Chemical Signals in Vertebrates 12

Marion L. East • Martin Dehnhard
Editors

Chemical Signals in Vertebrates 12

Springer

Editors
Marion L. East
Department of Evolutionary Ecology
Leibniz Institute for Zoo and Wildlife Research
Berlin, Germany

Martin Dehnhard
Department of Reproduction Biology
Leibniz Institute for Zoo and Wildlife Research
Berlin, Germany

ISBN 978-1-4614-5926-2 ISBN 978-1-4614-5927-9 (eBook)
DOI 10.1007/978-1-4614-5927-9
Springer New York Heidelberg Dordrecht London

Library of Congress Control Number: 2012955473

Printed on acid-free paper

Springer is part of Springer Science+Business Media (www.springer.com)

Preface

This volume reports the proceedings of the 12th meeting of the Chemical Signals in Vertebrates International Symposium (www.csiv.org), which was hosted by the Leibniz Institute for Zoo and Wildlife Research (IZW) in Berlin, Germany (www.izw-berlin.de). As chemical communication, particularly in social mammals, is a key subject of research interest at the IZW, we were very pleased to host the Congress this time.

The meeting took place 38 years after the first symposium in this series which was held in the USA in 1976. Previous meetings were held in Saratoga Springs, New York; Syracuse, New York; Sarasota, Florida; Laramie, Wyoming; Oxford, UK; Philadelphia, Pennsylvania; Tubingen, Germany; Ithaca, New York; Krakow, Poland; Corvallis, Oregon and Liverpool, UK.

The 12th meeting was well attended, attracting 104 scientists with a common interest in vertebrate chemical signalling, and its role in vertebrate behaviour from 25 countries. As in previous symposiums, the programme covered a broad range of topics including sexual selection, predator avoidance, migration, mother–offspring communication, social organisation, management of farm animals, olfactory receptors and neural processing. The talks and posters presented at the symposium were based on research findings in a wide diversity of vertebrate taxa and considered a broad range of chemical signals. The meeting was launched by a workshop on "Searching for chemical signals in vertebrates" which was ably run by Thomas Goodwin from Hendrix College, USA. This was followed by an interesting informal discussion of useful field and laboratory exercises for chemical ecology courses which was presented by Dietland Müller-Schwarze from Syracuse University, USA, and a welcome reception at the IZW. The main programme was held in the splendid Schloss Friedrichsfelde, located in the extensive grounds of the Tierpark, which is the larger of the two zoos located in Berlin. Posters were displayed throughout the conference and were the focus of attention during the "poster party" which provided ample opportunity to discuss each poster with its author over a drink and food from a buffet. During the closing ceremony the authors of the posters with the most votes from participants were presented with prizes by Martin Dehnhard. The symposium banquet in the Berlin Tierpark also provided a relaxing venue for extensive discussion and an exchange of ideas.

We are particularly grateful for financial support from the German Research Foundation (DFG), the sponsors of the poster prizes: Oxford Journals (Chemical Senses) and Cambridge University Press, and the Tierpark Berlin for providing free access for participants throughout the conference.

Finally, it is a pleasure to fully acknowledge the tremendous effort of Anke Schumann throughout the many months of preparation for the symposium, during the symposium and the handling of manuscripts for this volume. We are sure that all those who attended would agree that her tireless efforts made the meeting a success. We would like to thank all colleagues from the IZW who helped with a range of tasks during the running of the symposium, particularly Malvina Andris, Steffen Berthold, Dagmar Boras, Katrin Paschmionka, Marlies Rohleder, Karen Schneeberger, Caterina Wiedemann, Wolfgang Tauche, Dagmar Thierer and Kerstin Wilhelm.

Berlin, Germany

Marion L. East
Martin Dehnhard

Contents

Part I Olfaction: Receptors and Neural Processing

1 **Neural Substrate to Associate Odorants and Pheromones: Convergence of Projections from the Main and Accessory Olfactory Bulbs in Mice** 3
Bernardita Cádiz-Moretti, Fernando Martínez-García, and Enrique Lanuza

2 **Acceleration of Puberty Onset in Female Mice by Male Urinary Odours: Chemosensory, Physiological and Neuroendocrine Mechanisms** 17
Mélanie Jouhanneau and Matthieu Keller

3 **The Recessus Olfactorius: A Cryptic Olfactory Organ of Anuran Amphibians** 37
Christine Nowack, Sabrina Jordan, and Carolin Wittmer

Part II Mother-Offspring Communication

4 **Smell, Suck, Survive: Chemical Signals and Suckling in the Rabbit, Cat, and Dog** 51
Lourdes Arteaga, Amando Bautista, Daniel González, and Robyn Hudson

5 **Neonatal Recognition in Sheep** 61
Barend V. Burger, Marlize Z. Viviers, Niel J. le Roux, John Morris, Jan P.I. Bekker, and Maritha le Roux

6 **Suckling Odours in Rats and Mice: Biological Substrates that Guide Newborns to the Nipple** 77
Bruno Patris, Syrina Al Aïn, and Benoist Schaal

7 **The Role of Olfaction at Birth in Marsupial and Monotreme Mammals** 87
Nanette Y. Schneider, Geoff Shaw, and Marilyn B. Renfree

Part III Pheromones and Social Organisation

8 **Individual Variation in Pheromone Isoform Ratios of the Red-Legged Salamander, *Plethodon shermani*** 99
Adam J. Chouinard, Damien B. Wilburn, Lynne D. Houck, and Richard C. Feldhoff

9 **Rodent Urinary Proteins: Genetic Identity Signals and Pheromones** 117
Jane L. Hurst and Robert J. Beynon

10 **From Exploration to Selective Information Gathering: The Development of Chemosensory Investigation in Male African Elephants (*Loxodonta africana*)** 135
Bruce A. Schulte, Kathryn R. Bagley, Stacie Castelda, Helen Loizi, Nabil Nasseri, Dhaval K. Vyas, and Thomas E. Goodwin

Part IV Odour Profiles and Social Organisation

11 **Sex Differences in Olfactory Behaviours Reflect the Importance of Scent Marking for Social Integration in Adult Females and Competition Between Reproductively Active Males in the Spotted Hyena** 149
Marion L. East, Nicole Gusset-Burgener, and Heribert Hofer

12 **Is the Perception of Dietary Odour Cues Linked to Sexual Selection in Humans?** 161
Jitka Fialová, S. Craig Roberts, and Jan Havlíček

13 **Hemiterpenoids and Pyrazines in the Odoriferous Urine of the Maned Wolf (*Chrysocyon brachyurus*)** 171
Thomas E. Goodwin, Nucharin Songsasen, Laura J. Broederdorf, Blake A. Burkert, C. Joi Chen, Stephen R. Jackson, K. Bryan Keplinger, Margaret E. Rountree, Zach J. Waldrip, Margaret E. Weddell, Linda P. Desrochers, William K. Baker Jr., and Johannes P.F.G. Helsper

14 **The Perfume-Body Odour Complex: An Insightful Model for Culture–Gene Coevolution?** 185
Jan Havlíček and S. Craig Roberts

15 **A Historical Perspective on the Identification of Substances in the Territorial Scent Marks of Male Klipspringer Antelope *Oreotragus oreotragus*** 197
Hans E. Hummel

16 **Do Urinary Chemosignals Code for Sex, Age, and Season in the Giant Panda, *Ailuropoda melanoleuca?*** 207
Dingzhen Liu, Hong Yuan, Zhipeng Wang, Rongping Wei, Guiquan Zhang, and Lixing Sun

17 **On the Body Odour of Wild-Caught Mangrove-Marsh Bengal Tiger of Sundarban** 223
Mousumi Poddar-Sarkar, Sibdas Ray, Subrata Pal Chowdhury, Gopal Samanta, Paramaa Raha, and Ratan L. Brahmachary

18 **Ultrastructure and Chemical Composition of Elephant Hair in the Context of Chemical Signals in the Asian Elephant *Elephas maximus*** 227
Paramaa Raha, Mousumi Poddar-Sarkar, Utpal K. Nag, Radhaballabh Bhar, and Ratan L. Brahmachary

19 **Methyl Ketone Production in Juvenile Red-Sided Garter Snakes** 235
Emily J. Uhrig, Michael P. LeMaster, Deborah I. Lutterschmidt, and Robert T. Mason

Part V Predator Avoidance and Migration

20 **The Sophistication of Predator Odour Recognition by Minnows** 247
Douglas P. Chivers, Grant E. Brown, and Maud C.O. Ferrari

21 **Adaptive Responses of Embryonic Amphibians to Predation Risk** 259
Maud C.O. Ferrari and Douglas P. Chivers

22 **Response of Lumholtz' Tree-Kangaroos (*Dendrolagus lumholtzi*) to Odours from Native Arboreal and Introduced Terrestrial Predators: A Preliminary Study** 269
Sigrid Heise-Pavlov, Elizabeth Forbes, Clare Andersen, and Michelle Prince

23 **Voles in Space: Effects of Predator Odour on *Microtus arvalis* Breeding Females** 277
Mateusz Jochym and Stefan Halle

24 **Knowing Your Behaviour: The importance of Behavioural Assays in the Characterisation of Chemical Alarm Cues in Fishes and Amphibians** 295
Reehan S. Mirza, Chantale A. Laraby, and Ashley M. Marcellus

25 **Avoiding Predation: The Importance of Chemical and Visual Cues in Poison Frog Reproductive Behaviour** 309
Lisa M. Schulte, Rainer Schulte, and Stefan Lötters

Part VI Chemical Signals in Birds

26 **Besides Colours and Songs, Odour is the New Black of Avian Communication** 325
Francesco Bonadonna and Jerome Mardon

27 **Intraspecific Olfactory Communication in Zebra Finches (*Taeniopygia guttata*): Potential Information Apart from Visual and Acoustic Cues** 341
Barbara A. Caspers and E. Tobias Krause

28 **Male European Starlings Use Odorous Herbs as Nest Material to Attract Females and Benefit Nestlings** 353
Helga Gwinner

29 **Embryonic Domestic Chickens Can Detect Compounds in an Avian Chemosignal Before Breathing Air** 363
Julie C. Hagelin, Jacqueline C. Simonet, and Tyler R. Lyson

Part VII Reproduction in Domestic Animals

30 **Measurement of 16-Androstenes (5α-Androst-16-en-3-One, 5α-Androst-16-en-3α-ol, 5α-Androst-16-en-3β-ol) in Saliva of German Landrace and Göttingen Minipig Boars** 381
Martin Dehnhard, Hartmut Rohrmann, and Johannes Kauffold

31 **The Influence of Experimentally Reduced Vaginal Flora in Oestrus Females on the Mating Behaviour of Male Domestic Dogs (*Canis familiaris*)** 391
Michał Dzięcioł, Wojciech Niżański, Roland Kozdrowski, Małgorzata Ochota, and Ewa Stańczyk

32 **Socio-Sexual Stimuli and Reproductive Function: Emerging Perspectives of the Male Effect in Sheep and Goats** 397
Penny A.R. Hawken and Graeme B. Martin

Part VIII New Directions in Semiochemistry

33 **A Reverse-Engineering Approach to Identifying Which Compounds to Bioassay for Signalling Activity in the Scent Marks of African Wild Dogs (*Lycaon pictus*)** 417
Peter Apps, Lesego Mmualefe, and J. Weldon McNutt

34 **The Application of Proteomics to the Discovery and Quantification of Proteins in Scent Signals** 433
Robert J. Beynon, Stuart D. Armstrong, Sarah A. Roberts, Amanda Davidson, Deborah M. Simpson, Michael J. Turton, Jennifer Unsworth, and Jane L. Hurst

Index 449

Contributors

Syrina Al Aïn Developmental Ethology and Cognitive Psychology Group, Center for Smell, Taste and Food Science, CNRS, Université de Bourgogne, Dijon, France

Clare Andersen Wildlife Management, Wildlife Habitat, Port Douglas, QLD, Australia

Peter Apps Paul G. Allen Family Foundation Laboratory for Wildlife Chemistry, Botswana Predator Conservation Trust, Maun, Botswana

Stuart D. Armstrong Protein Function Group, Institute of Integrative Biology, University of Liverpool, Liverpool, UK

Lourdes Arteaga Centrol Tlaxcala de Biología de la Conducta, Universidad Autónoma de Tlaxcala, Tlaxcala, Mexico

Kathryn R. Bagley Department of Biology, Georgia Southern University, Statesboro, GA, USA

William K. Baker Jr. Little Rock Zoo, Little Rock, AR, USA

Panthera Research, Rockwall, TX, USA

Amando Bautista Centrol Tlaxcala de Biología de la Conducta, Universidad Autónoma de Tlaxcala, Tlaxcala, Mexico

Jan P.I. Bekker Institute for Plant Biotechnology, Stellenbosch University, Matieland, South Africa

Robert J. Beynon Protein Function Group, Institute of Integrative Biology, University of Liverpool, Liverpool, UK

Radhaballabh Bhar Department of Instrumentation Science, Jadavpur University, Kolkata, India

Francesco Bonadonna Behavioral Ecology Group, CEFE-CNRS, Montpellier Cedex 5, France

Ratan L. Brahmachary Embryology Unit, Indian Statistical Unit, Kolkata, India

Laura J. Broederdorf Department of Chemistry, Hendrix College, Conway, AR, USA

Grant E. Brown Department of Biology, Concordia University, Montreal, QC, Canada

Barend V. Burger Laboratory for Ecological Chemistry, Stellenbosch University, Matieland, South Africa

Blake A. Burkert Department of Chemistry, Hendrix College, Conway, AR, USA

Bernardita Cádiz-Moretti Department of Biologia Cel·lular, Laboratori de Neuroanatomia Funcional Comparada, Facultat de Ciències Biològiques, Universitat de València, Burjassot (València), Spain

Barbara A. Caspers Department of Animal Behaviour, University of Bielefeld, Bielefeld, Germany

Stacie Castelda Department of Biology, Georgia Southern University, Statesboro, GA, USA

C. Joi Chen Department of Chemistry, Hendrix College, Conway, AR, USA

Douglas P. Chivers Department of Biology, University of Saskatchewan, Saskatoon, SK, Canada

Adam J. Chouinard Department of Zoology, Oregon State University, Corvallis, OR, USA

Subrata Pal Chowdhury Department of Forest (Wildlife wing), Government of West Beangal, Bikash Bhavan, Calcutta, India

Amanda Davidson Mammalian Behaviour and Evolution Group, Institute of Integrative Biology, University of Liverpool, Neston, UK

Martin Dehnhard Department of Reproduction Biology, Leibniz Institute for Zoo and Wildlife Research, Berlin, Germany

Linda P. Desrochers Department of Chemistry, Hendrix College, Conway, AR, USA

Michał Dzięcioł Department of Reproduction and Clinic of Farm Animals, Wrocław University of Environmental and Life Sciences, Wrocław, Poland

Marion L. East Department of Evolutionary Ecology, Leibniz Institute for Zoo and Wildlife Research, Berlin, Germany

Richard C. Feldhoff Department of Biochemistry, University of Louisville, Louisville, KY, USA

Maud C.O. Ferrari Department of Biomedical Sciences, WCVM, University of Saskatchewan, Saskatoon, SK, Canada

Jitka Fialová Department of Anthropology, Charles University, Prague, Czech Republic

Elizabeth Forbes Vassar College, Poughkeepsie, NY, USA

Daniel González Instituto de Investigacionés Biomédicas, Universidad Nacional Autónoma de México, Ciudad Universitaria, México D.F., Mexico

Thomas E. Goodwin Department of Chemistry, Hendrix College, Conway, AR, USA

Nicole Gusset-Burgener Evolutionary Ecology Research Group, Leibniz Institute for Zoo and Wildlife Research, Berlin, Germany

Helga Gwinner Max-Planck-Institute for Ornithology, Eberhard-Gwinner-Straße, Seewiesen, Germany

Julie C. Hagelin Department of Biology, Swarthmore College, Swarthmore, PA, USA

Institute of Arctic Biology, University of Alaska, Fairbanks, AK, USA

Stefan Halle Institute of Ecology, Friedrich Schiller University, Jena, Germany

Jan Havlíček Department of Anthropology, Charles University, Prague, Czech Republic

Penny A.R. Hawken School of Animal Biology, The University of Western Australia, Crawley, WA, Australia

UWA Institute of Agriculture (Animal Production), The University of Western Australia, Crawley, WA, Australia

Sigrid Heise-Pavlov Center for Rainforest Studies, School for Field Studies, Yungaburra, QLD, Australia

Johannes P. F. G. Helsper Business Unit Bioscience, Plant Research International, Wageningen, The Netherlands

Heribert Hofer Evolutionary Ecology Research Group, Leibniz Institute for Zoo and Wildlife Research, Berlin, Germany

Lynne D. Houck Department of Zoology, Oregon State University, Corvallis, OR, USA

Robyn Hudson Instituto de Investigaciones Biomédicas, Universidad Nacional Autónoma de México, Ciudad Universitaria, México D.F., Mexico

Hans E. Hummel Justus-Liebig-University Giessen, Organic Agriculture, Giessen, Germany

University of Illinois Urbana-Champaign, Prairie Research Institute, Champaign, IL, USA

Jane L. Hurst Mammalian Behaviour and Evolution Group, Institute of Integrative Biology, University of Liverpool, Neston, UK

Stephen R. Jackson Department of Chemistry, Hendrix College, Conway, AR, USA

Mateusz Jochym Institute of Ecology, Friedrich Schiller University, Jena, Germany

Sabrina Jordan Department of Zoology/Developmental Biology, University of Kassel, WG Functional Vertebrate Anatomy, Kassel, Germany

Mélanie Jouhanneau UMR Physiologie de la Reproduction et des Comportements, UMR 0085 INRA—UMR 7247 CNRS—Université de Tours, INRA Tours, Nouzilly, France

Johannes Kauffold Large Animal Clinic for Theriogenology and Ambulatory Services, University of Leipzig, Leipzig, Germany

Matthieu Keller UMR Physiologie de la Reproduction et des Comportements, UMR 0085 INRA—UMR 7247 CNRS—Université de Tours, INRA Tours, Nouzilly, France

K. Bryan Keplinger Department of Chemistry, Hendrix College, Conway, AR, USA

Roland Kozdrowski Department of Reproduction and Clinic of Farm Animals, Wrocław University of Environmental and Life Sciences, Wrocław, Poland

E. Tobias Krause Department of Animal Behaviour, University of Bielefeld, Bielefeld, Germany

Enrique Lanuza Department of Biologia Cel·lular, Laboratori de Neuroanatomia Funcional Comparada, Facultat de Ciències Biològiques, Universitat de València, Burjassot (València), Spain

Chantale A. Laraby Department of Biology and Chemistry, Nipissing University, North Bay, ON, Canada

Michael P. LeMaster Department of Biology, Western Oregon University, Monmouth, OR, USA

Stefan Lötters Department of Biogeography, Trier University, Trier, Germany

Helen Loizi Department of Biology, Georgia Southern University, Statesboro, GA, USA

Dingzhen Liu Ministry of Education, Key Laboratory of Biodiversity Science and Ecological Engineering & Institute of Ecology, Beijing Normal University, Beijing, China

Deborah I. Lutterschmidt Department of Biology, Portland State University, Portland, OR, USA

Tyler R. Lyson Department of Biology, Swarthmore College, Swarthmore, PA, USA

Department of Geology and Geophysics, Yale University, New Haven, CT, USA

Ashley M. Marcellus Department of Biology and Chemistry, Nipissing University, North Bay, ON, Canada

Jerome Mardon Behavioral Ecology Group, CEFE-CNRS, Montpellier Cedex 5, France

Graeme B. Martin School of Animal Biology, The University of Western Australia, Crawley, WA, Australia

UWA Institute of Agriculture (Animal Production), The University of Western Australia, Crawley, WA, Australia

Fernando Martínez-García Department of Biologia Funcional i Antropologia Física, Laboratori de Neuroanatomia Funcional Comparada, Facultat de Ciències Biològiques, Universitat de València, Burjassot (València), Spain

Robert T. Mason Department of Zoology, Oregon State University, Corvallis, OR, USA

J. Weldon McNutt Paul G. Allen Family Foundation Laboratory for Wildlife Chemistry, Botswana Predator Conservation Trust, Maun, Botswana

Reehan S. Mirza Department of Biology and Chemistry, Nipissing University, North Bay, ON, Canada

Lesego Mmualefe Paul G. Allen Family Foundation Laboratory for Wildlife Chemistry, Botswana Predator Conservation Trust, Maun, Botswana

John Morris Mariendahl Experimental Farm, Faculty of AgriSciences, Stellenbosch University, Matieland, South Africa

Utpal K. Nag Department of Forest, Government of West Bengal, ADFO Wildlife Division I, Darjeeling, India

Nabil Nasseri Department of Biology, Georgia Southern University, Statesboro, GA, USA

Wojciech Niżański Department of Reproduction and Clinic of Farm Animals, Wrocław University of Environmental and Life Sciences, Wrocław, Poland

Christine Nowack Department of Zoology/Developmental Biology, University of Kassel, WG Functional Vertebrate Anatomy, Kassel, Germany

Małgorzata Ochota Department of Reproduction and Clinic of Farm Animals, Wrocław University of Environmental and Life Sciences, Wrocław, Poland

Bruno Patris Developmental Ethology and Cognitive Psychology Group, Center for Smell, Taste and Food Science, CNRS, Université de Bourgogne, Dijon, France

Michelle Prince Wildlife Management, Wildlife Habitat, Port Douglas, QLD, Australia

Mousumi Poddar-Sarkar Department of Botany, University of Calcutta, Kolkata, India

Paramaa Raha Department of Zoology, Bethune College, Kolkata, India

Sibdas Ray Department of Chemistry, University of Calcutta, Kolkata, India

Marilyn B. Renfree Department of Zoology, The University of Melbourne, Melbourne, VIC, Australia

S. Craig Roberts Department of Psychology, School of Natural Sciences, University of Stirling, Stirling, Scotland, UK

Sarah A. Roberts Mammalian Behaviour and Evolution Group, Institute of Integrative Biology, University of Liverpool, Neston, UK

Hartmut Rohrmann Rövershagen, Germany

Margaret E. Rountree Department of Chemistry, Hendrix College, Conway, AR, USA

Maritha le Roux Laboratory for Ecological Chemistry, Stellenbosch University, Matieland, South Africa

Niel J. le Roux Department of Statistics and Actuarial Science, Stellenbosch University, Matieland, South Africa

Gopal Samanta ABAHC, Alipurdwar I, Jalpaiguri, West Bengal, India

Benoist Schaal Developmental Ethology and Cognitive Psychology Group, Center for Smell, Taste and Food Science, CNRS, Université de Bourgogne, Dijon, France

Nanette Y. Schneider Department of Molecular Genetics, German Institute for Human Nutrition, Potsdam-Rehbrücke, Nuthetal, Germany

Department of Zoology, The University of Melbourne, Melbourne, VIC, Australia

Bruce A. Schulte Department of Biology, Western Kentucky University, Bowling Green, KY, USA

Department of Biology, Georgia Southern University, Statesboro, GA, USA

Lisa M. Schulte Department of Biogeography, Trier University, Trier, Germany

Rainer Schulte Instituto de Investigación Biológica de las Cordilleras Orientales—INIBICO, Tarapoto, Peru

Geoff Shaw Department of Zoology, The University of Melbourne, Melbourne, VIC, Australia

Jacqueline C. Simonet Department of Biology, Swarthmore College, Swarthmore, PA, USA

Perelman School of Medicine, University of Pennsylvania, Philadelphia, PA, USA

Deborah M. Simpson Protein Function Group, Institute of Integrative Biology, University of Liverpool, Liverpool, UK

Nucharin Songsasen Department of Reproductive Sciences, Center for Species Survival, Smithsonian Conservation Biology Institute, Front Royal, VA, USA

Ewa Stańczyk Department of Reproduction and Clinic of Farm Animals, Wrocław University of Environmental and Life Sciences, Wrocław, Poland

Lixing Sun Department of Biological Sciences, Central Washington University Ellensburg, Washington, WA, USA

Michael J. Turton Protein Function Group, Institute of Integrative Biology, University of Liverpool, Liverpool, UK

Emily J. Uhrig Department of Zoology, Oregon State University, Corvallis, OR, USA

Jennifer Unsworth Protein Function Group, Institute of Integrative Biology, University of Liverpool, Liverpool, UK

Marlize Z. Viviers Laboratory for Ecological Chemistry, Stellenbosch University, Matieland, South Africa

Dhaval K. Vyas Department of Biology, Georgia Southern University, Statesboro, GA, USA

Zach J. Waldrip Department of Chemistry, Hendrix College, Conway, AR, USA

Zhipeng Wang Ministry of Education, Key Laboratory of Biodiversity Science and Ecological Engineering & Institute of Ecology, Beijing Normal University, Beijing, China

Margaret E. Weddell Department of Chemistry, Hendrix College, Conway, AR, USA

Rongping Wei China Research and Conservation Center for the Giant Panda, Sichuan, China

Damien B. Wilburn Department of Biochemistry, University of Louisville, Louisville, KY, USA

Carolin Wittmer Department of Zoology/Developmental Biology, University of Kassel, WG Functional Vertebrate Anatomy, Kassel, Germany

Hong Yuan Ministry of Education, Key Laboratory of Biodiversity Science and Ecological Engineering & Institute of Ecology, Beijing Normal University, Beijing, China

Guiquan Zhang China Research and Conservation Center for the Giant Panda, Sichuan, China

Part I
Olfaction: Receptors and Neural Processing

Chapter 1
Neural Substrate to Associate Odorants and Pheromones: Convergence of Projections from the Main and Accessory Olfactory Bulbs in Mice

Bernardita Cádiz-Moretti, Fernando Martínez-García, and Enrique Lanuza

Abstract In rodents olfactory stimuli are essential for socio-sexual behaviour. Volatile stimuli mainly activate the main olfactory system, whereas non-volatile ones, some of them considered as pheromones, activate the accessory olfactory system. Traditionally, it has been considered that the efferent projections of the main and the accessory olfactory bulbs innervate different telencephalic areas. Recent studies performed in rats and mice have challenged this view. In this work we explore in mice, through the use of anterograde tracer injections in the olfactory bulbs, the brain centres where olfactory and vomeronasal information might converge.

Our results show that the olfactory and vomeronasal projections overlap mainly in the cortex-amygdaloid transition area, the anterior cortical amygdala and the medial amygdala (anterior and posterodorsal divisions), as well as, to a minor extent, in the anterior amygdaloid area and a restricted zone of the anterior piriform cortex. This indicates the existence of amygdaloid areas where the association between the olfactory and vomeronasal stimuli coming from the main and accessory olfactory bulb can take place. The chemosensory associative areas described in the amygdala might play a role in emotional learning involving the association of odours and pheromones.

B. Cádiz-Moretti • E. Lanuza (✉)
Department of Biologia Cel·lular, Laboratori de Neuroanatomia Funcional Comparada, Facultat de Ciències Biològiques, Universitat de València, C. Dr. Moliner, 50, 46100 Burjassot (València), Spain
e-mail: Enrique.Lanuza@uv.es

F. Martínez-García
Department of Biologia Funcional i Antropologia Física, Laboratori de Neuroanatomia Funcional Comparada, Facultat de Ciències Biològiques, Universitat de València, C. Dr. Moliner, 50, 46100 Burjassot (València), Spain

M.L. East and M. Dehnhard (eds.), *Chemical Signals in Vertebrates 12*,
DOI 10.1007/978-1-4614-5927-9_1,

1.1 Introduction

Rodents possess two long-distance chemosensory organs, the nasal or main olfactory epithelium (MOE) and the vomeronasal organ (VNO). Chemosensory stimuli detected by them play a key role in socio-sexual (Brennan and Kendrick 2006) and antipredatory behaviour (Papes et al. 2010). The MOE mainly detects air-borne volatiles, while the VNO seems specialised (with some exceptions) in detecting non-volatile chemicals (Krieger et al. 1999) possessing an intrinsic biological value, such as sexual pheromones or predator-derived chemosignals. Thus, in the context of socio-sexual behaviour, vomeronasal stimuli apparently trigger innate behavioural responses (Martínez-García et al. 2009; Keller et al. 2009).

The MOE and VNO project to the main (MOB) and accessory (AOB) olfactory bulbs, respectively. In the 1970s, tract-tracing experiments described that the main and accessory olfactory bulbs (OB) projected to adjacent but non-overlapping telencephalic nuclei (Scalia and Winans 1975; Skeen and Hall 1977). Traditionally, the structures considered as receiving only main olfactory projections are the piriform cortex (Pir), the olfactory tubercle (Tu), cortex-amygdala transition zone (CxA), anterior cortical amygdala (ACo), the nucleus of the lateral olfactory tract (Nlot) and the posterolateral cortical amygdala (PLCo). The nuclei traditionally considered as vomeronasal recipients are the bed nucleus of the accessory olfactory tract (BAOT), the posteromedial part of the medial division of the bed nucleus of the stria terminalis (BSTMPM), the anterior amygdaloid area (AA), the medial amygdala (anterior, MeA, and posterior, MeP, subdivisions) and the posteromedial cortical amygdala (PMCo). These observations led to the dual olfactory hypothesis, which stated that these two pathways act independently in the chemical information processing (see, for a review, Halpern 1987).

This view has been seriously challenged by recent data. Thus, tract-tracing experiments conducted in rats (Mohedano-Moriano et al. 2007; Pro-Sistiaga et al. 2007) and mice (Kang et al. 2009, 2011a) have demonstrated the existence of convergent projections from the MOB and AOB to several cortical and non-cortical amygdaloid structures (Gutiérrez-Castellanos et al. 2010). Functional experiments also indicate that both chemosensory systems play a complementary role in eliciting adequate behavioural responses (Martínez-García et al. 2009; Keller et al. 2009). This interaction can be observed in a particular case of emotional learning in which airborne olfactory stimuli are associated with sexual pheromones detected by the VNO. Male-derived involatile pheromones elicit an unconditioned attraction in female mice (even if the females have no previous experience with males or male-derived chemical signals; Roberts et al. 2010). In contrast, male-derived volatile odorants do not attract inexperienced females. Following experience with male-soiled bedding, which contains involatile pheromones and volatile odorants, females acquire a learned attraction to volatile male odours (Moncho-Bogani et al. 2002). Therefore, in this case, male pheromones act like appetitive unconditioned stimuli for females and, following chemosensory experience, become associated with male-derived volatiles, what results in a conditioned attraction for those volatiles (Moncho-Bogani et al. 2002; Martínez-Ricós et al. 2008).

In this context, we have started a neuroanatomical study of the neural centres receiving direct projections from both the main and accessory olfactory bulbs in the mouse brain, as a way to explore the brain regions where olfactory-vomeronasal associative learning might first occur.

1.2 Materials and Methods

For this work, we used 12 CD-1 adult female mice *Mus musculus*, which were treated throughout according to the EEC guidelines for European Communities Council Directives of 24th November 1986 (86/609/EEC). The Committee of Ethics on Animals Experimentation of the University of Valencia approved all the procedures.

To study the projections arising from the AOB and the MOB, iontophoretic injections of dextranamine (10,000 MW, lysine fixable, Invitrogen, Carlsbad, CA, USA) conjugated with tetramethylrhodamine and biotin (TBDA, 5% in PB 0.01 M, pH 8.0) were aimed at these structures. The tracer was delivered from glass micropipettes (10–50 μm diameter tips) by means of positive current pulses (7on/7off seconds, 3–5 μA, 10–15 min). Two minutes after the termination of each injection, the pipette was withdrawn while passing a retention current (–0.8 μA).

Animals were anaesthetised with intraperitoneal (IP) injections of a 3:2 ketamine (75 mg/kg, Merial laboratories, Barcelona, Spain) and medotomedine (1 mg/kg, Pfizer, Alcobendas, Madrid, Spain) solution, complemented with atropine (Sigma, St. Louis, MO, USA; 0.04 mg/kg, IP) to reduce cardio-respiratory depression and butorfanol (Fort Dodge Veterinaria, Girona, SPAIN; 5 mg/kg, subcutaneous) as analgesic. After fixing the mouse head in the stereotaxic apparatus (David Kopf, 963-A, Tujunga CA, USA), a small hole was drilled above the target zone. Tracer injections were performed at the following coordinates relative to Bregma: 4.25 mm AP; 0.95 L and 3.22 mm deep for the ventral MOB; 3.57 mm AP; 0.95 mm L and 0.90 mm deep for the AOB. After surgery, intramuscular injections of atipamezol (Pfizer, Alcobendas, Madrid, Spain; 1 mL/kg, intramuscular) were used to revert the medotomedine effects.

Three to five days later, animals were deeply anaesthetised with an overdose of sodium pentobarbital (Sigma) (90 mg/kg) and perfused with saline solution (0.9%) followed by 4% paraformaldehyde diluted in phosphate buffer (PB 0.1 M, pH 7.6). Brains were removed from the skull, postfixed for 4 h in the same fixative and cryoprotected with 30% sucrose buffered. Using a freezing microtome, we obtained parasagittal sections through the olfactory bulbs (30 μm) and frontal (40 μm) sections of the brain. For the detection of the TBDA, endogenous peroxidase was inactivated with 1% H_2O_2 (in 0.05 M Tris buffer saline pH 7.6) for 15 min and then sections were incubated for 90 min in ABC complex (Vectastain ABC kit,

Vector Labs, PK-6100, Burlingame, CA, USA) diluted 1:50 in TBS-Tx (Triton X-100 0.3% in TBS 0.05 M pH 7.6). After rinsing thoroughly with buffer, peroxidase activity was developed with 0.025% diaminobenzidine in PB (0.1 M, pH 8.0) and 0.01% H_2O_2 and 0.1% nickel ammonium sulphate. Series were mounted onto gelatinised slides, dehydrated with graded alcohols, cleared with xylene and coverslipped with Entellan. The sections were observed and photographed using an Olympus microscope equipped with a digital camera (XC-50). Line drawings and picture were arranged and labelled with Photoshop 7.0 (AdobeSystems, MountainView, CA, USA).

1.3 Results

1.3.1 Injections in the Accessory Olfactory Bulb

Six injections involved the AOB. Three of them were centred in the AOB (with no MOB involvement), one of which was restricted to the glomerular and mitral layer (Fig. 1.1a). This injection is described as a representative case. Importantly, in this AOB injection no labelled fibres appeared in the dorsolateral olfactory tract (dlot) (Fig. 1.1b, c), and no retrogradely labelled cells were observed.

Anterogradely labelled fibres originated by the AOB injection ran caudally in the inner part of the lateral olfactory tract (lot) to reach the anterior piriform cortex (Pir), nuclei of the lateral (Nlot) and accessory (Baot) olfactory tracts, anterior amygdala (AA), medial amygdala (Me) and several nuclei of the cortical amygdala.

In the anterior Pir labelled fibres were observed in the edge between the lot and layer IA (Fig. 1.2a, b). Labelled fibres display abundant varicosities and most of them are oriented parallel to the pial surface. In addition, in the anterior ventral Pir (where the lot runs below the pial surface, see Martínez-García et al. 2012) a number of fibres were observed crossing the layers II and III (Fig. 1.2c). In the posterior Pir (at the level of the AA and CxA), labelled fibres are not present anymore. At the medioventral edge of the Pir, in the limit with the adjacent striato-pallidal territory, some labelled fibres were observed running towards the ventral external capsule (Fig. 1.2d).

In the rostral CxA, next to the ventral and dorsal AA (anterior ventral amygdala (AAV) and anterior dorsal amygdala (AAD)), varicose fibres were seen in layers IA and IB (Fig. 1.2d–f), and a number of fine fibres crossed layer II (Fig. 1.2d). At this level, vertically oriented fine and thick fibres crossed through the AAD, entering the ventral tip of the external capsule (Fig. 1.2d, e). Labelled, fine fibres were seen scattered throughout the AAV and AAD (Fig. 1.2d, e).

At the level of the ACo, labelled varicose fibres appeared in its layer I (sublayers IA and IB; Fig. 1.2g, h). Some fibres crossed layer II (Fig. 1.2i) tangentially to the pial surface. In the Nlot, scattered varicose fibres were present in the layers IA and IB, whereas in the Baot a compact bundle of labelled fibres was observed (not shown).

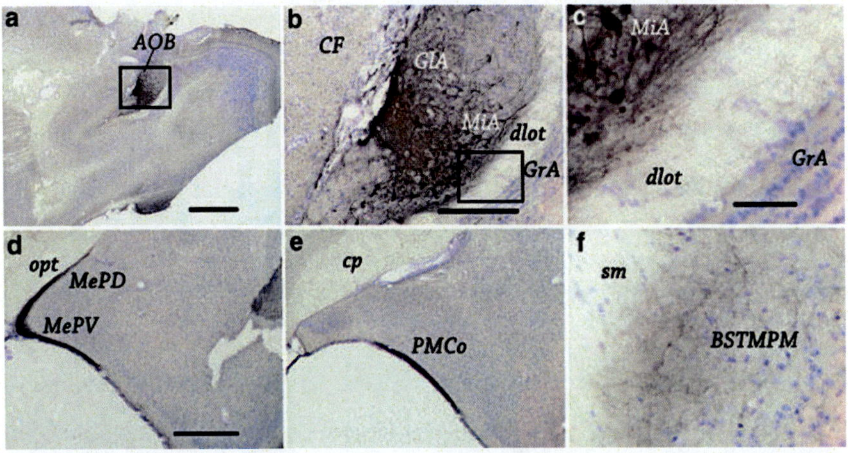

Fig. 1.1 Photomicrographs of Nissl counterstained parasagittal (**a–c**) and frontal (**d–f**) sections, illustrating the injection site of TBDA in the AOB (**a–c**) and the resulting anterograde labelling (**d–f**) in some of the vomeronasal nuclei. *Rectangles* in (**a, b**) correspond to (**b, c**), respectively. For abbreviations, see list. Scale bar: 1 mm for (**a**); 500 μm for (**b**); 250 μm for (**d, e**); 50 μm for (**c, f**)

In the anterior Me (MeA) we observed a dense plexus of terminal (fine and varicose) fibres in layer I (Fig. 1.2j–l), some of which reached layer II. Dorsal to the MeA, a few labelled fibres took a dorsal course to reach the ventral zone of the caudoputamen (CPu) (Fig. 1.2j). In the posterior Me (MeP) a dense field of anterogradely labelled, fine, varicose fibres occupied layer I (Fig. 1.1d), with a few fibres reaching layer II. From the dorsal MeP (MePD) labelled fibres were seen to enter the stria terminalis (st) and ran rostrally up to the medial bed nucleus of the stria terminalis (BST), where they gave rise to a terminal field specifically innervating its posteromedial division (Fig. 1.1f).

Finally, a dense terminal plexus of labelled fibres was observed in layers IA and IB of the PMCo (Fig. 1.1e).

1.3.2 Main Olfactory Bulb Injections

We obtained six injections in the MOB. Two of them were restricted to the ventral MOB and were located in the glomerular, external plexiform and mitral layers (as shown in a representative case in Fig. 1.3a, b), and are described here as examples.

From the injection site, labelled fibres entered the lateral olfactory tract (Fig. 1.3c) to reach the Pir, olfactory tubercle (Tu), AA, parts of the cortical amygdala, Nlot and Me. Occasionally, some retrogradely labelled cells were observed in the Pir, horizontal diagonal band (HDB) and magnocellular preoptic nucleus (MCPO).

Throughout the Pir, a dense plexus of beaded and varicose labelled fibres was observed in layer IA (Fig. 1.3d, e). Only occasionally labelled fibres were observed

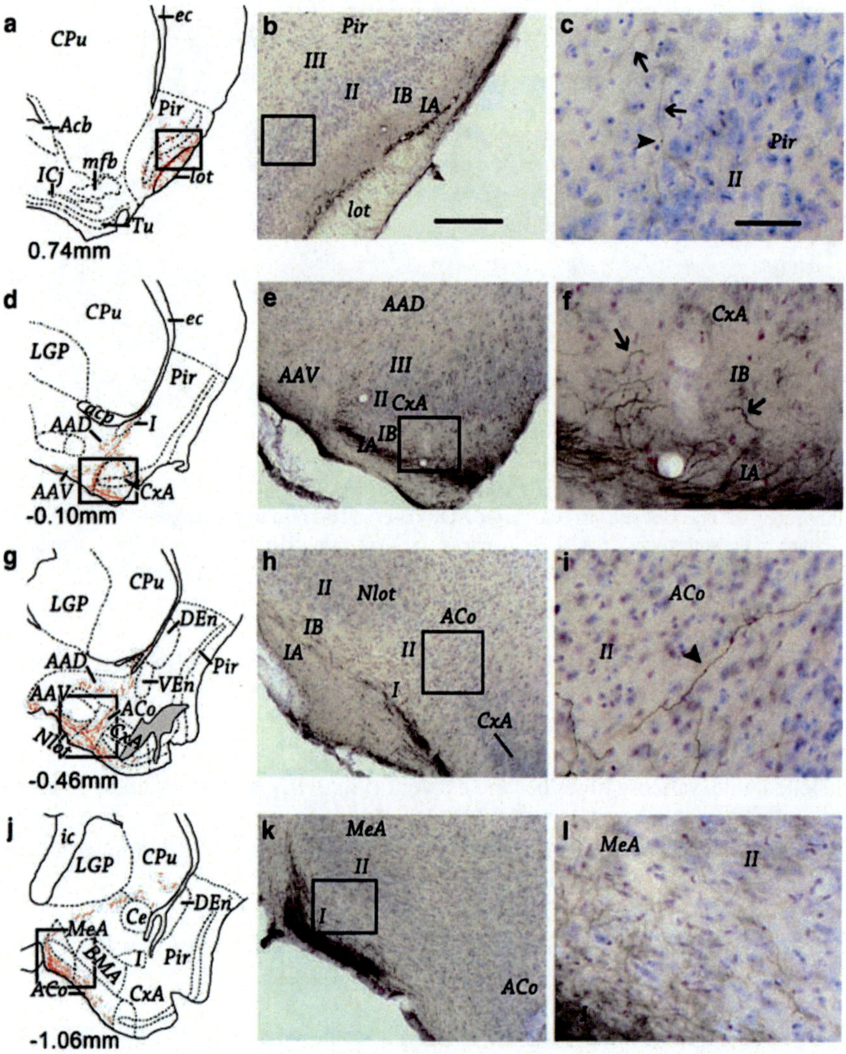

Fig. 1.2 Line drawings and photomicrographs of Nissl counterstained frontal sections illustrating the anterogradely labelled fibres in the telencephalon resulting from the AOB injection shown in Fig. 1.1. The drawings (**a**, **d**, **g**) illustrate the labelling in traditional olfactory recipients (with indication of the coordinate relative to bregma). For abbreviations, see list. *Rectangles* in (**a**, **d**, **g**, **j**) correspond to (**b**, **e**, **h**, **k**), respectively. *Rectangles* in (**b**, **e**, **h**, **k**), correspond to (**c**, **f**, **i**, **l**) respectively. Scale bar = 250 μm for (**b**, **e**, **h**, **k**); 50 μm for (**c**, **f**, **i**, **l**). *Arrows* indicate fibres and *arrowheads* varicosities

in layer IB. No labelled fibres appeared in layers II and III. In the Tu, anterogradely labelled fibres ran parallel to the pial surface (Fig. 1.3f), most of which restricted to layer IA, with some occasional fibres reaching the layer IB.

When the lateral olfactory tract reaches the amygdaloid formation, labelled varicose fibres were observed in layer IA of the CxA and in the superficial portion

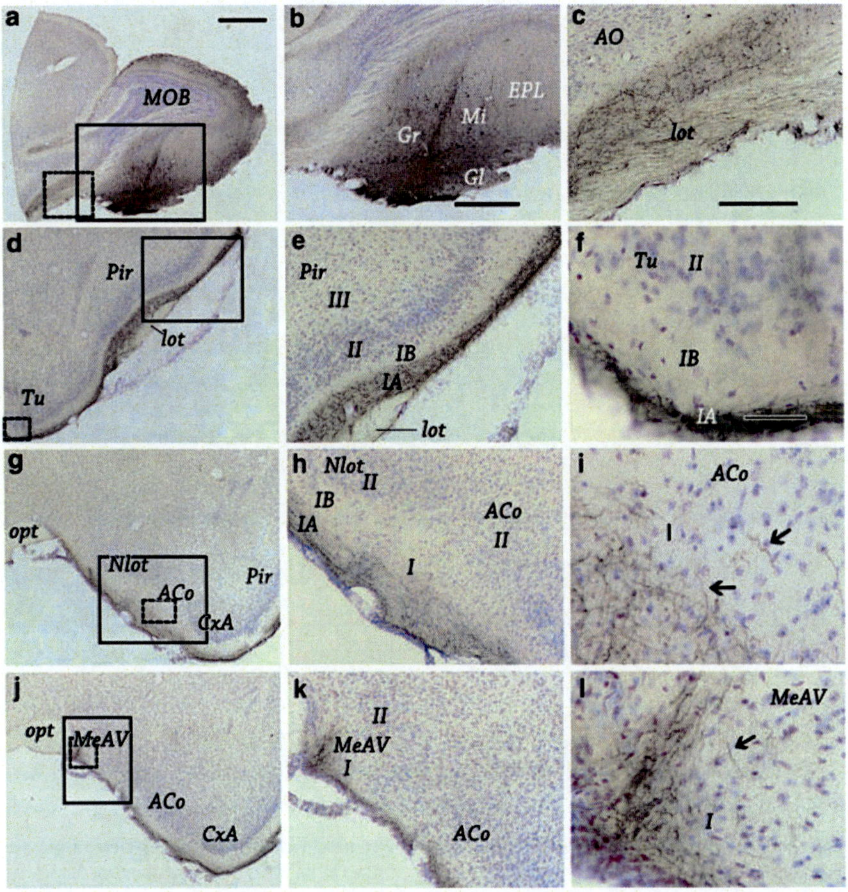

Fig. 1.3 Photomicrograph of Nissl counterstain of parasagittal (**a–c**) and frontal sections (**d–l**), illustrating the injection site in the ventral MOB (**a**, **b**) and the anterograde labelling in the traditionally considered vomeronasal (**d–i**) and olfactory (**j–l**) nuclei. *Continuous rectangles* in (**a**, **d**, **g**, **j**), correspond to (**b**, **e**, **h**, **k**), respectively. *Discontinuous rectangles* in (**a**, **d**, **g**, **j**), correspond to (**c**, **f**, **i**, **l**), respectively. For abbreviations, see list. Scale bar = 1 mm for (**a**); 500 μm for (**b**, **d**, **g**, **j**); 250 μm for (**c**, **e**, **h**, **k**); 50 μm to (**f**, **i**, **l**). *Arrows* indicate fibres

of the AAV (Figs. 1.3g and 1.4a–c). At the medial edge of the CxA, in the limit with the AAV, a few fibres were observed to run dorsally into the AAD (Fig. 1.4a). Within the cortical amygdaloid nuclei, thick varicose labelled fibres formed dense plexuses in layer IA of the ACo and the Nlot (Fig. 1.3g–i), where some fibres occasionally entered layer IB. Caudally in the cortical amygdala, both fine and thick beaded and varicose labelled fibres were present in layer IA of the PLCo, restricted to the most superficial zone (Fig. 1.4h).

In the medial amygdala, a number of thin labelled fibres were observed to reach the superficial aspect of the ventral part of the MeA (Figs. 1.3j–l and 1.4d–f). This antero-

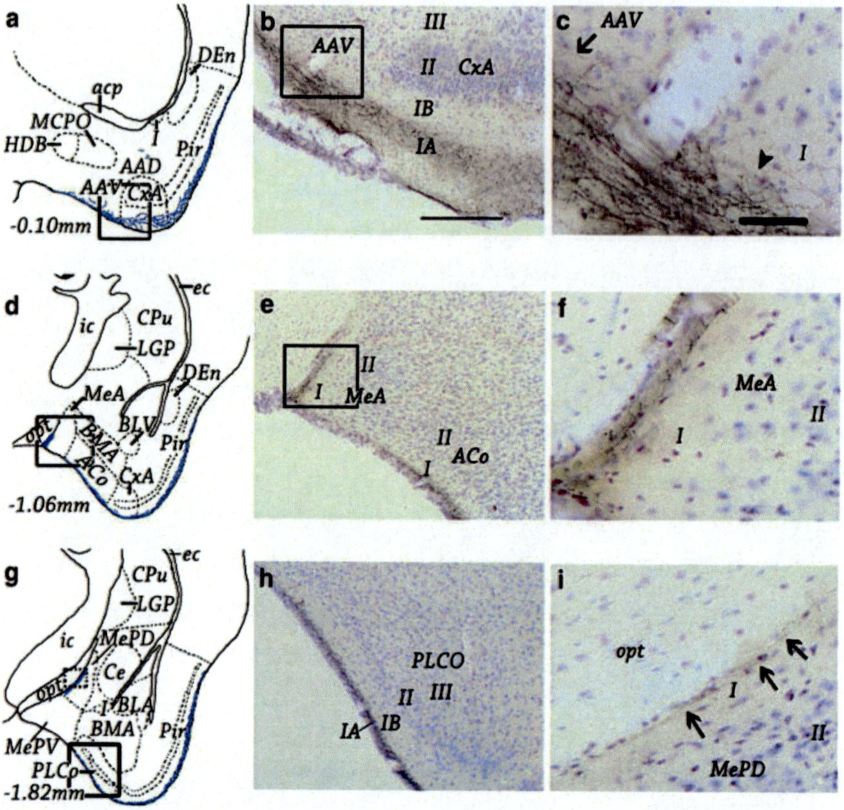

Fig. 1.4 Line drawings and photomicrographs of Nissl-stained frontal sections illustrating the anterogradely labelled fibres in the telencephalon of a MOB injection (same animal of Fig. 1.3). Photomicrographs show the labelling in traditionally considered vomeronasal recipients (except for (**h**) that is a traditionally olfactory recipient). *Continuous rectangles* in (**a**, **d**, **g**) correspond to (**b**, **e**, **h**), respectively. *Discontinuous rectangle* in (**g**) corresponds to (**i**). *Rectangles* in (**b**, **e**) correspond to (**c**, **f**). For abbreviations, see list. Scale bar: 250 μm in (**b**, **e**, **h**); 50 μm in (**c**, **f**, **i**). *Arrows* indicate fibres and *arrowheads* varicosities

grade labelling is particularly dense at the rostralmost levels of the MeA. The plexus of anterograde labelling in the MeA did not show continuity with the fibres present in the ACo (Fig. 1.4d, e). In the posterior medial amygdala, a few fine fibres were observed in layer I of the MePD (Fig. 1.4g, i), whereas no labelling was found in the MePV.

1.4 Discussion

Traditionally, the vomeronasal and olfactory systems have been considered to process chemosensory information independently, given their differential and non-convergent projections to the basal telencephalon (Skeen and Hall 1977;

Scalia and Winans 1975). This idea is known as the dual olfactory hypothesis (reviewed in Halpern 1987). The results of the present study confirm and extend previous results reported in mice and rats of convergent projections of the main and accessory olfactory bulbs to portions of the amygdala and olfactory cortex, thus contradicting the dual olfactory hypothesis. Our results indicate that the AOB projects to areas traditionally considered being olfactory recipients, such as the CxA, ACo and Nlot in the amygdala and a particular zone of the Pir, while the MOB projects to areas traditionally considered as vomeronasal, such as the AAV, MeA and MePD.

The Pir is the main projection target of the MOB, and therefore is usually viewed as the primary olfactory cortex (but see Martínez-García et al. 2012). Our results show fibres originated in the AOB innervating layers I, II and III in a small restricted area of the Pir: the anterior ventral Pir (surrounding the lot). It is important to note that the distribution of anterograde labelling in the Pir following AOB injections (fibres oriented perpendicularly to the brain surface that entered layers II and III) is very different from that observed after MOB injections (fibres oriented parallel to the brain surface concentrated in layer IA), and therefore it is unlikely that these fibres might arise from inadverted contamination of the MOB. In addition, these results agree with schematic drawings shown by Mohedano-Moriano et al. (2007) in rats (Table 1.1), as well as with the description of collaterals fibres entering the Pir after AOB injections in mice (Kang et al. 2011a; Table 1.1).

In the latter study only the fibres in layer I were described probably because the authors used flat-mount preparations of the telencephalon. In any case, these authors did not discuss the observed fibres in the Pir. However, this evidence suggests that this particular zone of the Pir is the first cortical area where vomeronasal and olfactory information converge.

With regard to the projections of the AOB to the amygdala, our results fully agree with those reported by Kang et al. (2009, 2011a), the only differences being that they did not report fibre labelling in the CxA, and that they described fibres in deep MeA. In rats, Pro-Sistiaga et al. (2007) and Mohedano-Moriano et al. (2007) reported AOB-originated projections in the same structures that we found in the present study (with the exception of a few fibres in the PLCo that we could not observe), thus suggesting that rats and mice have similar patterns of olfactory projections, although some differences might exist regarding the layers innervated, as summarised in Table 1.1.

We have also observed fibre labelling through the AAD that apparently reached the ventral tip of the external capsule (Fig. 1.2d). We could not identify the target structures of these fibres. In rats, similar anterograde labelling has been reported (Mohedano-Moriano et al. 2007) and was interpreted as part of the stria terminalis pathway to the BSTMPM. However, at the levels where this labelling appears the st is not present yet. An alternative explanation would be that these fibres are actually retrogradely labelled axons from neurons of the diagonal band that project back to the bulb (Shipley and Adamek 1984), although we did not observed retrograde labelled cell bodies in this nucleus.

Table 1.1 Comparison of the present results with recent studies of AOB and MOB projections performed in rats (Mohedano-Moriano et al. 2007; Pro-Sistiaga et al. 2007) and mice (Kang et al. 2009, 2011a)

	Layers where it can be seen labelling from AOB			Layers where it can be seen labelling from MOB		
Nucleus	Rat	Mouse	Present work	Rat	Mouse	Present work
Pir	Line drawing	Collaterals	IA-IB II-III	IA	IA	IA-IB
CxA	IA-II	–	IA-IB	IA-II	–	IA
AAV	I	I[a]	I	I	I[a]	I
Nlot	I-II-III	I[a]	IA-IB	IA-IB	I[a]	IA-IB
ACo	IB-III	Collaterals	IA-IB II-III	IA	I[a]	IA
Baot	I-II	I-II	I-II	IA	I[a]	IA
MeA	I-II-III	I-II	I-II	I	IA-II	I
MePV	I-II	–	I	No	No	No
MePD	I-II	I-II	I	No	IA-II	I
PMCo	IA-IB	IA-IB	IA-IB	No	No	No
PLCo	IA-IB	No	No	IA	I[a]	IA-IB

Collaterals refer to reported axon collaterals entering the structure in a flat-mount view

[a]Layers IA and IB not specified in the original paper—not mentioned in the original paper. *MOB* main olfactory bulb; *AOB* accessory olfactory bulbs; *MeA* anterior medial amygdale; *MePD* posterodorsal medial amygdala; *MePV* posteroventral medial amygdale; *Nlot* nucleus of the lateral olfactory tract; *Pir* piriform cortex; *PMCo* posteromedial cortical amygdale; *PLCo* posterolateral cortical amygdale

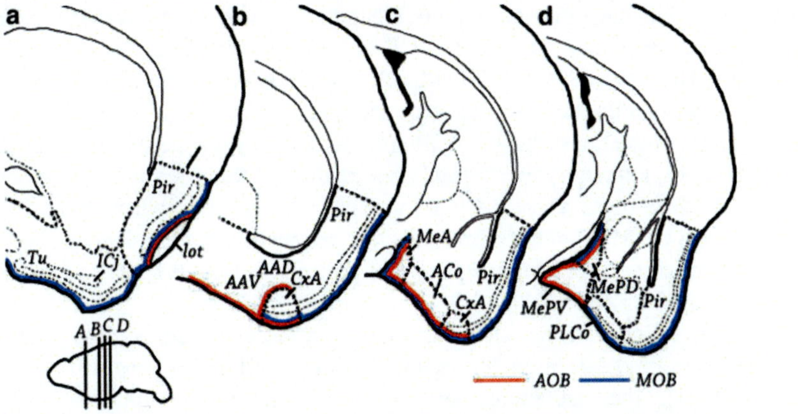

Fig. 1.5 Semi-schematic diagram showing the convergence from the AOB (*red*) and MOB (*blue*) projections

The amygdaloid targets of the MOB projection described in the present report are fully consistent with the results of previous works (Kang et al. 2009; Pro-Sistiaga et al. 2007), again with minor differences regarding the layers innervated in each structure, as summarised in Table 1.1. The only result of our work not previously reported is the presence of a few labelled fibres located deep in the AAV and in the AAD. Since this labelling is similar to that found in injections that only affected the granular layer of the MOB (not shown) that gave rise to abundant retrograde labelling, it is possible that it actually corresponds to retrogradely labelled axons.

Therefore, according to our results and other previous works, the amygdala contains several associative areas where chemosensory information coming from the AOB and MOB converge, including the CxA, ACo, MeA and MePD (Fig. 1.5). The CxA and ACo are structures receiving predominantly olfactory projections and a minor input from the AOB in layers I and II (also reaching layer III in the ACo) (Table 1.1).

The MeA and MePD are areas receiving predominantly vomeronasal projections and also a minor afferent from the MOB (layers I and II). Since the inputs to the inner layers are closest to the cell body, they presumably have a higher modulatory strength than those reaching the distal dendrites. Taking these consideration into account, we consider the four mentioned areas more relevant for the association of olfactory and vomeronasal information than the AAV, Nlot and Baot, where fibres coming from the MOB and AOB are superficial and may also be axons en passant (Table 1.1).

Although the present studies were performed in female mice, previous works showed no differences in the pattern of the AOB projections between male and female mice (von Campenhausen and Mori 2000). Regarding the MOB efferent projections, most of the previous works in mice were performed in males (e.g. Shipley and Adamek 1984; Kang et al. 2011a), and our results have not revealed any difference with these previous reports. Consistent with these observations, no sexual differences have been found in the AOB or MOB projections in rats (Dr. Martinez-Marcos, University of Castilla-LaMancha, personal communication). In addition, the reported similarities between mice and rats make it unlikely that the described convergent projections from the AOB and MOB to amygdaloid structures are different in males, although experimental confirmation of this possibility awaits further study. Noteworthy, a recent study has revealed a sexual dimorphism in the distribution within the MOB of the mitral cells that project to the medial amygdala, but no sex difference was found in the pattern of projections from the MOB to the medial amygdala (Kang et al. 2011b).

The direct convergence of projections from the main and accessory olfactory bulbs in the amygdala, as well as in a particular area of the anterior piriform cortex, suggests that these structures are likely implicated in the complementary role played by these two kinds of chemosensory information (Keller et al. 2009). In particular, it has been shown that non-volatile molecules detected by the vomeronasal system, which elicit unconditioned responses (e.g. sexual attraction), become associated by means of chemosensory experience with volatile molecules detected by the main olfactory system (Martínez-García et al. 2009). By this learning mechanism, animals may learn to respond to volatiles, which can be detected at a distance, and consequently track the source of the volatile molecules (e.g. a possible mate). A similar learning process might take place in the case of vomeronasal-detected predator signals (Papes et al. 2010) and associated volatiles, allowing the animals to avoid predators at a distance. This learning process has obvious advantages for the animal.

The medial nucleus of the amygdala is involved in sex-specific and sex-steroid olfactory preference (Dibenedictis et al. 2012). Studies of c-fos expression performed in female mice have shown that the cell bodies of this area are activated after exposure to male-derived chemicals (Moncho-Bogani et al. 2005; Kang et al. 2009), and electrolytic lesions of the MeP abolish the preference of females for male-derived urinary chemicals (Dibenedictis et al. 2012). Similar results have been reported in male hamsters (Maras and Petrulis 2006). This evidence, as well as the convergent projections of the vomeronasal (pheromones) and olfactory (volatiles) information in the medial amygdala, makes this structure a likely candidate to be involved in the association between odours and pheromones. With regard to the CxA and ACo, no information is available about their possible roles in pheromone detection, although it has been shown that the ACo expresses c-fos in female mice following chemoinvestigation of male-derived volatiles (Moncho-Bogani et al. 2005; Kang et al. 2009). Further studies are needed to clarify the role of these associative chemosensory structures.

Acknowledgments Funded by the Spanish Ministry of Science-FEDER (BFU2010-16656) and the Junta de Comunidades de Castilla-La Mancha (PEIC11-0045-4490). B.C.-M. is a predoctoral fellow of the "Becas Chile" program of the Government of Chile, and the present work is part of her doctoral thesis.

References

Brennan PA, Kendrick KM (2006) Mammalian social odours: attraction and individual recognition. Philos Trans R Soc Lond B Biol Sci 361:2061–2078

Dibenedictis BT, Ingraham KL, Baum MJ, Cherry JA (2012) Disruption of urinary odor preference and lordosis behavior in female mice given lesions of the medial amygdala. Physiol Behav 105:554–559

Gutiérrez-Castellanos N, Martínez-Marcos A, Martínez-García F, Lanuza E (2010) Chemosensory functions of the amygdala. Vitam Horm 83:165–193

Halpern M (1987) The organization and function of the vomeronasal system. Annu Rev Neurosci 10:325–362

Kang N, Baum KJ, Cherry JA (2009) A direct main olfactory bulb projection to the "vomeronasal" amygdala in female mice selectively responds to volatile pheromones from males. Eur J Neurosci 29:624–634

Kang N, Baum KJ, Cherry JA (2011a) Different profiles of main and accessory olfactory bulb mitral/tufted cell projections revealed in mice using an anterograde tracer and a whole-mount, flattened cortex preparation. Chem Senses 36:251–260

Kang N, McCarthy EA, Cherry JA, Baum KJ (2011b) A sex comparison of the anatomy and function of the main olfactory bulb–medial amygdala projection in mice. Neuroscience 172:196–204

Keller M, Baum MJ, Brock O, Brennan PA, Bakker J (2009) The main and accessory olfactory systems interact in the control of mate recognition and sexual behavior. Behav Brain Res 200:268–276

Krieger J, Schmitt A, Löbel D, Gudermann T, Schultz G, Breer H, Boekhoff I (1999) Selective activation of G protein subtypes in the vomeronasal organ upon stimulation with urine-derived compounds. J Biol Chem 274:4655–4662

Maras PM, Petrulis A (2006) Chemosensory and steroid-responsive regions of the medial amygdala regulate distinct aspects of opposite-sex odor preference in male Syrian hamsters. Eur J Neurosci 24:3541–3552

Martínez-García F, Martinez-Ricós J, Agustín-Pavón C, Martínez-Hernández J, Novejarque A, Lanuza E (2009) Refining the dual olfactory hypothesis: pheromone reward and odour experience. Behav Brain Res 200:277–286

Martínez-García F, Novejarque A, Gutiérrez-Castellanos N, Lanuza E (2012) Piriform cortex and amygdala. In: Watson C, Paxinos G, Puelles L (eds) The mouse nervous system. Academic, Amsterdam, pp 140–220

Martínez-Ricós J, Agustín-Pavón C, Lanuza E, Martínez-García F (2008) Role of vomeronasal system in intersexual attraction in female mice. Neuroscience 153:383–395

Mohedano-Moriano A, Pro-Sistiaga P, Ubeda-Bañon I, Crespo C, Insausti R, Martinez-Marcos A (2007) Segregated pathways to the vomeronasal amygdala: differential projections from the anterior and posterior divisions of the accessory olfactory bulb. Eur J Neurosci 25:2065–2080

Moncho-Bogani J, Lanuza E, Hernández A, Novejarque A, Martínez-García F (2002) Attractive properties of sexual pheromones in mice: innate or learned? Physiol Behav 77:167–176

Moncho-Bogani J, Martínez-García F, Novejarque A, Lanuza E (2005) Attraction to sexual pheromones and associated odorants in female mice involves activation of the reward system and basolateral amygdala. Eur J Neurosci 21:2186–2198

Papes F, Logan DW, Stowers L (2010) The vomeronasal organ mediates interspecies defensive behaviors through detection of protein pheromone homologs. Cell 141:692–703

Pro-Sistiaga P, Mohedano-Moriano A, Ubeda-Bañon I, Arroyo-Jimenez M, Marcos P, Artacho-Pérula E, Crespo C, Insausti R, Martinez-Marcos A (2007) Convergence of olfactory and vomeronasal projections in the rat basal telencephalon. J Comp Neurol 504:346–362

Roberts SA, Simpson DM, Armstrong SD, Davidson AJ, Robertson DH, McLean L, Beynon RJ, Hurst JL (2010) Darcin: a male pheromone that stimulates female memory and sexual attraction to an individual male's odour. BMC Biol 8:75

Scalia F, Winans SS (1975) The differential projections of the olfactory bulb and accessory olfactory bulb in mammals. J Comp Neurol 161:31–56

Shipley MT, Adamek GD (1984) The connections of the mouse olfactory bulb: a study using orthograde and restrograde transport of wheat germ agglutinin conjugated to horseradish peroxidase. Brain Res Bull 12:669–688

Skeen LC, Hall WC (1977) Efferent projections of the main and the accessory olfactory bulb in the tree shrew (*Tupaia glis*). J Comp Neurol 172:1–36

Von Campenhausen H, Mori K (2000) Convergence of segregated pheromonal pathways from the accessory olfactory bulb to the cortex in the mouse. Eur J Neurosci 12:33–46

Chapter 2
Acceleration of Puberty Onset in Female Mice by Male Urinary Odours: Chemosensory, Physiological and Neuroendocrine Mechanisms

Mélanie Jouhanneau and Matthieu Keller

Abstract Among many environmental factors, social factors, especially olfactory cues, can profoundly influence the timing of puberty onset. This is particularly well established in the house mouse (*Mus musculus*) where male urinary odours have been shown to accelerate puberty onset in prepubertal females. The urinary chemosignal(s) accelerate vaginal opening, the first oestrus and uterine growth, three parameters indicative of puberty onset. Several puberty-accelerating chemosignals, the majority of which are small volatile ligands bound to the major urinary proteins (MUPs), have been isolated from male urine; however when they are efficient, it is usually to a lower extent than male urine. These puberty-accelerating chemosignals are detected by the accessory olfactory system and the olfactory information is then transmitted to the hypothalamus. Among various potential neurological targets of the olfactory information, we focus on the hypothalamic neurons expressing neuropeptide kisspeptin, identified as a major regulator of the hypothalamic–pituitary–gonadal axis governing puberty onset. Preliminary results indicate that male odours seems to accelerate the structural changes usually observed in this neuronal population at puberty onset, thus suggesting that it could be a central hypothalamic relay of pheromonal information.

2.1 Introduction

In many mammalian species, male olfactory stimuli play an important role in the control of female reproductive function. A classical example of such a role is the acceleration of puberty onset that occurs in prepubertal female mice when they

M. Jouhanneau • M. Keller (✉)
UMR Physiologie de la Reproduction et des Comportements, UMR 0085 INRA—UMR 7247 CNRS—Université de Tours, INRA Tours, Nouzilly, France
e-mail: mkeller@tours.inra.fr

M.L. East and M. Dehnhard (eds.), *Chemical Signals in Vertebrates 12*,
DOI 10.1007/978-1-4614-5927-9_2,

are exposed following weaning to male pheromones; this effect is also known as the "Vandenbergh effect" (Vandenbergh 1967). Indeed, in the late 1960s and early 1970s, several authors described that juvenile female mice cohabiting with a sexually mature male showed precocious puberty onset, measured through an earlier occurrence of vaginal opening or first oestrus (Castro 1967; Vandenbergh 1969; Drickamer 1974; Eisen 1975; Bronson and Maruniak 1976). This acceleration of puberty occurred in females deprived of vision and hearing suggesting that these stimuli were not required (Bronson and Maruniak 1975). An adult male placed behind a wire mesh barrier also induces puberty acceleration (Drickamer 1974, 1975; Bronson and Maruniak 1975), suggesting that olfactory cues are mostly responsible for puberty acceleration. This acceleration of puberty has been reported in many other rodent species, including rats (Vandenbergh 1976), lemmings (Hasler and Banks 1975) or voles (Hasler and Nalbandov 1974; Baddaloo and Clulow 1981; Frankiewicz and Marchlewska-Koj 1985; Spears and Clarke 1986; Sawrey and Dewsbury 1991; Liang and Shi 2007). The acceleration of puberty onset in mice has been extensively investigated and the researches have been directed at describing (1) the required exposure timing, (2) the pattern of endocrine responses, (3) the production of chemosignal(s) and (4) their chemical identification. Nevertheless, data related to the neuroendocrine mechanisms involved are still scarce. In this context, this chapter reviews the current knowledge regarding the acceleration of puberty onset by male chemosignal(s) in female mice and suggests new perspectives of investigation.

2.2 Puberty Onset in Female Mice

2.2.1 Neuroendocrine Mechanisms

Puberty is the transitional period between the juvenile life-stage and adulthood during which reproductive function becomes functional. This process begins with the activation of the hypothalamic–pituitary–gonadal axis (Fig. 2.1; Sisk and Foster 2004). The pubertal emergence of the pulsatile secretion of gonadotropin-releasing hormone (GnRH) leads to the release of the pituitary gonadotropins: luteinizing hormone (LH) and follicle-stimulating hormone (FSH), into the blood circulation. In female mammals, these hormones act on ovaries to direct oocyte maturation leading up to ovulation as well as the production of steroid hormones, notably estradiol. In turn, high levels of estradiol induce a proliferative expansion of the mammary gland, the growth of the genital apparatus (ovaries, uterus and vagina; Alonso and Rosenfield 2002) and the preovulatory LH surge through a positive feedback loop on hypothalamus and pituitary gland.

There are two main hypotheses proposed to explain the mechanism behind the onset of puberty: (1) the "gonadostat" hypothesis based on steroid-dependent mechanisms and (2) the "central drive" hypothesis based on steroid-independent mechanisms (Ojeda and Skinner 2006). The "gonadostat" hypothesis of puberty states that

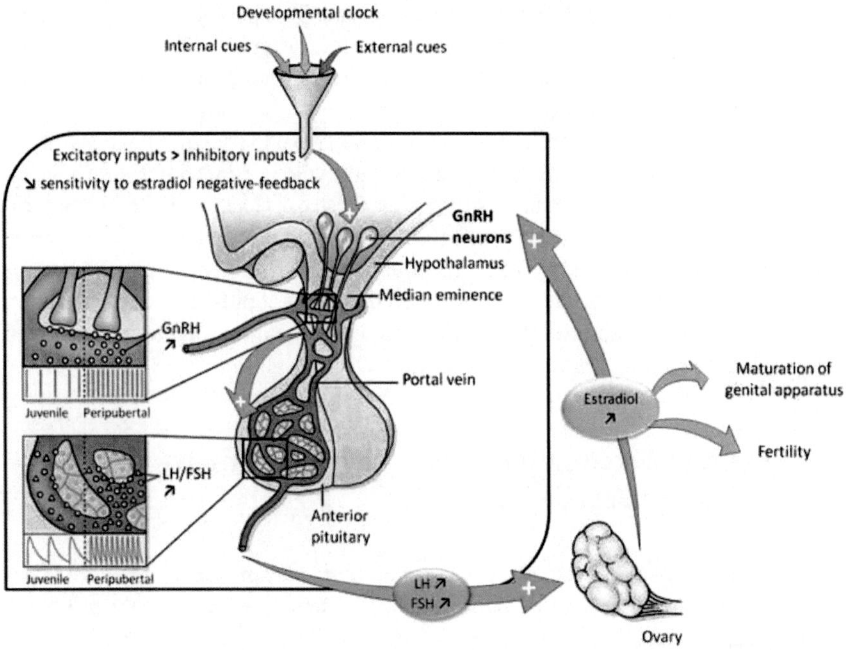

Fig. 2.1 Activation of hypothalamic–pituitary–gonadal axis during puberty onset. Puberty begins with an increase in gonadotropin-releasing hormone (GnRH) neuronal activity under influences of both sex steroid-dependent and -independent mechanisms. In response to pulsatile GnRH secretion, blood-borne luteinizing hormone (LH) and folliculo-stimulating hormone (FSH) act on ovaries which, in turn, secrete oestradiol. Under high level of oestradiol, the maturation of female genital apparatus starts as well as the establishment of oestrous cycle. Adapted by permission from Macmillan Publishers Ltd: Nature Neuroscience 7(10): 1040–1047

a decrease in the sensitivity of the central GnRH pulse generator to the negative feedback exerted by low levels of estradiol in the juvenile period occurs at puberty onset. This sensitivity change would lead to an increase of gonadotropins levels and, in turn, stimulate further release of estradiol. The "central drive" hypothesis proposes that changes in the upstream regulatory inputs to GnRH neurons are responsible for the onset of puberty. An increase in excitatory inputs over inhibitory inputs during the peripubertal period should initiate the pulsatile secretion of GnRH independently of steroid feedback. Currently, it is assumed that both steroid-dependent and -independent mechanisms underlie the onset of puberty.

2.2.2 *The Evaluation of Puberty Onset in Female Mice*

Puberty onset normally occurs around 28–35 days of age depending on the strains or environmental conditions. The emergence of puberty is usually determined by the measurement of parameters which are dependent upon high estradiol levels produced by ovaries. (1) The opening of the vagina: This parameter is evaluated by daily monitoring the vagina from weaning until the vaginal orifice is fully open. (2) The first oestrus: This parameter is evaluated by doing daily vaginal smears from vaginal opening. Indeed, the oestrous cycle may be conveniently monitored by examining the cytological content of vaginal smears reflecting changes in the structure of the vaginal epithelium. The oestrous cycle is divided into four phases: pro-oestrus, oestrus, metoestrus and dioestrus. Oestrus is characterised by the presence of cornified cells (i.e. hexagonal anucleated epithelial cells) and is the period during which ovulation and sexual behaviour occur. (3) Uterine weight: The uterus is highly sensitive to estradiol and increases in weight at puberty. An earlier entry into puberty should therefore be characterised by an earlier vaginal opening and first oestrus as well as a higher uterine weight.

2.3 Puberty Acceleration Induced by Male Urinary Odours in Female Mice

In the literature, two different olfactory stimuli derived from sexually mature males have been shown to accelerate puberty onset in female mice, namely urine and soiled bedding.

2.3.1 Methods of Stimulation

Usually, females are stimulated by a daily application of male urine to the oronasal groove of females or by introducing soiled bedding from males to the homecages of females. Generally, olfactory stimuli are applied from weaning (usually on postnatal day (PD) 21) until females reach puberty. A daily amount of 0.1 μL male urine is sufficient to produce the same degree of acceleration than the much larger amounts normally used (i.e. 10–50 μL urine; Drickamer 1982, 1984). While studies generally use urine or bedding pooled from a group of sexually mature males to ensure an effective stimulation of the females, it has also been shown that acceleration of puberty onset can be obtained by using olfactory stimuli coming from a single male, thus suggesting that the chemosignal(s) originating from a single male is as efficient as a mix from different individuals (Drickamer 1983). Finally, studies have shown that a short period of stimulation may be as effective as a continuous stimulation. A minimum exposure time of 30 min/day to the odour of male bedding is necessary to induce puberty acceleration in females, but longer durations are more effective (>3 h/day; Drickamer 1983). Moreover, a 3-day stimulation accelerated puberty onset when applied from PD 24 or later (Colby and Vandenbergh 1974; Schellinck et al. 1993). This result was confirmed by exposing females to males (Wilson et al. 1980). Interestingly, puberty was also advanced by preweaning exposure to olfactory stimuli (Vandenbergh 1967; Fullerton and Cowley 1971; Cowley and Wise 1972; Drickamer 1988a).

Table 2.1 Comparison of acceleration of vaginal opening and first oestrus by male urinary odours

		Vaginal opening		First oestrus		
Stimuli	Strain	Age	Difference in days from controls	Age	Difference in days from controls	
Male urine	ICR/Alb	–	–	31.5±0.3	5	Drickamer (1982)
		–	–	30.3±0.6	5	Drickamer (1983)
		–	–	31.3±0.8	3	Drickamer (1983)
		–	–	29.2±0.6	4	Drickamer (1984)
		–	–	29.1±0.6	6	Drickamer (1986)
	Swiss-Webster	28.8	1	30.8	5	Colby and Vandenbergh (1974)
Male-soiled bedding	Wild	–	–	48.9±1.3	8	Drickamer (1979)
	Albino	32.7±0.5	NS	45.4±1.5	9	Vandenbergh (1969)
	ICR/Alb	–	–	30.4±0.7	4	Drickamer (1983)
	Laboratory	26.2±0.5	2	30.7±0.6	3	Drickamer (1974)
		26.2±0.3	2	31.5±0.6	4	Drickamer (1974)

NS = Not significant ($p > 0.05$)

2.3.2 Magnitude of the Accelerating Effect

Exposure to male urinary odours results in an advance in vaginal opening by approximately 2 days. A similar advance was observed with male-soiled bedding (Table 2.1). The advance of first oestrus using these methods was 3–9 days in exposed females compared to non-exposed (Table 2.1). Finally, the acceleration effect is even more heterogeneous when this effect is assessed by measuring uterine weights (with uterine weight increases between 50 and 270% in comparison to control females (Table 2.2)). Although the puberty acceleration is a quite robust model, a large degree of variability in the advancement of puberty is observed between different studies. Many factors can account for these differences in the magnitude of the effect, including the stimulation with male odours, the strains of mice used or the measures used to evaluate puberty onset.

2.3.3 Hormonal Responses

Bronson and Desjardins (1974) were able to describe the pattern of hormone changes underlying puberty acceleration by measuring plasma LH, FSH and estradiol levels in CF-1 females cohabiting between PD 25 and 28 with an adult male. On the day of exposure to the male, a rapid increase in LH pulsatility is observed in the first 3 h, followed by a dramatic 15- to 20-fold increase in estradiol levels 12 h later. Estradiol increase induces uterine growth from 24 h of exposure. A second estradiol surge

Table 2.2 Comparison of uterine weight by male urinary odours

				Uterus			
Stimulus	Strain	Exposure time	Exposure frequency	Age at removal	Weight (mg)	Weight gain relative to controls	
Male urine	CF-1	3 PDs	3 per day	Third PD	26 ± 2	NS	Bronson and Maruniak (1975)
		3 PDs	Continuous		43 ± 10	+52%	Bronson and Maruniak (1975)
		3 PDs	Continuous		44	+56%	Bronson and Maruniak (1976)
	SJL/JxSWR/J	PD 25	Once	PD 27	50	+100%	Wilson et al. (1980)
		PD 25–PD 27	1 per day	PD 29	80.0 ± 6.8	+156%	Novotny et al. (1999a)
	ICR/Alb	PD 21–PD 27	2 per day	PD 27	95	+138%	Novotny et al. (1999b)
		PD 21–PD 28	1 per day	PD 28	56.9 ± 9.25	+65%	Price and Vandenbergh (1992)
		PD 25–PD 27	1 per day	PD 29	102.9 ± 12	+178%	Novotny et al. (1999a)
		PD 28–PD 36	1 per day	PD 37	60.5 ± 8.74	+93%	Vandenbergh et al. (1975)
	Swiss-Webster	PD 28–PD 35	1 per day	PD 36	70	+250%	Lombardi et al. (1976)
	Swiss	PD 30–PD 32	1 per day	PD 33	50 ± 10	+100%	Mucignat-Caretta et al. (1995)
	BALB/c	PD 22(24)–28(30)	2 per day	PD 28(30)	90	+200%	Flanagan et al. (2011)
	C57BL/6	PD 22(24)–28(30)	2 per day	PD 28(30)	Data not shown	NS	Flanagan et al. (2011)
Male-soiled bedding	Swiss	PD 28–PD 30	Continuous	PD 29	90	+260%	Schellinck et al. (1993)

NS = Not significant ($p > 0.05$)

occurs after 36 h of exposure. At this time, uterus is twofold larger in male-exposed compared to isolated females (Bronson and Stetson 1973; Bronson and Maruniak 1975). On day 3, 75% of female mice express a pro-oestrus-like uterus (ballooned uterus with intraluminal fluid and a weight over 100 mg). Normal adult-like preovulatory changes in circulating concentrations of FSH and LH are detected. In conclusion, adult male exposure elicits an immediate release of subovulatory levels of LH which produces enough estradiol to initiate an adult-like preovulatory surge of LH associated with a uterine growth.

Although the endocrine mechanisms of puberty acceleration in females associated with male olfactory stimuli have not been investigated in detail, two studies have reported a rapid increase of serum LH levels in the first 2 h after exposure to male urine (Bronson and Maruniak 1976; Darney et al. 1992). This response is however less strong than in the case of exposure to a male.

2.4 Mechanisms of Production of the Puberty-Accelerating Chemosignal(s)

2.4.1 Hormonal Regulation

The puberty-accelerating activity disappears when castrated male urine is used, while the effect is restored when animals receive testosterone injection (Vandenbergh 1969; Bronson and Maruniak 1975; Vandenbergh et al. 1975; Lombardi et al. 1976; Pandey and Pandey 1988), thus demonstrating that the urinary chemosignal(s) production is androgen-dependent.

2.4.2 Source of Production

Urine from testosterone-injected ovariectomised females accelerates uterine growth similarly to male urine (Lombardi et al. 1976). This indicates that neither testes nor any male-specific organs, such as epididymes, seminal vesicles, coagulating gland and prostate, are necessary sites of the puberty-accelerating chemosignal(s) synthesis. Moreover, females showed a hastened first oestrus when exposed to male urine without preputial secretions (bladder urine) or urine of preputialectomised or adrenalectomised males (Colby and Vandenbergh 1974; Drickamer and Murphy 1978; Drickamer 1983; Khan et al. 2009). Therefore, preputial and adrenal glands are not involved in puberty-accelerating chemosignal(s) production. To date, no study has identified the organ where the male chemosignal(s) is produced; however Lombardi et al. (1976) suggested that the liver and/or the kidneys may be likely candidates involved in chemosignal(s) metabolism and its excretion. It must be noticed that this question is rendered difficult by the fact that the chemical nature of the chemosignal(s) is a matter of debate (see Sect. 2.5).

2.4.3 Excretion

As mentioned above, the puberty-accelerating olfactory signal(s) are excreted in male urine. The excretion of this chemosignal(s) is exclusively found in adults (Drickamer and Murphy 1978) and is still present in 15-months-old animals (Drickamer 1988b). The chemosignal(s) begins to be produced around PD 80 in wild mice and PD 50 in laboratory ICR/Alb mice (Drickamer 1988b, 1992).

2.4.4 Social Influences on the Efficiency of the Chemosignal(s)

When prepubertal males are housed with an adult male, there is a retarded excretion of the puberty-accelerating chemosignal(s) (no effect at least until an age of 72 days vs. 50 days in the absence of adult, Drickamer 1992). Pooled urine collected from grouped males (up to five per cage) continues to accelerate puberty as well as urine from isolated males (Drickamer 1983). Dominance relationships seem also to influence the expression of the puberty-accelerating chemosignal(s) because subordinate males did not produce any puberty-accelerating effect, in contrast to dominant (Lombardi and Vandenbergh 1977; Drickamer 1983). This could be related to the fact that dominant males tend to have higher testosterone levels and presumably higher puberty-accelerating chemosignal(s) levels since their production is androgen-dependent (Zielinski and Vandenbergh 1993) and that dominant male urinary odours, contrary to those of subordinate male, preferentially activate a neural signalling pathway leading to the hypothalamus (i.e. the accessory olfactory pathway including the accessory olfactory bulb (AOB), the medial amygdala, the bed nucleus of the stria terminalis, the preoptic nucleus and the ventromedial hypothalamus; Veyrac et al. 2011; see Sect. 2.6).

2.5 Identification of Puberty-Accelerating Chemosignal(s)

Until now, several compounds have been isolated in male urine from various laboratory strains. However, none of these compounds seems to have a comparable effectiveness to male urine.

The early studies by Vandenbergh et al. (1975) suggested that the puberty-accelerating chemosignal(s) is present in the high-molecular-weight (HMW) male urine fraction and that it belonged to a family of hydrophilic proteins. However, as animal or urine placed behind a wire mesh barrier induces a significant effect on female sexual maturation, a volatile nature of the chemosignal(s) can not be excluded (Vandenbergh 1969; Bronson and Maruniak 1975; Wilson et al. 1980). This was reinforced by the fact that male low-molecular-weight (LMW) urine fraction was shown to accelerate puberty measured by uterine weight (Vandenbergh et al. 1976).

In the 1990s, several authors isolated different compounds acting potentially on puberty acceleration (Table 2.3): the major urinary proteins complex (MUPs; in HMW fraction; Mucignat-Caretta et al. 1995), some of their volatile ligands (2-*sec*-butyl-4,5-dihydrothiazole, 3,4-dehydro-*exo*-brevicomin and 6-hydroxy-6-methyl-3-heptanone in HMW fraction; Novotny et al. 1999b) and two volatile amines (isobutylamine and isoamylamine in LMW fraction; Nishimura et al. 1989).

2.5.1 *MUPs*

MUPs belong to a larger family of proteins known as lipocalins. MUPs are small hydrophilic proteins (18 kDa) which bind volatile hydrophobic molecules through their central hydrophobic pocket. MUPs are synthesised in the liver under androgen control and are present in adult male urine at very high concentration (1–5 mg/mL; Cavaggioni and Mucignat-Caretta 2000; Beynon and Hurst 2004; Zhou and Rui 2010). Mucignat-Caretta et al. (1995) demonstrated that MUPs depleted from their volatile ligands through organic extraction or by competition displacement with a high-affinity ligand (IBMP: 2-isobutyl-3-methoxypyrazine) induce a precocious puberty by comparing the uterine weight between females exposed or not. Moreover, a synthetic hexapeptide, called MUP-related hexapeptide (*N*-Glu-Glu-Ala-Arg-Ser-Met), which contains part of the amino acid sequence of the *N*-terminal sequence of MUP (four residues in common) also induces puberty acceleration. In contrast, the volatile ligands separated from MUP were inactive.

These results, however, were challenged by Novotny et al. (1999b) who suggested that volatile ligands were still persistent in the MUPs fraction and argued for the key role of these volatile ligands in puberty acceleration. Indeed, the MUPs fraction retained the typical odour of male mouse (which is known to be due to the presence of volatiles) even after extensive purifications and lyophilisations. In addition, the recombinant MUP-1, which is devoid of any volatile ligand, had no biological activity in the context of puberty acceleration. The efficiency of the MUP-related hexapeptide is also questionable since neither Novotny et al. (1999b) nor Flanagan et al. (2011) found an accelerating effect when retesting its effect.

2.5.2 *Volatile Ligands Bound to MUPs*

Volatile ligands bound to MUPs are small, hydrophobic, androgen-dependent molecules (Flower 1996). The advantages of such ligand–protein binding are that it provides a slow release of volatile chemosignal(s) into the environment and thus is a protection against degradation (Hurst et al. 1998; Beynon and Hurst 2003). Five synthetic analogues of volatile ligands exhibiting a strong affinity to male MUPs (2-*sec*-butyl-4,5-dihydrothiazole, 3,4-dehydro-*exo*-brevicomin, α- and β-farnesene and 6-hydroxy-6-methyl-3-heptanone; Bacchini et al. 1992; Robertson et al. 1993;

Table 2.3 Urine fractions and synthetic compounds responsible for puberty acceleration

	Puberty-acceleration bioassay							
	Concentration	Donor strain	Recipient strain	Exposure mode	Vaginal opening	First oestrus	Uterine weight	
MUPs complex		Swiss	Swiss	Oronasal	–	–	✓	Mucignat-Caretta et al. (1995)
		ICR/Alb	ICR/Alb	Oronasal	–	–	✓	Novotny et al. (1999b)
		Swiss	BALB/c	Oronasal	–	–	NS	Flanagan et al. (2011)
MUP-1 recombinant	4 mg/mL	–	ICR/Alb	Oronasal	–	–	NS	Novotny et al. (1999b)
MUP-related hexapeptide	500 μg/mL	–	Swiss	Oronasal	–	–	✓	Mucignat-Caretta et al. (1995)
		–	ICR/Alb	Oronasal	–	–	NS	Novotny et al. (1999b)
		–	BALB/c	Oronasal	–	–	NS	Flanagan et al. (2011)
NP-SPE fraction		Swiss	BALB/c	Oronasal	–	–	✓	Flanagan et al. (2011)
		BALB/c	BALB/c	Oronasal	–	–	✓	Flanagan et al. (2011)
		C57BL/6	BALB/c	Oronasal	–	–	✓	Flanagan et al. (2011)
Isobutylamine	0.1 M	–	ICR/Alb	Airborne	NS	✓	–	Nishimura et al. (1989)
Isoamylamine	0.1 M	–	ICR/Alb	Airborne	NS	✓	–	Nishimura et al. (1989)
Isobutylamine + Isoamylamine	2×0.05 M	–	ICR/Alb	Airborne	✓	✓	–	Nishimura et al. (1989)
		–	ICR/Alb	Airborne	–	NS	–	Price and vandenbergh (1992)
		–	ICR/Alb	Oronasal	–	–	NS	Price and vandenbergh (1992)
		–	BALB/c	Oronasal	–	–	NS	Flanagan et al. (2011)
2-*sec*-butyl-4,5-dihydrothiazole	1.3 ppm	–	ICR/Alb	Oronasal	–	–	✓	Novotny et al. (1999b)
	2.6 ppm	–	BALB/c	Oronasal	–	–	NS	Flanagan et al. (2011)
3,4-Dehydro-*exo*-brevicomin	1.3 ppm	–	ICR/Alb	Oronasal	–	–	✓	Novotny et al. (1999b)
β-Farnesene	250 ppm	–	ICR/Alb	Oronasal	–	–	✓	Novotny et al. (1999b)
		–	BALB/c	Oronasal	–	–	NS	Flanagan et al. (2011)
6-Hydroxy-6-methyl-3-heptanone	2,000 ppm	–	SJL/J×SWR/J	Oronasal	–	–	✓	Novotny et al. (1999a)
		–	ICR/Alb	Oronasal	–	–	✓	Novotny et al. (1999a)
		–	ICR/Alb	Oronasal	–	–	✓	Novotny et al. (1999b)
		–	BALB/c	Oronasal	–	–	NS	Flanagan et al. (2011)

NS = Not significant ($p > 0.05$)

Lehman-McKeeman et al. 1998; Sharrow et al. 2002; Marchlewska-Koj et al. 2000) showed individually puberty-accelerating effects in ICR/Alb mice (Novotny et al. 1999b). 6-hydroxy-6-methyl-3-heptanone produced also a significant effect in SJL/JxSWR/J mice, but, like in ICR/Alb, the effect was lower compared to male urine (Novotny et al. 1999a). However, Flanagan et al. (2011) tested three out of these compounds (2-*sec*-butyl-4,5-dihydrothiazole, β-farnesene and 6-hydroxy-6-methyl-3-heptanone) on BALB/c females but did not found any accelerating effect.

2.5.3 *Volatile Amines*

Nishimura et al. (1989) isolated from LMW urinary fraction two volatile amines that were much more concentrated in urine of intact than castrated males: isobutylamine and isoamylamine. Synthetic analogues of these compounds accelerate vaginal opening and the first oestrus in ICR/Alb females but only when mixed together. It is important to note that the concentrations used were much higher than the biological concentrations of these compounds found in male urine (10^{-1} vs. $10^{-8,-9}$ M). However, Price and Vandenbergh (1992) were unable to reproduce this experiment; neither male urine nor the mixture of two amines accelerated the first oestrus when placed over the cages. Moreover, when stimuli were applied to the oronasal groove, the mixture did not increase uterine weight whereas male urine did. The lack of effect could be explained by the results of Flanagan et al. (2011) who reported that females are behaviourally aversive to direct application of these two stimuli. Therefore, the role of volatile amines in puberty acceleration is questionable.

2.5.4 *New Lines of Investigation: Identification of Chemosignal(s) Using Solid Phase Extraction*

Since any of the identified molecules described above accelerate puberty in BALB/c females to a similar extent than did male urine, Flanagan et al. (2011) recently attempted to identify the chemosignal(s) by using solid phase extraction (SPE). The fraction containing LMW compounds induces puberty acceleration effect in BALB/c females contrary to the fraction containing MUPs which shows no biological effect. Further fractionation of the LMW urine indicated that the chemosignal(s) is hydrophilic and was similarly effective than male urine in bioassay measuring uterine weight. Moreover, BALB/c females exposed to the corresponding hydrophilic fractions from BALB/c, Swiss and C57BL/6 males showed equivalent effects on uterine weight gain. This suggests that urine of these three strains expresses similar chemosignal(s). Among the identified puberty-accelerating chemosignal(s), BALB/c, Swiss and C57BL/6 male urine shared 2-*sec*-butyl-4,5-dihydrothiazole, 3,4-dehydro-*exo*-brevicomin and isoamylamine. However, these three compounds were not found in the hydrophilic fraction from each strain suggesting that the hydrophilic fraction contains a chemosignal(s) that remains to be identified.

To date, a consensual puberty-accelerating pheromone has not been yet identified. Studies demonstrate rather that various compounds, presumably being volatile ligands, seem to be involved.

2.6 Neural Pathways Mediating the Integration of the Puberty-Accelerating Chemosignal(s)

2.6.1 Chemosensory Perception

In mice, olfactory cues are detected and analysed by the main and accessory olfactory systems. The main olfactory system (MOS) detects airborne odorants whereas the accessory olfactory system requires direct physical contact with the source of odour. Both systems detect wide and partly overlapping range of molecular cues, from small organic volatiles to non-volatile peptides and proteins. Therefore, both systems respond to complex natural stimuli such as urine or soiled bedding (Baum and Kelliher 2009; Keller et al. 2009).

The role of these two olfactory systems in the puberty acceleration effect has been investigated. An approach to test the role of MOS and AOS is either to destroy the main olfactory epithelium (MOE) by zinc sulphate or to surgically ablate the vomeronasal organ (VNO) respectively (McBride et al. 2003; Keller et al. 2006a, b). Neuronal activity assessed by immediate-early gene expression (e.g. c-fos) in olfactory bulbs can also be used (Kovacs 2008).

In response to male-soiled bedding, strong neuronal activity was reported in the AOB (Schellinck et al. 1993). The ablation of the VNO or the section of the vomeronasal nerves abolished the rapid increase of uterine growth in females (Lomas and Keverne 1982; Kaneko et al. 1980). Moreover, TRPC2 mutant mice, which lack the primary sensory channel of the VNO, are unresponsive to the puberty-accelerating chemosignal(s) (Flanagan et al. 2011). By contrast, few neuronal activity is reported in the main olfactory bulb (MOB) in response to male-soiled bedding, contrary to what is observed in the AOB (Schellinck et al. 1993). Moreover, females with bilateral MOB lesions showed an uterine weight increase in the same range than what is observed in females housed with a male (Kaneko et al. 1980). Females with a destruction of the MOE had a hastened first oestrus (Drickamer 1983) confirming the major role of the AOS. Nevertheless, the uteri of Swiss females devoid of MOE and stimulated with male-soiled bedding did not grow after 10 days of stimulation in comparison to intact females also stimulated with male chemosignal(s) (Slotnick et al. 2010). To conclude, it seems that the puberty-accelerating chemosignal(s) is mainly detected by the AOS. However, a role for the MOS probably needs further investigations. The neural signalling pathway leading chemosignal information from olfactory systems to the hypothalamus has not been yet studied. The medial amygdala receiving strongly inputs from AOB and MOB and projecting to hypothalamus could relay this information (Kevetter and Winans 1981; Kang et al. 2009; Boehm 2006; Boehm et al. 2005; Yoon et al. 2005).

2.6.2 *Hypothalamic Integration*

Changes in LH levels in response to the puberty-accelerating chemosignal(s) sign an activation of the hypothalamic–pituitary–gonadal axis and especially of hypothalamic GnRH neurons (Ebling 2005). Unfortunately, very few studies have been performed to dissect the mechanisms involved in puberty acceleration. Darney et al. (1992) measured GnRH concentration by radioimmunoassay in hypothalamus samples after an 8-day exposure to male urine; however, despite differences in uterine growth, GnRH concentration in urine-treated females was similar to control animals.

Recently, a new population of neurons expressing the peptide kisspeptin has been shown to act as a major regulator of GnRH neurons during the puberty onset and more generally during reproductive life (Colledge 2008; d'Anglemont de Tassigny and Colledge 2010). The importance of kisspeptin relies on the fact that: (1) up to 90% of GnRH neurons express receptors of kisspeptin (also known as GPR54; Han et al. 2005; Irwig et al. 2004; Messager et al. 2005) and GPR54 expression increases along puberty onset (Herbison et al. 2010); (2) the number of neurons expressing kisspeptin and the percent of GnRH neurons showing close appositions with kisspeptin fibres increase dramatically in peripubertal mice (Clarkson and Herbison 2006); (3) humans and mice with *kiss1* or *GPR54* mutations showed hypogonadotropic hypogonadism characterised by the absence of pubertal sexual development as well as low gonadal steroid and gonadotropin levels (de Roux et al. 2003; Funes et al. 2003; d'Anglemont de Tassigny et al. 2007); (4) exogenous administration of kisspeptin peptide in prepubertal female rats leads to a precocious activation of the hypothalamic–pituitary–gonadal axis assessed by an earlier vaginal opening and a larger uterine weight (Navarro et al. 2004).

In addition, kisspeptin neurons can be regulated by different signals coming from metabolic, physical or social environment. Notably, it appears that kisspeptin neurons can be sensitive to male sexual odours. In adult female goat, the hypothalamic arcuate nuclei (ARC) containing a population of kisspeptin neurons show an instantaneous rise of electrical activity in presence of buck odours (Hamada et al. 1996; Okamura et al. 2010; Murata et al. 2011). The periodic increase of the electrical activity in this hypothalamic region has been previously shown as being associated with discrete pulses of LH (Wakabayashi et al. 2010). Moreover, in adult female mice, it was shown that c-fos is expressed in kisspeptin-immunoreactive neurons in the rostral periventricular area of the third ventricle of hypothalamus (RP3V) in the presence of male-soiled bedding (Bakker et al. 2010). Given the critical role that kisspeptin plays in puberty onset and that kisspeptin neurons respond to male urinary odours in adult female mice, it is imaginable that kisspeptin neurons could be involved in precocious neuroendocrine changes occurring in prepubertal females in responses to male urinary chemosignal(s).

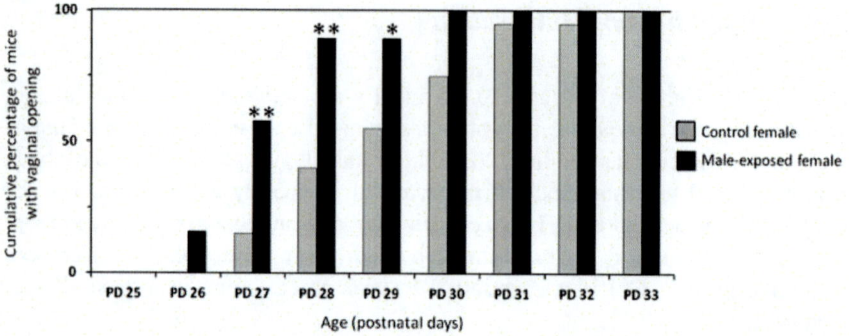

Fig. 2.2 Vaginal opening pattern according to age by male urinary odours. Cumulative percent of female mice showing a vaginal opening according to age and exposure or not to male-soiled bedding from weaning (PD 21). *$p<0.05$, **$p<0.01$

2.6.3 Regulation of Kisspeptin Neurons by Male Urinary Odours

On PD 21, 39 females (Swiss strain) were weaned and housed in individual cages. Then, females were assigned to one of two treatments: (1) clean bedding exposure (control) or (2) male-soiled bedding exposure. Females were examined each morning for vaginal opening and, once the vagina opened, a vaginal lavage was made until a fully cornified vaginal smear appeared (first oestrus). Females exposed to male-soiled bedding reached vaginal opening and first oestrus significantly earlier than control females (Fig. 2.2). Indeed, vaginal opening (VO) and first oestrus (FE) were advanced of about 2 days (VO: PD 27.5±0.3 vs. PD 29.3±0.3, $U=71$, $p<0.001$; FE: PD 28.3±0.4 vs. PD 30.3±0.5, $U=84$, $p<0.01$). The strongest difference in the proportion of animals with vaginal opening was observed on PD 28; 90% of male-exposed females showed a vagina fully open vs. 40% of control females ($X^2_1=10.36$, $p=0.001$).

Then, using the same protocol, exposed and non-exposed females were sacrificed at PD 28; their perfused brains were removed and their uteri weighed ($N=17$ and 14 respectively). As expected, male-exposed females showed a larger uterine weight than non-exposed females with a weight increase by 50% (97.7 mg±12.2 vs. 65.6 mg±9.1 respectively; $U=60$, $p=0.02$). In hypothalamus, two parameters of kisspeptin network development known to change at puberty in females were measured (Clarkson and Herbison 2006): (1) the number of kisspeptin neurons in the rostral periventricular area of the third ventricle (RP3V) and (2) the percent of GnRH neurons showing close contacts with kisspeptin fibres in the rostral preoptic area (rPOA).

Although the study is now still in progress, preliminary results showed an increase of GnRH neurons showing appositions with kisspeptin fibres in rPOA ($t_{29}=2.61$, $p=0.014$), while no difference in the number of kisspeptin cells in RP3V was observed (Fig. 2.3). These preliminary results suggest that the communication between kisspeptin and GnRH neurons is modulated during puberty onset by sexual odours.

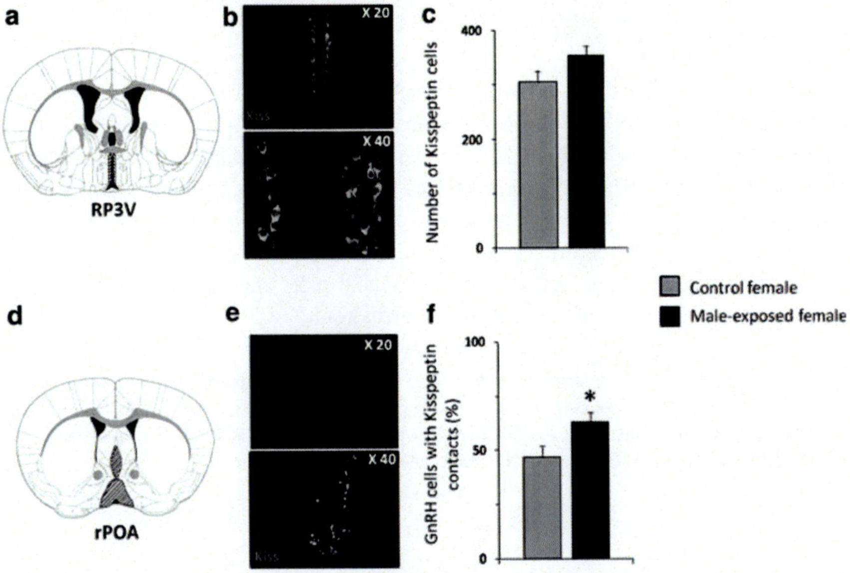

Fig. 2.3 Expression of Kisspeptin by male urinary odours. (**a**) Coronal mice brain sections showing the rostral periventricular area of the third ventricle of hypothalamus (RP3V, hatched zone; adapted from figure 32 of Paxinos and Franklin 2001). (**b**) Confocal images (×20 and ×40 magnification) with kisspeptin-immunoreactive neurons. (**c**) Numbers of RP3V kisspeptin-immunoreactive cell bodies on PD 28 in females exposed or not to male-soiled bedding from PD 21. (**d**) Coronal mice brain sections showing the rostral preoptic area of the hypothalamus (rPOA, hatched zone; adapted from figure 27 of Paxinos and Franklin 2001). (**e**) Confocal images (×20 and ×40 magnification) with GnRH neurons (in *red*) and kisspeptin fibres (in *green*). (**f**) Percent of rPOA GnRH neurons with kisspeptin fibre appositions on PD 28 in females exposed or not to male-soiled bedding from PD 21. *$p<0.05$

2.7 Ecological Validity and Potential Applications

Acceleration of puberty onset in female mice is not just a laboratory curiosity. Under natural conditions, wild male mice produce urinary chemosignal(s) capable of accelerating puberty in wild as well as laboratory females (Drickamer 1979; Pandey and Pandey 1988; Massey and Vandenbergh 1981), thus validating the use of laboratory mice as a model to study the mechanisms sustaining puberty acceleration by male odours in wild mice.

In the field of agronomic sciences, olfactory cues have been also shown to impact the reproductive function of farm animals. Indeed, the acceleration of puberty in females by male chemosignal(s) has been found in several important domestic farm animals (Vandenbergh 1989); among these are pigs (Brooks and Cole 1970; Kirkwood et al. 1981) or heifers (Izard and Vandenbergh 1982). Therefore, the use of social odours to modulate farm animal reproduction can represent an attractive strategy to replace the use of exogenous hormonal treatments which have damaging side effects for the environment.

2.8 Conclusions

This review highlighted the extensive researches carried out on the acceleration of puberty onset in female house mice by male urinary odours. Although many experiments have been done to elucidate the chemosensory, physiological or endocrine aspects of this effect, there is a clear lack of investigation regarding the neuroendocrine consequences of male odour exposure in females. This aspect will undoubtedly be explored in the coming years, especially thanks to the identification of new regulators of the hypothalamic–pituitary–gonadal axis, such as kisspeptin, which provide potential targets of pheromonal information.

Acknowledgements The experiments reported in this chapter were performed with the support of the French National Research Agency (ANR) grants: ANR-PHEROSEX, -MALE-EFFECT, -BIMUCE; Mélanie Jouhanneau is a PhD student supported by INRA PHASE Department and Région Centre. Matthieu Keller is a CNRS permanent research fellow.

References

Alonso LC, Rosenfield RL (2002) Oestrogens and puberty. Best Pract Res Clin Endocrinol Metab 16:13–30

Bacchini A, Gaetani E, Cavaggioni A (1992) Pheromone binding proteins of the mouse, *Mus musculus*. Experientia 48:419–421

Baddaloo EGY, Clulow FV (1981) Effects of the male on growth, sexual-maturation, and ovulation of young female meadow voles, microtus-pennsylvanicus. Can J Zool 59:415–421

Bakker J, Pierman S, Gonzalez-Martinez D (2010) Effects of aromatase mutation (ArKO) on the sexual differentiation of kisspeptin neuronal numbers and their activation by same versus opposite sex urinary pheromones. Horm Behav 57:390–395

Baum MJ, Kelliher KR (2009) Complementary roles of the main and accessory olfactory systems in mammalian mate recognition. Annu Rev Physiol 71:141–160

Beynon RJ, Hurst JL (2003) Multiple roles of major urinary proteins in the house mouse, *Mus domesticus*. Biochem Soc Trans 31:142–146

Beynon RJ, Hurst JL (2004) Urinary proteins and the modulation of chemical scents in mice and rats. Peptides 25:1553–1563

Boehm U, Zou Z, Buck LB (2005) Feedback loops link odor and pheromone signaling with reproduction. Cell 123:683–695

Boehm U (2006) The vomeronasal system in mice: from the nose to the hypothalamus- and back! Semin Cell Dev Biol 17:471–479

Bronson FH, Stetson MH (1973) Gonadotropin release in prepubertal female mice following male exposure: a comparison with the adult cycle. Biol Reprod 9:449–459

Bronson FH, Desjardins C (1974) Circulating concentrations of FSH, LH and oestradiol and progesterone associated with acute, male induced puberty in female mice. Endocrinology 94:548–558

Bronson FH, Maruniak JA (1975) Male-induced puberty in female mice: evidence for a synergistic action of social cues. Biol Reprod 13:94–98

Bronson FH, Maruniak JA (1976) Differential effects of male stimuli on follicle-stimulating hormone, luteinizing hormone, and prolactin secretion in prepubertal female mice. Endocrinology 98:1101–1108

Brooks PH, Cole JA (1970) The effect of the presence of a boar on the attainment of puberty in gilts. J Reprod Fertil 23:435–440
Castro BM (1967) Age of puberty in female mice: relationship to population density and the presence of adult males. An Acad Bras Cienc 39:289–292
Cavaggioni A, Mucignat-Caretta C (2000) Major urinary proteins, alpha(2U)-globulins and aphrodisin. Biochim Biophys Acta 1482:218–228
Clarkson J, Herbison AE (2006) Postnatal development of kisspeptin neurons in mouse hypothalamus; sexual dimorphism and projections to gonadotropin-releasing hormone neurons. Endocrinology 147:5817–5825
Colby DR, Vandenbergh JG (1974) Regulatory effects of urinary pheromones on puberty in the mouse. Biol Reprod 11:268–279
Colledge WH (2008) GPR54 and kisspeptins. Results Probl Cell Differ 46:117–143
Cowley JJ, Wise DR (1972) Some effects of mouse urine on neonatal growth and reproduction. Anim Behav 20:499–506
d'Anglemont de Tassigny X, Fagg LA, Dixon JP, Day K, Leitch HG, Hendrick AG, Zahn D, Franceschini I, Caraty A, Carlton MB, Aparicio SA, Colledge WH (2007) Hypogonadotropic hypogonadism in mice lacking a functional Kiss1 gene. Proc Natl Acad Sci U S A 104:10714–10719
d'Anglemont de Tassigny X, Colledge WH (2010) The role of kisspeptin signaling in reproduction. Physiology 25:207–217
Darney KJ, Goldman JM, Vandenbergh JG (1992) Neuroendocrine responses to social regulation of puberty in the female house mouse. Neuroendocrinology 55:434–443
de Roux N, Genin E, Carel JC, Matsuda F, Chaussain JL, Milgrom E (2003) Hypogonadotropic hypogonadism due to loss of function of the KiSS1-derived peptide receptor GPR54. Proc Natl Acad Sci U S A 100:10972–10976
Drickamer LC (1974) Contact stimulation, androgenized females and accelerated sexual maturation in female mice. Behav Biol 12:101–110
Drickamer LC (1975) Contact stimulation and accelerated sexual maturation of female mice. Behav Biol 15:113–115
Drickamer LC, Murphy RX (1978) Female mouse maturation: effects of excreted and bladder urine from juvenile and adult males. Dev Psychobiol 11:63–72
Drickamer LC (1979) Acceleration and delay of first estrus in wild *Mus musculus*. J Mammal 60:215–216
Drickamer LC (1982) Acceleration and delay of first vaginal oestrus in female mice by urinary chemosignals: dose levels and mixing urine treatment sources. Anim Behav 30:456–460
Drickamer LC (1983) Male acceleration of puberty in female mice (*Mus musculus*). J Comp Psychol 97:191–200
Drickamer LC (1984) Effects of very small doses of urine on acceleration and delay of sexual maturation in female house mice. J Reprod Fertil 71:475–477
Drickamer LC (1986) Acceleration and delay of puberty in female mice via urinary chemosignals: age of the urine stimulus. Dev Psychobiol 19:155–161
Drickamer LC (1988a) Preweaning stimulation with urinary chemosignals and age of puberty in female mice. Dev Psychobiol 21:77–87
Drickamer LC (1988b) Effects of donor age on urinary chemosignals that influence the timing of first oestrus in young female house mice, Mus domesticus. J Reprod Fertil 83:829–834
Drickamer LC (1992) Pre- and postweaning excretion of puberty-influencing chemosignals in house mice. Dev Psychobiol 25:1–16
Ebling FJ (2005) The neuroendocrine timing of puberty. Reproduction 129:675–683
Eisen EJ (1975) Influence of the male's presence on sexual maturation, growth and feed efficiency of female mice. J Anim Sci 40:816–825
Flanagan KA, Webb W, Stowers L (2011) Analysis of male pheromones that accelerate female reproductive organ development. PLoS One 6:e16660
Flower DR (1996) The lipocalin protein family: structure and function. Biochem J 318:1–14

Frankiewicz J, Marchlewska-Koj A (1985) Effect of conspecifics on sexual maturation in female European pine voles (*Pitymys subterraneus*). J Reprod Fertil 74:153–156
Fullerton C, Cowley JJ (1971) The differential effect of the presence of adult male and female mice on the growth and development of the young. J Genet Psychol 3:179–189
Funes S, Hedrick JA, Vassileva G, Markowitz L, Abbondanzo S, Golovko A, Yang S, Monsma FJ, Gustafson EL (2003) The KiSS-1 receptor GPR54 is essential for the development of the murine reproductive system. Biochem Biophys Res Commun 312:1357–1363
Hamada T, Nakajima M, Takeuchi Y, Mori Y (1996) Pheromone-induced stimulation of hypothalamic gonadotropin-releasing hormone pulse generator in ovariectomized, estrogen-primed goats. Neuroendocrinology 64:313–319
Han SK, Gottsch ML, Lee KJ, Popa SM, Smith JT, Jakawich SK, Clifton DK, Steiner RA, Herbison AE (2005) Activation of gonadotropin-releasing hormone neurons by kisspeptin as a neuroendocrine switch for the onset of puberty. J Neurosci 25:11349–11356
Hasler JF, Banks EM (1975) The influence of mature males on sexual maturation in female collared lemmings (Dicrostonyx groenlandicus). J Reprod Fertil 42:583–586
Hasler MJ, Nalbandov AV (1974) The effect of weanling and adult males on sexual maturation in female voles (*Microtus ochrogaster*). Gen Comp Endocrinol 23:237–238
Herbison AE, de Tassigny X, Doran J, Colledge WH (2010) Distribution and postnatal development of Gpr54 gene expression in mouse brain and gonadotropin-releasing hormone neurons. Endocrinology 151:312–321
Hurst JL, Robertson DHL, Tolladay U, Beynon RJ (1998) Proteins in urine scent marks of male house mice extend the longevity of olfactory signals. Anim Behav 55:1289–1297
Irwig MS, Fraley GS, Smith JT, Acohido BV, Popa SM, Cunningham MJ, Gottsch ML, Clifton DK, Steiner RA (2004) Kisspeptin activation of gonadotropin releasing hormone neurons and regulation of KiSS-1 mRNA in the male rat. Neuroendocrinology 80:264–272
Izard MK, Vandenbergh JG (1982) The effects of bull urine on puberty and calving date in crossbred beef heifers. J Anim Sci 55:1160–1168
Kaneko N, Debski EA, Wilson MC, Whitten WK (1980) Puberty acceleration in mice. II. Evidence that the vomeronasal organ is a receptor for the primer pheromone in male mouse urine. Biol Reprod 22:873–878
Kang N, Baum MJ, Cherry JA (2009) A direct main olfactory bulb projection to the 'vomeronasal' amygdala in female mice selectively responds to volatile pheromones from males. Eur J Neurosci 29:624–634
Keller M, Douhard Q, Baum MJ, Bakker J (2006a) Destruction of the main olfactory epithelium reduces female sexual behavior and olfactory investigation in female mice. Chem Senses 31:315–323
Keller M, Pierman S, Douhard Q, Baum MJ, Bakker J (2006b) The vomeronasal organ is required for the expression of lordosis behaviour, but not sex discrimination in female mice. Eur J Neurosci 23:521–530
Keller M, Baum MJ, Brock O, Brennan PA, Bakker J (2009) The main and the accessory olfactory systems interact in the control of mate recognition and sexual behavior. Behav Brain Res 200:268–276
Kevetter GA, Winans SS (1981) Connections of the corticomedial amygdala in the golden hamster. I. Efferents of the "vomeronasal amygdala". J Comp Neurol 197:81–98
Khan A, Berger RG, Decatanzaro D (2009) Preputialectomised and intact adult male mice exhibit an elevated urinary ratio of oestradiol to creatinine in the presence of developing females, whilst promoting uterine and ovarian growth of these females. Reprod Fertil Dev 21:860–868
Kirkwood RN, Forbes JM, Hughes PE (1981) Influence of boar contact on attainment of puberty in gilts after removal of the olfactory bulbs. J Reprod Fertil 61:193–196
Kovacs KJ (2008) Measurement of immediate-early gene activation- c-fos and beyond. J Neuroendocrinol 20:665–672
Lehman-McKeeman LD, Caudill D, Rodriguez PA, Eddy C (1998) 2-sec-butyl-4,5-dihydrothiazole is a ligand for mouse urinary protein and rat alpha 2u-globulin: physiological and toxicological relevance. Toxicol Appl Pharmacol 149:32–40

Liang H, Shi D (2007) Tactical reduction of copulatory competition: effects of male urine odour on maturation rates of Brandt's voles, *Lasiopodomys brandtii*. Folia Zool Brno 56:144–152

Lomas DE, Keverne EB (1982) Role of the vomeronasal organ and prolactin in the acceleration of puberty in female mice. J Reprod Fertil 66:101–107

Lombardi JR, Vandenbergh JG, Whitsett JM (1976) Androgen control of the sexual maturation pheromone in house mouse urine. Biol Reprod 15:179–186

Lombardi JR, Vandenbergh JG (1977) Pheromonally induced sexual maturation in females: regulation by the social environment of the male. Science 196:545–546

Marchlewska-Koj A, Cavaggioni A, Mucignat-Caretta C, Olejniczak P (2000) Stimulation of estrus in female mice by male urinary proteins. J Chem Ecol 26:2355–2366

Massey A, Vandenbergh JG (1981) Puberty acceleration by a urinary cue from male mice in feral populations. Biol Reprod 24:523–527

McBride K, Slotnick B, Margolis FL (2003) Does intranasal application of zinc sulfate produce anosmia in the mouse? An olfactometric and anatomical study. Chem Senses 28:659–670

Messager S, Chatzidaki EE, Ma D, Hendrick AG, Zahn D, Dixon J, Thresher RR, Malinge I, Lomet D, Carlton MB, Colledge WH, Caraty A, Aparicio SA (2005) Kisspeptin directly stimulates gonadotropin-releasing hormone release via G protein-coupled receptor 54. Proc Natl Acad Sci U S A 102:1761–1766

Mucignat-Caretta C, Caretta A, Cavaggioni A (1995) Acceleration of puberty onset in female mice by male urinary proteins. J Physiol 486:517–522

Murata K, Wakabayashi Y, Sakamoto K, Tanaka T, Takeuchi Y, Mori Y, Okamura H (2011) Effects of brief exposure of male pheromone on multiple-unit activity at close proximity to kisspeptin neurons in the goat arcuate nucleus. J Reprod Dev 57:197–202

Navarro VM, Castellano JM, Fernandez-Fernandez R, Barreiro ML, Roa J, Sanchez-Criado JE, Aguilar E, Dieguez C, Pinilla L, Tena-Sempere M (2004) Developmental and hormonally regulated messenger ribonucleic acid expression of KiSS-1 and its putative receptor, GPR54, in rat hypothalamus and potent luteinizing hormone-releasing activity of KiSS-1 peptide. Endocrinology 145:4565–4574

Nishimura K, Utsumi K, Yuhara M, Fujitani Y, Iritani A (1989) Identification of puberty-accelerating pheromones in male mouse urine. J Exp Zool 251:300–305

Novotny MV, Jemiolo B, Wiesler D, Ma W, Harvey S, Xu F, Xie TM, Carmack M (1999a) A unique urinary constituent, 6-hydroxy-6-methyl-3-heptanone, is a pheromone that accelerates puberty in female mice. Chem Biol 6:377–383

Novotny MV, Ma W, Wiesler D, Zidek L (1999b) Positive identification of the puberty-accelerating pheromone of the house mouse: the volatile ligands associating with the major urinary protein. Proc Biol Sci 266:2017–2022

Ojeda SR, Skinner MK (2006) Puberty in the rat. In: Neill JD, Plant TM, Pfaff DW, Challis JRG, de Kretser DM, Richards JS, Wassarman PM (eds) Knobil and Neill's physiology of reproduction, 3rd edn. Academic, St Louis, pp 2061–2126

Okamura H, Murata K, Sakamoto K, Wakabayashi Y, Ohkura S, Takeuchi Y, Mori Y (2010) Male effect pheromone tickles the gonadotrophin-releasing hormone pulse generator. J Neuroendocrinol 22:825–832

Pandey SD, Pandey SC (1988) Male-induced puberty acceleration in young female wild mice: hormonal regulation and source of pheromonal cue. Acta Physiol Hung 72:165–170

Paxinos G, Franklin KBJ (2001) The mouse brain in stereotaxic coordinates, 2nd edn. Academic, San Diego

Price MA, Vandenbergh JG (1992) Analysis of puberty-accelerating pheromones. J Exp Zool 264:42–45

Robertson DHL, Beynon RJ, Evershed RP (1993) Extraction, characterization, and binding analysis of two pheromonally active ligands associated with major urinary protein of house mouse (*Mus musculus*). J Chem Ecol 19:1405–1416

Sawrey DK, Dewsbury DA (1991) Males accelerate reproductive development in female montane voles. J Mammal 72:343–346

Schellinck HM, Smyth C, Brown R, Wilkinson M (1993) Odor-induced sexual maturation and expression of c-fos in the olfactory system of juvenile female mice. Brain Res Dev Brain Res 74:138–141

Sharrow SD, Vaughn JL, Zidek L, Novotny MV, Stone MJ (2002) Pheromone binding by polymorphic mouse major urinary proteins. Protein Sci 11:2247–2256

Sisk CL, Foster DL (2004) The neural basis of puberty and adolescence. Nat Neurosci 7:1040–1047

Slotnick B, Restrepo D, Schellinck H, Archbold G, Price S, Lin W (2010) Accessory olfactory bulb function is modulated by input from the main olfactory epithelium. Eur J Neurosci 31:1108–1116

Spears N, Clarke JR (1986) Effect of male presence and of photoperiod on the sexual maturation of the field vole (*Microtus agrestis*). J Reprod Fertil 78:231–238

Vandenbergh JG (1967) Effect of the presence of a male on the sexual maturation of female mice. Endocrinology 81:345–349

Vandenbergh JG (1969) Male odor accelerates female sexual maturation in mice. Endocrinology 84:658–660

Vandenbergh JG, Whitsett JM, Lombardi JR (1975) Partial isolation of a pheromone accelerating puberty in female mice. J Reprod Fertil 43:515–523

Vandenbergh JG (1976) Acceleration of sexual maturation in female rats by male stimulation. J Reprod Fertil 46:451–453

Vandenbergh JG, Finlayson JS, Dobrogosz WJ, Dills SS, Kost TA (1976) Chromatographic separation of puberty accelerating pheromone from male mouse urine. Biol Reprod 15:260–265

Vandenbergh JG (1989) Coordination of social signals and ovarian function during sexual development. J Anim Sci 67:1841–1847

Veyrac A, Wang G, Baum MJ, Bakker J (2011) The main and accessory olfactory systems of female mice are activated differentially by dominant versus subordinate male urinary odors. Brain Res 1402:20–29

Wakabayashi Y, Nakada T, Murata K, Ohkura S, Mogi K, Navarro VM, Clifton DK, Mori Y, Tsukamura H, Maeda K, Steiner RA, Okamura H (2010) Neurokinin B and dynorphin A in kisspeptin neurons of the arcuate nucleus participate in generation of periodic oscillation of neural activity driving pulsatile gonadotropin-releasing hormone secretion in the goat. J Neurosci 30:3124–3132

Wilson MC, Beamer WG, Whitten WK (1980) Puberty acceleration in mice. I. Dose–response effects and lack of critical time following exposure to male mouse urine. Biol Reprod 22:964–972

Yoon H, Enquist LW, Dulac C (2005) Olfactory inputs to hypothalamic neurons controlling reproduction and fertility. Cell 123:669–682

Zhou Y, Rui L (2010) Major urinary protein regulation of chemical communication and nutrient metabolism. Vitam Horm 83:151–163

Zielinski WJ, Vandenbergh JG (1993) Testosterone and competitive ability in male house mice, *Mus musculus*: laboratory and field studies. Anim Behav 45:873–891

Chapter 3
The Recessus Olfactorius: A Cryptic Olfactory Organ of Anuran Amphibians

Christine Nowack, Sabrina Jordan, and Carolin Wittmer

Abstract The recessus olfactorius is a presumably olfactory structure found on the rostral floor of the main nasal cavity of many anuran amphibians. Nevertheless, it has rarely been investigated or even recognized in the past. The aim of the present study was to broaden the knowledge regarding this enigmatic sensory organ. Therefore, we examined the nasal epithelia of the cane toad *Bufo marinus* and the fire-bellied toad *Bombina orientalis*, using histology, scanning electron microscopy and lectin histochemistry.

Histologically, the recessus olfactorius largely resembles a typical pseudostratified olfactory epithelium. However, we found significant differences when comparing the recessus olfactorius with the main olfactory organ and the vomeronasal organ. The surface differentiations of the recessus olfactorius can be assigned to four different cell types: sensory cells covered with either cilia located on protruding knobs or with long microvilli, and supporting cells bearing either short microvilli or cilia without any surface elevation. Additionally, the lectin binding patterns indicate the presence of a unique mucous constitution of the recessus olfactorius.

Taken together our results strongly support the idea, that the recessus olfactorius has a discrete olfactory function. Furthermore, a possible homology between the recessus olfactorius and the middle chamber epithelium of clawed frogs like *Xenopus laevis* is discussed.

C. Nowack (✉) • S. Jordan • C. Wittmer
Department of Zoology/Developmental Biology, University of Kassel, WG Functional Vertebrate Anatomy, Heinrich-Plett-Straße 40, 34132 Kassel, Germany
e-mail: nowack@uni-kassel.de; sabrina.jordan@uni-kassel.de; c.wittmer@uni-kassel.de

M.L. East and M. Dehnhard (eds.), *Chemical Signals in Vertebrates 12*,
DOI 10.1007/978-1-4614-5927-9_3, © Springer Science+Business Media New York 2013

3.1 Introduction

The recessus olfactorius is a distinct area of sensory epithelium within the nasal cavity of many anuran amphibians (frogs and toads). It is clearly isolated from both the main olfactory organ and the vomeronasal organ. Since its first descriptions about 80 years ago (Du Toit 1933; Rowedder 1937; Helling 1938; Slabbert 1945) few investigations have addressed this subject, and most details about the recessus olfactorius remain undiscovered. The only current study touching the recessus olfactorius describes an immunoreactivity of this "novel sensory epithelium" for the neuronal marker NCAM (Jungblut et al. 2011), suggesting it functions in sensing water-borne olfactory stimuli. Further data on the ultrastructure or functional properties of this organ remain unknown. Occasional discussions consider whether the recessus olfactorius is homologous to the so-called "middle chamber epithelium" of pipid frogs like *Xenopus laevis* (Helling 1938; Paterson 1951).

In the present study, we analyse different aspects of the recessus olfactorius of the cane toad *Bufo marinus* and the fire-bellied toad *Bombina orientalis* using histology, scanning electron microscopy, and lectin histochemistry. The results are compared to findings regarding the main olfactory and the vomeronasal system.

3.2 Methods and Materials

3.2.1 Animal Material

Adult animals of the two anuran species *B. marinus* and *B. orientalis* were examined. Series of histological cross sections of the nasal regions of both species were available from the anatomical collection of our work group. This material consisted of 10 μm paraffin sections stained with either Azan or Delafield's haematoxylin. Size and sex of these animals were not known. Sections were analysed and documented using microscopic digital photographs and computer-assisted schematic drawings.

For electron microscopy and lectin histochemistry, in each case we used one female animal of *B. orientalis* from our own breeding group (SVL 3.0 cm). Additionally, one female *B. marinus* (SVL 5.5 cm) was purchased from a commercial supplier for electron microscopy. The animals were anaesthetised using an aqueous solution of MS-222, decapitated under anaesthesia, and the heads fixed according to the methods described below.

3.2.2 Scanning Electron Microscopy

For electron microscopy the material was fixed in a 2.5% solution of glutaraldehyde in 0.05 M phosphate buffer for at least 60 h. After rinsing in the same buffer, the tissue was dehydrated with graded steps of ethanol up to a concentration of 70%. At this stage the olfactory epithelia were dissected under a microscope. Dehydration was continued, then the pieces of epithelia were dried using a critical point drying apparatus, mounted on pin stubs, and sputter coated with platinum. Coated samples were analysed using a Hitachi S-4000 scanning electron microscope and documented with an attached digital imaging scanning system (DISS 5).

3.2.3 *Lectin Histochemistry*

For lectin histochemistry, the animal material was fixed in a mixture of 96% ethanol and 35% formaldehyde (7:3) for four days. After rinsing with tap water, the tissue was embedded in paraffin using standard protocols. Ten micrometre thick coronal paraffin sections of the nasal region were produced and mounted on chromalum–gelatin-coated glass slides. After that, the tissue was deparaffinised and rehydrated, followed by an endogenous peroxidase blocking step. The slides were then incubated with the lectin *Dolichos biflorus* agglutinin (DBA) at a concentration of 10 μg/mL. The lectin used in the present study has a specificity for *N*-α-D-galactosaminyl-residues (α-GalNAc) (Etzler and Kabat 1970). To enable detection of the binding sites, the DBA used was horseradish peroxidase-conjugated. After an incubation time of 48 h at 4°C, the sections were rinsed in PBS (phosphate-buffered saline, pH 7.4). Localisation of DBA was visualised with the chromogen NovaRED®. After coverslipping with Entellan®, the staining results were analysed and documented by light microscopy (Leica DM E with camera EC 3). Additional slides were treated with 20 μg/mL rhodamine-conjugated DBA, and results visualised using a confocal laser scanning microscope (CLSM, Leica TCS SP 5 II). The confocal images obtained were modified using the programme Adobe® Photoshop® to enhance contrast and are presented as negatives. As a control for both experiments, DBA was omitted from the incubation medium.

3.3 Results

3.3.1 *Histology*

The position of the recessus olfactorius within the nasal cavity is almost identical in *B. marinus* and *B. orientalis*. In both species it is located at the rostral bottom of the main nasal chamber, at the level of the external naris (Fig. 3.1a). It is located laterally from the sensory epithelium of the main olfactory organ, which covers most of the internal surfaces of the main chamber (Fig. 3.1a–c). The vomeronasal organ is located in the lower parts of the nasal cavity, within the inferior chamber (Fig. 3.1b).

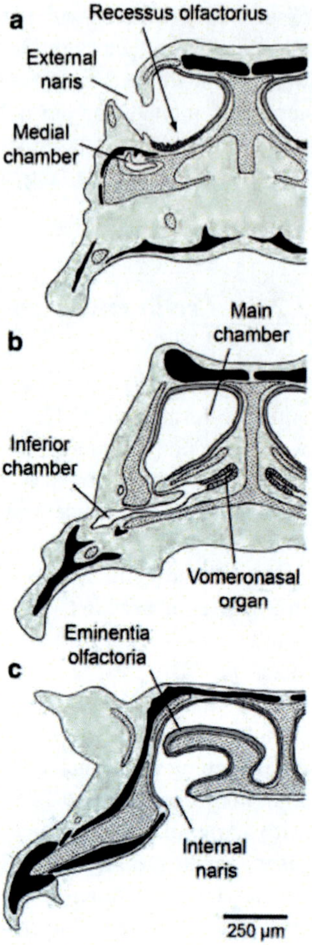

Fig. 3.1 Schematic drawings of histological transverse sections of the nasal region of *B. marinus*. (**a**) Rostral nasal region, demonstrating the position of the recessus olfactorius near to the external naris. (**b**) More posteriorly located section running through the vomeronasal organ. (**c**) Caudal region of the nasal cavity, which opens into the oral cavity via the internal naris. *Dark grey*: sensory epithelium of the main olfactory organ; *hatched dark grey*: sensory epithelium of the vomeronasal organ; *crosshatched dark grey*: recessus olfactorius; *stippling*: cartilage; *black*: bone

The histology of the sensory epithelia of the vomeronasal and the main olfactory organ of *B. marinus* and *B. orientalis* corresponds to the situation generally found in anuran amphibians (Moulton and Beidler 1967). Both epithelia exhibit three different types of cells: olfactory receptor cells, supporting cells, and basal cells. The arrangement of these cells and their respective nuclei is demonstrated in Fig. 3.2. The surface of the main olfactory organ is covered by long cilia of the receptor cells (Fig. 3.2a), whereas the cilia present in the vomeronasal organ belong to the supporting cells of the epithelium (Fig. 3.2b).

The histological architecture of the sensory epithelium of the recessus olfactorius of both species resembles that found in the main olfactory and the vomeronasal organ. Again, there are three different types of cells or cell nuclei visible within the sensory epithelium, subdivided into three layers (Fig. 3.2c). In the apical area of the epithelium, a more or less level row of nuclei is embedded into large amounts of

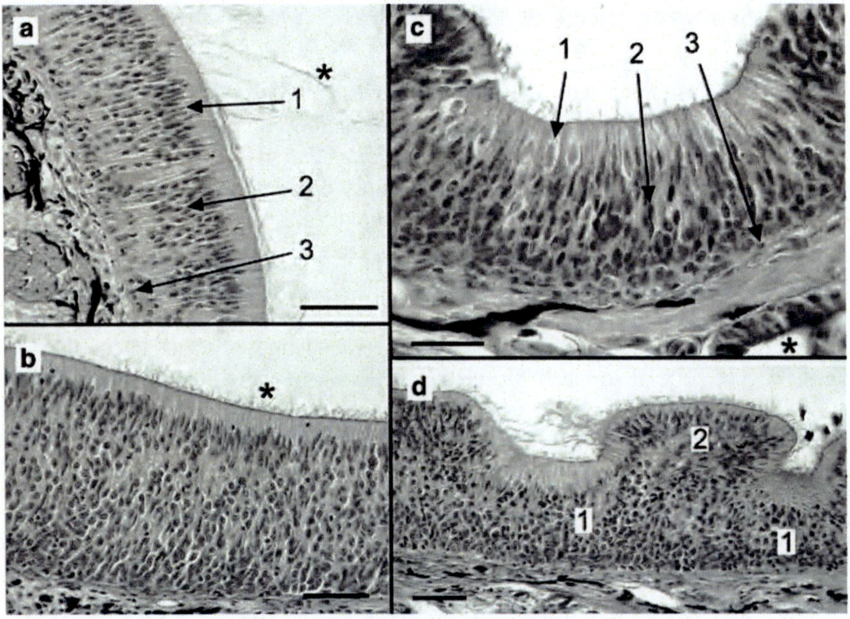

Fig. 3.2 Histological cross sections through the sensory epithelia of the main olfactory organ (**a**), the vomeronasal organ (**b**), and the recessus olfactorius (**c**, **d**). (**a**) Main olfactory organ of *B. marinus*. The nuclei of the different cell types (1: supporting cells; 2: sensory cells; 3: basal cells) are clearly distinguishable. *Asterisk*=cilia of the sensory cells. (**b**) Vomeronasal organ of *B. marinus*. *Asterisk*=cilia of the supporting cells. (**c**) Recessus olfactorius of *B. orientalis* (1: nucleus embedded into large amount of cytoplasm; 2: zone of more closely packed nuclei; 3: smaller nucleus of the basal area). *Asterisk*=alveolus of the glandula nasalis oralis interna. (**d**) Recessus olfactorius of *B. marinus*. Note the irregular surface. The sensory epithelium (1) is located between protruding buds, covered with non-sensory epithelium (2). Scale bars=50 μm; staining: Delafield's haematoxylin

cytoplasm. Below this apical area lies a broad zone with a large number of closely packed nuclei, with little cytoplasm in between. In the basal area of the epithelium, several isolated, considerably smaller nuclei are found, again surrounded by rather small amounts of cytoplasm. The surface of the epithelium is covered with vertically directed, short cilia.

In *B. orientalis* the recessus olfactorius has a uniform epithelial structure, forming a single small depression with a smooth surface (Fig. 3.2c). In contrast, the recessus olfactorius of *B. marinus* displays a rather irregular surface area, with the actual sensory epithelium mainly located in lowered furrows between bud-like protrusions, which themselves are covered with nonsensory epithelium (Fig. 3.2d). Both *B. orientalis* and *B. marinus* possess a glandula nasalis oralis interna, which is closely associated with the recessus olfactorius (Fig. 3.2c) and opens at its medial border.

3.3.2 Scanning Electron Microscopy

Scanning electron microscopy enabled us to investigate the epithelial surface structures of the different olfactory systems. In *B. orientalis* and *B. marinus*, the surface of the recessus olfactorius differs considerably from that of the main olfactory organ and the vomeronasal organ. In the main olfactory organ, the sensory epithelium is evenly covered by the dendritic knobs of the receptor cells, bearing six to ten olfactory cilia (Fig. 3.3a, e) with a maximum length of up to 180 μm in *B. marinus*. Between the protruding knobs, a dense carpet of microvilli emanates from the supporting cells. The surface of the vomeronasal organ is mainly coated by the considerably shorter cilia of the supporting cells (with an average length of about 10 μm in both species). The cellular surface between these cilia forms individual small microvilli (Fig. 3.3b). Furthermore there are interspersed bundles of the notably long microvilli (up to 5 μm in length), which have been reported to be protrusions of the vomeronasal sensory cells (Trotier et al. 1994).

The sensory epithelium of the recessus olfactorius exhibits a more complicated surface arrangement. Figure 3.3c shows an overview of the recessus olfactorius of *B. marinus*. As observed in the histological sections, the recessus in this species has an irregular, sulcated epithelial surface, mainly covered with round buds (Fig. 3.3d). The apical surface of these buds is covered with nonsensory ciliated epithelium, whereas sensory epithelium is found between the buds as well as on their lateral flanks. The sensory epithelium possesses up to four different types of surface differentiations (Fig. 3.3f). We were able to observe areas with very short microvilli and patches with long microvilli (approximately 5 μm in length). Additionally, some cells are covered with cilia, either situated on distinctly protruding knobs, or on a flat surface without any elevation. Both types of cilia are less than 15 μm in length. The cellular surface between the cilia (with and without knobs) is also covered with short microvilli.

3.3.3 Lectin Histochemistry

The lectin binding study with DBA resulted in a distinct staining pattern for the epithelium of the recessus olfactorius, when compared to the main olfactory and the vomeronasal epithelium. Furthermore, there were differences observed in the general binding characteristics of DBA, depending on whether it was linked to horseradish peroxidase (HRP) or rhodamine.

The sensory epithelium of the main olfactory organ (Fig. 3.4a, d) developed intensely stained structures in the apical third of the epithelium and at the luminal surface. These highlighted structures involved secretory vesicles of the supporting cells in the area above the nuclei, as well as a second group of vesicles directly beneath the cell membrane. In contrast to the fluorescence detection method, the peroxidase detection method revealed some additional labelled structures including

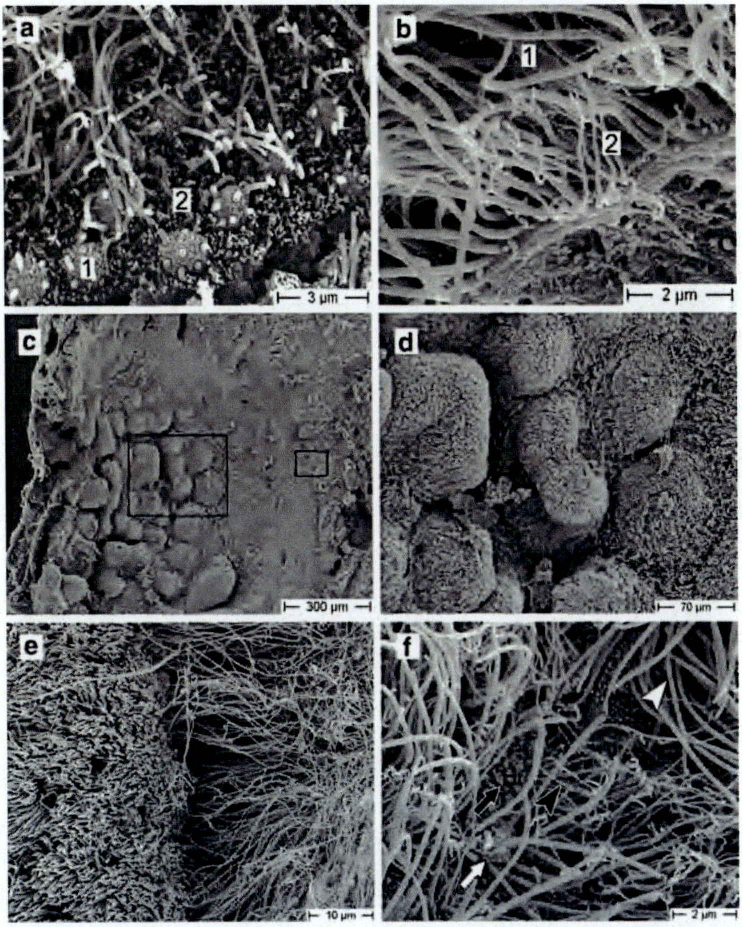

Fig. 3.3 Scanning electron microscopic pictures showing the surface structures of the sensory epithelia of the main olfactory organ (**a**), vomeronasal organ (**b**), and recessus olfactorius (**c–f**). (**a**) Main olfactory organ of *B. orientalis*. 1: Olfactory knobs with sensory cilia (partly cut off due to the preparation); 2: microvilli covering supporting cells. (**b**) Vomeronasal organ of *B. orientalis*. 1: cilia of the supporting cells; 2: microvilli of sensory cells. (**c**) Recessus olfactorius of *B. marinus*. *Boxed* areas magnified in (**d**, **e**). (**d**) Detailed view of the sulcated epithelial surface of the recessus olfactorius. (**e**) *Borderline* between recessus olfactorius (*left*) and main olfactory epithelium (*right*). (**f**) Magnified view of the recessus olfactorius sensory epithelium of *B. marinus*. The epithelium is covered with four different types of surface differentiations: short microvilli (*black arrow*), long microvilli (*black arrowhead*), cilia protruding from knobs (*white arrow*), and cilia without knobs (*white arrowhead*)

the nuclei, cytoplasm, the cilia of olfactory receptor neurons, and the covering mucus.

Unlike the main olfactory organ, the epithelium of the vomeronasal organ (Fig. 3.4b, e) displayed no detectable staining when using rhodamine-labelled DBA. HRP-linked DBA produced a weak staining at the nuclei and short cilia of the supporting cells, as

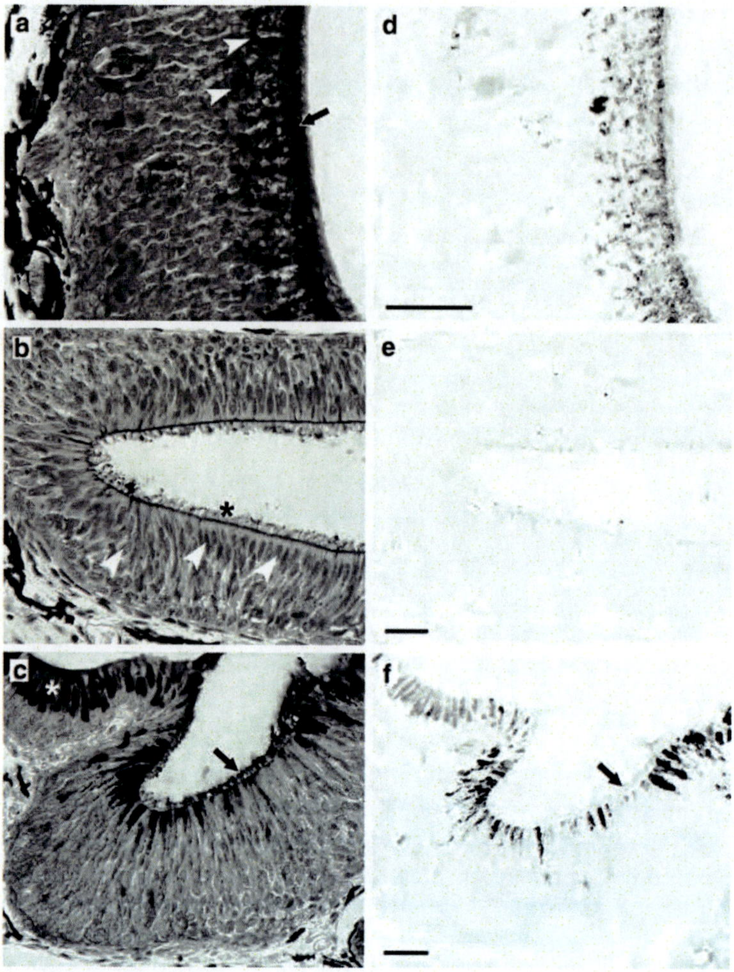

Fig. 3.4 Histological cross sections through olfactory sensory epithelia of *B. orientalis* after lectin histochemistry with DBA-HRP (**a–c**) and DBA-rhodamine (**d–f**), the latter presented as negatives. (**a, d**) Main olfactory epithelium with intensely labelled apical structures above supporting cell nuclei (*white arrowheads*) and directly under the surface (*black arrow*). (**b, e**) Vomeronasal epithelium, only HRP-linked DBA produced detectable staining at the cilia (*asterisk*) and the nuclei (*white arrowheads*). (**c, f**) Recessus olfactorius demonstrating heavily stained vesicular structures within its apical third. Note the different staining of the cilia (*black arrows*). *White* (*asterisk*) = adjacent respiratory epithelium with elongated goblet cells. Scale bars = 50 μm

well as a thin line bordering the distal edge of the epithelium. Additionally, some olfactory receptor neurons contained labelled vesicles above their nuclei; these neurons also possessed darkly coloured dendrites of varying diameters.

The epithelium of the recessus olfactorius (Fig. 3.4c, f) presented an almost identical staining pattern with both the HRP- and the rhodamine-labelled lectin. In either case, large parts of the epithelium remained almost unstained. The only significantly labelled elements were conspicuous vesicular structures, proceeding in cone-shaped arrays from the apical border into deeper levels of the epithelium. Moreover, HRP-linked DBA labelled the short cilia and the apical epithelial surface intensely.

3.4 Discussion

The results presented in this study provide exciting new insights and details of a so far disregarded sensory organ found in anuran amphibians. The recessus olfactorius exhibits unique, characteristic features and is hence clearly distinguishable from the main olfactory organ and the vomeronasal organ in a variety of aspects.

3.4.1 Histology

Our histological results indicate that the recessus olfactorius seems to resemble a typical olfactory epithelium. Therefore, it is likely that the nuclei found in the apical sector of the epithelium belong to supporting cells which contain large amounts of cytoplasm in this upper epithelial region (Fig. 3.2c). Beneath this area, a broad layer of densely packed receptor cell nuclei is located, above the basal cells. To unequivocally confirm this interpretation, further studies on the cellular ultrastructure of the epithelium (e.g. transmission electron microscopy) are needed.

3.4.2 Scanning Electron Microscopy

Despite the histological similarities, the recessus olfactorius differs considerably from the main olfactory and the vomeronasal organ in its surface differentiations. There seem to be at least four distinct cell types viewable at the epithelial surface (Fig. 3.3f). In comparison to already known surface structures from the other two olfactory systems, the following cell types seem to be present: two different types of sensory cells—one developing an olfactory knob with cilia, the other notably long microvilli; and two types of nonsensory cells—with short microvilli and cilia without a knob respectively. Intriguingly, this cytological arrangement is similar to the one described for the middle chamber epithelium of *Xenopus* (Reiss and Burd 1997; Hansen et al. 1998), an accessory olfactory organ only found in pipid frogs. Again transmission electron microscopic investigations are needed to confirm our observations.

3.4.3 Lectin Histochemistry

Lectin binding studies often allow a clear discrimination between olfactory (sub-) systems (e.g. Meyer et al. 1996). In the present study, we found some major differences in the binding properties for DBA when comparing the three investigated olfactory organs. This indicates varying amounts of *N*-α-D-galactosaminyl-residues (Etzler and Kabat 1970), in particular in the distal parts of the epithelia.

A variant distribution of labelled secretory vesicles was observed especially in the recessus olfactorius and the main olfactory organ. Considering our histological results, it is likely that the labelled vesicles in the recessus olfactorius, like in the main olfactory organ, should be elements of supporting cells. These cells are known to produce large amounts of mucus, also containing odorant-binding proteins (Moulton and Beidler 1967; Snyder et al. 1991). Therefore, we suggest that the observed lectin binding patterns resemble differences in the mucus environment of the three olfactory epithelia and possibly differences in odorant perception. As a consequence, the recessus olfactorius is likely to have both a unique mucous constitution and individual function.

We also observed considerable differences depending on whether DBA was linked to HRP or rhodamine. This variation in staining pattern is most likely based on the property of the HRP-molecule to interfere physically with potential binding partners of DBA (Brooks et al. 1996), resulting in an enhanced number of binding sites. For a less ambiguous detection of the desired sugar residues, we recommend the use of fluorescent-labelled lectins.

3.4.4 Concluding Remarks

The recessus olfactorius of the two investigated species—at least from an anatomical point of view—has to be considered as a true olfactory organ. Most of the characteristics described above also indicate that this organ covers a range of functions which differ from those of the main olfactory and the vomeronasal organ. What specific olfactory function the recessus olfactorius mediates remains unclear. The apparent similarity to the middle chamber epithelium of pipid frogs may indicate a functional homology between the two systems. As the middle chamber epithelium seems to be specialised in the detection of water-borne odorants ("water-nose", Altner 1962; Meyer et al. 1996; Reiss and Eisthen 2008), this may also be the case for the recessus olfactorius. On the other hand, the recessus olfactorius is also developed in exclusively terrestrial species like Mozambique rain frogs, *Breviceps mossambicus* (Helling 1938), where water-borne stimuli most likely do not play a role in the life history of these animals. In either case, the glandular secretions discharged on the surface of the recessus olfactorius by epithelial cells or the glandula nasalis interna should be important for the selective binding of the appropriate olfactory molecules, a mechanism which seems to be common in anuran olfaction (Nowack and Wöhrmann-Repenning 2010; Nowack 2011).

At present, the basic question of a phylogenetic homology between the recessus olfactorius and the middle chamber epithelium remains to be determined. As indicated above, Helling (1938)—who studied the olfactory systems of a vast variety of anurans in great detail—was convinced that the middle chamber epithelium and the recessus olfactorius are homologous. Our anatomical results seem to support this hypothesis. On the contrary, it was claimed by Paterson (1951) that a discrete recessus olfactorius is observable for a short period during the larval development of the pipid *Xenopus laevis*, in addition to an already developed middle chamber epithelium. This result clearly speaks against a homology between the two systems, but rather supports the idea of convergent evolution. As other authors have not described comparable observations, or even addressed the question, the situation remains unclear. Further studies on the embryological development, the functional properties, as well as the presence or absence of olfactory receptor molecules in the recessus olfactorius should help to unravel this issue.

Acknowledgements The authors would like to sincerely thank Harald Rühling, Julia Schulze, and Andreas Arendt for their very helpful technical support. Parts of this research were supported by the *Zentrale Forschungsförderung* (ZFF) of the University of Kassel (grant number 1415, P/249).

References

Altner H (1962) Untersuchungen über Leistungen und Bau der Nase des südafrikanischen Krallenfrosches *Xenopus laevis* (Daudin, 1803). Z vergl Physiol 45:272–306

Brooks SA, Lymboura M, Schumacher U, Leathem AJ (1996) Histochemistry to detect *Helix pomatia* lectin binding in breast cancer: methodology makes a difference. J Histochem Cytochem 44:519–524

Du Toit CA (1933) Some aspects of the cranial morphology of *Rana grayi* Smith. Proc Zool Soc Lond 103:715–734

Etzler ME, Kabat EA (1970) Purification and characterization of a lectin (plant hemagglutinin) with blood group A specificity from *Dolichos biflorus*. Biochemistry 9:869–877

Hansen A, Reiss JO, Gentry CL, Burd GD (1998) Ultrastructure of the olfactory organ in the clawed frog, *Xenopus laevis*, during larval development and metamorphosis. J Comp Neurol 398:273–288

Helling H (1938) Das Geruchsorgan der Anuren, vergleichend-morphologisch betrachtet. Z Anat 108:587–643

Jungblut LD, Pozzi AG, Paz DA (2011) Larval development and metamorphosis of the olfactory and vomeronasal organs in the toad *Rhinella* (*Bufo*) *arenarum* (Hensel, 1867). Acta Zool 92:305–315

Meyer DL, Jadhao AG, Bhargava S, Kicliter E (1996) Bulbar representation of the 'water-nose' during *Xenopus* ontogeny. Neurosci L 220:109–122

Moulton D, Beidler L (1967) Structure and function in the peripheral olfactory system. Physiol Rev 47:1–52

Nowack C (2011) Functional anatomy of the lateral nasal gland in anuran amphibians and its relation to the vomeronasal organ. J Herpetol 45:511–515

Nowack C, Wöhrmann-Repenning A (2010) The nasolacrimal duct of anuran amphibians: suggestions on its functional role in vomeronasal perception. J Anat 216:510–517

Paterson NF (1951) The nasal cavities of the toad *Hemipipa carvalhoi* Mir.-Rib. and other Pipidae. Proc Zool Soc Lond 121:381–415

Reiss JO, Burd GD (1997) Cellular and molecular interactions in the development of the *Xenopus* olfactory system. Sem Cell Dev Biol 8:171–179

Reiss JO, Eisthen HL (2008) Chemical senses: comparative anatomy and physiology of chemical senses in amphibians. In: Thewissen JGM, Nummela S (eds) Sensory evolution on the threshold: adaptions in secondarily aquatic vertebrates. University of California Press, California, pp 43–63

Rowedder W (1937) Die Entwicklung des Geruchsorgans bei *Alytes obstetricans* und *Bufo vulgaris*. Zeitschr f Anat u Entwicklungsgesch 107:91–123

Slabbert GK (1945) Contributions to the cranial morphology of the European anuran *Bombina variegata* (Linné). Ann Univ Stell 23:67–89

Snyder DA, Rivers AM, Yokoe H, Menco BPM, Anholt RRH (1991) Olfactomedin: purification, characterization, and localization of a novel olfactory glycoprotein. Biochemistry 30:9143–9153

Trotier T, Døving KB, Rosin JF (1994) Functional properties of frog vomeronasal receptor cells. In: Kurihana K, Suzuki N, Ogawa H (eds) Olfaction and taste XI, Springer. Tokyo, Japan, pp 188–191

Part II
Mother-Offspring Communication

Chapter 4
Smell, Suck, Survive: Chemical Signals and Suckling in the Rabbit, Cat, and Dog

Lourdes Arteaga, Amando Bautista, Daniel González, and Robyn Hudson

Abstract All mammalian young depend for their survival for at least some period on the mother's milk. Locating the mammary region, attaching to a nipple or teat, and being motivated to suck until milk flows is one of the first and most vital challenges they face. For altricial species such as the European rabbit (*Oryctolagus cuniculus*), the cat (*Felis silvestris catus*) and the dog (*Canis lupus familiaris*) born without vision, not directly aided by the mother, and having to compete with littermates, the challenge is particularly great. All three species use chemical cues apparently emitted under hormonal control by lactating females to achieve this. From birth they quickly orient to the mother's ventrum, rapidly locate nipples, and respond with nipple-search behavior and nipple attachment to other lactating females but not to non-lactating females. Kittens, but not rabbits or puppies, also quickly establish a nipple order in which each kitten uses primarily one or sometimes two particular nipples. Recognition of own nipples appears to depend on learned olfactory cues, possibly contained in each kitten's own saliva. These three species illustrate similarities and differences in the use of chemical cues by mammalian young in the suckling context and raise questions warranting further investigation: are there commonalities in the origin and chemical composition of the suckling signals emitted by the mothers of such taxonomically different species? Are there commonalities in the neural processing of such signals, for example, in the participation of the main and accessory olfactory systems? To what extent are inborn responses augmented or even replaced by learned chemical cues? And does the early experience of such cues affect the response to chemical signals in later life?

L. Arteaga (✉) • A. Bautista
Centrol Tlaxcala de Biología de la Conducta, Universidad Autónoma de Tlaxcala,
Tlaxcala 90070, Mexico
e-mail: mlac328@gmail.com

D. González • R. Hudson
Instituto de Investigaciones Biomédicas, Universidad Nacional Autónoma de México,
Ciudad Universitaria, México D.F. Mexico
e-mail: rhudson@biomedicas.mx

M.L. East and M. Dehnhard (eds.), *Chemical Signals in Vertebrates 12*,
DOI 10.1007/978-1-4614-5927-9_4,

4.1 Introduction

All mammalian young depend for their survival for at least some period of early postnatal life on the mother's milk. Locating the mammary region, attaching to a nipple or teat, and being sufficiently motivated to suck until milk flows is one of the first and most vital challenges they face. For altricial species born without vision or audition, often with poor motor coordination, not directly aided by the mother, and often having to compete with littermates, the challenge is particularly great (Hudson and Trillmich 2008; Hudson et al. 2011a). We can therefore expect newborn mammals to have evolved specialised sensory, motor, and cognitive mechanisms to meet this challenge. Additionally, given differences in mammalian lifestyles and accompanying patterns of maternal behaviour, we can further expect variation in the mechanisms the young of different species have evolved to manage different suckling conditions. Indeed, there is strong evidence from an increasing range of mammals that highly specific chemical cues on the lactating female's ventrum play a vital role in the suckling context. Here we give three examples which we use to compare the role of inborn (pheromonal) and learned chemical cues (cf. Wyatt 2010) in ensuring successful suckling in three taxonomically very different but well known and experimentally accessible species: the European rabbit (a lagomorph), the domestic cat (a felid), and the domestic dog (a canid) (Hudson et al. 2011b).

4.2 The European Rabbit

The European rabbit (*Oryctolagus cuniculus*), the ancestor of all domestic and laboratory breeds, is arguably the best-studied mammalian species with regard to the role of chemical stimuli in the regulation of suckling behaviour. This is in part due to the rabbit's unusually limited pattern of maternal care, which makes it particularly easy to study the behaviour of mothers and young in the nursing context. Briefly, female rabbits give birth in a nursery burrow, leave the young almost immediately, and only return to nurse them for a few minutes once approximately every 24 h. During nursing the mother stands virtually motionless over the litter, and the pups compete vigorously for access to nipples in order to obtain sufficient milk to survive the next 24 h (Hudson and Distel 1983; Drummond et al. 2000; Bautista et al. 2005, 2008).

Various studies using a variety of behavioural and surgical methods have shown that species-specific pheromonal cues on the mother's ventrum, which increase in concentration towards the nipples, are essential for the release of the pups' stereotyped nipple-search behaviour, nipple attachment, and sucking (Hudson and Distel 1983, 1986, 1987; Distel and Hudson 1984, 1985; reviews in Hudson and Distel 1995; Hudson et al. 2008). Emission of these cues is under hormonal control. Pups do not respond to anestrous females with nipple-search behaviour, but respond weakly to oestrous females, more strongly to pregnant females, and very strongly to lactating females (Hudson and Distel 1984, 1990). This graded response across the female reproductive cycle can be readily simulated by sequentially administering the hormones estradiol (oestrus), progesterone (pregnancy), and prolactin (lactation) to ovariectomised, non-pheromone producing females (Hudson et al. 1990; González-Mariscal et al. 1994). The response to these chemical cues is inborn in the sense that even without postnatal experience pups respond appropriately to lactating females. Indeed, they often attach to nipples and start sucking while parturition is still in progress (Hudson et al. 1999a), and pups delivered by caesarean section respond on their first contact with a lactating female in the same manner as normally delivered young (Hudson 1985). Cues releasing nipple-search movements and nipple grasping are contained in fresh rabbit milk (Schley 1981; Keil et al. 1990) and a compound in the milk, 2-methylbut-2-enal (2 MB2), has been found to release this behaviour (Schaal et al. 2003). However, as pups respond with nipple-search behaviour to hormonally primed non-lactating females and 2 MB2 is apparently not present on the female's ventral skin (Moncomble et al. 2005), neither the identity nor the origin of the chemical cue(s) eliciting nipple-search behaviour in the functionally relevant natural suckling context is clear (review in Hudson et al. 2008).

Although rabbit pups do not need to learn to respond to the nipple-search pheromone, they rapidly learn to associate novel odours with suckling and to respond to them as to the pheromone itself. If the mother's ventrum is perfumed with a novel odorant and the pups are then nursed by her, the following day they will respond to the novel odour with the full nipple-search sequence when the odorant is presented on the ventrum of an ovariectomised rabbit, a female cat, or a warmed rabbit fur (Hudson 1985; Kindermann et al. 1991, 1994; Allingham et al. 1999). This rapid associative learning is dependent on the presence of the pheromone and somatosensory perioral stimulation rather than on milk intake, at least in young pups (Hudson et al. 2002), and occurs most strongly within a few days of birth (Hudson 1993; Kindermann et al. 1994). Although the functional significance of such learning, if any, is not known, it may contribute to increased suckling efficiency (Kindermann et al. 1994; Coureaud et al. 2002), to the recognition of littermates and later kin recognition, and to recognition of the mother when pups leave the nursery burrow (cf. Mykytowycz and Ward 1971). It may also contribute to the acquisition of food preferences via association with compounds from the mother's diet in her milk, on her skin, or in the faecal pellets she deposit in the nest (Bilkó et al. 1994; Altbäcker et al. 1995; Hudson et al. 1996; review in Hudson et al. 1999b).

Perception of the nipple-search pheromone and learning of suckling-associated odours are mediated by the main rather than by the accessory olfactory system. When made anosmic by removing the olfactory bulbs or irrigating the nasal mucosa with zinc sulphate, newborn rabbits do not respond with nipple-search behaviour to lactating females and are unable to suckle (Schley 1977, 1981; Distel and Hudson 1985), and even when the accessory olfactory system is left intact (Hudson and Distel 1986, 1987). However, olfactory cues alone are not sufficient to ensure suckling. If the subopthalmic branches of the trigeminal nerve innervating the muzzle are transected and so eliminating somatosensory input, pups search for nipples but are unable to grasp them and suck (Distel and Hudson 1985).

4.3 The Domestic Cat

The domestic cat (*Felis silvestris catus*) has a very different pattern of maternal behaviour to the rabbit. Whereas parturition in the rabbit usually takes less than 10 min (Hudson et al. 1999a), in the cat it may take several hours (Hudson et al. 2009). Unlike the rabbit, after giving birth the mother stays with her kittens almost continuously during the first few days, spending most of this time lying on her side in a posture that exposes her nipples to the young. Although she frequently licks the kittens, she does not directly help them to find and attach to nipples (Hudson et al. 2009). Like rabbit pups, newborn kittens quickly orient to the mother's ventrum, show nipple-search behaviour, and may also attach to nipples and suckle while parturition is still in progress. Later, when the mother returns to the litter and adopts the nursing posture, the kittens usually attach to nipples and start sucking within 1 or 2 min although milk letdown may take 15 min or more (Hudson et al. 2009). Unlike rabbit pups, kittens show a gradient in nipple use, and from birth (within the first 12 h) they preferentially orient and attach to the more posterior nipples (Fig. 4.1 upper left panel). The reason for this pattern is not clear as kittens' weight gain seems largely independent of the location of their most used nipples (Hudson et al. 2009).

As in the rabbit, chemical cues on the mother's ventrum are important for the release of the characteristic nipple-search behaviour (cf. Prechtl 1952; Rosenblatt 1972), nipple attachment and sucking, and also appear to be under hormonal control. Thus, kittens fail to respond to non-reproducing females with nipple-search behaviour, respond weakly to pregnant females, and strongly to lactating females (Raihani et al. 2009). However, hormone substitution experiments similar to those conducted in the rabbit are now needed to confirm this and to identify the hormones involved. Furthermore, nothing is known regarding the chemical nature of these cues other than that they differ from the cues used by rabbits since kittens do not respond to lactating rabbits with nipple-search behaviour, nor rabbits to lactating cats (Hudson 1985; unpublished observations).

Unlike rabbits, kittens rapidly establish a nipple order in which each kitten uses mainly one or two particular nipples (Ewer 1959; Hudson et al. 2009; Fig. 4.1 lower left panel). This order is established within 1 or 2 days postpartum (Hudson et al. 2009), and even when kittens are hungry and tested singly in the absence of competition from littermates, they usually fail to attach to nipples other than their "own" (Raihani et al. 2009). When they do so, they typically show lower arousal, weaker sucking behaviour, and soon release the nipple (Raihani et al. 2009). Identification of own nipples appears to depend on chemical rather than on topographical or thermal cues. When tested on lactating females other than their mother, kittens fail to attach to nipples corresponding in location to their particular nipple(s) on their own mother, even after eye-opening (Raihani et al. 2009). Although the nature of chemical "signatures" (cf. Wyatt 2010) distinguishing individual nipples is not known, kittens possibly mark them with their own saliva and/or learn subtle differences in chemical profiles characterising each nipple. While this needs to be investigated

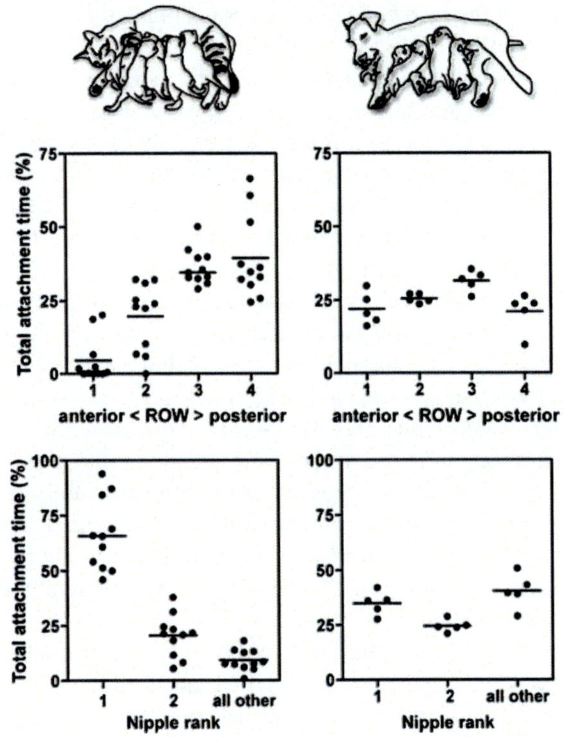

Fig. 4.1 *Left*: In kittens of the domestic cat, preferential use of posterior nipples during the first postnatal month (*above*), and individual use of particular nipples expressed as ranks in which 1 is the highest rank (*below*). Adapted from Hudson et al. (2009). *Right*: In puppies of the domestic dog, no anterior–posterior gradient in preferential nipple use (*above*) and no individual use of particular nipples (*below*). Each *dot* represents a litter mean (cats: $n = 11$ litters; dogs: $n = 5$ litters)

experimentally, it is clear that as in the rabbit, kittens are able to rapidly learn chemical cues associated with suckling (see also Rosenblatt 1972).

That newborn kittens indeed depend on chemical cues to locate and attach to nipples is confirmed by reports of their failure to do so when made anosmic by removing the olfactory bulbs or irrigating the nasal mucosa with zinc sulphate (Kovach and Kling 1967; Larson and Stein 1984; Schuleikina-Turpaeva 1986). However, it is not known whether perception of these cues is mediated by the main and/or by the accessory olfactory system. Nevertheless, as in the rabbit, chemical cues alone are not sufficient to ensure suckling. If the muzzle and lips of kittens are anaesthetised, they are unable to grasp nipples and suckle (Larson and Stein 1984; Blass et al. 1988).

4.4 The Domestic Dog

Maternal behaviour in the domestic dog (*Canis lupus familiaris*) is similar to that in the cat. Parturition may also take several hours and after giving birth the mother usually stays with the litter almost continuously during the first few days, spending most of this time lying on her side in a posture that exposes her nipples to the young.

As for the rabbit and cat, she does not directly help the young to find and attach to nipples (unpublished observations). Nevertheless, like rabbit pups and kittens, puppies orient to and respond to the mother's ventrum with nipple-search behaviour and may also attach to nipples and suckle while parturition is still in progress (unpublished observations). Although our study of suckling behaviour in dogs is still in progress (presently five litters), the pattern of behaviour of the mothers and litters we have so far observed was very similar even though the litters were from different breeds and of different sizes (four to nine puppies). All were observed and filmed in private homes, either as house pets or yard dogs.

Although results from only five litters are presently available, these suggest that, as in the rabbit and cat, chemical cues on the mother's ventrum are important for the release of nipple-search behaviour and nipple attachment. Puppies of the one litter we have so far tested did not respond to two non-reproducing females with nipple-search behaviour, even after several days' suckling experience, but did so immediately they were returned to their mother. Further litters need to be tested and, as in the rabbit, also on pregnant and on ovariectomised, hormone-treated females. At present, nothing is known regarding the chemical nature of suckling cues in the dog. Indeed, it is surprising how little is known about the early development of an animal as familiar and accessible as the dog and particularly in relation to this vital developmental context.

It can nevertheless be stated that, in contrast to the cat, in none of the five litters we have observed across the first postnatal month was there any form of nipple order. There was no apparent anterior–posterior gradient in nipple use (Fig. 4.1 upper right panel) or an individual preference for particular nipples (Fig. 4.1 lower right panel).

4.5 Interspecific Similarities and Differences

The similarities and differences between the rabbit, cat, and dog in olfactory-guided suckling behaviour described above are summarised in Table 4.1. Common to all three (although information for the dog is still limited) is the importance of pheromonal cues on the mother's ventrum for the release of nipple-search behaviour, nipple-attachment, and sucking. Pheromonal gradients on the mother's ventrum exist in the rabbit and probably the cat but apparently not in the dog. The hormonal control of emission of these cues still needs to be examined in more detail in the cat and dog. Also lacking for the cat and dog is information as to whether pheromonal cues are contained in milk and whether they are mediated by the main or by the accessory olfactory system.

Surprising and still needing to be accounted for is why kittens should develop an individual order in nipple use and puppies not. The lack of a nipple order in rabbits is understandable given the short duration of each nursing event and pups' scramble competition for nipple access (Bautista et al. 2005). We previously proposed (Hudson et al. 2009) that in the cat a nipple order might help ensure that by regularly sucking

Table 4.1 Similarities and differences in the role of pheromonal and learned chemical cues in facilitating suckling in the newborn rabbit, cat, and dog

	Rabbit	Cat	Dog
Pheromonal suckling cues	+	+	(+)
Gradient on mother's ventrum	+	+	(–)
Emission under hormonal control	+	+	(+)
Contained in milk	+	?	?
Mediation by main olfactory system	+	?	?
Learned suckling cues	+	+	?
Mediation by main olfactory system	+	?	?
"Teat order"	–	+	–

the same nipple each kitten maintains its own milk supply. However, considering the information now available for the dog, this by itself is not a completely satisfactory explanation. Also important might be the difference between feline and canid mothers in their life style. Mothers of most feline species (obligatory carnivores) are solitary hunters while social canids such as the dog (or more precisely, its ancestor the wolf; a partial omnivore) get their share from the communal kill. In addition, in contrast to dogs, felines—particularly smaller species such as the domestic cat—largely depend for defence on their great agility and notable climbing ability. Feline mothers may therefore have specialised in reducing weight and clumsiness by rapidly shutting down superfluous mammary glands while canid mothers can afford to keep all glands functional. Thus, the pressure on cat mothers to keep the number of functional mammary glands to a minimum may have driven kittens to limit suckling to one or two particular nipples so as to secure their own milk supply.

Acknowledgements We thank Marise Lilian Trejo Elzalde and Luis Daniel Cárdenas Vargas for video analysis, Joel Flores Bonilla for recruiting owners of newborn puppies, Carolina Rojas for technical support, and Hans Distel for valuable discussion. Financial support was provided by CONACYT (48692-Q) and DGAPA (IN223910).

References

Allingham K, Brennan PA, Distel H, Hudson R (1999) Expression of c-fos in the main olfactory bulb of neonatal rabbits in response to garlic as a novel and conditioned odour. Behav Brain Res 104:157–167

Altbäcker V, Hudson R, Bilkó Á (1995) Rabbit mothers' diet influences pups' later food choice. Ethology 99:107–116

Bautista A, Mendoza-Degante M, Coureaud G, Martínez-Gómez M, Hudson R (2005) Scramble competition in newborn domestic rabbits for an unusually restricted milk supply. Anim Behav 70:1011–1021

Bautista A, Martínez-Gómez M, Hudson R (2008) Mother-young and within-litter relations in the European rabbit *Oryctolagus cuniculus*. In: Alves PC, Ferrand N, Hackländer K (eds) Lagomorph biology: evolution, ecology, and conservation. Springer, Berlin, pp 211–223

Bilkó Á, Altbäcker V, Hudson R (1994) Transmission of food preference in the rabbit: the means of information transfer. Physiol Behav 56:907–912

Blass EM, Shuleikina-Turpaeva K, Luschekin V (1988) Sensory determinants of nipple attachment behavior in 2-4-day-old-kittens. Dev Psychobiol 21:365–370

Coureaud G, Schaal B, Hudson R, Orgeur P, Coudert P (2002) Transnatal olfactory continuity in the rabbit: behavioral evidence and short-term consequences of its disruption. Dev Psychobiol 40:372–390

Distel H, Hudson R (1984) Nipple-search performance by rabbit pups: changes with age and time of day. Anim Behav 32:501–507

Distel H, Hudson R (1985) The contribution of the olfactory and tactile modalities to the nipple-search behavior of newborn rabbits. J Comp Physiol A 157:599–605

Drummond H, Vázquez E, Sánchez-Colón S, Martínez-Gómez M, Hudson R (2000) Competition for milk in the domestic rabbit: survivors benefit from littermate deaths. Ethology 106:511–526

Ewer RF (1959) Suckling behaviour in kittens. Behaviour 15:146–162

González-Mariscal G, Chirino R, Hudson R (1994) Prolactin stimulates emission of nipple pheromone in ovariectomized New Zealand white rabbits. Biol Reprod 50:373–376

Hudson R (1985) Do newborn rabbits learn the odor stimuli releasing nipple-search behavior? Dev Psychobiol 18:575–585

Hudson R (1993) Olfactory imprinting. Curr Opin Neurobiol 3:548–552

Hudson R, Distel H (1983) Nipple location by newborn rabbits: behavioural evidence for pheromonal guidance. Behaviour 85:260–277

Hudson R, Distel H (1984) Nipple-search pheromone in rabbits: dependence on season and reproductive state. J Comp Physiol A 155:13–17

Hudson R, Distel H (1986) Pheromonal release of suckling in rabbits does not depend on the vomeronasal organ. Physiol Behav 37:123–129

Hudson R, Distel H (1987) Regional autonomy in the peripheral processing of odor signals in newborn rabbits. Brain Res 421:85–94

Hudson R, Distel H (1990) Sensitivity of female rabbits to changes in photoperiod as measured by pheromone emission. J Comp Physiol A 167:225–230

Hudson R, Distel H (1995) On the nature and action of the rabbit nipple-search pheromone: a review. In: Apfelbach D, Müller-Schwarze D, Reutter K, Weiler E (eds) Chemical signals in vertebrates VII. Elsevier Science, Oxford, pp 223–232

Hudson R, Trillmich F (2008) Sibling competition and cooperation in mammals: challenges, developments and prospects. Behav Ecol Sociobiol 62:299–307

Hudson R, González-Mariscal G, Beyer C (1990) Chin marking behavior, sexual receptivity, and pheromone emission in steroid-treated, ovariectomized rabbits. Horm Behav 24:1–13

Hudson R, Bilkó Á, Altbäcker V (1996) Nursing, weaning and the development of independent feeding in the rabbit (*Oryctolagus cuniculus*). Z Saeugetierkund (now Mammal Biol) 61:39–48

Hudson R, Cruz Y, Lucio RL, Ninomiya J, Martínez-Gómez M (1999a) Temporal and behavioral patterning of parturition in rabbits and rats. Physiol Behav 66:599–604

Hudson R, Schaal B, Bilkó Á (1999b) Transmission of olfactory information from mother to young in the European rabbit. In: Box HO, Gibson KR (eds) Mammalian social learning: comparative and ecological perspectives. Cambridge University Press, UK, pp 141–157

Hudson R, Labra-Cardero D, Mendoza-Soylovna A (2002) Suckling, not milk, is important for the rapid learning of nipple-search odors in newborn rabbits. Dev Psychobiol 41:226–235

Hudson R, Rojas C, Arteaga L, Martínez-Gómez M, Distel H (2008) Rabbit nipple-search pheromone versus rabbit mammary pheromone revisted. In: Hurst JL, Beynon RJ, Roberts SC, Wyatt TD (eds) Chemical signals in vertebrates 11. Springer, Berlin, pp 315–324

Hudson R, Raihani G, González D, Bautista A, Distel H (2009) Nipple preference and contests in suckling kittens of the domestic cat are unrelated to presumed nipple quality. Dev Psychobiol 51:322–332

Hudson R, Bautista A, Reyes-Meza V, Morales Montor J, Rödel HG (2011a) The effect of siblings on early development: a potential contributor to personality differences in mammals. Dev Psychobiol 53:564–574

Hudson R, Bautista A, González D, Arteaga L (2011b) Similarities and differences in the use of suckling odors by newborn rabbits, cats, and dogs. Dev Psychobiol 53:752

Keil W, von Stralendorff F, Hudson R (1990) A behavioral bioassay for analysis of rabbit nipple-search pheromone. Physiol Behav 47:525–529

Kindermann U, Gervais R, Hudson R (1991) Rapid odor conditioning in newborn rabbits: amnesic effects of hypothermia. Physiol Behav 50:457–460

Kindermann U, Hudson R, Distel H (1994) Learning of suckling odors by newborn rabbits declines with age and suckling experience. Dev Psychobiol 27:111–122

Kovach JK, Kling A (1967) Mechanisms of neonate sucking behaviour in the kitten. Anim Behav 15:91–101

Larson MA, Stein BE (1984) The use of tactile and olfactory cues in neonatal orientation and localization of the nipple. Dev Psychobiol 17:423–436

Moncomble RS, Coureaud G, Quennedey B, Langlois D, Perrier G, Schaal B (2005) The mammary pheromone of the rabbit: from where does it come? Anim Behav 69:29–38

Mykytowycz R, Ward MM (1971) Some reactions of nestlings of the wild rabbit, *Oryctolagus cuniculus* (L.), when exposed to natural rabbit odours. Forma Functio 4:137–148

Prechtl HFR (1952) Angeborene Bewegungsweisen junger Katzen. Experientia 8:220–221

Raihani G, González D, Arteaga L, Hudson R (2009) Olfactory guidance of nipple attachment and suckling in kittens of the domestic cat: inborn and learned responses. Dev Psychobiol 51:662–671

Rosenblatt JS (1972) Learning in newborn kittens. Sci Am 227:18–25

Schaal B, Coureaud G, Langlois D, Giniès C, Sémon E, Perrier G (2003) Chemical and behavioural characterization of the rabbit mammary pheromone. Nature 424:68–72

Schley P (1977) Die Ausschaltung des Geruhsvermögens und sein Einfluß auf das Saugverhalten von Jungkaninchen. Berl Münch Tierärztl Wochenschr 90:382–385

Schley P (1981) Geruchssinn und Saugverhalten bei Jungkaninchen. Kleintier Praxis 26:197–264

Schuleikina-Turpaeva KV (1986) Sensory organization of alimentary behavior in the kitten. In: Rosenblatt JS, Busnel MC, Hinde RA (eds) Advances in the study of behavior, vol 16. Academic, New York, pp 1–37

Wyatt TD (2010) Pheromones and signature mixtures: defining species-wide signals and variable cues for identity in both invertebrates and vertebrates. J Comp Physiol A 196:685–700

Chapter 5
Neonatal Recognition in Sheep

Barend V. Burger, Marlize Z. Viviers, Niel J. le Roux, John Morris, Jan P.I. Bekker, and Maritha le Roux

Abstract The strong bond formed between ewe and lamb shortly after parturition is an important factor in lamb survival. Evidence exists that the ewe can distinguish her lamb by its unique smell, but the constituents of such a putative pheromone have not yet been identified. We have identified 133 volatile organic compounds in the cranial wool of Döhne Merino lambs that are presumably constituents of the neonatal recognition cue of this sheep race. Quantitative analysis and comparison of the odour profiles of the twins of 16 ewes of a flock of 165 twin-bearing ewes (9.69% sample group) revealed that the wool volatiles of twins are qualitatively as well as quantitatively practically identical, but differ from those of other twins or non-twin lambs in the flock. The 88 constituents present in at least 20% of the analysed wool samples were considered as variables for multivariate analysis. A *P*-value <0.0001 was calculated, indicating that the pairing of twins according to the qualitative and quantitative composition of the wool is statistically highly significant. Bioassays carried out during the lambing seasons of 2009 and 2010 confirmed the previously established role of the odour of lambs in ewe-lamb recognition. However, ewes rejected alien lambs dressed in jackets that were sprayed with mixtures formulated with synthetic analogues

B.V. Burger (✉) • M.Z. Viviers • M. le Roux
Laboratory for Ecological Chemistry, Stellenbosch University,
Private Bag X1, Matieland 7602, South Africa
e-mail: LECUS@sun.ac.za

N.J. le Roux
Department of Statistics and Actuarial Science, Stellenbosch University,
Private Bag X1, Matieland 7602, South Africa

J. Morris
Mariendahl Experimental Farm, Faculty of AgriSciences, Stellenbosch University,
Private Bag X1, Matieland 7602, South Africa

J.P.I. Bekker
Institute for Plant Biotechnology, Stellenbosch University,
Private Bag X1, Matieland 7602, South Africa

M.L. East and M. Dehnhard (eds.), *Chemical Signals in Vertebrates 12*,
DOI 10.1007/978-1-4614-5927-9_5, © Springer Science+Business Media New York 2013

of the identified wool volatiles according to the qualitative and quantitative compositions of the experimental ewes' own lambs. This is probably due to the volatiles not being released into the atmosphere in quantitative ratios emulating the odour of the lambs accurately enough to satisfy the experimental ewes.

5.1 Introduction

Ever since the Greek philosopher Aristotle had reportedly (Lynch et al. 1992) first described skinning a dead lamb, placing the skin on a foster lamb and confining the ewe with the foster lamb as the most successful way of achieving fostering in sheep, *Ovis aries*, this method is still practiced by many sheep farmers. Sheep are seasonal breeders and large numbers of ewes give birth in a relatively short lambing season. For the ewe, the recognition of her lamb is of crucial importance in selectively providing nourishment to her offspring. Although auditory and visual cues are also used, ewes appear to rely on olfactory cues as final assurance before allowing lambs at the udder (Lindsay 1988). For a short period around the time of parturition ewes become exceptionally interested in neonatal lambs and may even steal each other's lambs. The bond between ewe and lamb is established by contact with the new-born lamb within 4–6 h after birth. If this fails to happen, maternal interest wanes and the ewe will not accept any lamb. However, once the bond has been formed, ewe and lamb can be separated for relatively long periods without disruption of the bond (Lindsay 1988).

The common factor that characterises most post-partum lamb deaths is impediment or disturbance of bond formation between ewes and their offspring (Lindsay 1988). In order to curtail the financial losses resulting from the rejection of new-born lambs, a well-known phenomenon in all sheep races, farmers and scientists have examined methods of facilitating adoption of rejected lambs by ewes with an ample supply of milk. For example, transfer of the odour of the foster mother's own lamb to the rejected or to an orphaned lamb is practiced with limited success by sheep farmers; this also has been subjected to scientific scrutiny (Alexander 1978; Alexander et al. 1989a, b; Price et al. 1984, 2003). However, the cue involved in ewe-lamb bonding has not been identified. There are several sources from which the wool-associated volatile organic compounds (wool VOCs) constituting each lamb's unique olfactory cue could conceivably be released. The odour signature of a lamb could be a maternal label originating from the amniotic fluid, milk, saliva, or some other secretion or excretion of the ewe. Three possible sources of maternal labels in sheep, colostrum, amniotic fluid and inguinal gland secretion, have recently been investigated by Rietdorf (2002). Because ewes become strongly attracted to amniotic fluid for a few hours after parturition (Lévy et al. 1983; Vince et al. 1985), farmers often smear the new-born lamb with the ewe's amniotic fluid in attempting to strengthen the bond with a rejected lamb. However, the origin of the amniotic fluid had no effect on maternal acceptance, which is an indication that the fluid contains cues for general attractiveness rather than for individual recognition (Lévy et al. 2004).

Alternatively, each lamb could be born with its own individual odour profile, with the amniotic fluid as a supporting role player in the greater recognition process, initiating the ewe's interest in her lamb and inducing licking and grooming behaviour so that she can become acquainted with the lamb's olfactory profile. The interest that ewes display in their lambs' hindquarters, head and neck areas is well documented (Alexander et al. 1983; Houpt 2005; Lynch et al. 1992; Price et al. 1984). However, our observations have shown the ewe has no preference for a specific area on her lamb's body during the initial identification process; she simply inspects that part of her lamb closest to her. These observations are in agreement with the results of the experiments carried out by Alexander and Stevens (1981), Brennan and Kendrick (2006) and Lynch et al. (1992), who found that the discrimination of ewes between their own and alien lambs is mediated primarily by olfactory cues from the lamb's wool and skin, not a specific area on the lamb's body, and also not from the amniotic fluid. Furthermore, Kendrick et al. (1992) found that lamb odours had no influence on the activity of the olfactory bulb neurons of ewes during the period before birth, but when tested in the period after birth these odours were a most potent olfactory stimulus for the ewes with the smell of the lamb's wool being almost as effective a stimulus as the smell of the whole lamb. Subsequently, Lévy et al. (2004) have shown that ewes can recognise their lambs without any direct contact and with volatile molecules as the only cues available for recognition. Romeyer et al. (1993) and Poindron et al. (2006) showed that ewes use phenotype matching to recognise their lambs, since recognition was not influenced when the genetic make-up of the ewes differed from that of the lambs, neither was recognition influenced by the ewe's secretions. A ewe implanted with monozygotic twin embryos that was allowed to learn the odour of one of the twin lambs after birth could recognise the second monozygotic twin lamb without being previously exposed to this lamb. However, when this experiment was repeated with dizygotic twin lambs, the ewe could not identify the non-identical twin lamb after being exposed to the other dizygotic twin lamb.

The object of this study was to identify and quantify the VOCs present in the cranial wool of Döhne Merino lambs that could be constituents of the neonatal recognition cue in this species, to subject the results to statistical analyses and to carry out preliminary field tests with synthetic analogues of the identified VOCs.

5.2 Methods and Materials

5.2.1 Sample Collection and Preparation

For statistical purposes, 16 of a flock of 165 twin-bearing ewes were used as a sample group (9.69%). Using a pair of scissors cranial wool samples (ca. 200–400 mg) were collected from 32 lambs on the morning after they were born. The samples were stored in glass vials (25 mL) with Teflon-faced septa at –20°C until analysed.

5.2.2 Analytical Methods

Gas chromatographic (GC) analyses were carried out with a Carlo Erba HRGC gas chromatograph (Milan, Italy) with a flame ionization detector (FID) and a Grob split/splitless sample inlet. Gas chromatographic-low resolution electron impact mass spectrometric analyses (GC-LRMS) of material isolated from the wool samples were carried out on a Carlo Erba QMD 1000 GC-MS instrument (Milan, Italy) operated in the split mode. The following columns were used: column A [glass, 40 m × 0.25 mm i.d., coated with 0.25 μm of the apolar phase PS-089-OH (DB-5 equivalent)], column B [glass, 40 m × 0.25 mm i.d., coated with 0.25 μm of the polar stationary phase AT-1000 (FFAP equivalent)], enantioselective column C [glass, 30 m × 0.3 mm i.d., coated with 0.25 μm of OV-1701-OH containing 10% heptakis(2,3-di-*O*-methyl-6-*O*-tert-butyldimethylsilyl)-β-cyclodextrin] and enantioselective column D [glass, 30 m × 0.3 mm i.d., coated with 0.25 μm of OV-1701-OH containing 10% heptakis (2,3-di-*O*-acetyl-6-*O*-tert-butyldimethylsilyl)-β-cyclodextrin]. Hydrogen was used as carrier gas at a linear velocity of 50 cm/s (column temperature 40°C). The injector was operated at 220°C and the FID at 280°C. Columns A and B were programmed at 2°C/min from 40 to 280°C and from 40 to 250°C, respectively. The enantioselective columns C and D were programmed from 40 to 220°C at 1°C/min. Electron impact (EI) mass spectral data were acquired at 70 eV from *m/z* 25 to 350. Helium was used as carrier gas at a linear velocity of 28.6 cm/s (column temperature 40°C). Inlet and interface temperatures were 220 and 250°C, respectively.

GC-MS analysis was used to confirm the identification of the wool VOCs by retention time comparison with commercially available synthetic compounds, compounds that were available from previous research projects or compounds that were synthesised during the present investigation.

Headspace sampling of the wool VOCs was carried out by sample enrichment probe (SEP) using a second generation SEP50 sample enrichment probe (MasChrom Analisetegniek, Stellenbosch, South Africa) consisting of a fused silica stalk or shaft carrying a 50-mm sleeve (47 mg) of polydimethylsiloxane (PDMS) tubing (Burger et al. 2006, 2011a). The wool sample (70 mg) was placed in an 8-mL vial; the SEP50 was installed in this vial with the lower tip of the SEP clearing the sample. After tightening the vial cap, the SEP was exposed to the headspace vapour of the wool at 70°C for 1 h. The volatile compounds retained in the PDMS were subjected to GC-MS analysis. The SEP was left in the injector until completion of the analysis (Burger et al. 2006).

5.2.3 Statistical Analysis

The peak areas of the target analytes were used to construct data matrices for the wool collected from day-old lambs in 2007 and 2009. All analytical data were normalised to produce variables with zero means and unit standard deviation (Kowalski and Bender 1972). Principal component analysis (PCA) biplots, permutation tests and investigations into predictivity were performed using R (Vienna, Austria) as described in Aldrich et al. (2004) and Gardner-Lubbe et al. (2008).

5.2.4 *Bioassays*

Bioassays were performed with single-bearing Döhne Merino ewes during the lambing season at Mariendahl from March to April 2009 and 2010 in a 0.2-ha paddock. Similar bioassays with twin-bearing ewes were carried out in lambing pens inside a naturally illuminated and ventilated shed, or outside in an enclosed concrete-surfaced arena.

These bioassays were carried out by dressing alien lambs in appropriately treated cotton jackets and presenting the alien lambs to the experimental ewes in experiments similar to those carried out by Price et al. (1984). For example, the lambs of the experimental ewes were dressed in jackets and spent 24–48 h with their respective mothers before the bioassays were carried out. In experiments in which alien lambs were dressed in these jackets, precautions were taken against odour transfer from the alien lamb to the jacket or odour diffusion through the jacket by fitting the lamb with disposable diapers and wrapping it from hind quarters to shoulder blades, as well as its legs in several layers of low density polyethylene film.

5.3 Results and Discussion

As a working hypothesis it was presumed that the odour profiles of twins lambs were identical, or at least similar, and different from those of other lambs in the flock and that it should thus be possible to identify the chemical compounds constituting the ewe-lamb recognition cue by intra- and inter-twin comparison of the qualitative and quantitative composition of the effluvia of a representative sample of lambs. Because of its high sensitivity and simplicity, the SEP technique (Burger et al. 2006, 2011a) was used to sample the wool VOCs from the collected cranial wool samples. Qualitative and quantitative analyses of the VOCs were thus carried out by SEP-GC-MS.

A total of 133 constituents (Table 5.1), comprising several compound classes, were unambiguously identified by GC-LRMS and GC-HRMS analysis, retention index (RI) determination and by GC-MS comparison with authentic synthetic analogues of these compounds (Burger et al. 2011b). The enantiomeric compositions of the chiral VOCs were determined by GC-MS comparison with the corresponding synthetic compounds using enantioselective columns C and D, and data published by Maas et al. (1994) and Burger et al. (2008).

The quantitative composition of the volatile organic material present in the headspace of the individual wool samples was determined by GC-MS analysis.

Table 5.1 VOCs of the cranial wool of Döhne Merino lambs collected during the lambing seasons of 2007 and 2009

No	Compound	Remarks	No	Compound	Remarks
C10	Nonane	b, c, d, k, m	**C99**	2-Heptadecanone	b, c, d, o
C20	Decane	b, c, d, k, m, o	**C14**	6-Methyl-2-heptanone	b, c, d, k, o
C28	Undecane	b, c, d, o	**C47**	3-Methyl-2-undecanone	b, c, d, e, g
C36	Dodecane	b, c, d, o	**C16**	3-Octanone	b, c, d, o
C46	Tridecane	b, c, d, j, o	**P4**	3-Decanone	b, c, d
C58	Tetradecane	b, c, d, j, n, o	**P3**	Acetic acid	b, c, d
C72	Hexadecane	b, c, d, j, o	**P5**	Propanoic acid	b, c, d
C80	Heptadecane	b, c, d, j, o	**P6**	Butanoic acid	b, c, d
C89	Octadecane	b, c, d, j, o	**P7**	Hexanoic acid	b, c, d
C100	Nonadecane	b, c, d, j, o	**P8**	Heptanoic acid	b, c, d
C44	1-Tridecene	b, c, d, f, o	**P9**	Octanoic acid	b, c, d, j
C55	1-Tetradecene	b, c, d, f, m, o	**C42**	Nonanoic acid	b, c, d, j
C66	1-Pentadecene	b, c, d, f, o	**C53**	Decanoic acid	b, c, d, j
C118	1-Pentacosene	b, c, e	**C69**	Dodecanoic acid	b, c, d, j, m
C120	1-Hexacosene	b, c, e	**C85**	Tetradecanoic acid	b, c, d, j, o
C2	1-Pentanol	b, c, d, k	**C94**	Pentadecanoic acid	b, c, d, j, n, o
C15	1-Heptanol	b, c, d, o	**C104**	Hexadecanoic acid	b, c, d, j, m, o
C25	1-Octanol	b, c, d, j, o	**C107**	Heptadecanoic acid	b, c, d, j
C33	1-Nonanol	b, c, d, o	**C112**	Octadecanoic acid	b, c, d, j
P10	1-Tridecanol	b, c, d	**C115**	Eicosanoic acid	b, c, d, j
C75	1-Tetradecanol	b, c, d, j, l	**C117**	Heneicosanoic acid	b, c, d, j
C87	1-Pentadecanol	b, c, d, j, l, o	**C119**	Docosanoic acid	b, c, d, j
C98	1-Hexadecanol	b, c, d, j, l, m	**C82**	12-Methyltridecanoic acid	b, c, d, e, j
C108	1-Octadecanol	b, c, d, j, l	**C102**	14-Methylpentadecanoic acid	b, c, d, e, j
C114	1-Eicosanol	b, c, d, j	**C106**	15-Methylhexadecanoic acid	b, c, d, e, j
C116	1-Heneicosanol	b, c, e, j	**C113**	17-Methyloctadecanoic acid	b, c, d, e, j
P1	3-Methyl-1-butanol	b, c, d	**C103**	(*Z*)-9-Hexadecenoic acid	b, c, d, f

C6	4-Methyl-1-pentanol	b, c, d	**C110**	(*Z,Z*)-9,12-Octadecadienoic acid	b, c, d
P2	3-Octanol	b, c, d	**C111**	(*Z*)-9-Octadecenoic acid	b, c, d, f
C4	Hexanal	b, c, d, k	**C32**	Benzoic acid	b, c, d
C8	Heptanal	b, c, d, k, o	**C88**	Ethyl tetradecanoate	b, c, d, o
C19	Octanal	b, c, d, o	**C92**	Isopropyl tetradecanoate	b, c, d, e, o
C27	Nonanal	b, c, d, o	**C48**	Nonan-4-olide	b, c, d, g, o
C35	Decanal	b, c, d, o	**C61**	Decan-4-olide	b, c, d, g
C45	Undecanal	b, c, d, o	**C76**	Dodecan-4-olide	b, c, d, g
C57	Dodecanal	b, c, d, o	**C109**	Hexadecan-4-olide	b, c, d, g
C67	Tridecanal	b, c, d, o	**C56**	6,10-Dimethyl-2-undecanone	b, c, d, h, p
C73	Tetradecanal	b, c, d, o, p	**C60**	(5*E*)-6,10-Dimethyl-5,9-undecadien-2-one	b, c, d, o, p
C81	Pentadecanal	b, c, d, o, p	**C91**	2,6,10,14-Tetramethylhexadecane	b, c, d, h
C90	Hexadecanal	b, c, d, o, p	**C93**	6,10,14-Trimethyl-2-pentadecanone	b, c, d, h, o
C101	Heptadecanal	b, c, d, o, p	**C95**	6,10,14-Trimethyl-2-pentadecanol	b, c, d, h
C105	Octadecanal	b, c, d	**C122**	Squalene	b, c, d
C1	3-Methylpentanal	b, c, d, h, p	**C5**	2-Methylpyrimidine	b, c, d
C24	7-Methyloctanal	b, c, d, e, m, o	**C11**	2,5-Dimethylpyrimidine	b, c, d
C31	8-Methylnonanal	b, c, d, e, o, p	**C29**	*N*-Methyl-2-piperidinone	b, c, d
C41	9-Methyldecanal	b, c, d, e, o	**C37**	3-Ethyl-4-methyl-1H-pyrrole-2,5-dione	b, c, i, n, o
C51	10-Methylundecanal	b, c, d, e, o	**C38**	3-Ethyl-4-methylpyrrolidine-2,5-dione	b, i, g, n
C68	12-Methyltridecanal	b, c, d, e, p	**C39**	3-Methyl-4-vinyl-1H-pyrrole-2,5-dione	b, c, i, n
C77	13-Methyltetradecanal	b, c, d, e, o, p	**C9**	Dimethyl sulphone	b, c, d, n, o
C86	14-Methylpentadecanal	b, c, d, e, m, o, p	**C18**	2-Pentylfuran	b, c, d, o
C97	15-Methylhexadecanal	b, c, d, e, m, p	**C124**	Cholest-5-en-3β-ol	b, c, d, j
C23	(*E*)-2-Octenal	b, c, d, e, o	**C3**	Unidentified	
C30	(*E*)-2-Nonenal	b, c, d, e, o	**C7**	Unidentified	
C40	(*E*)-2-Decenal	b, c, d, e, o	**C12**	Unidentified	n
C50	(*E*)-2-Undecenal	b, c, d, e, f, o	**C21**	Unidentified	

(continued)

Table 5.1 (continued)

No	Compound	Remarks	No	Compound	Remarks
C62	(*E*)-2-Dodecenal	b, c, d, e, m, o	**C49**	Unidentified	n, o
C78	(*E*)-2-Tetradecenal	b, c, d, e	**C52**	Unidentified	o
C13	Benzaldehyde	b, c, d, n, o	**C59**	Unidentified	
C22	Phenylacetaldehyde	b, c, d, n, o	**C63**	Unidentified	
C17	2-Octanone	b, c, d, k, o	**C64**	Unidentified	o
C26	2-Nonanone	b, c, d, o	**C70**	Unidentified	o
C34	2-Decanone	b, c, d, o	**C74**	Unidentified	m
C43	2-Undecanone	b, c, d, o	**C83**	Unidentified	m
C54	2-Dodecanone	b, c, d, o	**C84**	Unidentified	b, m
C65	2-Tridecanone	b, c, d, o	**C96**	Unidentified	b, c, o
C71	2-Tetradecanone	b, c, d, p	**C121**	Unidentified	
C79	2-Pentadecanone	b, c, d, o, p	**C123**	Unidentified	

a Constituents identified using apolar column A are numbered from **C1** to **C124** and those identified using polar column B are numbered from **P1** to **P10**; *b* Low-resolution EI mass spectrum; *c* Library spectrum (NBS and/or NIST); *d* Retention time comparison with synthetic compound; *e* Kovats retention index; *f* Double bond localisation by DMDS derivatisation and GC-MS analysis; *g* Absolute configuration determined (Burger et al. 2011b); *h* Absolute configuration not determined; *i* Tentative identification; *j* Compounds previously identified in lanolin (Schlossman and McCarthy 1979; Motiuk 1979a, b, 1980); *k* Compounds previously identified in wool (Lisovac and Shooter 2003); *l* Compounds previously identified in inguinal gland of ewes (Rietdorf 2002); *m* Compounds found in the wool during the lambing season of 2007 that were not identified in the lambing season of 2009; *n* Compounds found in the wool during the lambing season of 2009 that were not identified in the lambing season of 2007; *o* Compounds found in cloth jackets left on lamb for 3 days; *p* Compounds synthesised in this investigation

Constituents are grouped according to compound class

Approximately 50% of the total average peak area of the volatile compounds present in the headspace of the wool of the 32-day-old Döhne Merino lambs consisted of only five compounds: ethyl tetradecanoate (20%), nonanal (12%), 6,10,14-trimethyl-2-pentadecanone (9%), tetradecanal (4%) and Pentadecanal (4%). The role of genetic influences on the odour profile of lambs was investigated by comparison of the quantitative composition of the wool volatiles of twin lambs and that of randomly selected lambs. This amounted to an evaluation of the significance of the intra-twin similarities and the inter-twin differences in the composition of the wool volatiles.

The results of our analyses showed that the odours in twins are remarkably similar, in contrast to the inter-twin differences that were found between the odours of lambs born to different ewes. This is illustrated by comparison of examples of the results of a few of these analyses in Figs. 5.1 and 5.2.

Of the 133 constituents identified as wool volatiles only those present in 20% or more of the wool samples were considered as variables for statistical analysis, reducing the number of constituents included in the analysis to 88. The chromatographic data used for multivariate analyses nevertheless still contained a large number of variables (peak areas for these identified constituents, $n \leq 88$) relative to the number of sample units (wool samples, $n \leq 32$). The peak areas were normalised across all samples to produce comparable variables (Kowalski and Bender 1972). Permutation tests were carried out to determine whether the wool of twin lambs can be grouped according to the qualitative and quantitative composition of the volatile organic material present in their wool. The following null hypothesis was tested: twin lambs cannot be grouped according to the qualitative and quantitative composition of their wool volatiles. Thus, the calculated P-value provides an indication of whether the similarities found between twin lambs can be attributed to the fact that they are twins or to random variation in the data. Using this principle, the similarity in the odour profile of the wool of a pair of twin lambs was assessed by executing 10,000 permutation tests. A P-value <0.0001 was calculated, indicating that the pairing of the twin pairs of lambs according to the qualitative and quantitative composition of the wool did not occur by chance, and is statistically significant. Twin lambs thus possess odour profiles that are unique and their odour profiles are more similar than those of other randomly selected non-twin lambs.

The 88 constituents, and the quantities in which they are found in the wool of day-old lambs born during the lambing season of 2007, were used to construct a PCA biplot in order to obtain insight into the multivariate character of the results (Fig. 5.3). The biplot depicted in Fig. 5.3 is a multivariate scatterplot that gives a graphical presentation of samples (lambs), displayed as points on the graph, and variables (identified constituents), displayed as linear axes. Each axis of the biplot is used similar to the usage of an axis in an ordinary scatterplot: a line is drawn from any point in a biplot perpendicular to a biplot axis and the value of the point for that variable is read from the axis. In addition, the angle between any two axes is an approximation of the correlation between the respective variables. Constituents displayed as axes that are 90° in relation to one another have no correlation with each other, and those constituents lying on axes close to one another have a high level of

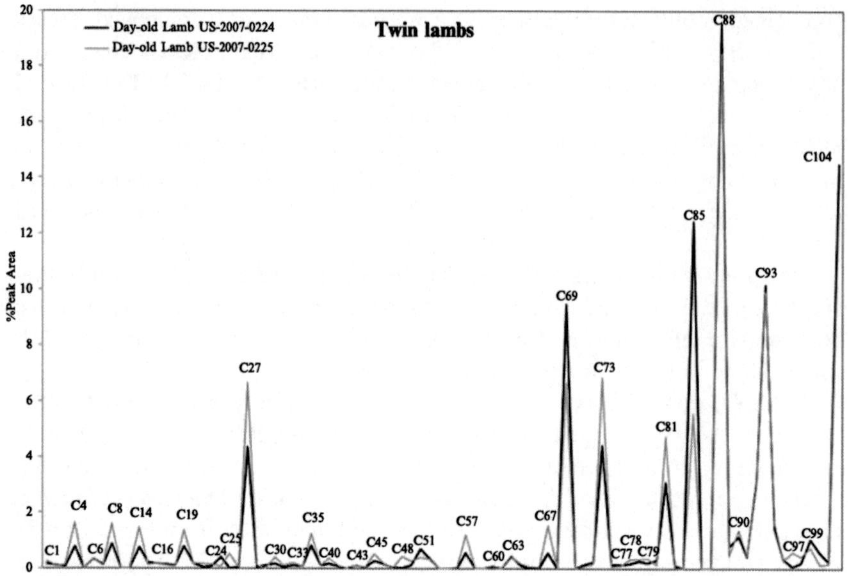

Fig. 5.1 Quantitative comparison (percentage peak area) of the SEP-enriched wool volatiles of day-old twin lambs US-2007-0224 (*black*) and US-2007-0225 (*grey*) by superimposing the reconstructed gas chromatograms with *grey* on *top*. The peaks are numbered according to the information given in Table 5.1. Note that the concentrations of compounds 87–96 are so similar that the *black* chromatogram of lamb US-2007-0224 is only visible at the apexes of the peaks

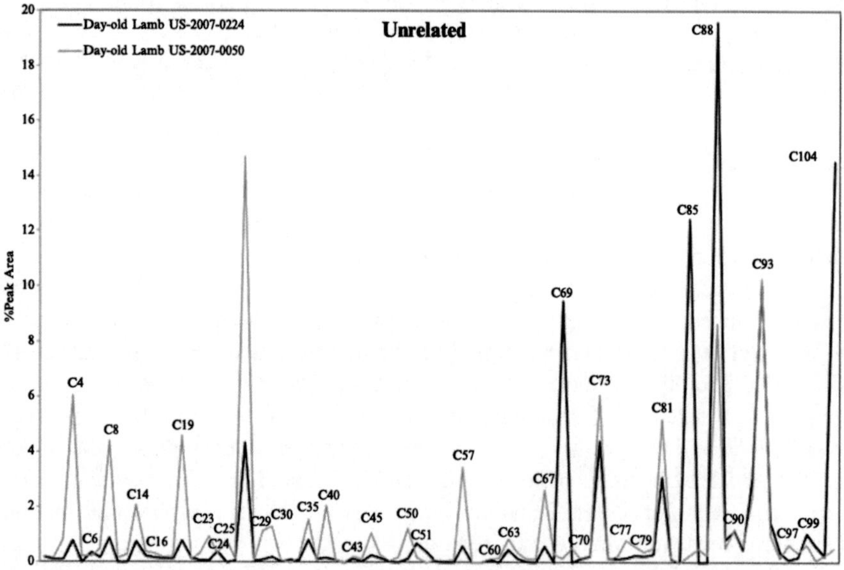

Fig. 5.2 Quantitative comparison (percentage peak area) of the SEP-enriched wool volatiles of the unrelated lambs US-2007-0224 (*black*) and US-2007-0050 (*grey*)

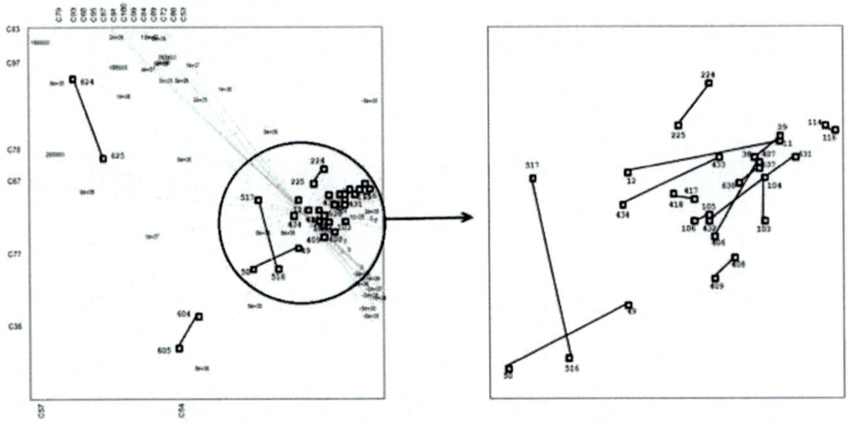

Fig. 5.3 (**a, b**) PCA biplot for the wool volatiles collected from day-old lambs during the lambing season of 2007, and a magnified view on the right of the crowded area in Fig. 5.3a. The PCA component loadings are listed in Table 5.2. The first and second principal components (i.e. the component loadings) are used as scaffolding for constructing the PCA biplot. Following Gower et al. (2011) the scaffolding axes are not calibrated; instead the axes representing the variables are calibrated to be used similar to the axes in an ordinary scatterplot. This usage leads to a figure that is easier to interpret

correlation with each other. An axis is labelled at the positive value of its calibration. The quality of a biplot is an overall measure of the accuracy of the 2D approximation of the data matrix. Furthermore, sample predictivities and axis predictivities provide detailed information about how accurately each data point is represented in the biplot and the degree of accuracy in the predictions made from a biplot axis (Gardner-Lubbe et al. 2008). The predictivity values range from 0 to 1, with a value of 1 representing the best predictivity. Figure 5.3 provides the optimal 2D presentation of the data matrix. Not all 32 samples and 88 variables are equally well represented in the biplot, as can be judged from their respective predictivities (Gardner-Lubbe et al. 2008). Only the 21 axes with predictivities higher than 0.800 are displayed in Fig. 5.3. The quality of the display in the PCA biplot was measured and this value reflects the proportion of the variation in the data accounted for in the 2D display (Gower and Hand 1996; see also Gower et al. 2011). The quality of display for the PCA biplot in Fig. 5.3 is 54%. This means that 54% of the variation of the data is explained in the first two dimensions of the 2D biplot; the other 46% is explained in the remaining 30 dimensions. In Fig. 5.3(b) the axes are not displayed in order to draw attention to the positions of the twin lambs. Each square represents a sample of wool collected from a day-old lamb, and the last three digits of the lamb's number are indicated on the biplot. The closer the samples are placed in the PCA biplot, the more similar their odour profiles. The lambs of each of the 16 twins are connected with black lines. The group of lambs concentrated in the encircled area indicates that they possess odour profiles that are generally similar.

Table 5.2 Component loadings for PCA biplot depicted in Fig. 5.3

Cmpd	FirstPC	ScondPC	Cmpd	FirstPC	ScondPC	Cmpd	FirstPC	ScondPC	Cmpd	FirstPC	ScondPC
C1	−0.136	−0.057	C27	−0.133	−0.091	C55	−0.120	−0.141	C79	−0.150	0.113
C2	−0.099	−0.076	C28	−0.084	−0.124	C56	−0.128	0.116	C80	−0.122	0.151
C3	−0.097	−0.183	C29	−0.003	−0.108	C57	−0.139	−0.097	C81	−0.141	−0.007
C4	−0.077	−0.100	C30	−0.131	−0.071	C58	−0.014	0.007	C83	−0.142	0.088
C5	0.011	−0.076	C31	−0.059	−0.087	C59	−0.009	0.013	C84	−0.133	0.144
C6	−0.117	−0.025	C33	−0.058	0.121	C60	−0.145	0.128	C85	−0.060	0.140
C7	−0.083	−0.164	C34	−0.065	−0.063	C62	0.002	−0.066	C86	−0.055	−0.166
C8	−0.086	−0.110	C35	−0.152	−0.060	C63	−0.019	0.052	C87	−0.143	0.133
C10	−0.045	0.050	C36	−0.088	−0.133	C64	−0.120	−0.075	C88	−0.001	−0.021
C11	0.017	−0.042	C40	−0.086	−0.141	C65	−0.108	0.127	C89	−0.135	0.154
C14	−0.132	−0.110	C41	−0.133	−0.112	C66	−0.118	−0.120	C90	−0.148	−0.013
C15	−0.054	0.015	C42	−0.102	0.147	C67	−0.157	0.014	C91	−0.142	0.138
C16	−0.075	−0.038	C43	−0.114	0.049	C68	−0.056	−0.209	C92	−0.026	−0.009
C17	−0.033	−0.081	C44	−0.112	−0.132	C69	−0.021	0.075	C93	−0.136	0.115
C18	−0.135	0.043	C45	−0.078	−0.087	C70	−0.096	−0.126	C95	−0.141	0.120
C19	−0.067	−0.131	C46	−0.100	−0.158	C71	−0.141	−0.045	C96	−0.122	0.128
C20	−0.020	−0.044	C47	−0.048	−0.144	C72	−0.122	0.154	C97	−0.156	0.080
C21	−0.150	−0.032	C48	−0.081	0.056	C73	−0.121	−0.099	C98	−0.118	0.124
C23	−0.068	−0.120	C50	−0.104	−0.103	C74	−0.118	0.135	C99	−0.127	0.139
C24	−0.136	−0.087	C51	−0.065	−0.012	C75	−0.132	0.081	C100	−0.131	0.129
C25	−0.119	−0.118	C53	−0.115	0.174	C77	−0.155	−0.023	C101	−0.056	−0.002
C26	−0.087	−0.089	C54	−0.117	−0.171	C78	−0.158	0.030	C104	−0.078	0.136

Cmpd compound; *FirstPC* first principal component loadings; *ScondPC* second principal component loadings

The 21 constituents with predictivities higher than 0.800 in order of decreasing predictivity are 2,6,10,14-tetramethylhexadecane (**C91**), 2-pentadecanone (**C79**), octadecane (**C89**), 1-pentadecanol (**C87**), (5*E*)-6,10-dimethyl-5,9-undecadien-2-one (**C60**), 15-methylhexadecanal (**C97**), unidentified constituent **C84**, (*E*)-2-tetradecenal (**C78**), 2-dodecanone (**C54**), tridecanal (**C67**), decanoic acid (**C53**), decanal (**C35**), 6,10,14-trimethyl-2-pentadecanol (**C95**), 13-methyltetradecanal (**C77**), hexadecane (**C72**), heptadecane (**C80**), nonadecane (**C100**), 2-heptadecanone (**C99**), unidentified constituent **C83**, 6,10,14-trimethyl-2-pentadecanone (**C93**) and dodecanal (**C57**). Interestingly, all the aldehydes have their maxima on the left hand side of the plot, with dodecanal at the bottom right hand side of the plot. This suggests that all the methyl ketones, except 2-dodecanone, and all the hydrocarbons and decanoic acid, are highly correlated with one another.

Price et al. (1984) have shown that lambs can be fostered by means of odour transfer from a prospective foster mother's own lamb to an alien lamb. In their experiment, lambs were dressed in nylon stockinettes for 72 h, after which the stockinettes were placed on alien lambs and these lambs were presented to the ewes. A total of 38% of the alien lambs were immediately accepted and another 50% were accepted over a further period of 36 h. We carried out similar experiments in order to develop and standardise our bioassays. In this phase of the study the experimental ewes' own lambs were dressed in hoodless jackets made from white or pale grey cotton fleece cloth. Taking precautions to prevent odour diffusion from alien lambs to the jackets, alien lambs wearing the jackets were then presented to the respective experimental ewes (videos available at http://scholar.sun.ac.za/handle/10019.1/3980). Drastic methods were not applied to force the experimental ewe to accept an alien lamb. An alien lamb was considered accepted if it was allowed to suckle and if the ewe directed no butts or butt attempts at it for a period of 5 min after their first contact. If an alien lamb was rejected by the ewe and she had displayed aggressive behaviour towards the lamb, the experiment was immediately terminated and the lamb was removed from the experimental arena. The alien lambs were accepted in 74% ($n=19$) of these experiments. In similar experiments carried out with single-bearing ewes, a 100% acceptance rate ($n=6$) was obtained. The volatile constituents present in the headspace of one of the jackets used in a successful fostering experiment were analysed and no compounds were identified in the headspace of the jacket in addition to those listed in Table 5.1.

The possibility of using extracts of the wool volatiles in similar experiments was considered. Approximately 8 g wool was sheared from the bodies of day-old lambs and Soxhlet-extracted with either pentane ($n=15$), diethyl ether ($n=4$), dichloromethane (DCM) ($n=4$) or ethyl acetate ($n=1$) for 2 h, after which the extracts were sprayed uniformly on new washed fleece cloth jackets. The respective solvents were allowed to evaporate before bioassays were carried out within 24 h after the wool had been harvested. In five of these experiments (20%) (pentane $n=4$; DCM $n=1$) the alien lambs were accepted by the experimental ewes.

Disappointing results were obtained when mixtures of synthetic analogues of wool volatiles were subjected to bioassays in a few exploratory experiments. It was concluded that, due to selective co-evaporation of some of the more volatile compounds, the quantitative composition of the mixture of the extracted or synthetic volatile compounds that were left on the jackets after the solvent had evaporated probably did not emulate the odour profiles of the experimental ewes' own lambs closely enough to satisfy the ewes. Headspace analyses of jackets treated with the wool extracts mentioned above or with mixtures of the synthetic VOCs revealed that the VOCs are, in fact, not released from the treated jackets in the quantitative ratio in which they are present in the original wool or in the synthetic mixtures. It is clear that especially the volatile and apolar compounds are prone to co-evaporation with the solvents used for extraction of the VOCs or for the dilution of the synthetic mixtures. It is also possible that the dissemination of the VOCs from cotton cloth could differ from that of natural lamb's wool.

Acknowledgements Financial support by the NRF and Stellenbosch University is gratefully acknowledged. We thank the experimental farm of Stellenbosch University, Mariendahl, for making experimental sheep available for this research and for invaluable advice and assistance.

References

Aldrich C, Gardner S, Le Roux NJ (2004) Monitoring of metallurigal process plants using biplots. AIChE J 50:2167–2186

Alexander G (1978) Odour, and the recognition of lambs by Merino ewes. Appl Anim Ethol 4:153–158

Alexander G, Stevens D (1981) Recognition of washed lambs by Merino ewes. Appl Anim Ethol 7:77–86

Alexander G, Stevens D, Bradley LR (1983) Washing lambs and confinement as aids to fostering. Appl Anim Ethol 10:251–261

Alexander G, Goodrich BS, Stevens D, Bradley LR (1989a) Maternal interest in lambs smeared with polar and nonpolar substances. Austr J Exp Agric 29:513–515

Alexander G, Stevens D, Bradley LR (1989b) Fostering in sheep: an exploratory comparison of several approaches. Austr J Exp Agric 29:509–512

Brennan PA, Kendrick KM (2006) Mammalian social odours: attraction and individual recognition. Philos Trans R Soc Lond B Biol Sci 361:2061–2078

Burger BV, Le Roux M, Marx B, Herbert SA, Amakali KT (2011a) Development of a second-generation sample enrichment probe (SEP) for improved sorptive analysis of volatile organic compounds. J Chromatogr A 1218:1567–1575

Burger BV, Marx B, Le Roux M, Burger WJG (2006) Simplified analysis of organic compounds in headspace and aqueous samples by high-capacity sample enrichment probe. J Chromatogr A 1121:259–267

Burger BV, Viviers MZ, Bekker JPI, Le Roux M, Fish N, Fourie WB, Weibchen G (2008) Chemical characterization of territorial marking fluid of male Bengal tiger, *Panthera tigris*. J Chem Ecol 34:659–671

Burger BV, Viviers MZ, le Roux NJ, Morris J, Bekker JPI, Le Roux M (2011b) Olfactory cue mediated neonatal recognition in sheep, *Ovis aries*. J Chem Ecol 37:1150–1163

Gardner-Lubbe S, Le Roux NJ, Gower JC (2008) Measures of fit in principal component and canonical variate analyses. J Appl Stat 35:947–965

Gower J, Gardner-Lubbe S, Le Roux N (2011) Understanding biplots. Wiley, Chichester

Gower JC, Hand DJ (1996) Biplots. Chapman & Hall, London

Houpt KA (2005) Domestic animal behaviour for veterinarians and animal scientists, 4th edn. Blackwell, Oxford
Kendrick KM, Lévy F, Keverne EB (1992) Changes in the sensory processing of olfactory signals induced by birth in sheep. Science 256:833–836
Kowalski BR, Bender CF (1972) Pattern recognition. A powerful approach to interpreting chemical data. J Am Chem Soc 94:5632–5639
Lévy F, Poindron P, Le Neindre P (1983) Attraction and repulsion by amniotic fluids and their olfactory control in the ewe around parturition. Physiol Behav 31:687–692
Lévy F, Keller M, Poindron P (2004) Olfactory regulation of maternal behavior in mammals. Horm Behav 46:284–302
Lindsay D (1988) Breeding the flock: modern research and reproduction in sheep. Inkata, Melbourne
Lisovac AM, Shooter D (2003) Volatiles from sheep wool and the modification of wool odour. Small Ruminant Res 49:115–124
Lynch JJ, Hinch GN, Adams DB (1992) The behaviour of sheep. Biological principles and implications for production. C.A.B. International and CSIRO Australia, Melbourne
Maas B, Dietrich A, Mosandl A (1994) Collection of enantiomer separation factors obtained by capillary gas chromatography on chiral stationary phases. J High Resolut Chromatogr 17:109–115
Motiuk K (1979a) Wool wax acids: a review. J Am Oil Chem Soc 56:91–97
Motiuk K (1979b) Wool wax acids: a review. J Am Oil Chem Soc 56:651–658
Motiuk K (1980) Wool wax acids: a review. J Am Oil Chem Soc 57:145–146
Poindron P, Lévy F, Keller M (2006) Maternal responsiveness and maternal selectivity in domestic sheep and goats: the two facets of maternal attachment. Dev Psychobiol 49:54–70
Price EO, Dunn GC, Talbot JA, Dally MR (1984) Fostering lambs by odor transfer: the substitution experiment. J Anim Sci 59:301–307
Price EO, Dally MR, Hernandez L (2003) A note on the use of odor manipulation to facilitate the adoption of alien lambs by ewes bearing twins. Appl Anim Behav Sci 81:127–131
Rietdorf M (2002) Identifizierung und Synthese flüchtiger Substanzen aus Säugetieren. PhD thesis Institut für Organische Chemie, Universität Hamburg
Romeyer A, Porter RH, Poindron P, Orgeur P, Chesné P, Poulin N (1993) Recognition of dizygotic and monozygotic twin lambs by ewes. Behaviour 127:119–139
Schlossman ML, McCarthy JP (1979) Lanolin and derivatives chemistry: relationship to allergic contact dermatitis. Contact Dermatitis 5:65–72
Vince MA, Lynch JJ, Mottershead B, Green G, Elwin R (1985) Sensory factors involved in immediately postnatal ewe/lamb bonding. Behaviour 94:60–84

Chapter 6
Suckling Odours in Rats and Mice: Biological Substrates that Guide Newborns to the Nipple

Bruno Patris, Syrina Al Aïn, and Benoist Schaal

Abstract It is a general strategy for mammalian females to emit odour signals to direct their offspring to the mammae and to motivate their suckling. The survival of newborns depends on their own capacities to exploit the cues emitted by their mother, or by conspecific lactating females, and to direct their behaviour to the vital targets on the mother's body—the nipples. This chapter synthesises data on some natural substrates that contribute to nipple searching and grasping in the newborns of (laboratory strains of) two phylogenetically close species, *Rattus norvegicus* and *Mus musculus*.

6.1 Introduction

To survive, all mammalian neonates have to interact efficiently with the mother (and in rare cases with the father) to simultaneously obtain care and protection, as well as energetic, nutritional, protective and informational resources. To reach these resources, infants depend on their own neural and motor capacities to exploit the sensory cues emitted by maternal females. Once in physical contact with the mother, the targets of neonatal sensory and behavioural skills are the nipples (Schaal 2010). But great taxonomic differences in neonatal sensory and motor abilities are associated with different levels of offspring development at birth. Here, the discussion will be restricted to two representatives of the order *Rodentia*, family *Muridae*, sub-family *Murinae*, namely the domestic forms of *Rattus norvegicus* and *Mus musculus*. The rat and the mouse have in common that their pups are born altricial, in the sense that

B. Patris (✉) • S. Al Aïn • B. Schaal (✉)
Developmental Ethology and Cognitive Psychology Group, Center for Smell, Taste and Food Science, CNRS, Université de Bourgogne, Dijon, France
e-mail: bruno.patris@u-bourgogne.fr; benoist.schaal@u-bourgogne.fr

M.L. East and M. Dehnhard (eds.), *Chemical Signals in Vertebrates 12*,
DOI 10.1007/978-1-4614-5927-9_6, © Springer Science+Business Media New York 2013

they have limited locomotor and thermoregulatory abilities (Rosenblatt 1983; Blass et al. 1977; Bouslama et al. 2005). Accordingly, the pups of both species are raised in a nest where they are in quasi-permanent contact with their mother, and, in the case of mice where communal nursing is frequent, with other females (König and Markl 1987; Schultz and Lore 1993). Another correlate of neonatal altriciality is that the pups' sensorium is only partially functional at birth. Somesthesis and chemoreception are then already functional (although still maturing), audition comes into play only by postnatal day (PND) 9, and vision by days 14–15, in both the rat (Suckow et al. 2006) and the mouse (Ehret 1976; Weber and Olsson 2008). Neonate pups can thus only be guided to the mother's nipple by tactile, thermal and chemical cues. Although the contribution of somesthesis is admittedly considerable (e.g. Rosenblatt 1983), the emphasis will be here on olfaction which appears crucial in arousing, motivating and guiding the initial behaviour of neonates (Blass and Teicher 1980), and remains dominant in controlling behaviour during later development up to weaning (Breen and Leshner 1977; Hennessy et al. 1980).

The mouse and the rat may differ in the chemosensory systems that direct their behaviour. In both species, the main olfactory system is operative in the foetus several days before birth (Coppola and Millar 1997; Pedersen and Blass 1982; Smotherman and Robinson 1987), so that olfaction is in a good position to participate in the development of sensory processes and preferences. But rats and mice differ in the range of chemosensory subsystems governing neonatal behaviour. While the vomeronasal system functions in the rat foetus from the last days of gestation (Pedersen et al. 1983), it does not appear to do so before birth in mice (Coppola and O'Connell 1989; Coppola et al. 1993).

Nipple search and suckling in newborn animals is powerfully triggered, directed and regulated by olfactory cues. In this context, stimulus–response associations are determined by two types of mechanisms. First are mechanisms that apparently do not depend on postnatal experience for their development; these mechanisms rely on stimuli (signals) eliciting responses which morphology is quasi-complete at very first expression, i.e. apparently without need of prior direct exposure and training. Second are stimuli that are clearly experience-dependent, viz. involving direct exposure to give rise to an efficient response. These experiential processes can operate during gestation (e.g. Pedersen and Blass 1982; Smotherman 1982a, b; Todrank et al. 2011), during the natal process (Ronca et al. 1994), or, more obviously, postnatally in direct contact with the mother's body and mammae, which form an extremely potent context for the rapid learning of contingent odour cues (Alberts 1987).

Although the suckling interaction has been studied in many mammalian taxa from the ethological and ecological points of view, their chemical and chemosensory underpinnings remain unclear and still little investigated. This area of research has attracted a considerable surge of work in the 1970s and 1980s (reviewed, for example, in Rosenblatt 1983; Blass 1990), but conclusive results are still needed in many ways. Based on previous results and several reviews of them (Blass and Teicher 1980; Blass 1990; Raineki et al. 2010; Schaal et al. 2009; Schaal 2010), and based on our own recent work on the mouse, we intent here comparing the odour cues that likely guide rat and mouse newborns to the nipple and initiate suckling. From the outset, it may be posited that any comparison will be imperfect as rat and mouse pups have not been tested in exactly the same conditions with the same paradigms. However, homologous compounds or biological substrates (amniotic fluid, milk, maternal and pup salivas) have often been used in a functionally similar context (nipple searching/seizing) at similar developmental stages (i.e. in more or less nursing-experienced newborn pups). Accordingly, some comparative inferences and prospective studies may be proposed.

6.2 Lactating Females Are Uniquely Salient for Newborn Pups

Neonatal rats and mice are strongly aroused by the odour emitted by the whole body of conspecific lactating females when presented against the odour from non-lactating females. When put in contact with the abdomen of an anaesthetised conspecific lactating female, they express searching and oral grasping of a nipple (Hofer et al. 1976; Belin et al. 2011; Al Aïn et al. 2012b). Lactation correlates thus clearly with the release of attractive cues in both species (Galef and Muskus 1979; Al Aïn et al. 2012a).

The basic role of olfaction in suckling behaviour is attested through its perturbations either after genetic defection of the olfactory system (Hongo et al. 2000), by hampering olfactory abilities [lesion of olfactory bulbs (e.g. Kovach and Kling 1967); peripheral denervation (e.g. Hofer et al. 1976)], or by altering local stimuli by washing the source of emission with organic solvents (Hofer et al. 1976; Teicher and Blass 1976) or water (Hongo et al. 2000).

Thus, nipple-related odour cues are detectable to mouse and rat pups, and can be suppressed by mere washing. Although the precise bodily source(s) and nature of substrates involved are poorly understood, it is now clear that nipples of lactating females may constitute olfactory hotspots that attract inexperienced neonates and facilitate their suckling performance. The term "hotspot" is used here in all senses, as warmer skin regions overlaying mammary structures (Hofer et al. 1976) and affording at the same time thermal cues and thermally enhanced odour cues to newborn pups.

6.3 Prenatal Odour Cues in Initial Suckling

Right after being born, rat pups are endowed with a range of extra-mammary substrates brought to the nipple by the periparturient female's self-licking. These include blood, amniotic fluid (AF) and maternal saliva (see following section). AF was shown to elicit oral grasping of a nipple that has been painted with it (Teicher and Blass 1977). Rat pups on postnatal day 1 also display sustained crawling toward AF odour, but not toward an unfamiliar artificial odorant (Méndez-Gallardo and Robinson 2011). In *Balb-C* mice, nursing-inexperienced 5–8-h-old pups also initiated suckling when placed close to a nipple of a non-lactating female painted with AF taken from another conspecific female (Belin et al. 2011). The potency of AF odour was also noted in 2 day-old mice who preferred suckling a nipple coated with it, whereas saline and vanilla (as a novel odour) revealed ineffective (Kodama 1990).

The potency of AF to trigger appetitive behaviour in newly born rodents can be explained by some reinforcing value presumably assigned to it in utero. The reinforcing value of this odour experienced in aqueous conditions of reception is then well conserved in aerial conditions. But two volatile fractions of AF compounds may be considered: the one that is species-specific—common to all members of the species; and the other that is individual-specific—proper to each mother-foetus system. The above experiments in rats and mice used AF collected in females other than the mother, thus taping into the supra-individual fraction (at least within a given colony) of the AF odour. Individual-specific amniotic odorants may also govern pup behaviour, and it is through their systematic investigation that foetal olfactory learning was demonstrated in mammals. Prenatal odour learning was established in the rat concurrently by Pedersen and Blass (1982) and by Smotherman (1982a, b). These experimenters injected an artificial odorant into the amniotic sacs on embryonic day 19 or 20, and by testing newborn pups thereafter they showed an increased attraction to the target odorant. Hepper (1987) later demonstrated the ability of foetal rats to encode the individual characteristics of their own AF odour without addition of any extraneous odorant and suggested its involvement in kin recognition processes.

At parturition, rat and mouse females have their nipples endowed with cues from the amniotic environment that trigger activation, orientation and oral grasping in their newly born offspring. Accordingly, rat and mouse newborns are initially exposed to, and attracted by, the odour of the AF, even when it is not the neonate's own fluid (reviewed in Robinson and Méndez-Gallardo 2010); thus, the neonates' positive response to AF might be traced into some colony- or species-specific odour experience in utero.

6.4 Postnatal Odour Cues in Initial Suckling

AF is not the only releaser of initial nipple grasping response in newly born rat and mouse pups. Saliva from post-parturient females, but not from virgin females, was also shown to elicit nipple grasping in suckling-inexperienced rat pups (Teicher and Blass 1977). Olfactory correlates of lactation are thus also expressed in some extra-mammary sources.

The saliva-related responses of the mouse pup appear different from that of the rat pup, however. When placed at the nipple of a non-lactating female that was treated with saliva collected from a parturient dam, nursing-inexperienced mouse pups did *not* react by immediate grasping (Al Aïn et al. 2012b).

Thus, when examining the initial reactions of rat and mouse newborns to the saliva of lactating conspecifics, findings seem at variance in both species. Maternal saliva appeared indeed potent in neonatal rats (Teicher and Blass 1977), but not in neonatal mice (Belin et al. 2011). Whether this reflects a species difference in maternal saliva composition or differential disponibility of it on nipples is unknown. The maternal behaviour of rats and mice during and after parturition is indeed characterised by intense self- and pup licking (Hood et al. 2003; Weber and Olsson 2008). With advancing labour, between each delivery, females of both species alternate self- and pup licking, suggesting that maternal saliva is abundantly distributed over the ventral fur/nipples while pups are vigorously aroused by licking while they nose over the nipples. Thus, variation in direct exposure to maternal saliva odour may not explain the seeming species difference in the salience of saliva to elicit initial nipple grasping.

Regarding milk, the rat-mouse comparison is not possible as, surprisingly enough, it seems that no study has assessed the behavioural activity of conspecific milk in newly born rat pups. To the best of our knowledge, only one study has tested conspecific milk in rat pups, but these were aged PND 7–8. At this age, rat milk was ineffective in releasing the nipple grasping in rat pups (Singh and Hofer 1978). However, the potential implication of conspecific milk in the elicitation of nipple grasping was clearly demonstrated in newborn mice. Painting fresh mouse milk on the nipples of an anaesthetised non-lactating mouse had a strong positive effect on oral grasping in mouse pups (as compared with nipples painted with water) (Belin et al. 2011; Al Aïn et al. 2012b).

The behavioural activity of fresh milk on neonatal mice before any postnatal exposure may derive from prenatal learning. Neonate mice pups might thus encounter in milk a fraction of the compounds already met in AF, following the general process of transnatal chemosensory continuity noted in most mammals studied so far (reviewed in Schaal 2005), including the rat (e.g. Hepper 1987; Arias and Chotro 2006). However, whether newborn rats respond to rat milk is an open area for future research, as they have paradoxically been tested only with bovine milk (which is attractive for them; e.g. Méndez-Gallardo and Robinson 2011).

Maternal saliva and milk are, in part, transudates from maternal blood plasma, and as such may compositionally overlap with AF (Schaal 2005), facilitating the early learning of the odours they convey. But these different substrates may also just be present on the nipple at the very first suckling episode and become rapidly associated with suckling.

6.5 The Cueing Value of Nipple-Related Odour Substrates Changes as a Function of Nursing Experience

In the newborn mouse, the response to saliva changes over time and necessitates direct exposure before it emerges. The saliva of a lactating female begins to release nipple grasping in mice 6–8 h after birth (i.e. after about 1–4 h of suckling experience; Belin et al. 2011; Al Aïn et al. 2012b). After more extensive suckling experience during the first 24 h, pup saliva becomes effective in mice on PND 1

(Belin et al. 2011), as well as in rat pups on PND 2 and older (Blass and Teicher 1980). Accordingly, the cueing value of maternal saliva (in mouse pups) and of pup saliva (in mouse and rat pups) to promote nipple grasping depends on the amount of direct exposure to saliva odour in the context of suckling. But, in the mouse at least, current evidence indicates that pup saliva requires more suckling experience than maternal saliva to become behaviourally significant.

Thus, the cuing value of milk (see previous section) and of pup and maternal saliva appear to be shaped through learning. At PND 5–6, in mouse pups, all odour substrates presumed to be present on the nipples were separately as potent to elicit approach response as the whole nipple odour of lactating mice, and they were all equally attractive when paired (Al Aïn et al. 2012a). This suggests that all of these substrates may convey common compounds. It is indeed probable that maternal milk and saliva share volatiles derived from food or general metabolism (e.g. Galef and Henderson 1972; Bronstein et al. 1975). In the same way, suckling pups may impregnate nipples with their own saliva, as well as with a mixture of saliva and milk coagulate. Otherwise, all of these odour substrates may have got similar reinforcing value because they were encoded in contingence with the same rewarding endpoints, viz., comfort contact, warmth, and the mere action of sucking. Sucking-related odour learning is indeed a general phenomenon in mammalian perinates (e.g. Alberts 1987; Rosenblatt 1983; Schaal 2005), and it can result in an equivalent reinforcing value of functionally different odour cues. This interpretation is strengthened by the fact that the dorsal odour from lactating females, not associated with nursing, is less attractive to mouse pups than the ventral odour (Al Aïn et al. 2012a). Hence, being exposed for most of their interactions with the mother to the blend formed by mammary (milk) and extra-mammary odours (AF, salivas), mouse pups exhibit undifferentiated attraction to all these substrates presented separately. In addition, besides sucking-induced odour preference, newborn mice also develop a preference for odorants linked with aspects of maternal care that are specially arousing, such as ano-genital licking or its experimental simulation with a soft paintbrush (Durand et al. 2003; Bouslama et al. 2005; Armstrong et al. 2006).

Finally, maternal urine presumably spread on nipples during the avid licking of their own and their pups' ano-genital areas is also susceptible to create salient cues for pups. Nevertheless we have clearly observed in mice that maternal urine is not involved in the grasping-sucking response at birth and later (Belin et al. 2011; Al Aïn et al. 2012b), although it is later profoundly engaged in social relations (e.g. Hurst 1990).

6.6 Predisposed Mechanisms Implied in Suckling?

The above results explained the suckling response of rat and mouse neonates to odour substrates on the nipple through learning mechanisms. Nevertheless, some stimuli might initiate oral grasping before any previous exposure. In that line, Pedersen and Blass (1981) have tested rat pups' responsiveness to dimethyldisulfide (DMDS), a compound that was targeted by extrapolation if its involvement in the sexual attraction of male hamsters to the female. DMDS was indeed traced in the saliva of rat pups, and shown to be effective in nipple grasping of 3–5 day-old pups, but with a much lower potency than nipples of lactating rats. Nonetheless as the rat pups were already aged PND 3–5, they could have acquired DMDS odour as a cue because of its regular association with suckling. Nevertheless, DMDS appears singular because it has the power to elicit oral and general responsiveness already in the foetus, where it mimics the response to bovine milk (Smotherman and Robinson 1992). DMDS is reported not to be chromatographically (GC) traceable in AF (Blass 1990), leading to the logical conclusion that its behavioural salience might not depend on prenatal exposure. Overall, however, before DMDS can be conclusively defined as an unconditional stimulus triggering oral grasping in the rat newborn, its effect should be tested at the very first postnatal exposure.

Compounds emitted in the mammary tract (either in milk or in nipple-skin secretions) carrying similar effects have not been evidenced so far in the rat or in the mouse. Nevertheless, it cannot be excluded that mouse pups' initial response to conspecific milk could derive from an unconditional release of searching-grasping responses, in a way comparable to the rabbit mammary pheromone (Schaal et al. 2003, 2009). But this needs more investigation.

6.7 Conclusion

Studies on neonatal behaviour in the rat and the mouse have shown similarities, but also apparent species specificities, in the elicitation and regulation of nipple searching and grasping. Right at birth, neonate pups appear to rely on cues already encountered prenatally, AF, and that females spread on their ventrum and nipples in the course of their normal maternal behaviour. Rat and mouse pups are thus exposed to an already familiar odour substrate while rooting for, seizing and sucking, a nipple for the first time. But, in addition, other substrates are present due to the normal physiology and behaviour of the dam, such as secretions from nipple glands, milk, maternal saliva or pup saliva. All these compounds, at least when they are presented separately, do not seem to bear the same cueing value for neonatal rats and mice. While rat pups are attracted to saliva right from birth, mouse pups are not. Whether this disparity in neonatal dispositions reflects a species difference needs to be ascertained, as it will direct the search for chemicals that govern the vital responses of nursing-inexperienced rat and mouse pups towards nipples and milk.

References

Al Aïn S, Chraïti A, Schaal B, Patris B (2012a) Orientation of newborn mice to lactating females: biological substrates of semiochemical interest. Dev Psychobiol. doi:10.1002/dev.21003

Al Aïn S, Belin L, Schaal B, Patris B (2012b) How a newly born mouse gets to the nipple? Odor substrates eliciting first nipple grasping and sucking responses. Dev Psychobiol. doi: 10.1002/dev.21082

Alberts JR (1987) Early learning and ontogenetic adaptation. In: Krasnegor NA, Blass EM, Hofer MA, Smotherman WP (eds) Perinatal development: a psychobiological perspective. Academic, Orlando, FA, pp 12–38

Arias C, Chotro MG (2006) Interactions between prenatal ethanol exposure and postnatal learning about ethanol in rat pups. Alcohol 40:51–59

Armstrong CM, DeVito LM, Cleland TA (2006) One-trial associative odor learning in neonatal mice. Chem Senses 31:343–349

Belin L, Al Aïn S, Schaal B, Patris B (2011) Which odor substrates induce nipple grasping at birth in newborn mice? Dev Psychobiol 53:741

Blass EM (1990) Suckling: determinants, changes, mechanisms, and lasting impressions. Dev Psychol 26:520–533

Blass EM, Teicher MH (1980) Suckling. Science 210:15–22

Blass EM, Teicher MH, Cramer CP, Bruno JP, Hall WG (1977) Olfatory, thermal, and tactile controls of suckling in preauditory and previsual rats. J Comp Physiol Psychol 9:1248–1260

Bouslama M, Durand E, Chauvière L, Van den Bergh O, Gallego J (2005) Olfactory classical conditioning in newborn mice. Behav Brain Res 161:102–106

Breen MF, Leshner AI (1977) Maternal pheromone: a demonstration of its existence in the mouse (Mus musculus). Physiol Behav 18:527–529

Bronstein PM, Levine MJ, Marcus M (1975) A rat's first bite: the non-genetic, cross-generational transfer of information. J Comp Physiol Psychol 89:295–298

Coppola DM, Millar LC (1997) Olfaction in utero: behavioural studies of the mouse fetus. Behav Processes 39:53–68

Coppola DM, O'Connell RJ (1989) Stimulus access to olfactory vomeronasal receptors in utero. Neurosci Lett 106:241–248

Coppola DM, Budde J, Millar L (1993) The Vomeronasal duct has a protracted postnatal development in the mouse. J Morphol 218:59–64

Durand E, Dauger S, Vardon G, Gressens P, Gaultier C, De Schonen S et al (2003) Classical conditioning of breathing pattern after two acquisition trials in 2-day-old mice. J Appl Physiol 94:812–818

Ehret G (1976) Development of absolute auditory thresholds in the house mouse (Mus musculus). J Am Audiol Soc 1:179–184

Galef BG, Henderson PW (1972) Mother's milk: a determinant of the feeding preferences of weaning rat pups. J Comp Physiol Psychol 78:213–219

Galef BG, Muskus PA (1979) Olfactory mediation of mother-young contact in Long-Evans rats. J Comp Physiol Psychol 93:708–716

Hennessy MB, Li J, Levine S (1980) Infant responsiveness to maternal cues in mice of 2 inbred lines. Dev Psychobiol 13:77–84

Hepper PG (1987) The amniotic fluid: an important priming role in kin recognition. Anim Behav 35:1343–1346

Hofer MA, Shair H, Singh P (1976) Evidence that maternal ventral skin substances promote suckling in infant rats. Physiol Behav 17:131–136

Hongo T, Hakuba A, Shiota K, Naruse I (2000) Suckling dysfunction caused by defects in the olfactory system in genetic arhinencephaly mice. Biol Neonate 78:293–299

Hood KE, Dreschel NA, Douglas A (2003) Maternal behavior changes after immune challenge of neonates with developmental effects on adult social behavior. Dev Psychobiol 42:17–34

Hurst JL (1990) Urine marking in populations of wild house mice Mus domesticus Rutty. III Communication between the sexes. Anim Behav 40:233–243

Kodama N (1990) Preference for amniotic fluid in newborn mice. Annual Meeting of the International Society for Developmental Psychobiology, Cambridge, UK

König B, Markl H (1987) Maternal care in house mice. Behav Ecol Sociobiol 20:1–9

Kovach JK, Kling A (1967) Mechanisms of neonate sucking behaviour in the kitten. Anim Behav 15:91–101

Méndez-Gallardo V, Robinson RR (2011) Amniotic fluid and milk odor evoke crawling locomotion in the newborn rat. Dev Psychobiol 53:757

Pedersen PA, Blass EM (1981) Olfactory control over suckling in albino rats. In: Aslin RN, Alberts R, Petersen MR (eds) Development of perception, psychobiological perspectives, Vol. 1. Audition somatic perception and the chemical senses. Academic, New York, pp 359–381

Pedersen PA, Blass EM (1982) Prenatal and postnatal determinants of the 1st suckling episode in albino rats. Dev Psychobiol 15:349–355

Pedersen PE, Stewart WB, Greer CA, Shepherd GM (1983) Evidence for olfactory function in utero. Science 221:478–480

Raineki C, Pickenhagen A, Roth TL, Babstock DM, McLean JH, Harley CW, Lucion AB, Sullivan RM (2010) The neurobiology of infant maternal odor learning. Braz J Med Biol Res 43:914–919

Robinson SR, Méndez-Gallardo V (2010) Amniotic fluid as an extended milieu intérieur. In: Hood KE, Halpern CT, Greenberg G, Lerner RM (eds) Handbook of developmental science, behaviour, and genetics. Blackwell, New York, pp 234–284

Ronca AE, Kamm K, Thelen E, Alberts JR (1994) Proximal control of fetal rat behavior. Dev Psychobiol 27:23–38

Rosenblatt JS (1983) Olfaction mediates developmental transitions in the altricial newborn of selected species of mammals. Dev Psychobiol 16(5):347–375

Schaal B (2005) From amnion to colostrum to milk: bridging in early developmental transitions. In: Hopkins B, Johnson S (eds) Prenatal development of postnatal functions. Praeger, Westport, CT, pp 52–102

Schaal B (2010) Mammary odor cues and pheromones: mammalian infant-directed communication about maternal state, mammae, and milk. Vitam Horm 83:83–136

Schaal B, Coureaud G, Langlois D, Ginies C, Sémon E, Perrier G (2003) The mammary pheromone of the rabbit: chemical and behavioural characterisation. Nature 424:68–72

Schaal B, Coureaud G, Doucet S, Delaunay-El Allam M, Moncomble AS, Montigny D et al (2009) Mammary olfactory signalisation in females and odor processing in neonates: ways evolved by rabbits and humans. Behav Brain Res 200(2):346–358

Schultz LA, Lore RK (1993) Communal reproductive success in rats (Rattus norvegicus)—effects of group composition and prior social experience. J Comp Psychol 107:116–122

Singh PJ, Hofer MA (1978) Oxytocin reinstates maternal olfactory cues for nipple orientation and attachment in rat pups. Physiol Behav 20(4):385–389

Smotherman WP (1982a) Odor aversion learning by the rat fetus. Physiol Behav 29:769–771

Smotherman WP (1982b) In utero chemosensory experience alters taste preferences and corticosterone responsiveness. Behav Neural Biol 36:61–68

Smotherman WP, Robinson SR (1987) Psychobiology of fetal experience in the rat. In: Krasnegor NA, Blass EM, Hofer MA, Smotherman WP (eds) Prenatal development: a psychobiological perspective. Academic, Orlando, FL, pp 39–60

Smotherman WP, Robinson SR (1992) Dimethyl disulfide mimics the effects of milk on fetal behavior and responsiveness to cutaneous stimuli. Physiol Behav 52(4):761–765

Suckow MA, Weisbroth SH, Franklin CL (2006) The laboratory rat, 2nd edn. Elsevier-Academic, San Diego, CA

Teicher MH, Blass EM (1976) Suckling in newborn rats: eliminated by nipple lavage, reinstated by pup saliva. Science 193:422–425

Teicher MH, Blass EM (1977) First suckling response of the newborn albino rat: the roles of olfaction and amniotic fluid. Science 198:635–636

Todrank J, Heth G, Restrepo D (2011) Effects of in utero odorant exposure on neuroanatomical development of the olfactory bulb and odour preferences. Proc Soc Lond B 278:1949–1955

Weber EM, Olsson IAS (2008) Maternal behaviour in *Mus musculus* sp.: an ethological review. Appl Anim Behav Sci 114:1–22

Chapter 7
The Role of Olfaction at Birth in Marsupial and Monotreme Mammals

Nanette Y. Schneider, Geoff Shaw, and Marilyn B. Renfree

Abstract At birth all mammalian neonates depend on finding the milk source in order to survive. The development of mammalian young ranges from precocial to highly altricial, but olfaction plays an essential role in the milk-finding process in all mammalian species studied, along with other sets of cues depending on the development of the neonate's sensory systems. Of all mammals studied, so far marsupials and monotreme are the least developed at birth and have the least developed olfactory systems. Nonetheless the neonates are able to reach the pouch and the teat or milk patches unaided by their mothers. Behavioural studies confirm that olfaction pays a role in guiding the young to the pouch and teat and may be responsible for the switch in direction at the lip of the pouch in the tammar wallaby to climb down into the pouch and attach to a teat. The switch in direction at the pouch entrance may be triggered by olfactory cues acting via this the main olfactory system, as it is further developed than the accessory olfactory system. However, the latter may serve as a backup if the main olfactory system fails. How the information is processed by the poorly developed neonatal brain to control the movements and behaviour of marsupial and monotreme young remains to be investigated.

N.Y. Schneider (✉)
Department of Molecular Genetics, German Institute for Human Nutrition, Potsdam-Rehbrücke, Arthur-Scheunert-Allee 114-116, 14558 Nuthetal, Germany

Department of Zoology, The University of Melbourne, Melbourne, VIC 3010, Australia
e-mail: nanette@finistere-virtuell.de

G. Shaw • M.B. Renfree
Department of Zoology, The University of Melbourne, Melbourne, VIC 3010, Australia
e-mail: g.shaw@unimelb.edu.au; m.renfree@unimelb.edu.au

M.L. East and M. Dehnhard (eds.), *Chemical Signals in Vertebrates 12*,
DOI 10.1007/978-1-4614-5927-9_7,

7.1 Introduction

All mammalian neonates are dependent on milk to survive and grow and so must locate the milk source immediately after birth. Milk not only provides a specialised nutrition, but it also is critical to provide immune protection because most mammalian neonates have only an innate immune system at birth (Goldman 2007). The overall developmental stage of the neonate varies greatly between species. Some species such as the horse, cow and elephant produce one or two highly developed precocial young, while others such as shrews, marsupials and monotremes give birth to altricial young that are totally dependent on their mother. Some young are protected within a nest or a burrow while the mother is foraging, or in the case of most marsupials, within a pouch. Both strategies are found in monotremes. While the platypus leaves its young in a borrow, the echidna has a pouch and will carry its young while foraging until it starts developing its spines, then the female will leave the young in a borrow. The smallest and most altricial young are found in marsupials and monotremes whose young are delivered at a stage similar to mid-gestational eutherian embryos. At birth, marsupial young have to find their way from the urogenital sinus to the teat. This path varies in length with that of the red kangaroo (*Macropus rufus*) being the longest, at about 20 cm, to species such as the eastern quoll (*Dasyurus viverrinus*), where the young are deposited directly into the pouch from the urogenital opening during the birth process. The development of the pouch varies between species, with teats not covered, partly covered or fully covered (Tyndale-Biscoe and Renfree 1987). Three birth positions have been described in marsupials (Gemmell et al. 2002): standing on all fours, lying on the side or sitting up. The young of macropodids (kangaroos and wallabies) face the longest journey in which the females will adopt a sitting position with her tail passed forward in between her hind legs (Renfree et al. 1989). While these species mainly have a single young per gestation, this young is more advanced in development. In other marsupial families the neonate's journey is shorter due to birth position and body size of the female. Many of these species are polytocous and some have a greater number of offspring than teats (Tyndale-Biscoe and Renfree 1987).

Monotremes are the only mammals that lay eggs. The young hatch around 10.5 days after egg-laying in short-beaked echidnas (*Tachyglossus aculeatus*); the time of hatching is unknown for the platypus (*Ornithorhynchus anatinus*) (Griffiths 1978; Augee et al. 2006). The platypus female digs a nursing burrow in the river bank with the entrance above water level and a chamber filled with grass, leaves and roots at the end (Bennett 1835). The female retires to this burrow for some days, during which time she lays the eggs. There are no direct reports of egg-laying in platypus, but it is thought that the female curls up with the tail bent round and laid along her abdomen laying the egg into the 'incubatorium' formed between tail and abdomen (Burrell 1927). The eggs are covered with a sticky secretion of unknown origin, making them adhere to each other and presumably to hairs of the abdomen (Burrell 1927). Two young were found between tail and abdomen during the excavation of a burrow making this the likely nursing position (Burrell 1927). The female echidna also prepares a nursing borrow. During the egg laying she will curl up so that the cloaca reaches into the pouch making it possible to deposit the egg there (Griffiths 1978; Augee et al. 2006). The young (puggle) will stay in the pouch until it starts developing spines when it will then be left in the burrow.

Most mammalian young manage to find the teats (therian mammals) or mammary patches (monotremes) without much assistance from their mothers. Exceptions to this are primates including humans in which the mother needs to hold the neonate to her breast. So far, the sensory systems involved in this milk-finding process have been studied in a number of eutherian mammals and it is clear that olfaction plays a critical role essential for neonatal survival. New data on marsupials and monotremes suggests that they also depend on the same senses (Schneider 2011; Schneider et al. 2009). This chapter reviews our understanding of the use of odour cues and the olfactory systems in marsupials and monotremes that may be involved in the search for the teats or mammary patches.

7.1.1 Chemical Cues Are Important for Survival at Birth

At birth, most sensory organs are still developing and in some species some may not be functional until after birth. While the young are likely to use a variety of sensory cues such as touch, taste, warmth and visual cues in order to find the teat, it seems that all mammalian young use olfaction to some degree in this task. In order to lead the neonate to the milk, mammalian mothers have developed signals to guide the young. Odour cues have an important role as signals (e.g. Blass and Teicher 1980; Distel and Hudson 1985; Hudson and Distel 1983; Leon and Moltz 1971; Schaal et al. 1995; Schneider 2011; Schneider et al. 2009; Vince 1993; Vince and Ward 1984). Marsupial young are highly altricial at birth. The lung is still a sac-like structure and the majority of gas exchange occurs through the thin, highly vascular skin (Hughes and Hall 1988). The neonate moves itself unaided to the pouch using crawling movements of its powerfully developed forelimbs on which the paws are precocially developed with claws (Hughes and Hall 1988). At this time, the hindlimbs are still paddle-like structures and play no role in movement to the pouch. The sensory systems are still poorly developed. The eyes are poorly developed and the eyelids still fused, the auditory channel has not yet opened and the lips are fused laterally, leaving only an opening at the front to accept the teat. However, the nostrils are already open, Merkel cells are found underlying the skin of the snout (Hughes and Hall 1984; Jones and Munger 1985; Gemmell et al. 1988; Gemmell and Rose 1989) and otoliths are found in the vestibular system of the inner ear (Hughes and Hall 1984; Gemmell and Nelson 1988; Gemmell and Rose 1989; McCluskey et al. 2008). It has been long suggested (e.g. Hill and Hill 1955; Lyne 1974; Kratzing 1986; Gemmell and Nelson 1988) that olfactory cues play an important role in directing the young to the teat in marsupials, but until recently there were no direct studies to confirm this. Behavioural studies of the neonatal quokka (*Setonix brachyurus*), a small macropodid marsupial, showed that it uses gravity to orient its climb from the urogenital opening to the pouch entrance (Cannon et al. 1976). The first behavioural study to provide evidence that olfactory cues are important was done in the brushtail possums (*Trichosurus vulpecula*) (Veitch et al. 2000). In macropodids, where females adopt a sitting posture at birth, the neonate has to orient itself and climb upwards to the pouch. Once at the lip of the pouch, it has to change direction, to climb downwards into the pouch to attach to a teat (Schneider et al. 2009). We have recently shown that the neonate uses olfactory cues from its mother's pouch or teats to orientate this downwards climb (Schneider et al. 2009). The exact source of the cues and their possible chemical structure remain unknown. This chemical cue—a pheromone?—may also help direct the climb from the urogenital opening to the pouch lip, since during the characteristic parturient behaviour the mother energetically licks the pouch and the area between the pouch entrance and the urogenital opening in the minutes before the young is delivered, a behaviour which may well produce a well-defined odour trail to follow (Renfree et al. 1989; Schneider et al. 2009). In experimental tests, tammar wallaby (*Macropus eugenii*) neonates not only direct their climb towards a swab with pouch odour on it, but they also change direction from climbing up against gravity to climbing down once they have reached the odour source. It appears likely that a second chemical cue may be produced at the lip of the pouch, inducing the downward climb of the young into the pouch towards the teat (Schneider et al. 2009) or the mammary odours may be sufficiently strong to attract the young.

The young not only needs to locate the teat, but, in polytocous species, it may also have to compete for the best milk outlet, as has been well shown in eutherian mammals (for review Hudson et al. 2009). In marsupials, this race to the pouch may be a life or death competition, because many species have more young than they have teats (Tyndale-Biscoe and Renfree 1987). Since each young becomes permanently attached to the teat for the first weeks or months of lactation, any supernumerary young die (Tyndale-Biscoe and Renfree 1987). It is unlikely that successfully attached marsupial young compete for teats after they are no longer permanently attached but this is as yet unknown.

Newly hatched monotreme young are at a similar level of development to neonatal macropodids, which are the most advanced of the marsupials at birth.

An egg tooth is found on the upper lip which facilitates hatching (Griffiths 1978). Monotreme females secrete milk through the milk ducts onto the skin of areas called milk patches or milk fields. There are no teats. As monotreme young do not attach to a teat, their lips are not fused the way it is observed for marsupial young (Griffiths 1978). While most sensory systems appear to be non-functional at hatching, the nostrils are open suggesting that the olfactory organs are functional at this time (Griffiths 1978; Schneider 2011).

While the mothers provide the guiding cue, the neonates also need sensory systems and motor abilities to reach the milk source. Olfaction is likely to play a crucial role in detecting these cues. Mammals possess at least four chemosensory organs in their nasal cavity: the two major olfactory organs—the main olfactory epithelium and the vomeronasal organ (VNO)—and the septal organ (Rodolfo-Masera 1943) and Grüneberg's ganglion (Grüneberg 1973) (Fig. 7.1). The latter structure has, however, only been described in rodents.

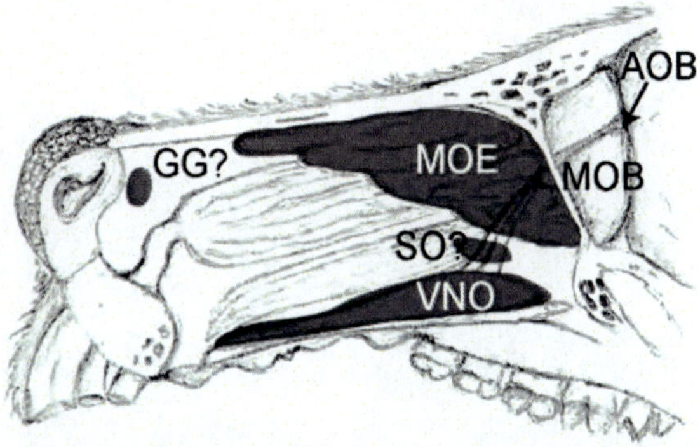

Fig. 7.1 Four olfactory sensory systems can be distinguished in mammals, shown here in a sketch of a tammar wallaby nose. While the vomeronasal organ (VNO) and main olfactory epithelium (MOE) have been described in the tammar wallaby, the septal organ (SO) has only been found in one marsupial, the bandicoot (Kratzing 1978). The Grünberg's ganglion (GG) near the nose tip in eutherians has not been identified in any marsupial. All these sensory organs project either to the main olfactory bulb (MOB) or to the accessory olfactory bulb (AOB)

7.2 Olfactory Systems Involved in Detection of Cues Produced by the Mother

While both major olfactory sensory organs, the main olfactory epithelium and VNO, could possibly be involved in detecting pheromonal cues produced by the mother to enable the neonate to find the teats, the main olfactory system may be of greater importance in most mammals than the VNO in this task. The only mammalian species that may be an exception are those who never leave their marine environment and suckle their young underwater (such as whales and dolphins). Olfactory receptor cells are differentiated before birth, e.g. in the mouse (Cuschieri and Bannister 1975; Graziadei et al. 1980), rat (Farbman and Margolis 1980; Gesteland et al. 1982), rabbit (Meisami et al. 1990) and human fetus (Nakashima et al. 1984) as well as in marsupials and monotremes (reviewed Schneider 2011; Fig. 7.2a, b).

However, the development of the main olfactory system varies between species. Marsupials and monotremes are the least developed. While the structure of the main olfactory system in neonatal marsupials is not yet mature, its development is comparable to that of an E14 rats fetus (Farbman and Menco 1986) for which electro-olfactograms of similar magnitude to those of adult rats can be recorded (Gesteland et al. 1982). Almost nothing is known about further processing of olfactory cues by the developing mammalian brain. The mammary pheromone of the rabbit activates the main olfactory bulb as well as brain regions involved in osmoregulation, thirst and motivation-guided motor responses (Charra et al. 2012). The pheromone selectively labels the anterior subdivision of the piriform cortex (figure-ground separation of target odorants), the organum vasculosum of the lamina terminalis

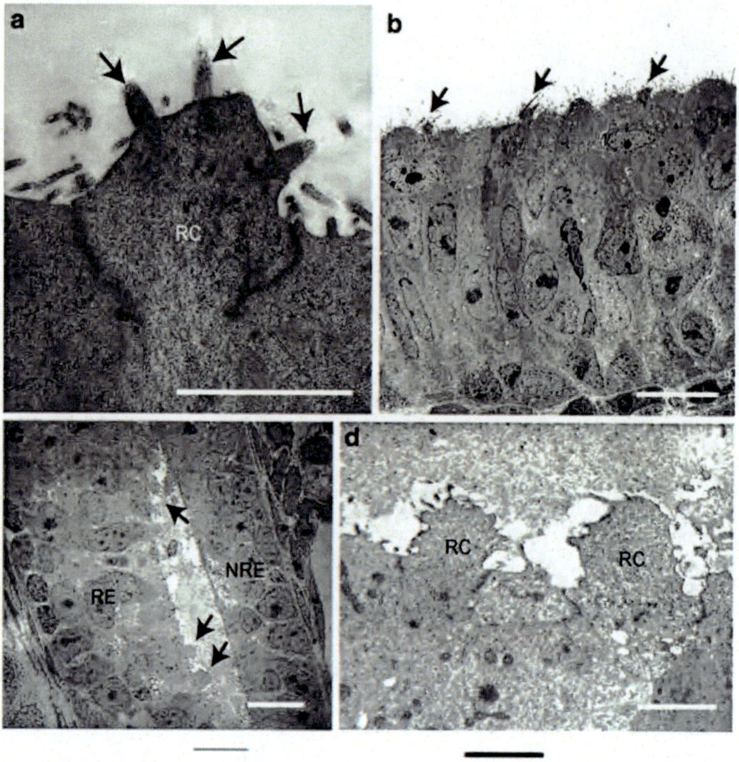

Fig. 7.2 Transmission electron micrograph of the neonatal tammar wallaby olfactory knobs ((**a**) cilia labelled with *arrows* and (**b**) olfactory knobs marked by *arrows*; scale 1 and 10 μm) on the surface of the receptor cells (RC) in the main olfactory epithelium. Similar olfactory knob-like structures as found in the adult receptor epithelium of the VNO are observed in the neonate ((**c**, **d**) *black arrows mark* olfactory knob-like structures on the receptor epithelium (RE); non-receptor epithelium (NRE); scale 10 and 2 μm)

(homeostatic and motor regulation; ingestive behaviour and regulation of thirst), the habenula (modulation of reward-based motor actions), and the lateral preoptic area (area of input from olfactory pathway). While only two layers, likely representing the olfactory nerve layer and the neuroepithelium, are present in the main olfactory bulb of the neonate kowari (*Dasyurides byrnei*) and eastern quoll, all other marsupial species studied have three layers including olfactory nerve layer and neuroepithelium (Schneider et al. 2009; Schneider 2011). In neonatal tammars, axons in the main olfactory epithelium project to the main olfactory bulb, but few axons were seen coursing from the main olfactory epithelium to the surface of the forebrain in the area of the future olfactory tubercle (Ashwell et al. 2008). Few fibres from the olfactory bulb pass through the anterior olfactory nuclei to the most rostral piriform cortex (Ashwell et al. 2008). It is still unclear how olfactory information could be processed by the neonatal brain as central projections of the olfactory bulb (along the lateral olfactory tract) seem not established until at least the second week of postnatal life and bulb output neurons cannot be retrogradely labelled from the lateral olfactory tract until after postnatal day 15 (Ashwell et al. 2008). Furthermore, Ashwell et al. (2008) were unable to find evidence of the forebrain nuclei projecting to the spinal cord at birth and found only few neurons projecting from the lateral hypothalamus to the brainstem.

The VNO and the nasopalatine duct are already open at birth in all investigated species (Schneider 2011). The VNO opens directly to the nasal cavity near the nasopalatine duct's opening in most species, although in the kowari the opening position so far could not be determined due to the orientation of the tissue sectioning. However, the VNO of the neonatal long-nosed bandicoot and brushtail possum opens to the nasopalatine duct. At birth, vomeronasal receptor cells with knob-like structures protruding into the lumen of the VNO are found in the tammar (Schneider et al. 2009; Fig. 7.2c, d). There is no accessory olfactory bulb in marsupials at birth. Some expression of transient receptor potential channel, subfamily C, member 2 (TRPC2) that is involved in the signal transduction cascade of vomeronasal receptor cells, is seen at birth, but only becomes specific to the VNO after day 30 postpartum in the tammar wallaby (Frankenberg et al. 2011; Schneider 2011). There are no behavioural studies that demonstrate the use of the accessory olfactory system to detect the teats in any species. While the vomeronasal receptor cells project via the ganglion terminale to the brain, it remains to be established if the accessory system has a role in teat location in marsupials. It may provide a back up for the main olfactory system in detecting signals to find the milk source in addition to other roles.

Our behavioural test with neonate tammar wallabies showed that they are able to direct their climb towards the pouch odour (Schneider et al. 2009). It is therefore likely that olfactory receptor cells of marsupials and monotremes are also capable of transducing odour signals to the brain. The behavioural response involves only a simple switch in direction so minimal development of neural connections between the olfactory epithelium and the brain may be sufficient. It remains to be determined how young of marsupials such as the kowari or eastern quoll, whose olfactory systems appear to be inadequately developed, and are likely non-functional, find the way to the teat (Schneider 2011). However, in both these species the neonates are deposited by the mother directly into the pouch and so the navigational challenge faced is less than for most other marsupials.

The main olfactory system of the monotremes seems to be further developed at hatching than in the neonatal marsupials studied. Bowman glands typical for the adult main olfactory epithelium are underlying the epithelium (Schneider 2011). Only goblet cells, single cell glands, were observed in the neonatal tammar wallaby (Schneider et al. 2009). The olfactory bulb already has three layers (Schneider 2011).

The palate of the monotreme is closed at hatching, apart from an opening in the palate that has the form of an equal-sided triangle whose apex is directed posterior (Seydel 1899). In newly hatched monotremes the VNO opens to the opening in the palate. This is distinctive compared with the adult platypus where it opens to the mouth cavity or in the adult echidna into the nasopalatine duct (Broom 1896; Symington 1891). The vomeronasal end gland at the caudal end of the VNO is already prominent in the late embryo in echidna eggs pre-hatching (Schneider 2011).

While nerves lead from the VNO to the brain, no tissues were available to examine the accessory olfactory bulb or ganglion terminal (Schneider 2011). Due to the similarly well-developed main olfactory systems, it seems likely that newly hatched monotreme have functional olfactory organs and may also use similar olfactory cues to find the milk patches by odours (Griffiths 1978; Schneider 2011; Schneider et al. 2009).

It is clear that olfaction plays an essential role in all mammalian neonates' quests for the milk source. While there are differences in the development of the olfactory systems, marsupials and monotremes have the least developed olfactory systems of all mammals so far investigated. However, the main olfactory system is functional in most studied species. The accessory olfactory system develops later than the main olfactory system and so far has not been shown to play a role in detecting the milk source. While it seems that the main olfactory system is detecting olfactory cues to assist the location of the teat, the details of the neural pathways involved in the neonatal tammar are still to be ascertained, an endeavour that will be facilitated by characterisation of the pouch odour cue. Depending on the species, different chemical signals are used to determine the location of the milk. So far the structure of only one mammary pheromone has been determined. Much more information in a greater number of species is needed to understand how olfactory communication and its role in suckling the young evolved.

References

Ashwell KWS, Marotte LR, Cheng G (2008) Development of the olfactory system in a wallaby (*Macropus eugenii*). Brain Behav Evol 71:216–230

Augee ML, Gooden B, Musser AM (2006). Echidna: extraordinary egg-laying mammal. CSIRO Publishing, Collingwood, p 136

Bennett G (1835) Notes on the natural history and habits of the *Omithorhynchus paradoxus*. Blum Trans Zool Soc Lond 1:229–258

Blass EM, Teicher MH (1980) Suckling. Science 210:15–22

Broom R (1896) On the organ of Jacobson in the monotremata. J Anat Physiol 30:70–80

Burrell H (1927) The platypus. Angus & Robertson, Sydney

Cannon JR, Bakker HR, Bradshaw SD, McDonald IR (1976) Gravity as the sole navigational aid to the newborn quokka. Nature 259:42

Charra R, Datiche F, Casthano A, Gigot V, Schaal B, Coureaud G (2012) Brain processing of the mammary pheromone in newborn rabbits. Behav Brain Res 226:179–188

Cuschieri A, Bannister LH (1975) The development of the olfactory mucosa in the mouse: electron microscopy. J Anat 119:471–498

Distel H, Hudson R (1985) The contribution of the olfactory and tactile modalities to the nipple-search behaviour of newborn rabbits. J Comp Physiol A 157:599–605

Farbman AI, Margolis FL (1980) Olfactory marker protein during ontogeny: immunohistochemical localization. Dev Biol 74:205–215

Farbman AI, Menco BP (1986) Development of olfactory epithelium in the rat. In: Breipohl W (ed) Ontogeny of olfaction. Springer, Berlin, pp 45–56

Frankenberg S, Schneider NY, Fletcher TP, Shaw G, Renfree MB (2011) Identification of two distinct genes at the vertebrate TRPC2 locus and their characterisation in a marsupial and a monotreme. BMC Mol Biol 12:39

Gemmell RT, Nelson J (1988) The ultrastructure of the olfactory system of three newborn marsupials species. Anat Rec 221:655–662

Gemmell RT, Rose RW (1989) The senses involved in movement of newborn Macropodoidea and other marsupials from cloaca to pouch. In: Grigg G, Jarman P, Hume I (eds) Kangaroos, wallabies and rat-kangaroos. Surrey Beatty & Sons Pty Limited, New South Wales, Australia, pp 349–354

Gemmell RT, Peters B, Nelson J (1988) Ultrastructure identification of Merkel cells around the mouth of newborn marsupials. Anat Embryol 177:403–408

Gemmell RT, Veitch C, Nelson J (2002) Birth in marsupials. Comp Biochem Physiol B Biochem Mol Biol 131:621–630

Gesteland RC, Yancey RA, Farbman AI (1982) Development of olfactory receptor neuron selectivity in the rat fetus. Neuroscience 7:3127–3136

Goldman AS (2007) The immune system in human milk and the developing infant. Breastfeed Med 2:195–204

Graziadei GAM, Stanley RS, Graziadei PPC (1980) The olfactory marker protein in the olfactory system of the mouse during development. Neuroscience 5:1239–1252

Griffiths M (1978). The biology of the monotremes. Academic, New York, p 367

Grüneberg H (1973) A ganglion probably belonging to the N. terminalis system in the nasal mucosa of the mouse. Z Anat Entwicklungsgesch 140:39–52

Hill JP, Hill WCO (1955) The growth stages of the native cat (*Dasyurus viverrinus*) together with observations on the anatomy of the newborn young. Trans Zool Soc Lond 28:349–452

Hudson R, Distel H (1983) Nipple location by newborn rabbits: behavioural evidence for pheromonal guidance. Behaviour 85:260–275

Hudson R, Raihani G, Gonzalez D, Bautista A, Distel H (2009) Nipple preference and contests in suckling kittens of the domestic cat are unrelated to presumed nipple quality. Dev Psychobiol 51:322–332

Hughes RL, Hall LS (1984) Embryonic development in the common brushtail possum *Trichosurus vulpecula*. In: Smith AP, Hume ID (eds) Possums and gliders. Sydney, Australian Mammal Society, pp 197–212

Hughes RL, Hall LS (1988) Structural adaptations of the newborn marsupial. In: Tyndale-Biscoe CH, Janssens PA (eds) The developing marsupial. Springer, Berlin, Heidelberg, pp 8–27

Jones TE, Munger BL (1985) Early differentiation of the afferent nervous system in glabrous snout skin of the opossum (*Monodelphis domesticus*). Somatosens Res 3:169–184

Kratzing JE (1978) The olfactory apparatus of the bandicoot (*Isoodon macrourus*): fine structure and the presence of the septal olfactory organ. J Anat 125:601–613

Kratzing JE (1986) Morphological maturation of the olfactory epithelia of Australian marsupials. In: Breipohl W (ed) Ontogeny of olfaction. Springer, Berlin, pp 57–70

Leon M, Moltz H (1971) Maternal pheromone: discrimination by pre-weanling albino rats. Physiol Behav 7:265–267

Lyne AG (1974) Gestation period and birth in the marsupial *Isoodon macrourus*. Aust J Zool 22:303–309

McCluskey SU, Marotte LR, Ashwell KW (2008) Development of the vestibular apparatus and central vestibular connections in a wallaby (*Macropus eugenii*). Brain Behav Evol 71:271–286

Meisami E, Louie J, Hudson R, Distel H (1990) A morphometric comparison of the olfactory epithelium of newborn and weanling rabbits. Cell Tissue Res 262:89–97

Nakashima T, Kimmelman CP, Snow JB (1984) Structure of human fetal and adult olfactory neuroepithelium. Arch Otolaryngol 110:614–646

Renfree MB, Fletcher TP, Blanden DR, Lewis PR, Shaw G, Gordon K, Short RV, Parer-Cook E, Parer D (1989) Physiological and behavioural events around the time of birth in macropodid marsupials. In: Grigg G, Jarman P, Hume I (eds) Kangaroos, wallabies and rat-kangaroos. New South Wales, Australia, Surrey Beatty & Sons Pty, pp 323–337

Rodolfo-Masera T (1943) Sur l'esistenza di un particolare organo olfacttivo nel setto nasale della cavia e di altri roditori. Arch Ital Anat Embryol 48:157–212

Schaal B, Orgeur P, Amould C (1995) Olfactory preferences in newborn lambs: possible influences of prenatal experience. Behaviour 132:351–365

Schneider NY (2011) The development of the olfactory organs in newly hatched monotremes and neonate marsupials. J Anat 219:229–242

Schneider NY, Fletcher TP, Shaw G, Renfree MB (2009) The olfactory system of the tammar wallaby is developed at birth and directs the neonate to its mother's pouch odours. Reproduction 138:849–857

Seydel O (1899) Ueber Entwicklungsvorgänge an der Nasenhöhle und am Mundhöhlendache von Echidna nebst Beiträgen zur Morphologie des peripheren Geruchsorgans und des Gaumens der Wirbelthiere. Denksch Med Naturwis Ges Jena 6:445–532

Symington M (1891) On the nose, the organ of Jacobson, and the dumb-bell-shaped bone in the *Ornithorhynchus*. Proc Zool Soc 39:575–585

Tyndale-Biscoe CH, Renfree MB (1987) Reproductive physiology of marsupials. Cambridge University Press, Cambridge, p 476

Veitch CE, Nelson J, Gemmell RT (2000) Birth in the brushtail possum, *Trichosurus vulpecula* (Marsupialia: Phalangerida). Aust J Zool 48:691–700

Vince M (1993) Newborn lambs and their dams: the interaction that leads to sucking. Adv Study Behav 22:239

Vince M, Ward T (1984) The responsiveness of newly born Clun Forest lambs to odour sources in the ewe. Behaviour 89:117–127

Part III
Pheromones and Social Organisation

Chapter 8
Individual Variation in Pheromone Isoform Ratios of the Red-Legged Salamander, *Plethodon shermani*

Adam J. Chouinard, Damien B. Wilburn, Lynne D. Houck, and Richard C. Feldhoff

Abstract For more than 15 years, pheromone signalling in plethodontid salamanders has served as an important amphibian system for studying how molecules can influence reproductive behaviour. In the red-legged salamander (*Plethodon shermani*), males utilise proteinaceous pheromones from a submandibular (mental) gland to affect female behaviour and reduce the duration of courtship. The two major proteins, Plethodontid Receptivity Factor (PRF) and Plethodontid Modulating Factor (PMF), exist as multiple isoforms within the gland of an individual male. While previous research has focused on biochemical characterisation of these proteins, this study explores the degree of intraspecific variability of pheromone isoform ratios among males of a single population. Biochemical analyses were performed on the mental gland extracts of individual male salamanders ($n = 108$) via high-performance liquid chromatography to quantify components of the pheromone mixture and establish the extent of individual variability. The results of these individual analyses revealed that the two main proteins (PMF and PRF) comprise over 80% of the total mixture, with a 5:3 ratio (PMF:PRF). Numerous minor peaks also appeared in the elution range of PRF, which could account for an excess of mRNA transcripts that have not been proteomically characterised. Lastly, these data demonstrate a remarkable diversity of isoform ratios among males. Taken together, these aspects of pheromone complexity and diversity could have profound functional effects on reproductive behaviour in this family of salamanders.

A.J. Chouinard (✉) • L.D. Houck
Department of Zoology, Oregon State University,
Corvallis, OR 97331, USA
e-mail: adam.chouinard@science.oregonstate.edu; houckl@science.oregonstate.edu

D.B. Wilburn • R.C. Feldhoff
Department of Biochemistry, University of Louisville,
Louisville, KY 40202, USA
e-mail: damien.wilburn@louisville.edu; rick.feldhoff@louisville.edu

M.L. East and M. Dehnhard (eds.), *Chemical Signals in Vertebrates 12*,
DOI 10.1007/978-1-4614-5927-9_8, © Springer Science+Business Media New York 2013

8.1 Introduction

Chemical signal profiles of many species are highly complex, often containing numerous components which may differ in their proportions between individuals (Wyatt 2003). A major challenge in studying these signals is overcoming the complexity in order to identify the functional role and chemical nature of each component. Generally, the first step in this process is to fractionate the mixture and evaluate these fractions through repeated bioassays. However, reducing a blend to its constituents can lead to negative assays due to a loss of essential multicomponent interactions. A complementary approach to overcoming the complexity of chemical signals is to consider the properties of the mixture as a whole, as this is what is perceived by the receiver in nature. The first step in this process is to define the composition of the whole chemical profile for the species of interest at the individual and population levels.

The aim of this research was to quantify the complexity and intraspecific diversity in signal composition of the male-specific courtship gland of the red-legged salamander, *Plethodon shermani*. This salamander is known for producing one of the first characterised vertebrate pheromones (Rollmann et al. 1999). Males of this species possess a submandibular secretory gland (the mental gland) that hypertrophies each year during the breeding season. The mental gland is used during courtship, which consists of a "tail-straddling walk" (TSW; Arnold 1976) where the male slowly walks forward and the female follows him with her front legs straddling his undulating tail. During TSW, the male will periodically turn around and quickly snap its mentum downward, contacting the mental gland to the snout of the female and delivering proteinaceous pheromones to the female's nares. These proteins ultimately arrive in the vomeronasal organ (VNO; see Dawley and Bass 1988, 1989), where they activate the sensory neurons of the vomeronasal epithelium (Wirsig-Wiechmann et al. 2002, 2006). These VNO receptor neurons send electrical signals to the accessory olfactory bulb and ultimately to several regions in the brain known to be involved in reproductive behaviour (Schmidt and Roth 1990; Laberge et al. 2008).

The secretion of the mental gland consists of multiple proteins, most of which exist in a variety of structural isoforms. Two proteins, Plethodontid Receptivity Factor (PRF) and Plethodontid Modulating Factor (PMF), have been well characterised at the biochemical level and comprise the majority of the glandular secretion (Feldhoff et al. 1999). PRF is a 22 kDa protein with sequence similarity to the IL-6 family of cytokines (Rollmann et al. 1999; Watts et al. 2004). In most males, PRF exists in a unique combination of three major isoforms (B, C1 and C2). The other major component (PMF) is a 7 kDa protein with sequence similarity to the three-finger protein superfamily (Palmer et al. 2007a). PMF persists as a large multigene complex with at least 99 mRNA haplotypes translated into at least 28 protein isoforms (Wilburn et al. 2012).

While the number of variants and degree of structural diversity between isoforms differ between protein types, PRF and PMF share two key features: (1) both protein types exist in multiple isoforms which appear to have arisen due to strong sexual selection (Watts et al. 2004; Palmer et al. 2007b, 2010), and (2) models of molecular evolution suggest that both PRF and PMF are some of the fastest evolving proteins (compared to examples from Swanson and Vacquier 2002). A third protein (C3) is also present, though less is known about this protein than its counterparts. Other than these three abundant proteins, the other components all consistently appear in very small quantities.

The proteinaceous secretions of the mental gland modulate the timing of courtship (Rollmann et al. 1999; Houck et al. 2007). In particular, PRF decreases the duration of the courtship ritual. Reduction in courtship duration may not be the sole function of this multicomponent signal, but may be a parallel effect of other underlying physiological roles. By studying the biochemical properties of the components, coupled with their neurophysiological modes of action, we expect to reveal the broader scope of functions that are facilitated by this variable multicomponent signal.

For practical reasons, most previous biochemical and behavioural analyses of these protein secretions have utilised pooled samples obtained from many individuals. This process permitted the use of standardised preparations for purification, biochemical characterisation, and use in behavioural assays. In contrast, the aim of this study was to investigate differences in pheromone composition between individual male salamanders of a single population. Because these proteins are so intimately involved in the reproductive behaviour of this species, variation among male pheromone profiles may have profound fitness effects as a result of the courtship interactions that they mediate.

8.2 Methods

8.2.1 Animal Collection and Protein Extraction

Male salamanders ($n=108$) were collected from Macon County, NC (35°10′38″N, 83°33′47″W) in August over the course of 3 years: 2007 ($n=22$), 2008 ($n=25$) and 2010 ($n=61$). A preliminary analysis of the 2007 data (Houck 2009) indicated that a larger sample size was required for the current study. Animals were housed in the laboratory in clear plastic containers (29 × 14 × 9 cm) at 16 °C on a 14:10 (light:dark) light cycle. Mental glands (MG) were surgically removed from each male for biochemical analysis. Males were anaesthetised in 7% diethyl ether for 10 min, and once fully anaesthetised, the MG was surgically removed from the underside of the chin with iridectomy scissors. Each gland was independently rinsed in Amphibian Ringer's solution to remove any residual mucus and then placed in 200 μL of 0.8 mM acetylcholine chloride in Amphibian Ringer's solution for extraction of protein from the tissue. After 30 min, glands in acetylcholine solution were centrifuged at >10,000 × g for 10 min. The supernatant was then subjected to a second centrifugation before storage at −80 °C. The MG for each male was extracted and processed independently to allow for biochemical analysis of individual-specific profiles.

8.2.2 *Biochemical Analysis*

Each sample was thawed and the protein concentration was determined by a bicinchoninic acid (BCA) assay (Smith et al. 1985). Samples were then analysed by reverse-phase high-performance liquid chromatography (RP-HPLC). For each male, 25 μg of MG protein was loaded on a C18 column (5 μm; 4.6×150 mm; Vydac, Hesperia, CA) equilibrated with 0.1% trifluoroacetic acid (TFA). Elution was accomplished on a Waters 2695 Alliance HPLC system (Waters Corp., Milford, MA) using a linear gradient (0–100%) of 70% acetonitrile/0.08% TFA at 1 mL/min for 70 min. Absorbance was measured at a wavelength of 220 nm.

8.2.3 *Data Analysis*

Previous analyses demonstrated that the protein secretions of the MG elute between 20 and 60 min (Feldhoff et al. 1999). The retention times for each component were consistent between runs and verified by the use of standards. Waters Empower™ Software was used to integrate and quantify peak areas. The absorbance values for each peak were converted into the percent per peak of the total protein content of the male MG extract. Each run was standardised to 25 μg, to allow for comparison of the relative protein composition of each male MG extract. For each male profile, the proportion of PMF, PRF and C3 was calculated and compared among all males to evaluate the average composition of the MG. The number of peaks within each of these components was also counted. In the case of PRF, the relative ratios of the three major components (B, C1 and C2) were calculated and compared among all males. These data were presented in a ternary plot for compositional analysis of three components ("vcd" package in R; Meyer et al. 2006). Kernel density estimates were also calculated for these data ("adehabitatHR" package in R; Calenge 2006). A least-squares cross validation (LSCV) method was used to increase the precision in detecting clusters within the total data. The probability function was assigned to a heat colour map with blue representing low probability and red demonstrating high probability. Relative composition of PMF was not quantitatively analysed because it is not possible to establish a reliable isoform count due to (1) the large number of isoforms, and (2) similarity of PMF isoform retention times.

8.3 Results

8.3.1 *Pheromone Composition*

The data from four individuals were omitted from the total sample ($n=108$) due to poor resolution of the RP-HPLC run, resulting in an effective sample size $n=104$. A representative chromatogram (Fig. 8.1) reveals retention times and absorbance

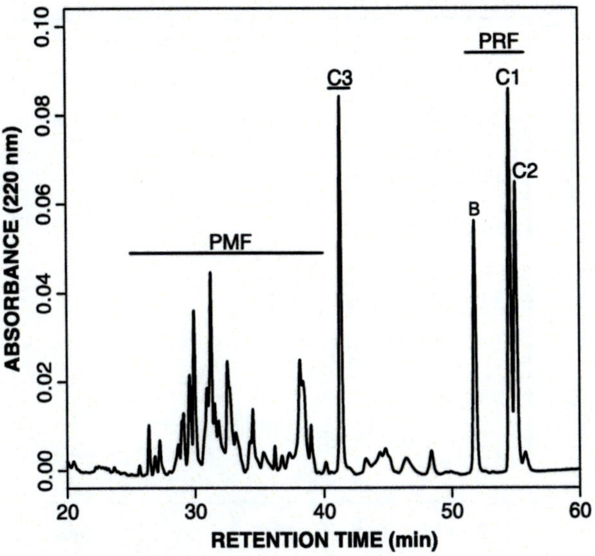

Fig. 8.1 Representative chromatogram of a complete mental gland profile from a single male *P. shermani*. The retention time and abundance can be seen for the numerous isoforms of PMF (20–40 min), component C3 (~41 min) and the three isoforms of PRF (50–60 min). The *y*-axis depicts absolute absorbance values, while the *x*-axis has been cropped to the range in which the proteinaceous components elute

values of the various proteinaceous components. The diversity of PMF can be observed as the wide range of peaks that appear between 20 and 40 min. The protein C3 eluted at ~41 min. All three isoforms of PRF (PRF-B, PRF-C1 and PRF-C2) eluted between 50 and 55 min. These three components of the MG secretion (PMF, PRF, C3) constituted the major proportion of the total mixture (Fig. 8.2). The combination of PMF and PRF comprised an average of 80.8% of the total profile. Of these two major components, PMFs ($\overline{x} = 48.8\%$) were consistently present in higher concentrations than PRFs ($\overline{x} = 32.0\%$). Of the remaining ~20% of the total composition, C3 made up an average of 10.3%. This leaves only 8.9% of the total MG secretion being composed of other minor components that were unidentified at the proteomic level.

The number of peaks of PMF varied greatly among individuals, ranging from 17 to 39 ($\overline{x} = 27.2$). As noted earlier, these data are likely to be an underestimate of PMF isoform numbers because peaks may contain multiple isoforms. Although there were three major components of PRF, the number of peaks in the range of PRF also varied between individuals, ranging from 3 to 6 ($\overline{x} = 4.2$). There were also occasionally other peaks in very close proximity to the protein C3, ranging from 1 to 3 ($\overline{x} = 1.6$). It should be noted that the protein C3 is not an additional PRF isoform. The current peak nomenclature refers to the fraction from which each protein was purified during strong anion-exchange chromatography (Wilburn et al. 2012). Three of the PRF isoforms identified by Rollmann et al. (1999) correspond to PRF-B

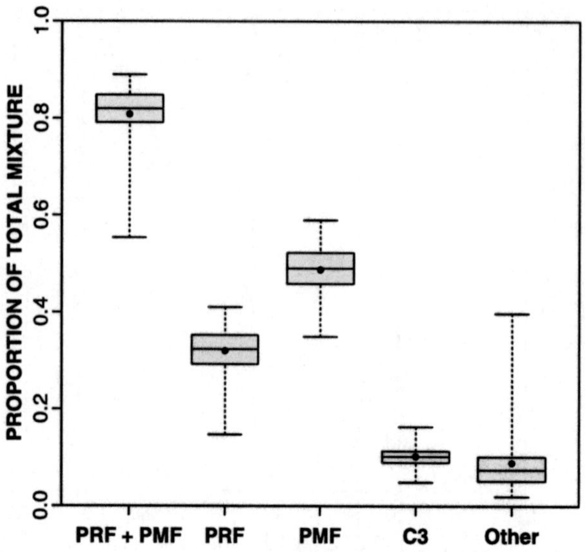

Fig. 8.2 Proportional composition of the complete mental gland extract. The two major proteins (PRF+PMF) make up over 80% of the entire secretion, with PMF regularly present in greater abundance. Of the remaining proteins, C3 makes up more than half (~10% of the whole extract). This leaves less than 10% of the extract being composed by unknown components. The *line* represents the median and the *circle* represents the mean. The *box* represents the interquartile range, while the *whiskers* depict the total range of the data (maximum–minimum)

(Isoform 1), PRF-C1 (Isoform 4) and PRF-C2 (Isoform 3). The remaining isoform (2) corresponds to a SNP variant of PRF-B.

8.3.2 Minor Components

Analysis of individual pheromone profiles revealed numerous minor components in the PRF range that would normally be unobservable in pooled samples, due to the abundance of the three major isoforms overshadowing these minor components. Several of these minor components consistently appeared in the PRF range of individual males (Fig. 8.3). The minor proteins were classified as such because they appeared in much lower proportion than the three major isoforms (with one exception described below), but also because they were observed less frequently. While the presence of these minor components has been established through RP-HPLC, no proteomic analyses have been conducted to ascertain the identity of these proteins. In all cases, an apostrophe after the name of a known component indicates an unknown peak eluting near that component and is not meant to imply that it is necessarily a sequence variant of that known protein.

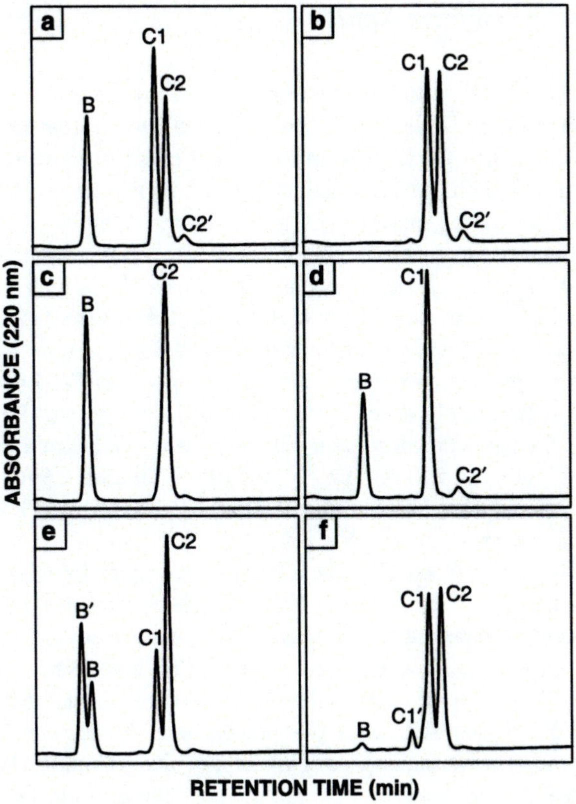

Fig. 8.3 Example chromatograms of PRF profiles that contain minor components. Depicted are data for males that: (**a**) have a complete MG profile, (**b**) are missing PRF-B, (**c**) are missing PRF-C1, (**d**) are missing PRF-C2, (**e**) have an extra peak before PRF-B, and (**f**) have an extra peak before PRF-C1. The *x*-axis depicts the range of retention times for PRF (50–60 min), but in some cases it has been slightly shifted to align the isoforms across all panels. The *y*-axis is scaled in arbitrary units

The most common of the minor components in the range of PRF was a peak that eluted just after PRF-C2 (C2′; Fig. 8.3a). This peak occurred in ~72% of all males, although always in small quantities (ranging from 0.4 to 5.9% of the total PRF for males possessing this peak). Another minor component eluted just before PRF-B (B′; Fig. 8.3e); this peak was the second most common of the minor peaks in the PRF range, appearing in ~13% of all males. This peak was the most prominent of the minor peaks in terms of abundance (ranging from 6.5 to 24.4% of the total PRF for males possessing this peak). In a single case, B′ was even more abundant than PRF-B (the example shown in Fig. 8.3e). Other minor peaks also occasionally eluted in the range before PRF-B, as well as between PRF-B and PRF-C1 (e.g. C1′; Fig. 8.3f); these peaks were less frequent than the minor isoforms described above.

8.3.3 Relative Isoform Ratios

While the relative ratios of PMF isoforms could not be calculated precisely, qualitative observation revealed that the PMF profile was different in essentially every male. In the case of PRF, the lower number of isoforms allowed for excellent resolution. As such, the relative ratios of the PRF isoforms could be quantitatively analysed. The majority of males in the sample (85/104; 81.7%) possessed all three major isoforms of PRF. The remaining males (19/104; 18.3%) had only two major components. In no case did a male in the sample have only one of the major isoforms. However, there were examples of each isoform missing from the profile of a male (Fig. 8.3b–d). In the 19 males lacking an isoform, the most common isoform missing was PRF-C1 (14/104 males; 13.5%). PRF-B was rarely absent (2/104 males; 1.9%), as was PRF-C2 (3/104 males; 2.9%). Of the males producing all three isoforms, the relative abundance of the isoforms present differed greatly between individuals. PRF-B and PRF-C2 are typically expressed in intermediate proportions, while PRF-C1 is often expressed in lower proportions (Fig. 8.4).

The above values for each of the isoforms do not convey information about the ratio of isoforms within a given male. The term "pheromone profile" will be used throughout to describe the pattern of relative component ratios that were specific to an individual male signal. The PRF profile of each male was summarised in a ternary plot (Fig. 8.5). This plot displays the relative ratios of each major isoform of the total PRF. Each point within the plot demonstrates the location of one male's profile within the three-variable space of the ternary plot, revealing a large degree of diversity in PRF profiles between males of the population. Almost every region of the plot was occupied by some male profiles (with the exception of the corners—see Sect. 8.4). However, while there was a great deal of diversity between males, some PRF profiles were more common in the population (i.e. a dense clustering of points). The highest density cluster (orange—red) occurred in males that possessed intermediate amounts of PRF-B (50–60%) and PRF-C2 (40–50%), with low amounts of PRF-C1 (0–10%). There was also a rather large region surrounding this cluster that had intermediate levels of density (aqua to yellow). This region was mostly dictated by an increasing proportion of PRF-C1, typically with PRF-B making up slightly less of the total composition (dropping to 20–50%).

8.4 Discussion

In these analyses we quantified several aspects of the mental gland composition of *Plethodon shermani*: (1) the proportional abundance of the major protein components, (2) the prevalence of minor components in the range of PRF, and (3) the intraspecific diversity of relative isoform ratios.

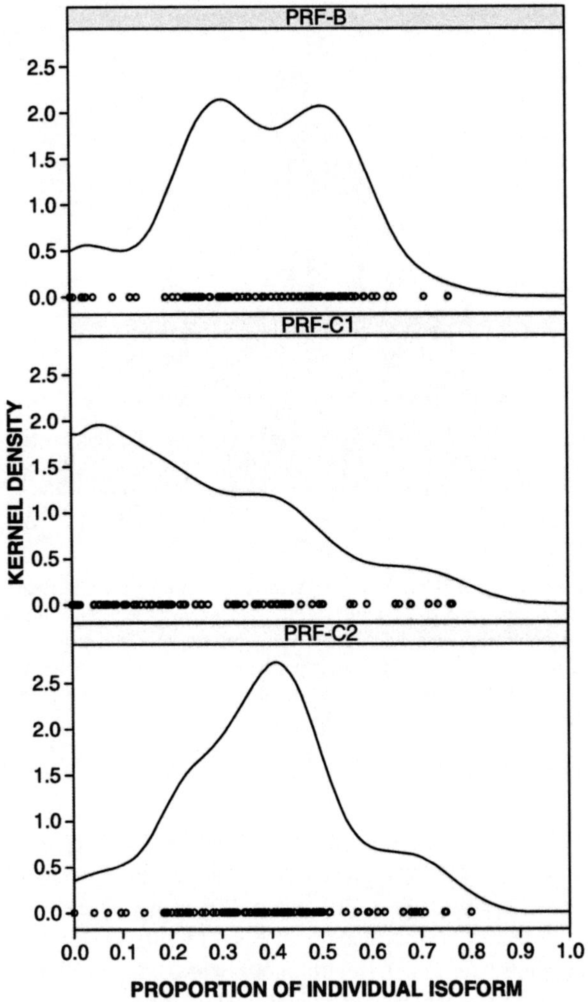

Fig. 8.4 Distribution of individual isoform proportions for the sample population. Isoforms PRF-B and PRF-C2 typically exist in intermediate proportions, while PRF-C1 is more likely to exist in lower proportions. Isoform PRF-C1 is also the most likely to be missing from a male's profile. The circles above the *x*-axis depict the individual data points, while the *y*-axis is a kernel function that estimates the density of the data for the corresponding *x*-value

8.4.1 Proportional Abundance of Major Components

These data allowed for quantitative analysis of the proportions of known components that constitute the total mental gland secretion. The overall pheromone mixture of this species is composed of mainly two protein types, PMF and PRF, constituting 80.8% of the total proteinaceous secretion at a ratio of 5:3. However, because PRF is roughly three times larger than PMF, the stoichiometry ratio is closer to ~5:1.

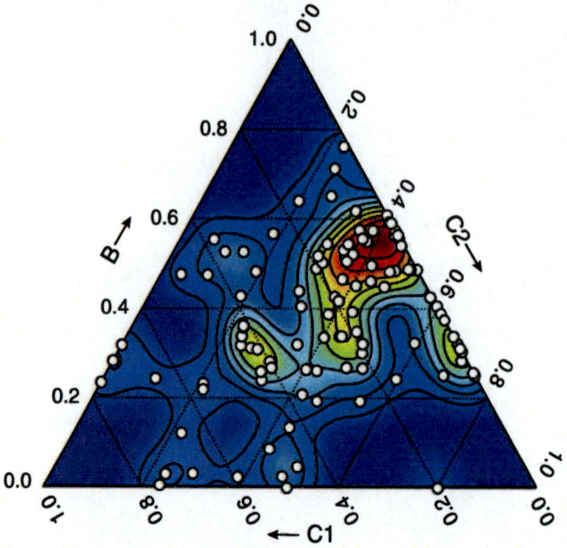

Fig. 8.5 Ternary plot depicting the intraspecific diversity in PRF isoform ratios of *P. shermani* ($n = 104$). Each point corresponds to the PRF profile of a single male. Each axis corresponds to the proportion of a single isoform and the gridlines for the appropriate isoform are in the same alignment as the units for its axis. To identify the proportions of each isoform in the total PRF composition, the axes for individual isoforms can be used to triangulate the position in the plot that corresponds to the profile of a given male. The coloured contours depict a kernel-density function that estimates the clustering of data in a given area of the plot. *Blue shading* indicates low density of points, with an increasing gradient to *red* indicating higher density

As the actual absorbance values for individual PMF peaks are typically less than or comparable to those of PRF, the greater abundance of PMF stems from the much greater number of isoforms (e.g. $n > 30$) than are typically found for PRF (e.g. $n \sim 3$).

Behavioural research has revealed that the protein PRF decreases the duration of courtship (Rollmann et al. 1999). Heterologous expression of PRF later showed that a single recombinant isoform was sufficient to elicit the full capacity of the behavioural response (Houck et al. 2008). When initially tested alone for behavioural effects, PMF increased the duration of courtship (Houck et al. 2007). These contradictory results are particularly confusing when considering that the effect of the whole mental gland secretion reduced courtship time similarly to PRF alone. If the whole extract elicited the same effect observed by PRF, yet the secretion is composed of more PMF than PRF, it seems difficult to explain the opposing behavioural effect of PMF. Recent research, however, suggests that the initial findings were due to an incomplete fraction of PMF isoforms being used in the first study (Wilburn and Houck, unpublished data). Thus, the presence of numerous isoforms may be essential for inducing an appropriate female response.

The third most abundant mental gland component was the (non-PRF) protein C3. This protein constitutes an average of 10.3% of the total extract. Little is currently known about C3 other than its (1) size (~18 kDa), (2) relatively negative charge, and

(3) relative hydrophobicity (RT ~ 41 min). The reason for this discrepancy is that the mRNA of C3 has not appeared in any of the previous analyses of mental gland cDNA (Kiemnec-Tyburczy et al. 2009). The lack of C3 mRNA is presumably a function of either the timing of its expression or the size of the transcript. Recent attempts at high-throughput sequencing of the mental gland transcriptome may reveal the identity of this third major component.

In summary, 91.1% of the mental gland is composed of three protein types. The remaining unknown proteins (8.9%) seem to be accounted for by several proteins expressed in low levels, rather than a few components with high expression. Kiemnec-Tyburczy et al. (2009) analysed expressed-sequence tags (ESTs) and identified several transcripts that were not classified as PRF or PMF. One of these transcripts was identified as sodefrin-like precursor factor (SPF). This protein is an ancestral plethodontid pheromone that is more abundant in the species that deliver pheromones via dermal scratching (Palmer et al. 2007b), rather than by direct contact of the male gland to the female's nares (as in *P. shermani*). SPF is thought to be undergoing evolutionary replacement by PRF, as it is greatly diminished in the mental gland mRNA of the more derived clade that utilises olfactory-delivery of pheromones, including *P. shermani* (Palmer et al. 2007b). Thus, low levels of SPF expression in *P. shermani* may account for some of the unidentified proteins found in the current analyses. Despite efforts to limit contamination, cytosolic and/or blood proteins may contribute to the unidentified fraction. Transcriptome sequencing and proteomic analyses will facilitate the identification of these uncharacterised proteins.

8.4.2 Minor Components in the Range of PRF

The second notable result of the individual male analysis was the appearance of numerous minor peaks in the elution range of PRF. As we have yet to characterise these peaks using mass-spectrometry, they may or may not represent PRF variants. However, the major PRFs presently account for only 3 of the 12 mRNA transcripts that have been identified (Watts et al. 2004; Rollmann et al. 1999). Not surprisingly, these minor components varied in their presence/absence, as well as in their proportions between individuals. Most minor components were expressed at low levels in less than 15% of the males. However, C2′ and B′ were present in ~72% and 13% of males, respectively, with B′ being the most highly expressed minor component (6.5–24.4% of the total PRF in males possessing B′). We speculate that B′ could be the SNP variant of PRF-B identified by Rollmann et al. (1999; Isoform 2).

The reason for differences in the distribution of PRF isoform between males could lie in the realm of (1) genetic, (2) transcriptional, or (3) translational variation. The results from the current study are consistent with past findings from our laboratories and suggest that PRF represents a highly duplicated multigene family. We further hypothesise that there are likely to be several gene copies of the major isoforms, with allelic variants at different loci resulting in the expression of the minor isoforms. A similar explanation for PMF diversity has been proposed by Wilburn et al. (2012), though with many more gene duplications resulting in the large number of observed haplotypes.

In contrast, the presumed differences in male isoform expression could be the result of either transcriptional or translational regulation. If all males possess more loci for PRF than are expressed as mRNA, differences in isoform presence between males could result from differences in transcriptional regulation. In the context of the multiple mRNA transcripts (Watts et al. 2004), variation between males could also result from differential translation rates. If there are a small number of loci or alleles that constitute the PRF gene family, it is also possible that alternative splicing could create mRNA variants, resulting in different isoforms being expressed between males. However, based on preliminary gene sequencing, alternative splicing is not a likely explanation for either PRF or PMF variability (Wilburn and Feldhoff, unpublished data). Lastly, it is possible that both genetic differences and regulatory mechanisms may be working in concert to produce the large degree of isoform diversity observed in PRF and PMF.

If additional PRF transcripts are occasionally translated in some males, the question becomes one of functional significance. Simply because a component is expressed in low levels does not mean that it could not have profound behavioural consequences. For example, the minor components in the pheromone blends of several species of moth are responsible for the entire signal's specificity and preserve its functionality (Linn et al. 1985, 1987). While it is not clear if these peaks are new isoforms of PRF, future proteomic characterisation of the minor peaks in the PRF range will address this question.

8.4.3 *Intraspecific Diversity of Relative Isoform Ratios*

Perhaps the most important aspect of this research is the demonstration of individual variation in pheromone isoform ratios among males of the same population (Figs. 8.4 and 8.5). Quantitative analysis of PRF profiles revealed that a subset of profiles was more prevalent in our study population (Fig. 8.5), but overall the males of this sample possessed a wide array of possible combinations of isoform proportions. In a graphical sense, the only regions of the ternary plot not occupied by any data are the corners (Fig. 8.5). The reason for this is that in order for a point to occupy a corner, a male must have very high levels of one isoform (e.g. >80%), with very low levels of the other two isoforms, and no instances of this were observed. The majority of male profiles (81.7%) had all three isoforms present, with the remaining males expressing two of the three isoforms. PRF-C1 was the most common isoform to be absent, with PRF-B and PRF-C2 each only missing from a small number of males. In summary, the mental gland profiles of male *P. shermani* generally exist as a well-balanced blend of all three PRF isoforms.

The amount of diversity observed in these pheromone proteins leads to three important questions pertaining to their function: (1) the efficacy of individual isoforms in eliciting sensory and behavioural responses, (2) interactive or synergistic effects stemming from combinations of isoforms, and (3) the potential role of isoform ratios acting as unique neurophysiological and behavioural signals. None of these concepts are mutually exclusive and they may all be occurring simultaneously. Future investigations should permit evaluation of the roles and functions of individual pheromones and combinations of pheromone isoforms.

Genes for PRF and PMF are thought to be under rapid gene duplication and pervasive positive selection (Watts et al. 2004; Palmer et al. 2007a, 2010). As these pheromones serve to facilitate reproduction, an intuitive explanation for this genetic pattern is sexual selection (Palmer et al. 2010). Thus, we hypothesise that there is a functional benefit to having multiple isoforms in eliciting sensory stimulation of the female. The presence of numerous variants of the pheromones may elicit the maximum effect in the peripheral and/or central nervous system (CNS), making females maximally affected by the pheromone during mating. Preliminary evidence for this hypothesis may come from the molecular evolution of vomeronasal receptors. In *P. shermani,* the VNO responds to both PRF and PMF (Wirsig-Wiechmann et al. 2002, 2006). In the rodent VNO, V2R receptors recognise proteinaceous pheromones (Haga et al. 2010). Kiemnec-Tyburczy et al. (2012) recently characterised cDNA sequences for numerous V2R receptor variants in the vomeronasal epithelium of *P. shermani*, though the specific receptor:ligand pairs for plethodontid pheromones remain to be determined.

A further rationale for pheromone complexity comes from the field of behaviour. Multicomponent pheromone mixtures are known to be present in communication scenarios that require greater specificity and complexity of information transfer. This concept has been demonstrated in insects (Hefetz and Graur 1988) as well as mammals (Hurst et al. 2001; Beynon et al. 2002; Hurst and Beynon 2004). If the pheromone mixtures of *P. shermani* are serving a signalling function in parallel with their known role in reproductive behaviour, the complexity of plethodontid chemical signals may allow for sex (Dawley 1984a), species (Dawley 1984b), population (Rollmann et al. 2000) and/or individual recognition (Jaeger 1981; Dawley 1984b; Simon and Madison 1984). Behavioural research has revealed that multiple plethodontid species are capable of sophisticated information transfer through chemical signals, which requires a high degree of complexity in the biochemical composition of the mixtures.

If males possess multiple components, with numerous isoforms of each pheromone, these variants may act in an additive or synergistic manner when perceived by females. The concept of different combinations having profound functional effects could be the case for different components (e.g. PMF+PRF), as well as between different isoforms of a single component (e.g. PRF-B+C1+C2). Interactive isoform effects may be perceived by the peripheral (VNO) or CNS in a manner that ultimately contributes to their appropriate functionality. Synergistic effects of pheromone components are also known to occur in insects and mammals (Wyatt 2003, Sect. 2.6) and a recent behavioural experiment suggests that such interactions between PMF isoforms may be necessary for appropriate functionality (Wilburn and Houck, unpublished data). Future research efforts will continue to investigate the potential for interactive effects of different pheromone components on reproductive behaviour, as well as address the mechanisms for this phenomenon at the levels of the peripheral and CNS.

The mental gland pheromone mixture of *P. shermani* is not only complex in terms of the number of components and the large number of structural variants; an additional level of signal complexity may arise from the actual proportions of the individual components. Several lines of evidence support the idea that relative pheromone ratios may serve an additional aspect of functionality in communication. The cells of the vertebrate VNO are known to be exceptionally sensitive to low ligand concentrations (Leinders-Zufall et al. 2000; Kikuyama et al. 1995). In addition, different subsets of cells are finely tuned to only specific ligands (Leinders-Zufall et al. 2000; Zufall et al. 2002). Importantly, a recent study has demonstrated that different ligand concentrations can elicit distinct signals via the VNO (He et al. 2010). Cumulatively, both the presence and concentration of plethodontid pheromones may be providing a highly integrated, complex signal to female salamanders. Thus, it may be that pheromone ratios could act as individual-specific signature mixtures (Wyatt 2010), indicator mechanisms of mate-quality (sensu Andersson 1994), or signals of genetic compatibility (Hurst 2009).

Even though there is a large degree of variability in mental gland profiles, some profiles occurred more frequently in the population. In the current sample, the highest density of profiles occurred with males that have intermediate levels of PRF-B and PRF-C2, with lower amounts of PRF-C1. Interestingly, the most common profile also appears to differ between populations (Rollmann et al. 2000). This type of distribution could be interpreted in terms of a profile that resulted from sexual selection in this population. This hypothesis assumes that females perceive and prefer certain combinations of isoform proportions. As such, this system could serve as an indispensible tool for understanding the nature of coevolution between male traits and female preference.

8.5 Conclusion and Future Directions

The presented biochemical analyses have provided a quantitative description of the characteristics and variability of the mental gland composition of the red-legged salamander, *Plethodon shermani*. Plethodontid salamanders provide an important non-mammalian system for studying the role of pheromones in reproduction and social behaviour. This amphibian family also serves as a useful study system in the field of molecular evolution: pheromone genes are under rapid and pervasive positive selection in contrast to stasis exhibited by most other elements of the functional complex that serves to facilitate reproduction (Watts et al. 2004). Overall, this complex revolves largely around pheromone transfer, which involves genes, proteins, physiology, sexually dimorphic anatomy, neurobiology and behaviour. Our future research aims are to investigate the physiological processes that regulate pheromone production in addition to the molecular and neurophysiological mechanisms that mediate pheromone reception. Ultimately, this study has identified several aspects of pheromone complexity that could have profound effects on the functionality of this multicomponent chemical signal. These aspects of signal complexity include: (1) the number of components, (2) the particular isoforms present and (3) the relative ratios of the individual components and their isoforms. This description of the remarkable individual variability in pheromone composition establishes the foundation for testing hypotheses regarding the way in which the complexity and diversity of the signal may affect sensory reception and ultimately behaviour. Future research efforts will address the potential fitness consequences to individuals possessing particular differences in pheromone composition, as well as the evolutionary pressures and mechanisms that may have led to the rapid diversification of structural variants in the protein pheromones present in this family of salamanders.

Acknowledgements We thank Kathleen Bowen, Pamela Feldhoff, Kari Leichty, Sarah Eddy and Josef Uyeda for help in the lab or field and/or for comments on the manuscript. We also thank James Costa and the staff of the Highlands Biological Station (Highlands, NC). Funding was supported by National Science Foundation grants to LDH (IOS-0818554) and RCF (IOS-0818649).

References

Andersson M (1994) Sexual selection. Princeton University Press, Princeton

Arnold SJ (1976) Sexual behavior, sexual interference, and sexual defense in the salamanders *Ambystoma maculatum*, *Ambystoma tigrinum* and *Plethodon jordani*. Z Tierpsychol 42:247–300

Beynon RJ, Veggerby C, Payne CE, Robertson DH, Gaskell SJ, Humphries RE, Hurst JL (2002) Polymorphism in major urinary proteins in mouse liver: a biochemical study. Biochem Genet 28(7):1429–1445

Calenge C (2006) The package adehabitat for the R software: a tool for the analysis of space and habitat use by animals. Ecol Model 197:516–519

Dawley EM (1984a) Identification of sex through odors by male red-spotted newts, *Notopthalmus viridescens*. Herpetologica 40(2):101–105

Dawley EM (1984b) Recognition of individual, sex, and species odours by salamanders of the *Plethodon glutinosus—P. jordani* complex. Anim Behav 32(2):352–361

Dawley EM, Bass AH (1988) Organization of the vomeronasal organ in a plethodontid salamander. J Morphol 198:243–355

Dawley EM, Bass AH (1989) Chemical access to the vomeronasal organs of a plethodontid salamander. J Morphol 200:163–174

Feldhoff RC, Rollmann SM, Houck LD (1999) Chemical analyses of courtship pheromones in a plethodontid salamander. In: Johnston RE, Muller-Schwarze D, Sorenson PW (eds) Advances in chemical signals in vertebrates. Kluwer Academic/Plenum Publishers, New York, pp 117–125

Haga S, Hattori T, Sato T, Sato K, Matsuda S, Kobayakawa R, Sakano H, Yoshihara Y, Touhara K (2010) The male mouse pheromone ESP1 enhances female sexual receptive behaviour through a specific vomeronasal receptor. Nature 466:118–122

He J, Ma L, Kim S, Schwartz J, Santilli M, Wood C, Durnin MH, Yu CR (2010) Distinct signals conveyed by pheromone concentrations to the mouse vomeronasal organ. J Neurosci 30(22):7473–7483

Hefetz A, Graur D (1988) The significance of multicomponent pheromones in denoting specific compositions. Biochem Sys Ecol 16(6):557–566

Houck LD (2009) Pheromone communication in amphibians and reptiles. Annu Rev Physiol 73:161–176

Houck LD, Palmer CA, Watts RA, Arnold SJ, Feldhoff PW, Feldhoff RC (2007) A new vertebrate courtship pheromone, PMF, affects female receptivity in a terrestrial salamander. Anim Behav 73:315–320

Houck LD, Watts RA, Arnold SJ, Bowen KE, Kiemnec KM, Godwin HA, Feldhoff PW, Feldhoff RC (2008) A recombinant courtship pheromone affects sexual receptivity in a plethodontid salamander. Chem Senses 33:623–631

Hurst JL (2009) Female recognition and assessment of males through scent. Behav Brain Res 200:295–303

Hurst JL, Beynon RJ (2004) Scent wars: the chemobiology of competitive signalling in mice. Bioessays 26:1288–1298

Hurst JL, Payne CE, Nevison CM, Marie AD, Humphries RE, Robertson DH, Cavaggioni A, Beynon RJ (2001) Individual recognition in mice mediated by major urinary proteins. Nature 414:631–634

Jaeger RG (1981) Dear enemy recognition and the costs of aggression between salamanders. Am Nat 117:962–974

Kiemnec-Tyburczy KM, Watts RA, Gregg RC, von Borstel D, Arnold SJ (2009) Evolutionary shifts in courtship pheromone composition revealed by EST analysis of plethodontid salamander mental glands. Gene 432:75–81

Kiemnec-Tyburczy KM, Woodley SK, Watts RA, Arnold SJ, Houck LD (2012) Expression of vomeronasal receptors and related signaling molecules in the nasal cavity of a caudate amphibian (*Plethodon shermani*). Chem Senses 37:335–346

Kikuyama S, Toyoda F, Ohmiya Y, Matsuda K, Tanaka S, Hayashi H (1995) Sodefrin: a female-attracting peptide pheromone in newt cloacal glands. Science 267:1643–1645

Laberge F, Feldhoff RC, Feldhoff PW, Houck LD (2008) Courtship pheromone induced c-Fos-like immunolabeling in the female salamander brain. Neuroscience 151:329–339

Leinders-Zufall T, Lane AP, Puche AC, Weidong M, Novotny MV, Shipley MT, Zufall F (2000) Ultrasensitive pheromone detection by mammalian vomeronasal neurons. Nature 405:792–796

Linn CE, Campbell MG, Roelofs WL (1985) Male moth sensitivity to multicomponent pheromones: critical role of female-released blend in determining the functional role of components and active space of the pheromone. J Chem Ecol 12(3):659–668

Linn CE, Campbell MG, Roelofs WL (1987) Pheromone components and active spaces: what do moths smell and where do they smell it? Science 237:650–652

Meyer D, Zeileis A, Hornik K (2006) The strucplot framework: visualizing multi-way contingency tables with vcd. J Stat Softw 17(3):1–48

Palmer CA, Hollis DM, Watts RA, Houck LD, McCall MA, Gregg RG, Feldhoff PW, Feldhoff RC, Arnold SJ (2007a) Plethodontid modulating factor, a hypervariable salamander courtship pheromone in the three-finger protein superfamily. FEBS J 274:2300–2310

Palmer CA, Watts RA, Houck LD, Picard AL, Arnold SJ (2007b) Evolutionary replacement of components in a salamander pheromone signaling complex: more evidence for phenotypic-molecular decoupling. Evolution 61(1):202–215

Palmer CA, Watts RA, Hastings AP, Houck LD, Arnold SJ (2010) Rapid evolution of Plethodontid Modulating Factor, a hypervariable salamander courtship pheromone, is driven by positive selection. J Mol Evol 70:427–440

Rollmann SM, Houck LD, Feldhoff RC (1999) Proteinaceous pheromone affecting female receptivity in a terrestrial salamander. Science 285:1907–1909

Rollmann SM, Houck LD, Feldhoff RC (2000) Population variation in salamander courtship pheromones. J Chem Ecol 26(12):2713–2724

Schmidt A, Roth G (1990) Central olfactory and vomeronasal pathways in salamanders. J Hirnforsch 31(5):543–553

Simon G, Madison DM (1984) Individual recognition in salamanders: cloacal odours. J Herpetol 12(3):397–406

Smith PK, Krohn RI, Hermanson GT, Mallia AK, Gartner FH, Provenzano MD, Fujimoto EK, Goeke NM, Olson BJ, Klenk DC (1985) Measurement of protein using bicinchoninic acid. Anal Biochem 150:76–85

Swanson WJ, Vacquier VD (2002) The rapid evolution of reproductive proteins. Nat Rev Gen 3:137–144

Watts RA, Palmer CA, Feldhoff RC, Feldhoff PW, Houck LD, Jones AG, Pfrender ME, Rollmann SM, Arnold SJ (2004) Stabilizing selection on behavior and morphology masks positive selection on the signal in a salamander pheromone signaling complex. Mol Biol Evol 21(6):1032–1041

Wilburn DB, Bowen KE, Gregg RG, Cai J, Feldhoff PW, Houck LD, Feldhoff RC (2012) Proteomic and UTR analyses of a rapidly evolving hypervariable family of vertebrate pheromones. Evolution 66:2227–2239

Wirsig-Wiechmann C, Houck LD, Feldhoff PW, Feldhoff RC (2002) Pheromonal activation of vomeronasal neurons in plethodontid salamanders. Brain Res 952:335–344

Wirsig-Wiechmann C, Houck LD, Wood JM, Feldhoff PW, Feldhoff RC (2006) Male pheromone protein components activate female vomeronasal neurons in the salamander *Plethodon shermani*. BMC Neurosci 7:26–34

Wyatt TD (2003) Pheromones and animal behaviour: communication by smell and taste. Cambridge University Press, Cambridge

Wyatt TD (2010) Pheromones and signature mixtures: defining species-wide signals and variable cues for identity in both invertebrates and vertebrates. J Comp Physiol A 196:685–700

Zufall F, Kelliher KR, Leinders-Zufall T (2002) Pheromone detection by mammalian vomeronasal neurons. Microsc Res Techniq 58:251–260

Chapter 9
Rodent Urinary Proteins: Genetic Identity Signals and Pheromones

Jane L. Hurst and Robert J. Beynon

Abstract Many rodents communicate through scent marks that contain a complex mixture of volatile components along with species-specific proteins of the lipocalin family. These proteins have an internal cavity that binds low molecular weight hydrophobic volatiles and can slow their release from scent marks, while the proteins themselves are detected through receptors in the vomeronasal organ. By far the best studied urinary lipocalins to date are the major urinary proteins (MUPs) of the house mouse (*Mus musculus domesticus*). MUPs are encoded by a cluster of >21 genes on mouse chromosome 4 that has undergone recent rapid expansion in this species. Mice of both sexes make substantial investment in these proteins, with considerable polymorphism in the patterns of MUPs expressed between individuals. This MUP pattern provides the main genetic identity signal in mouse urine scent used to recognise individual scent owners, close kin, and can also be used to assess genetic heterozygosity. A male-specific atypical MUP named darcin acts as a sex pheromone, responsible for eliciting instinctive female sexual attraction to spend time near male urinary scent marks. Importantly, darcin also stimulates rapid associative learning, such that females that contact darcin learn and are subsequently attracted to the associated airborne odour of that particular male but not to airborne odours of other males. This pheromone-induced mechanism of learned attraction targets and reinforces sexual attraction towards a particular male, allowing sexual attraction to males to be inherent but also selective. Comparative studies of other

J.L. Hurst (✉)
Mammalian Behaviour and Evolution Group, Institute of Integrative Biology, University of Liverpool, Leahurst Campus, Neston CH64 7TE, UK
e-mail: jane.hurst@liv.ac.uk

R.J. Beynon
Protein Function Group, Institute of Integrative Biology, University of Liverpool, Liverpool L69 7ZB, UK
e-mail: r.beynon@liv.ac.uk

M.L. East and M. Dehnhard (eds.), *Chemical Signals in Vertebrates 12*,
DOI 10.1007/978-1-4614-5927-9_9, © Springer Science+Business Media New York 2013

rodents are starting to reveal considerable diversity in concentration, sexual dimorphism and polymorphic complexity of lipocalins in scent signals; these differences are likely to reflect species-specific requirements for recognition and assessment through scent.

9.1 Rodent Urinary Proteins

Vertebrate scents typically are very complex compared to the much simpler pheromone signals used by invertebrates. Scents can contain several hundred low molecular weight volatile components, which can differ substantially between individuals and with an individual's current status. Added to this complexity of volatile molecules, it is becoming increasingly clear that proteins can also be very important components of vertebrate scents. Many rodents exhibit an obligate proteinuria, deliberately excreting a high concentration of species-specific specialised communication proteins in the urine that they use for scent marking. Communication proteins have also been found in other scent sources including tears (Shaw et al. 1983; Kimoto et al. 2007), saliva (Shaw et al. 1983; Talley et al. 2001) and vaginal secretions (Briand et al. 2004). Proteins of the lipocalin family, in particular, are widely used in scent communication (Flower 1996). There appear to be two distinct groups of lipocalins used as vertebrate semiochemicals based on sequence homology (Turton et al. 2010). Mice and rats (family Muridae) commonly excrete small lipocalins of 18–19 kDa called major urinary proteins or MUPs. Most of these are synthesised in the liver, secreted into serum and then rapidly excreted in urine. However, MUPs are also expressed in a number of secretory tissues including submaxillary, lachrymal, parotid, sublingual, nasal, mammary and preputial glands, with specific tissue expression varying between species (MacInnes et al. 1986; Shahan et al. 1987). Voles and hamsters (family Cricetidae) also excrete urinary lipocalins; those characterised to date are slightly smaller than MUPs at 16–17 kDa. Vole and hamster urinary lipocalins share greatest sequence homology with aphrodisin, a lipocalin in the vaginal secretion of female hamsters that promotes sexual behaviour in males (Briand et al. 2004). Vole and hamster urinary lipocalins share greater sequence homology with olfactory binding proteins (OBPs) from mice and rats than with MUPs, although the 3D structure of urinary proteins from murid and cricetid rodents is very similar (Fig. 9.1).

9.1.1 Ligand Binding

Lipocalins have a highly conserved eight-stranded β-barrel structure (Fig. 9.1) with an internal hydrophobic cavity suitable for the binding and transport of low molecular weight hydrophobic molecules (Flower 1996). Accordingly, mouse MUPs bind

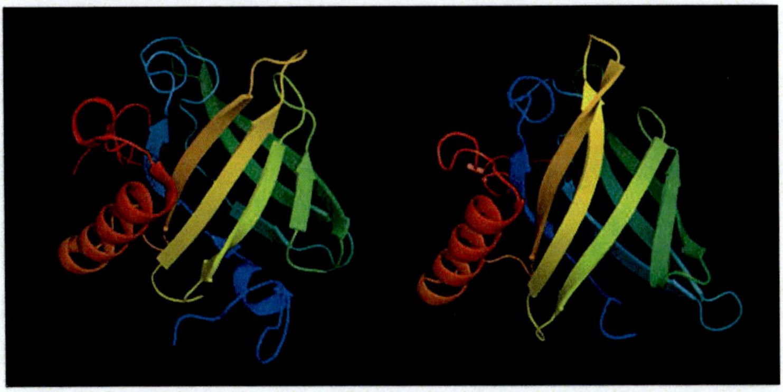

Fig. 9.1 Tertiary structure of rodent urinary lipocalins used in scent communication. *Left*: MUP11, encoded by one of the central *Mup* genes of the house mouse (M Phelan unpublished data). *Right*: homology model of the 3D structure of roborovskin excreted by the dwarf hamster *Phodopus roborovskii* determined by Phyre homology modelling software (after Turton et al. 2010)

small lipophilic molecules, including a number of identified volatile pheromones in male urine (Bacchini et al. 1992; Robertson et al. 1993; Novotny et al. 1999; Sharrow et al. 2002); these volatiles have specific receptors in the vomeronasal organ and act as priming pheromones that stimulate female oestrus cycling (Leinders-Zufall et al. 2000). In addition to known pheromones, urinary lipocalins sequester other volatile metabolites such as those derived from food (Robertson et al. 1998; Kwak et al. 2011) and urinary volatiles associated with the major histocompatibility complex (MHC) type of laboratory mice (Singer et al. 1993). Binding and transport of hydrophobic molecules by MUPs in scents may be an important mechanism for delivering these volatiles to vomeronasal receptors. Further, one specific MUP isoform, expressed in the nasal mucosa and in the vomeronasal organ of mice, has highly specific binding of the male mouse pheromone 2-*sec*-butyl-4,5-dihydrothiazole (Perez-Miller et al. 2010) and it has been proposed that this MUP may capture, concentrate and transport this pheromone to vomeronasal receptors (Sharrow et al. 2002). Ligand binding is also important for holding highly volatile scent components in scent marks, slowing their release. When not bound to MUPs, free volatiles are lost from urine scent marks within minutes (Robertson et al. 2001), while those bound to MUPs continue to be released gradually and can be detected by animals over at least 24 h (Hurst et al. 1998; Humphries et al. 1999). Different MUP isoforms vary in their binding affinity for volatile ligands according to amino acid substitution in the central cavity (Marie et al. 2001; Armstrong et al. 2005; Perez-Miller et al. 2010). Differential expression of specific MUP isoforms may thus contribute to controlling the concentration of different volatile ligands in scent and their release rates from scent marks as well as capture in the nose, although the effects of protein expression on the volatile scent profile are not yet well understood.

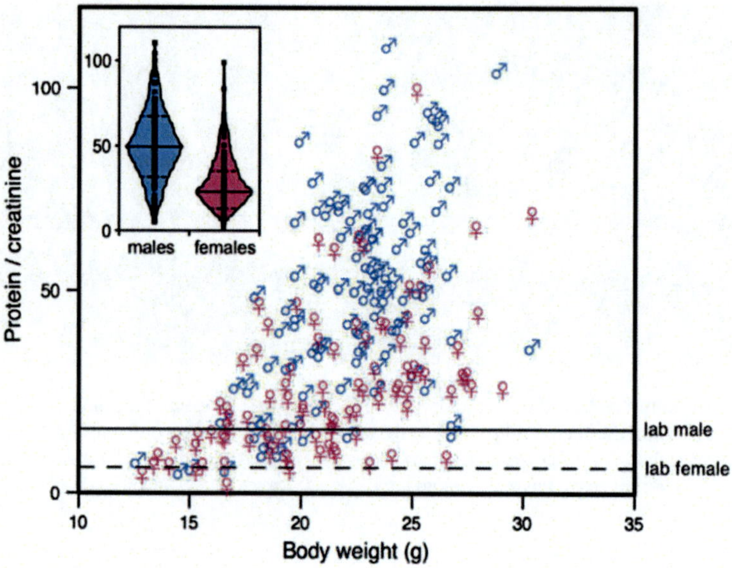

Fig. 9.2 Urinary protein concentration among house mice living in four populations in semi-natural enclosures (each 250 m²). *Blue*: male ($n=115$); *pink*: female ($n=104$). Protein concentration is expressed as mg/mg creatinine to correct for urine dilution (Beynon and Hurst 2004). Protein to creatinine ratio increases with body weight, even though the amount of creatinine excreted is likely to show a small increase with muscle mass. *Solid* and *dotted horizontal lines* indicate normal maximum urinary protein levels in adult male and female laboratory mice, respectively. Insert shows percentiles for each sex. SDS-PAGE confirmed that urinary protein consists almost entirely of MUP in both sexes

9.1.2 Protein Investment

Rodents often make considerable investment in urinary lipocalins, although this can vary substantially with age, sex, season and social conditions. Adult house mice, in particular, invest heavily in urinary MUPs. There is a clear sexual dimorphism in protein investment among laboratory mice, with adult males consistently expressing two to eightfold more urinary protein than females of the same strain (Cheetham et al. 2009). However, investment among laboratory animals may not be typical, influenced by the lack of normal sociosexual stimulation in the laboratory and artificial selection over many generations in captivity. Wild-stock house mice in semi-natural enclosure populations invest considerably more in the production of MUPs (Fig. 9.2). Urinary protein is usually expressed as a ratio of protein to creatinine to control for differences in urine dilution, as the rate of production of creatinine (a breakdown product of creatine phosphate in muscle) is fairly constant and equivalent between animals of similar muscle mass (Beynon and Hurst 2004). While the average protein to creatinine ratio of adult laboratory males is around 10 (range 2.5–15), this is five times higher among wild males in semi-natural populations (mean 54, range 9.5–113). The difference in female output is even greater, with an average protein to creatinine ratio of 3 (range 0.8–6) among laboratory females, but 30 (range 5.5–99) in semi-natural enclosure populations. Most notably, there is considerable overlap in the amount of MUP investment between males and females under semi-natural conditions, and substantial variation between individuals (Fig. 9.2). Among males, a significant correlation between MUP output and body weight ($r_{114}=0.54$, $P<0.0001$) is due to MUP output increasing during the development of subadult males (<20 g); among adult males (≥20 g), variation in output is not strongly related to body weight ($r_{89}=0.19$, $P=0.08$). However, female MUP output increases with body weight even among breeding females ($r_{104}=0.50$, $P<0.0001$). This is perhaps surprising given the very high investment that breeding females are making simultaneously in pregnancy and lactation. In some species, such as *Rattus norvegicus*, only males invest in urinary MUPs, although females express MUPs in other tissues including the preputial glands (MacInnes et al. 1986). In other species, such as the dwarf hamster *Phodopus roborovskii*, investment appears to be similarly high in both sexes (Turton et al. 2010).

Expression of MUPs in the liver is under complex multi-hormonal control, with expression of different MUP genes regulated by different hormones. The high output of MUPs by adult male mice correlates with the onset of sexual maturity and depends on testosterone (Knopf et al. 1983), which controls not only the high concentration of MUPs expressed by adult males but also expression of several male-specific MUPs (Mudge et al. 2008). Among laboratory rats, only males express MUPs in the liver for urinary excretion; these are induced by testosterone, growth hormone, thyroxine and glucocorticoid hormones, but repressed by estrogens (Kulkarni et al. 1985). Growth hormone and thyroxine also induce liver MUP expression in mice, acting synergistically but with each hormone differentially increasing expression of different but overlapping set of MUPs (Knopf et al. 1983). However, while MUP expression clearly changes during sexual development, as yet, little is known about the extent to which physiological variation in hormone levels among adult animals affects the amount or patterns of MUPs expressed. The total amount of MUP expressed by female laboratory mice varies through the oestrus cycle, with up-regulation of MUPs at the beginning of oestrus (Stopka et al. 2007). It is not yet known whether this up-regulation has any biological significance as a signal of approaching oestrus, as variation in MUP output through the oestrus cycle appears to be minor compared to the considerable individual variation in MUP output among females under more naturalistic conditions.

Among house mice, investment in MUP increases in response to social competition (Garratt et al. 2011b, 2012). MUP output is higher among mice housed in breeding pairs with territorial neighbours than among those housed in isolated breeding pairs or with no reproductive experience (Fig. 9.3). The rate of scent marking and scent replenishment also increases with social competition (Garratt et al. 2012). As MUPs bind volatiles and delay their release (see above), this means that both MUPs and their volatile ligands in scent marks are elevated when animals come under competitive pressure to defend their territories. However, MUP output decreases as males begin to senesce (from approximately 24 months old among

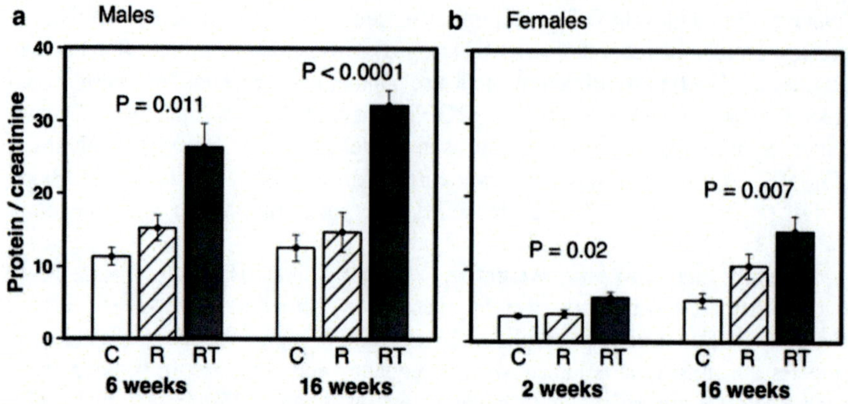

Fig. 9.3 Investment in urinary protein varies with social condition in both male and female house mice. Mice were housed in laboratory enclosures (1.2×1.2 m) in isolated breeding pairs (R, *hatched*) or in breeding pairs that defended their enclosure from a neighbour pair (RT, *solid*). Control non-breeding mice (C, *open*) were housed in same-sex dyads (females) or on their own (males). Urinary protein was sampled after 6 and 16 weeks (males, Garratt et al. 2012) or after 2 and 16 weeks in enclosures (females, Garratt et al. 2011b). Protein concentration is expressed as mg/mg creatinine to correct for urine dilution. ANOVAs assessed differences in protein concentration between groups

captive wild mice), with a matched reduction in the amount of male-specific urinary volatiles that bind to MUPs, although the pattern of MUPs does not change. This reduces the strength of urine scent mark signals from old adults compared to young adult and middle-aged males, reducing the longevity of volatile scent signals and female attraction (Garratt et al. 2011a).

9.1.3 *Individual Polymorphism*

While the total amount of urinary protein expressed can vary widely between individuals according to development and condition, the qualitative pattern of urinary proteins is often characteristic of the species or of the species and sex of the animal. For example, male and female *Phodopus roborovskii* express the same 17,144 Da urinary lipocalin, roborovskin, together with another 6 kDa protein that has not yet been identified (Turton et al. 2010). The bank vole *Myodes glareolus* expresses two urinary lipocalins that differ between males and females, but individuals of the same sex share the same isoforms (Beynon et al. 2008; Stopkova et al. 2010).

By contrast, in some species there can be considerable individual polymorphism in the patterns of isoforms expressed. In the house mouse *Mus musculus domesticus*, each individual expresses a complex pattern of multiple MUP isoforms, most of which differ from each other by only a few amino acids. These individual MUP patterns are genetically encoded and very stable once the adult pattern has developed,

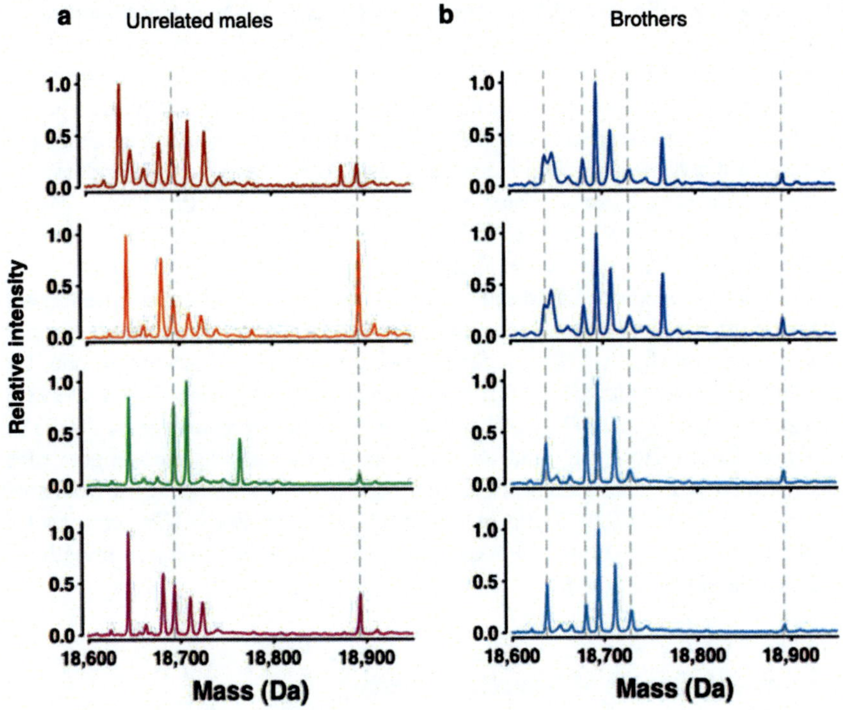

Fig. 9.4 Typical MUP patterns of individual male house mice revealed by electrospray ionisation mass spectrometry of intact urine samples. Intact mass spectra show the intensity of each mass peak relative to the highest peak in the spectrum for (**a**) four unrelated males, and (**b**) four full sib brothers. *Dashed lines* connect mass peaks shared between all four males in each set. Among unrelated males, two masses are shared: a peak at 18,694 Da is common to all mice, though may represent several MUP isoforms of similar mass (Mudge et al. 2008), and the 18,893 Da male-specific MUP darcin (Roberts et al. 2010). Brothers of the same MUP type (same colour) share all mass peaks with similar relative intensities of expression, while brothers with one common MUP haplotype share more mass peaks than unrelated males with two different MUP haplotypes

such that repeated samples from the same individual are very similar. In outbred populations, these individual MUP profiles are so highly polymorphic that unrelated individuals each express a different urinary MUP pattern (Fig. 9.4a). These MUPs are encoded by a tight cluster of more than 20 functional *Mup* genes on mouse chromosome 4 (Mudge et al. 2008; Logan et al. 2008). Most of the urinary MUPs (expressed in liver) are encoded by genes within the central region of this cluster; these central genes have undergone recent and rapid expansion through multiple gene duplication events and are responsible for this considerable polymorphism in individual MUP patterns (Mudge et al. 2008). However, *Mup* genes are inherited as a tightly linked cluster or haplotype and close relatives, such as full sibs, are likely to share much more similar or almost identical MUP patterns according to whether one or both of their MUP haplotypes are shared through common inheritance (Fig. 9.4b). Thus, MUP patterns in this species have the potential to provide signals both of individual identity and of close relatedness between individuals (see Sect. 9.2). This expansion of the *Mup* cluster in commensal house mice may reflect the need for more elaborate communication of genetic identity in a species that has become adapted to live at high density where food resources are abundant, leading to much greater social complexity than in species where individuals are spatially dispersed. Closely related aboriginal species of *Mus* that live in widely dispersed territories lack this individual polymorphism in urinary MUP patterns (Robertson et al. 2007; Beynon et al. 2008).

Completely independently, the *Mup* gene cluster has also undergone recent rapid expansion in the Norway rat *Rattus norvegicus* (Logan et al. 2008), another social species that typically lives in large family-based groups. Rat MUPs are encoded by a tight cluster of approximately 20 genes on rat chromosome 5 (Kurtz 1981; McFadyen and Locke 2000). Little is yet known about expression of rat MUPs in wild populations. However, initial analysis of urine from several wild-caught males has revealed individual variation but also consistent expression of rat MUPs of masses 18,714 and 18,730 Da, the two most strongly expressed MUPs across males (Beynon et al. 2008). Further work is underway in our laboratory to investigate MUP expression in wild rats.

9.2 Scent Signals of Genetic Identity

The direct encoding of urinary lipocalins in the genome, together with species, sex and individual specificity in expression, suggests that these proteins have the potential to provide genetic identity signals in scents. So far, the functional significance of these proteins as genetic identity signals has only been explored in house mice.

9.2.1 MUPs Are Signals of Individual Identity

While many genes influence the complex urinary volatile profiles of different individuals, mice use MUP pattern to identify the individual owners of scent marks including their own. When male mice encounter scent marks from other males within their territory rather than their own scent, they investigate the scent closely and then counter-mark by increasing their rate of scent marking nearby the intruder's scent (Hurst 1990). However, they fail to show this response if the other male's urine contains the same MUP pattern as their own, even when there are many other genetic differences between the males. Airborne volatiles emanating from urine scent marks attract attention to an intruder male's scent and induce close investigation, regardless of whether the intruder male has the same or different MUP type as the territory owner. However, after sniffing the scent closely, males fail to countermark urine that contains their own MUP type, depositing no more scent marks than in response to own urine or to water (Hurst et al. 2001; Hurst and Beynon 2008).

Direct evidence that this is due to MUP type comes from using a recombinant MUP to alter a male's own highly familiar MUP pattern; this then causes a male to counter-mark his own, altered scent (Hurst et al. 2001).

Female mice also use male MUP patterns to recognise the individual owners of male urinary scent marks, which are deposited to advertise their location and competitive ability to females. When females encounter the competitive scent marks of two males, they learn the scent of the male that deposited the freshest scent (the counter-marking male) and prefer this male when they subsequently meet the males themselves (Rich and Hurst 1999; Cheetham et al. 2007). However, females fail to show any such discrimination between two males if they share the same MUP pattern, even though there may be many other genetic differences between them that influence the males' scents including MHC type (Cheetham et al. 2007). Such MUP sharing between competing adult males is likely to be unusual in normal outbred mouse populations though, given the high degree of MUP polymorphism and dispersal of related male offspring. To confirm that the MUP pattern itself determines the individual signatures of male scent owners, rather than the influence of genes closely linked to MUPs, we have manipulated individual male MUP patterns directly with recombinant proteins. In each case, changing a male's MUP pattern disrupts recognition of the male's urinary odour while control manipulations do not (authors' unpublished data).

Similar studies of the effect of urinary proteins on individual recognition in other species are currently lacking. Notably, though, not all species show individual polymorphism in the urinary proteins expressed. It is possible that urinary proteins may still play a role in individual identity signals in these species through binding individual-specific patterns of volatiles and thus helping to retain an individual signature in scent marks.

9.2.2 MUPs Are Signals of Kinship

The ability to recognise close kin has many potential benefits to animals, and there is substantial behavioural evidence that many animals are able to recognise kin even when they are previously unfamiliar. Surprisingly little is known, though, about the genetic identity signals that allow animals to recognise unfamiliar kin (Gardner and West 2007). Gene clusters like *Mup* and MHC have been shown to induce discriminable differences in scent, while both are so highly polymorphic that only closely related individuals are likely to share the same type. Both MHC and MUP type are thus good candidates to provide genetic markers of kinship, although combinations of other genes that influence scents could also be involved. Studies of inbred laboratory mice have provided some evidence that MHC scents may be used as a marker of kinship to avoid inbreeding (Penn 2002). However, although some inbred strains show MHC-disassortative mate preferences when strains differ only in MHC type, such preferences differ between strains (Jordan and Bruford 1998). Laboratory mice are problematic models to investigate kin recognition though because they lack normal genetic variation in the signals that might be used; artificial selection in the laboratory has also created animals that are willing to inbreed and show abnormally low competitive aggression (Manning et al. 1992; Cheetham et al. 2009). Instead, we need to understand the genetic markers of kinship that can be recognised against the variable genetic and environmental backgrounds and behaviour typical of natural populations (Hurst and Beynon 2008).

Identifying which markers are used in natural populations is difficult because, by definition, markers used to recognise unfamiliar kin must correlate closely with relatedness; these markers will also correlate with the sharing of many other genes inherited through common descent. To establish which markers are used to detect kinship, study designs must break the natural correlation between the sharing of one or more candidate markers and other polymorphic genes that are shared due to relatedness. We took this approach to investigate whether house mice use sharing of MHC type, MUP type or both to avoid inbreeding with close kin. Wild stock house mice of the same degree of relatedness (sibs and paternal half-sibs), but which differed in their sharing of MHC and MUP haplotypes, were allowed to breed freely in large seminatural enclosures (Sherborne et al. 2007). This revealed that there was no avoidance of breeding between mice that shared the same MHC type when the correlation with MUP sharing and genetic background was controlled. However, there was a strong deficit of breeding between mice that shared the same MUP type. This suggests that mice are able to recognise MUP patterns that are shared between close kin and use this to avoid inbreeding. The considerable similarity of shared MUP patterns between close kin, despite differences in their genetic backgrounds (Fig. 9.4b), provides a direct genetic marker of kinship that can easily be detected through urine scent; thus, it is most likely that the deficit of breeding when mice shared the same MUP pattern resulted from a pre-copulatory preference for mates of different MUP type. However, as mating was not observed directly, the possibility of a post-copulatory selection mechanism cannot yet be ruled out. We are conducting further studies to define the mechanism underlying MUP-disassortative mate preferences. Notably, there was no deficit of breeding between related mice with partly shared MUP types (Sherborne et al. 2007). Partial sharing at a single polymorphic locus or cluster of genes such as MUP or MHC is likely to occur between many unrelated animals simply by chance; using only this as a marker of close relatedness could result in rejection of a large number of unrelated mates (Paterson and Hurst 2009). Instead, mice appear to use the more conservative signal of complete MUP sharing, although this will only allow recognition of a proportion of close relatives in outbred populations. Genetic signals used for kin recognition in other species have yet to be identified.

9.2.3 *MUPs May Have a Role in Signalling Genetic Heterozygosity*

Animals avoid inbreeding with close relatives to promote heterozygosity in their offspring, but they may also favour animals that are more heterozygous as mates or as cooperative breeding partners. While heterozygosity itself is not heritable, it increases individual vigour and success in intrasexual competition. There is some evidence that the polymorphic MUP patterns of house mice may be used as a direct signal of genetic heterozygosity. Females show a preference for associating with MUP heterozygous over MUP homozygous males when heterozygosity across the rest of the genome is controlled, choosing to spend more time co-habiting the nest site of MUP heterozygous males (Thom et al. 2008). There is no evidence as yet that females prefer to mate with MUP heterozygous males. Instead, females may benefit from choosing to nest with a male of greater vigour that is better able to defend the nest site than a more homozygous male, as this is an important factor in offspring survival in this species.

9.3 Urinary Proteins as Pheromones

Pheromones are chemical signals that are produced for communication between animals of the same species and elicit a specific instinctive behavioural or physiological response (Karlson and Luscher 1959; Wyatt 2010). Unlike polymorphic signatures that differ between individual conspecifics, allowing animals to adjust their behaviour towards different individuals, pheromones are typically produced consistently across individuals as they underpin important instinctive responses. In house mice, although most urinary MUPs differ in expression between individuals providing individual signatures of identity and kinship, one urinary MUP called darcin is produced consistently by adult males and has been identified to act as a male sex pheromone.

9.3.1 Darcin Is a Male Mouse Sex Pheromone

When female mice are presented with a choice between unfamiliar urine from an adult male vs. an adult female, they show an instinctive sexual attraction to spend more time near male urine regardless of any prior experience with males or male scents (Moncho-Bogani et al. 2002; Ramm et al. 2008). Although male mouse urine contains many androgen-specific components, including a number of volatiles that act as priming pheromones to stimulate female reproductive physiology (Leinders-Zufall et al. 2000), the instinctive sexual attraction to spend time near male scent is due to a single component of male mouse urine. Darcin is an atypical male-specific MUP, named after Jane Austen's romantic hero Mr Darcy in the novel *Pride and Prejudice* to recognise its distinct and specific role in stimulating female sexual attraction (Roberts et al. 2010). Darcin is phylogenetically very distinct from most other urinary MUPs, being encoded by one of the genes in the peripheral region of the *Mup* cluster (Mudge et al. 2008). In addition to its unique behavioural role, darcin has several unusual physical properties, including high mobility on reducing gel electrophoresis, and is responsible for binding most of the abundant volatile male pheromone 2-*sec*-butyl 4,5 dihydrothiazole (Armstrong et al. 2005). Despite this, its tertiary structure is quite similar to other MUPs (Phelan et al. 2010). Although darcin binds the volatile 2-*sec*-butyl 4,5 dihydrothiazole in male urine, it is darcin itself that stimulates female sexual attraction, not the volatile ligands that it binds and releases. Females show the same instinctive sexual attraction towards darcin alone, produced as a recombinant protein and containing no natural mouse ligands, as they do towards intact male urine. Further, adult male urine that contains no darcin (or only trace amounts) fails to stimulate any more attraction than female urine. Darcin is detected on close nasal contact with male urine, presumably through MUP receptors in the vomeronasal organ, although a specific receptor for darcin has yet to be identified.

9.3.2 Targeting Response to Specific Individuals

Besides stimulating instinctive female attraction to spend time near male urine on nasal contact, darcin induces a learned attraction to the airborne odour encountered at the same time, through *associative learning*. Airborne odours emanating from scents stimulate females to approach and briefly investigate the scent closely (particularly if the airborne scent is unfamiliar). This brings animals into contact with additional information that can be detected through the vomeronasal system, such as darcin. However, the airborne odours themselves stimulate no instinctive female attraction to spend time in the vicinity of male scent (Moncho-Bogani et al. 2002; Ramm et al. 2008). On contact with darcin, though, females learn to spend time near that male's airborne urinary odour, similar to their attraction to darcin itself (Roberts et al. 2010). The learned airborne odour is specific to that individual male; females that have contacted urine from one individual do not learn any attraction to airborne odours from other males (Ramm et al. 2008; Roberts et al. 2010). This learned attraction to the individual odour of a specific male is critically dependent on the presence of the MUP sex pheromone darcin: if male urine does not contain darcin, females fail to learn any attraction to the male's airborne urinary odour, but addition of recombinant darcin to the male's urine stimulates the normal learned attraction (Roberts et al. 2010).

The consequence of this pheromone-induced learning mechanism is that it targets and reinforces sexual attraction to a specific individual male. A common sex pheromone in male urine ensures that female recognition and attraction to spend time near male scent is an instinctive, reliable response that does not need to be learned. However, females also need to be selective in choosing the best mates available to them in the local population. Females can use other information in individual male scents to assess individual quality and genetic compatibility and will preferentially spend time closely sniffing scents from males that are uninfected (Penn et al. 1998; Kavaliers et al. 2005), socially dominant (Jones and Nowell 1974; Mossman and Drickamer 1996) and that are able to defend their territories successfully from competitor males (Rich and Hurst 1999). As detection of darcin in a male's scent requires close contact and, presumably, active delivery of scents to MUP receptors in the vomeronasal organ, females can control their exposure to this involatile sex pheromone through such active sniffing. However, once they have contacted darcin and learned a response to the associated airborne odour, females will then become attracted to spend time near airborne odours from that particular male as they move around the local area, and to the odour of the male himself. Thus, this pheromone-induced learning will reinforce attraction to particular individual males, integrating an instinctive response to the pheromone with flexible, learned behaviour. The induction of associative learning by the pheromone is itself an instinctive response to the pheromone, even though it results in learning.

9.3.3 Aggression

MUPs produced by male mice have also been described as aggression-promoting pheromones, as swabbing a pooled mixture of recombinant MUP isoforms on the body of a castrated adult male increases aggression against the male when it is introduced into a normal resident male's cage (Chamero et al. 2007). However, pheromones are communication signals that are deliberately produced by a signaller to induce a specific response in a receiver of the same species, to their mutual benefit. As animals would not evolve signals specifically to induce aggression against themselves, the concept of an 'aggression-promoting pheromone' to incite an instinctive hostile response against the signaler is misconceived. Even though one or more mouse MUPs induce an aggressive response in other males, MUPs cannot evolve as pheromones to promote aggression.

There are several reasons why aggression may be induced by MUPs. First, one of the MUPs (darcin) is a male-specific sex pheromone that plays an important role in female sexual attraction (see above). As males producing this signal are competitors for breeding opportunities, detection of darcin on another male is likely to induce a competitive aggressive response. However, it is not yet known whether aggression is a specific response to darcin rather than to all MUPs as aggression has been tested only towards a pooled mixture of recombinant MUP isoforms that included darcin (Chamero et al. 2007). Secondly, aggression may be induced by any androgen-dependent scent components that indicate that an intruder is another adult male. In support of this, male-specific volatile pheromones that prime oestrus in female mice stimulate strong aggression from competitor males (Novotny et al. 1985), while Chamero et al. (2007) found that the low molecular weight fraction of male urine stimulates as much aggression as the high molecular weight fraction containing MUPs and bound volatile ligands. Thirdly, mice discriminate familiar group members (tolerated in the territory) from unfamiliar intruders by their individual odour signatures. As MUP pattern provides the identity signature used to recognise individual mice in other contexts (Sect. 9.2.1), application of a subset of recombinant MUPs in a different ratio to the familiar pattern expressed by the strain (Mudge et al. 2008) could have altered the highly familiar identity signature used to recognise familiar cagemates. To test whether this learned familiarity of individual MUP signatures influences aggression, future studies need to assess whether aggression is induced by changes in the familiar profile of MUPs that are not male-specific. Of course, it is very possible that aggression is stimulated by darcin, other androgen-dependent MUPs and urinary volatiles, unfamiliar individual MUP signatures and MUP signatures that belong to familiar territorial opponents, as each of these scent stimuli would signify the presence of competitors or non-group members in a male's defended territory.

9.4 Conclusions

The urinary proteins of house mice are now known to play a number of different roles in chemical signalling. The polymorphism of individual MUP patterns provides a stable and persistent signal of the individual identity, kinship and heterozygosity of the scent owner, allowing animals to recognise and adjust their behaviour towards different individuals and kin. Selective binding and release of low molecular weight urinary volatiles by these proteins ensures that volatile signals are retained and slowly released from scent marks, allowing scents to be detected over a prolonged time period. Urinary proteins may also be involved in delivering volatiles to appropriate receptors in the receiving animal, although further work is needed to understand their importance in signal detection. Perhaps most excitingly, at least one of these proteins, darcin, acts as an important sex pheromone. This stimulates an instinctive sexual attraction in females to spend time near the pheromone, but also targets and reinforces sexual attraction to the odour of a specific male through associative learning. Such pheromone-induced associative learning is likely to be an important mechanism underpinning the ability of mammals and other vertebrates to respond flexibly towards pheromone signals from different individuals. The challenge now is to gain a much broader understanding of the functions of these important communication proteins in other species and explain why species differ so greatly in individual polymorphism, sexual dimorphism and levels of expression.

Acknowledgements We thank Amanda Davidson, Sarah Roberts, Duncan Robertson, Stuart Armstrong, Michael Thom, Amy Sherborne, John Waters, Felicity Fair and Rick Humphries for help in collecting data in Figs. 9.2 and 9.4 and additional data referred to in the text. This work was supported by research grants from the Biotechnology and Biological Sciences Research Council and from the Natural Environment Research Council.

References

Armstrong SD, Robertson DHL, Cheetham SA, Hurst JL, Beynon RJ (2005) Structural and functional differences in isoforms of mouse major urinary proteins: a male-specific protein that preferentially binds a male pheromone. Biochem J 391:343–350

Bacchini A, Gaetani E, Cavaggioni A (1992) Pheromone binding proteins of the mouse, *Mus musculus*. Experientia 48:419–421

Beynon RJ, Hurst JL (2004) Urinary proteins and the modulation of chemical scents in mice and rats. Peptides 25:1553–1563

Beynon RJ, Turton MJ, Robertson DHL, Armstrong SD, MacNicoll A, Humphries RE, Hurst JL (2008) Urinary lipocalins in Rodenta: is there a generic model? In: Hurst JL, Beynon RJ, Roberts SC, Wyatt TD (eds) Chemical Signals in Vertebrates 11. Springer, New York, pp 37–50

Briand L, Trotier D, Pernollet JC (2004) Aphrodisin, an aphrodisiac lipocalin secreted in hamster vaginal secretions. Peptides 25:1545–1552

Chamero P, Marton TF, Logan DW, Flanagan K, Cruz JR, Saghatelian A, Cravatt BF, Stowers L (2007) Identification of protein pheromones that promote aggressive behaviour. Nature 450:899–902

Cheetham SA, Thom MD, Jury F, Ollier WER, Beynon RJ, Hurst JL (2007) The genetic basis of individual recognition signals in the mouse. Curr Biol 17:1771–1777

Cheetham SA, Smith AL, Armstrong SD, Beynon RJ, Hurst JL (2009) Limited variation in the major urinary proteins of laboratory mice. Physiol Behav 96:253–261

Flower DR (1996) The lipocalin protein family: structure and function. Biochem J 318 (Pt 1):1–14

Gardner A, West SA (2007) Social evolution: the decline and fall of genetic kin recognition. Curr Biol 17:R810–R812

Garratt M, Stockley P, Armstrong SD, Beynon RJ, Hurst JL (2011a) The scent of senescence: sexual signalling and female preference in house mice. J Evol Biol 24:2398–2409

Garratt M, Vasilaki A, Stockley P, McArdle F, Jackson M, Hurst JL (2011b) Is oxidative stress a physiological cost of reproduction? An experimental test in house mice. Proc Biol Sci 278:1098–1106

Garratt M, McArdle F, Stockley P, Vasilaki A, Beynon RJ, Jackson M, Hurst JL (2012) Tissue dependent changes in oxidative damage with male reproductive effort in house mice. Funct Ecol 26(2):423–433

Humphries RE, Robertson DHL, Beynon RJ, Hurst JL (1999) Unravelling the chemical basis of competitive scent marking in house mice. Anim Behav 58:1177–1190

Hurst JL (1990) Urine marking in populations of wild house mice *Mus domesticus* Rutty. I. Communication between males. Anim Behav 40:209–222

Hurst JL, Beynon RJ (2008) Chemical communication in societies of rodents. In: D'Ettorre P, Hughes DP (eds) Sociobiology of communication: an interdisciplinary perspective. Oxford University Press, Oxford, pp 97–118

Hurst JL, Robertson DHL, Tolladay U, Beynon RJ (1998) Proteins in urine scent marks of male house mice extend the longevity of olfactory signals. Anim Behav 55:1289–1297

Hurst JL, Payne CE, Nevison CM, Marie AD, Humphries RE, Robertson DHL, Cavaggioni A, Beynon RJ (2001) Individual recognition in mice mediated by major urinary proteins. Nature 414:631–634

Jones RB, Nowell NW (1974) A comparison of the aversive and female attractant properties of urine from dominant and subordinate male mice. Anim Learn Behav 2:141–144

Jordan WC, Bruford MW (1998) New perspectives on mate choice and the MHC. Heredity 81:239–245

Karlson P, Luscher M (1959) Pheromones: a new term for a class of biologically active substances. Nature 183:55–56

Kavaliers M, Choleris E, Pfaff DW (2005) Recognition and avoidance of the odors of parasitized conspecifics and predators: differential genomic correlates. Neurosci Biobehav R 29:1347–1359

Kimoto H, Sato K, Nodari F, Haga S, Holy TE, Touhara K (2007) Sex- and strain-specific expression and vomeronasal activity of mouse ESP family peptides. Curr Biol 17:1879–1884

Knopf JL, Gallagher JF, Held WA (1983) Differential multihormonal regulation of the mouse major urinary protein gene family in the liver. Mol Cell Biol 3:2232–2240

Kulkarni AB, Gubits RM, Feigelson P (1985) Developmental and hormonal regulation of alpha 2u-globulin gene transcription. Proc Natl Acad Sci USA 82:2579–2582

Kurtz DT (1981) Rat alpha 2u globulin is encoded by a multigene family. J Mol Appl Genet 1:29–38

Kwak J, Josue J, Faranda A, Opiekun MC, Preti G, Osada K, Yamazaki K, Beauchamp GK (2011) Butylated hydroxytoluene is a ligand of urinary proteins derived from female mice. Chem Senses 36:443–452

Leinders-Zufall T, Lane AP, Puche AC, Ma W, Novotny MV, Shipley MT, Zufall F (2000) Ultrasensitive pheromone detection by mammalian vomeronasal neurons. Nature 405:792–796

Logan DW, Marton TF, Stowers L (2008) Species specificity in major urinary proteins by parallel evolution. PLoS One 3:e3280

MacInnes JI, Nozik ES, Kurtz DT (1986) Tissue-specific expression of the rat alpha2u globulin gene family. Mol Cell Biol 6:3563–3567

Manning CJ, Potts WK, Wakeland EK, Dewsbury DA (1992) What's wrong with MHC mate choice experiments? In: Doty RL, Muller-Schwarze D (eds) Chemical Signals in Vertebrates 6. Plenum, New York, pp 229–235

Marie AD, Veggerby C, Robertson DHL, Gaskell SJ, Hubbard SJ, Martinsen L, Hurst JL, Beynon RJ (2001) Effect of polymorphisms on ligand binding by mouse major urinary proteins. Protein Sci 10:411–417

McFadyen DA, Locke J (2000) High-resolution fish mapping of the rat alpha2u-globulin multigene family. Mamm Genome 11:292–299

Moncho-Bogani J, Lanuza E, Hernandez A, Novejarque A, Martinez-Garcia F (2002) Attractive properties of sexual pheromones in mice: innate or learned? Physiol Behav 77:167–176

Mossman CA, Drickamer LC (1996) Odor preferences of female house mice (*Mus domesticus*) in seminatural enclosures. J Comp Psychol 110:131–138

Mudge JM, Armstrong SD, McLaren K, Beynon RJ, Hurst JL, Nicholson C, Robertson DH, Wilming LG, Harrow JL (2008) Dynamic instability of the major urinary protein gene family revealed by genomic and phenotypic comparison between c57 and 129 strain mice. Genome Biol 9:R91

Novotny M, Harvey S, Jemiolo B, Alberts J (1985) Synthetic pheromones that promote inter-male aggression in mice. Proc Natl Acad Sci USA 82:2059–2061

Novotny MV, Ma W, Wiesler D, Zidek L (1999) Positive identification of the puberty-accelerating pheromone of the house mouse: the volatile ligands associating with the major urinary protein. Proc R Soc B 266:2017–2022

Paterson S, Hurst JL (2009) How effective is recognition of siblings on the basis of genotype? J Evol Biol 22:1875–1881

Penn DJ (2002) The scent of genetic compatibility: sexual selection and the major histocompatibility complex. Ethology 108:1–21

Penn D, Schneider G, White K, Slev P, Potts W (1998) Influenza infection neutralizes the attractiveness of male odour to female mice (*Mus musculus*). Ethology 104:685–694

Perez-Miller S, Zou Q, Novotny MV, Hurley TD (2010) High resolution x-ray structures of mouse major urinary protein nasal isoform in complex with pheromones. Protein Sci 19:1469–1479

Phelan MM, McLean L, Simpson DM, Hurst JL, Beynon RJ, Lian LY (2010) ^{1}H, ^{15}N and ^{13}C resonance assignment of darcin, a mouse major urinary protein. Biomol NMR Assign 4:239–241

Ramm SA, Cheetham SA, Hurst JL (2008) Encoding choosiness: female attraction requires prior physical contact with individual male scents in mice. Proc R Soc B 275:1727–1735

Rich TJ, Hurst JL (1999) The competing countermarks hypothesis: reliable assessment of competitive ability by potential mates. Anim Behav 58:1027–1037

Roberts SA, Simpson DM, Armstrong SD, Davidson AJ, Robertson DH, McLean L, Beynon RJ, Hurst JL (2010) Darcin: a male pheromone that stimulates female memory and sexual attraction to an individual male's odour. BMC Biol 8:75

Robertson DHL, Beynon RJ, Evershed RP (1993) Extraction, characterization and binding analysis of two pheromonally active ligands associated with major urinary protein of house mouse (*Mus musculus*). J Chem Ecol 19:1405–1416

Robertson DHL, Hurst JL, Hubbard S, Gaskell SJ, Beynon RJ (1998) Ligands of urinary lipocalins from the mouse: uptake of environmentally derived chemicals. J Chem Ecol 24:1127–1140

Robertson DHL, Marie AD, Veggerby C, Hurst JL, Beynon RJ (2001) Characteristics of ligand binding and release by major urinary proteins. In: Marchlewska-Koj A, Muller-Schwarze D, Lepri J (eds) Chemical Signals in Vertebrates 9. Plenum Press, New York, pp 169–176

Robertson DHL, Hurst JL, Searle JB, Gunduz I, Beynon RJ (2007) Characterization and comparison of major urinary proteins from the house mouse, *Mus musculus domesticus*, and the aboriginal mouse, *Mus macedonicus*. J Chem Ecol 33:613–630

Shahan K, Denaro M, Gilmartin M, Shi Y, Derman E (1987) Expression of six mouse major urinary protein genes in the mammary, parotid, sublingual, submaxillary, and lachrymal glands and in the liver. Mol Cell Biol 7:1947–1954

Sharrow SD, Vaughn JL, Zidek L, Novotny MV, Stone MJ (2002) Pheromone binding by polymorphic mouse major urinary proteins. Protein Sci 11:2247–2256

Shaw PH, Held WA, Hastie ND (1983) The gene family for major urinary proteins: expression in several secretory tissues of the mouse. Cell 32:755–761

Sherborne AL, Thom MD, Paterson S, Jury F, Ollier WER, Stockley P, Beynon RJ, Hurst JL (2007) The genetic basis of inbreeding avoidance in house mice. Curr Biol 17:2061–2066

Singer AG, Tsuchiya H, Wellington JL, Beauchamp GK, Yamazaki K (1993) Chemistry of odortypes in mice—fractionation and bioassay. J Chem Ecol 19:569–579

Stopka P, Janotova K, Heyrovsky D (2007) The advertisement role of major urinary proteins in mice. Physiol Behav 91:667–670

Stopkova R, Zdrahal Z, Ryba S, Sedo O, Sandera M, Stopka P (2010) Novel OBP genes similar to hamster aphrodisin in the bank vole, *Myodes glareolus*. BMC Genomics 11:45

Talley HM, Laukaitis CM, Karn RC (2001) Female preference for male saliva: implications for sexual isolation of *Mus musculus* subspecies. Evolution 55:631–634

Thom MD, Stockley P, Beynon RJ, Hurst JL (2008) Scent, mate choice and genetic heterozygosity. In: Hurst JL, Beynon RJ, Roberts SC, Wyatt TD (eds) Chemical Signals in Vertebrates 11. Springer, New York, pp 291–301

Turton MJ, Robertson DHL, Smith JR, Hurst JL, Beynon RJ (2010) Roborovskin, a lipocalin in the urine of the Roborovski hamster, *Phodopus roborovskii*. Chem Senses 35:675–684

Wyatt TD (2010) Pheromones and signature mixtures: defining species-wide signals and variable cues for identity in both invertebrates and vertebrates. J Comp Physiol A Neuroethol Sens Neural Behav Physiol 196:685–700

Chapter 10
From Exploration to Selective Information Gathering: The Development of Chemosensory Investigation in Male African Elephants (*Loxodonta africana*)

Bruce A. Schulte, Kathryn R. Bagley, Stacie Castelda, Helen Loizi, Nabil Nasseri, Dhaval K. Vyas, and Thomas E. Goodwin

Abstract Male African and Asian elephants exhibit a lengthy extended developmental period from weaning to social reproductive maturity. Chemical signals play an important role in elephant society, and male Asian elephants (*Elephas maximus*) have been shown to go through developmental changes in chemical signal sending, reception and social interactions. We examined the chemosensory behaviour of male African elephants (*Loxodonta africana*) over four life stages (calf, juvenile, adolescent and adult). We observed the investigatory behaviour of the trunk by wild male African elephants, as well as their responses to potential chemical signals (i.e. urine, control and individual compounds from elephant urine). One of our objectives was the search for compounds in the urine of African elephant females that might serve as chemical signals. We conducted biological assays with elephants in South Africa and Tanzania. No single compound evoked responses indicative of a chemical signal. The chemosensory response to these signals showed an intriguing pattern of increasing interest by males from the calf to adolescent stages with a significant drop off in chemosensory behaviour by adult males. Chemosensory behaviour in male African elephants appears to go through developmental refinement from exploration to selective information gathering once they mature.

B.A. Schulte (✉)
Department of Biology, Western Kentucky University, Bowling Green, KY, USA

Department of Biology, Georgia Southern University, Statesboro, GA, USA
e-mail: bruce.schulte@wku.edu

K.R. Bagley • S. Castelda • H. Loizi • N. Nasseri • D.K. Vyas
Department of Biology, Georgia Southern University, Statesboro, GA, USA

T.E. Goodwin
Department of Chemistry, Hendrix College, Conway, AR, USA

M.L. East and M. Dehnhard (eds.), *Chemical Signals in Vertebrates 12*,
DOI 10.1007/978-1-4614-5927-9_10,

10.1 Introduction

In group-living, polygynous mammals, offspring are raised in a social environment, dominated in the early years by the presence of their mother and related caregivers (Clutton-Brock 2009). The young have the opportunity to interact with other immature individuals of both sexes as well as adult females. Contact with adult males may be more limited, especially in matriarchal societies such as those experienced by African elephants (*Loxodonta africana*) with which adult males interact infrequently. Females are likely to spend their entire lives in social groups containing close and distant kin, whereas males disperse, leaving the natal group for a largely bachelor or solitary lifestyle (Clutton-Brock 1989). Exposure to an adult male is more frequent in mammalian harem societies, where an adult male typically defends his harem against other males. However, young males are likely to be exposed to primarily agonistic interactions between adult males. In contrast, young females witness cooperative behaviours among adult females in the group (Wrangham 1980; Clutton-Brock 2009).

While females develop within the social framework of their adult existence, males undergo a relatively rapid and extreme alteration in their social landscape over ontogeny (Chepko-Sade and Halpin 1987). Male reproductive success is dependent on successful outcomes of intrasexual competition, requiring large size, ornaments, competitive skills and high social status. Female reproductive success is resource limited, and investment in offspring and social group dynamics is emphasised (Darwin 1871; Clutton-Brock et al. 1982). Traits that are considered genetically controlled adaptations, such as social strategies and predator–prey behaviours, are likely to be influenced by social learning (Thornton and Clutton-Brock 2011). In matriarchal societies, females have ample opportunity to learn from same-sexed adults, but males have fewer chances to observe and apply the lessons from such learning. For instance, young male elephants may witness the interaction between adult females, but these females (social group members) may be intolerant of similar type contacts by a younger male of their own social group. Because male offspring lack regular role models (i.e. adult males) for appropriate adult behaviour, they may be more investigative of their physical and social environment than similarly aged females. After weaning around 4–5 years of age, male elephants wander further from their mother more so than similarly aged females (Lee and Moss 1999). Males also tend to play more than females (Lee 1986; Meyer 2006). Such exploration may provide opportunities for social learning as a prerequisite for successful dispersal when males interact with similarly aged individuals regularly (Chiyo et al. 2011).

Following dispersal from the natal herd, adolescent males often travel in bachelor groups. Typically these bachelor herds are loosely knit assemblages of teenage males, although both younger and older males may be a part of the group (Poole 1989). In a study of male elephants in Botswana, all males of 10 years and older preferred to be in the company of fully adult males (>35 years old) (Evans and Harris 2008). The reason for this association is not clear, although the authors conjecture that younger males may learn from the older individuals, or older males may be trying to suppress competitors. In either case, males have the potential to communicate, interact and begin to establish themselves in a male social hierarchy.

During the years of adolescence, male Asian elephants (*Elephas maximus*) experience a phenomenon known as moda musth (Rasmussen et al. 2002). Adult male Asian and African elephants go through an annual period of high testosterone, temporal gland secretion, urine dribbling and heightened aggressiveness termed musth (Jainudeen et al. 1972; McKay 1973; Poole 1987), indicated chemically for Asian elephants by the compound frontalin in the temporal gland secretion (Rasmussen and Greenwood 2003, 2005). Musth male African elephants are dominant to other males and preferred as mates (Hollister-Smith et al. 2007). In addition, male Asian elephants in musth are better at detecting the oestrous pheromone, Z-7-dodecen-1-yl acetate (Z7:12Ac), released by female Asian elephants that are approaching ovulation (Rasmussen et al. 2005). The discovery of moda musth indicates that the development of adult musth goes through stages. In fact, the change of the male temporal gland secretion from a sweet-smelling honey odour to a pungent, malodourous scent reflects the alteration in the enantiomeric ratio (+ and -) of frontalin. In moda musth frontalin exists mainly as the (+) enantiomer, while in adult musth it is approximately racemic (Greenwood et al. 2005). Asian elephants show no or low interest in the moda musth signal, but male and female Asian elephants, except follicular phase females, are repulsed by the adult mid-phase musth scent. Thus, at least in the Asian species, male elephants go through developmental changes in signal delivery, signal reception and social interaction. The chemical composition of urine from wild male African elephant urine changes over development (Rasmussen and Wittemyer 2002). Furthermore, adult male African elephants in captivity distinguish luteal and follicular urine from conspecific females, indicating the presence of an oestrous chemical signal (Bagley et al. 2006). The identity of this signal is not known, but of potential relevance is the fact that both known chemical signals in Asian elephants (frontalin and Z7:12Ac) are used by insects as chemical signals alone or in blends, which has focused our attention on such compounds in African elephants (Castelda et al. 2008; Goodwin and Schulte 2009). In our search, we have used bioassays as a means of examining the development of chemosensory behaviour in African elephants (Loizi et al. 2009; Merte et al. 2010).

10.2 Objectives

The objectives of the present study were to examine the chemosensory behaviour of male African elephants at different developmental stages while also screening for potential chemical signals. We hypothesised that male elephants would exhibit increasing rates of investigative chemosensory behaviour as they developed from calves to adolescents. By adulthood, male elephants should be refining their behaviour to track signals of greatest importance to their reproductive success.

10.3 Methods

Our observations for the present study occurred from 2003 to 2008 at Addo Elephant National Park (AENP) in South Africa and at Ndarakwai Ranch in northeastern Tanzania. AENP is located 72 km northeast of Port Elizabeth. Elephant-proof fencing encloses some 14,000 ha with over 400 African elephants in the main reserve (Whitehouse and Schoeman 2003). The vegetation is composed of sub-tropical succulent thicket and grassland (Whitehouse and Hall-Martin 2000). Five main waterholes supply pumped water throughout the year, while numerous natural pans are created during rainy periods. The average rainfall for the region is 445 mm annually (Paley and Kerley 1998). Ndarakwai Ranch is an unfenced, privately owned area of 4,300 ha with a mix of grassland and acacia woodland. This region experiences short and long wet seasons separated by dry seasons (Vyas 2006), similar to nearby Amboseli National Park where rainfall averages 350 mm annually (Poole 1999).

At AENP, we know the identity and age (usually to at least the year) of each elephant (see Loizi et al. 2009 for details). We have assisted in the maintenance of a photographic database for each elephant (Whitehouse and Hall-Martin 2000). At Ndarakwai Ranch, we used morphological features (e.g. shoulder height and tusk length approximations) to estimate ages (Moss 1996) as follows: calves (<5 y), juveniles (5–9 y), adolescents, pubescents or subadults (10–19 y), and adults (≥20 y). These ages reflect the approximate time of major developmental changes for elephants (calf: nursing; juvenile: weaning; adolescent: capability of producing viable gametes and dispersal [males] or first offspring production [females]; adult: full reproductive status and rise in rank). This classification also was used for the elephants at AENP for comparisons.

At both locations, we observed elephants and set up bioassays at waterholes. Observations at waterholes permit an unobstructed view of elephants and their trunk movements. Waterholes serve as points of congregation for elephants, so chemical signals in their urine and faeces are readily available. We used continuous focal observations that started when an elephant moved within a body length of a placed sample and ended when the elephant departed this same distance. Multiple visits by the same elephant were combined so that an individual elephant was represented only once in the dataset. For the present study, we focused our data analysis on male elephants over the four age classes. At AENP, we observed from a vehicle because the elephants were acclimated to the presence of vehicles by tourists and game drive personnel. At Ndarakwai Ranch, we observed elephants from a 6 m high platform adjacent to the waterhole. Elephants ignored observers who remained quiet and still while recording data.

A double-blind procedure was implemented for bioassays by coding samples such that the observer did not know their identities. Twelve compounds from African elephant urine were selected for assays. The compounds were selected because they appeared often in African elephant urine, were somewhat unusual and yet available commercially (i.e. α-damascone, β-damascone, damascenone), some were known signals in insects (i.e. (1S)-(-)-verbenone, (S)-*cis*-verbenol, *endo*-brevicomin, *exo*-brevicomin and *E,E*-α-farnesene), and some also were considered signals or potential signals in Asian elephants (2-decanone, 2-nonanone, cyclohexanone, and racemic frontalin; Perrin and Rasmussen 1994; Rasmussen et al. 1996; Goodwin et al. 2007; Goodwin and Schulte 2009). Based on the convergence between insects and elephants on the two known Asian elephant chemical signals (frontalin and *Z*7:12Ac), we were especially interested in compounds found in the urine of African elephants that had signal value in insects. Hence, our discovery of brevicomins and frontalin in female African elephant urine provided highly anticipated candidate signal compounds (Goodwin et al. 2006). The oestrous pheromone (*Z*7:12Ac) found in Asian elephant urine was not assayed because it has not been identified in female African elephant urine. We used vanilla extract or vanillin as a low response control (Schulte and Rasmussen 1999). When available, we used female elephant urine to evoke a higher level of response. In our many years of conducting such bioassays, there has been no evidence of the responses to one substance influencing the responses to others (i.e. assaying elephant urine does not increase the responses to single compounds). We could only obtain non-oestrous urine. The compounds were presented in concentrations of 100 μL of the chemical compound to 500 mL water (Rasmussen et al. 1997). Samples were placed a few metres back from the edge of a waterhole and poured on the natural substrate. We avoided placing samples near elephant faeces or wet areas upon which elephants may have urinated. Samples were placed no closer than 3 m apart so elephants could not be in proximity to two samples at one time. Typically, they were further apart (e.g. 5 m or more). Samples were poured when elephants were in sight, but were far enough away that we could safely return to our observation point (tower or vehicle) without altering the approach trajectory of the elephants.

We recorded the responses of wild male elephants to placed samples of elephant urine and single chemical compounds that occur in elephant urine. State behaviours (main behaviours: drink, eat, mud, stand, walk) were recorded as durations, while short-lasting behaviours, termed events (e.g. chemosensory responses with trunk, head-shakes) were converted from frequency to rate (frequency/proximity). We used state behaviours to check that behaviours at waterholes were similar over time and location within and between sites. We do not report the state behaviour data herein because we have found no important differences in state behaviours at the two locations (Schulte et al. 2008). For the analysis of chemosensory behaviour, we grouped the main ones (sniff, check, place and flehmen) as a single response variable (SCPF) (see ethograms in Schulte and Rasmussen 1999; Bagley et al. 2006; Meyer 2006; Meyer et al. 2008). These main chemosensory responses have been shown to correlate well with accessory trunk behaviours (Bagley et al. 2006), so we report only the main responses. In the field, we have determined that standardising response frequencies by the number of proximities (movements within one body length of a sample) is more appropriate than per unit time because the duration nearby a sample can be relatively short (<1 min). For the developmental question, we combined all event behaviours relevant to the investigation to incorporate more data on overall investigatory behaviour. Our primary interest was to examine whether responses changed by age class. We set up an orthogonal set of linear

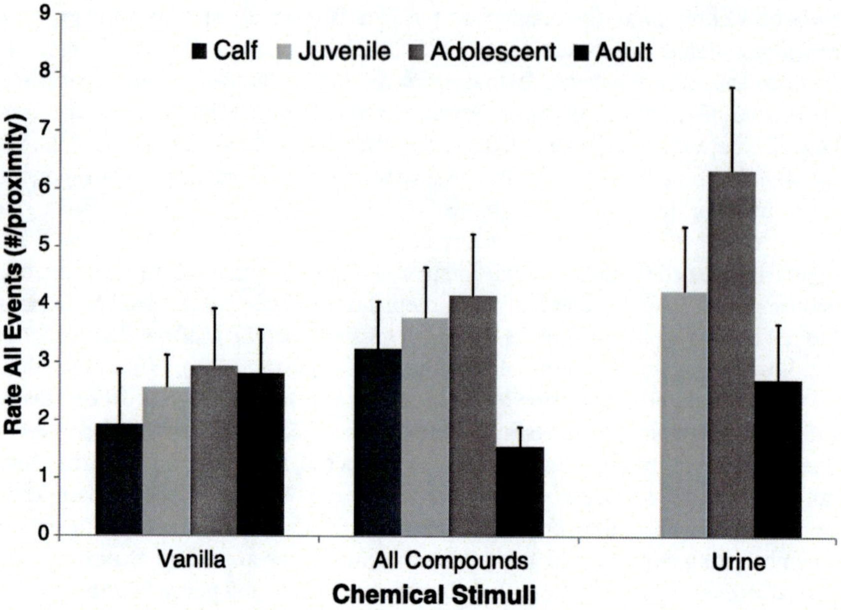

Fig. 10.1 Rate of relevant event behaviours (SCPF and accessory trunk behaviours, see Bagley et al. (2006) for definitions, +1 SEM) by male African elephants in four age classes: calf (<5 y), juvenile (5–9 y), adolescent (10–19 y) and adult (≥20 y)

contrasts in which we compared the responses as follows: adolescents to adults, calves to juveniles, and the average of calves and juveniles to the average of adolescents and adults. Analyses were performed in Statistica (StatSoft, Inc.). We reported all rates on a per proximity basis with ± 1 SEM.

10.4 Results

To examine developmental patterns in the response to chemical stimuli, we calculated the mean responses to all 12 compounds and analysed the results by age class in comparison to the responses to the two controls (Fig. 10.1). Our interest was in examining the response across age classes and not between the samples by age class. When comparing the responses of the four age classes of male elephants to the combined 12 single compounds, the chemosensory events did not differ by age class ($F_{3,164}=2.05$, $P=0.11$). In our a priori set of linear contrasts, the responses of adolescents were significantly different from adults ($F_{1,164}=5.37$, $P=0.02$); whereas, calves did not differ from juveniles ($F_{1,164}=0.17$, $P=0.68$) nor did younger elephants (calves + juveniles) differ in their response from older elephants (adolescents + adults) ($F_{1,164}=0.57$, $P=0.45$). As shown in Fig. 10.1, adult males were less responsive than any other age class to the 12 single compounds.

Table 10.1 Rate of sniff, check, place and flehmen (SCPF) responses (number per approach to within body length, termed proximity) to chemical samples (given in millimolar [mM] concentrations for the experimental compounds) by all age classes of male African elephants at Addo Elephant National Park South Africa and Ndarakwai Ranch Tanzania

Compound	mM	Location	Mean SCPF	SEM
Cyclohexanone	1.93	Tanzania	2.52	0.91
Damascenone	0.98	South Africa	1.03	0.18
α-Damascone	0.97	South Africa	0.35	0.06
β-Damascone	0.97	South Africa	0.30	0.05
2-Decanone	1.06	Tanzania	1.75	0.46
E,E-α-Farnesene	0.82	Tanzania	0.73	0.36
endo-Brevicomin	1.24	Tanzania	3.33	2.30
exo-Brevicomin	1.24	Tanzania	1.87	0.71
Frontalin	1.39	Tanzania	2.17	0.75
2-Nonanone	1.15	Tanzania	1.87	0.56
(1S)-(-)-Verbenone	1.30	South Africa	1.53	0.26
(S)-*cis*-Verbenol	1.31	South Africa	0.91	0.16
Urine (luteal or pre-puberty)		Tanzania	2.47	0.24
Vanilla		Both	1.57	0.14

To determine whether any of the single compounds showed promise as chemical signals, we analysed the chemosensory responses to the 12 compounds separately. The rate of sniff, check, place and flehmen responses by all age classes of elephant to the 12 placed compounds did not differ from their responses to vanilla extract or elephant urine (Table 10.1). Three of the compounds, cyclohexanone, *endo*-brevicomin and frontalin, had mean response levels comparable to that of non-oestrous urine, but the variation in responses was quite high across elephants (and even within age classes). For example, *endo*-brevicomin evoked higher responses than any other substance from an adolescent male elephant, but several other males in this and other age classes did not respond at all despite being in close proximity to the sample.

10.5 Discussion

Male elephants diverge morphologically and behaviourally from females early in life and while females remain with their natal herd, males disperse, banding into bachelor groups or travel about singly. As shown in Fig. 10.1, chemosensory behaviours by males followed the predicted pattern with increasing rates through adolescence and then refinement in adulthood. If we had assayed a chemical with reproductive signal value (i.e. a pheromone), such as an oestrous or musth signal, then we would have expected to witness the highest rates of response by adult male elephants.

In a previous examination of chemosensory behaviour by male and female African elephants, Loizi et al. (2009) showed that adult males were the most likely age and sex class to investigate conspecific urine and faeces. A similar finding was evident in a study comparing the four age classes of elephants as they approached waterholes at AENP; adult male elephants exhibited the highest rate of sniff behaviours to conspecific faeces (Fig. 2 in Merte et al. 2010). Yet, when sniff responses to all substrates were considered, Merte et al. (2010) documented a pattern of rising responses for calves through adolescents with a drop-off by adult males (compare Fig. 3 in Merte et al. 2010 to Fig. 10.1 herein). The current study suggests that the response rate of these males is driven by the signal value of the substance investigated. Over development, male African elephants appear to expand the chemosensory exploration of their environment in terms of rate of response, and then as adults, refine their investigative skills, focusing their assessments more specifically on signal sources and specific signals of reproductive and social importance.

Experimental trials with captive male (Bagley et al. 2006) and female (Meyer et al. 2008) African elephants as well as field studies that have involved stimuli detected by chemosenses (e.g. Bates et al. 2007, 2008; Nasseri 2009) suggest that chemical signalling is vital to African elephant communication (Rasmussen and Schulte 1998). Two single compounds have been identified in Asian elephants that function in the signalling of estrus (*Z*7:12Ac) and musth (frontalin) (Rasmussen and Greenwood 2005). Each of these compounds performs signal functions in insects as well. Thus, for our studies on African elephants, we pursued the single compound approach and were especially interested in compounds found in African elephant urine with previously known signal value in insects and Asian elephants. However, to date, we have not documented responses to 12 candidate compounds, indicating that as single compounds they play a signal role in African elephants. We know from the studies with Asian elephants that *Z*7:12Ac is bound to a urinary protein and so activity drops quickly over time when it is assayed just in an aqueous buffer solution (Lazar et al. 2004). Nevertheless, fresh samples in buffer do evoke flehmen and other chemosensory responses by male Asian elephants that clearly distinguish this compound as bioactive (Rasmussen et al. 1996, 1997). While in the present study we were unable to assay a positive control that evokes high responses (i.e. oestrous urine), the finding that responses to the individual chemicals were not greater than non-oestrous urine suggests that none of these compounds alone have full signal value to male African elephants. We continue to search for the identity of chemical signals in African elephants, expanding our candidate pool of compounds while also incorporating a variety of approaches (Goodwin and Schulte 2009). If a single compound serves as a chemical signal to male African elephants, then it is unlikely that any of these 12 compounds serve such a role, although we did not examine ratios of enantiomers. These compounds may be a part of a multi-compound signal. Understanding the variation in responses to such compounds may reveal personality traits, an area that we are exploring currently. Despite not identifying a bioactive compound, the bioassays served as a valuable tool, providing insight on the developmental pattern of chemosensory behaviour in male African elephants.

Acknowledgements We thank our respective colleges and universities for support of our scholarly activities and the Leibniz Institute for Zoo and Wildlife Research for hosting the 12th Meeting on Chemical Signals in Vertebrates. Funding was provided by the National Science Foundation (B.A.S., L.E.L. Rasmussen and T.E.G.: NSF-DBI-02-17062, -16862, -17068, respectively), Biospherics Research Corporation, Georgia Southern University and Hendrix College. We appreciate the assistance of personnel at Addo Elephant National Park, Nelson Mandela Metropolitan University and especially the Centre for African Conservation Ecology and South Africa National Parks (permit 200(2-11)-12-11 BSCH). We are grateful to Peter & Margot Jones and the staff at Ndarakwai Ranch, and COSTECH and TAWIRI (permit No. 2004-170-N-2004-32) for permission to work in Tanzania. The research met with IACUC approval at Georgia Southern University.

References

Bagley KR, Goodwin TE, Rasmussen LEL, Schulte BA (2006) Male African elephants (*Loxodonta africana*) can distinguish oestrous status via urinary signals. Anim Behav 71:1439–1445

Bates LA, Sayialel KN, Njiraini NW, Moss CJ, Poole JH, Byrne RW (2007) Elephants classify human ethnic groups by odor and garment color. Curr Biol 17:1938–1942

Bates LA, Sayialel KN, Njiraini NW, Poole JH, Moss CJ, Byrne RW (2008) African elephants have expectations about the locations of out-of-sight family members. Biol Lett 4:34–36

Castelda SM, Goodwin TE, Schulte BA (2008) Investigating chemical signals in African elephants for convergence with insects and similarities with Asian elephants. In: Proceedings of the 2007 International Elephant Conservation & Research Symposium. Orlando, Florida, p 81–91

Chepko-Sade BD, Halpin ZT (eds) (1987) Mammalian dispersal patterns: the effects of social structure on population genetics. Universitu of Chicago Press, Chicago, p 342

Chiyo PI, Archie EA, Hollister-Smith JA, Lee PC, Poole JH, Moss CJ, Alberts SC (2011) Association patterns of African elephants in all-male groups: the role of age and genetic relatedness. Anim Behav 81:1093–1099

Clutton-Brock TH (1989) Mammalian mating systems. Proc R Soc Lond B 236:339–372

Clutton-Brock TH (2009) Structure and function in mammalian societies. Philos Trans R Soc B 364:3229–3242

Clutton-Brock TH, Guinness FE, Albon SD (1982) Red Deer: the behavior and ecology of two sexes. University of Chicago Press, Chicago

Darwin C (1871/1958) The descent of man and selection in relation to sex. The Modern Library, New York, NY

Evans KE, Harris S (2008) Adolescence in male African elephants, *Loxodonta africana*, and the importance of sociality. Anim Behav 76:779–787

Goodwin TE, Schulte BA (2009) Prospecting for mammalian chemical signals via solventless extraction techniques: an elephantine task. ChemoSense 11(2):9–15

Goodwin TE, Eggert M, House SJ, Weddell ME, Schulte BA, Rasmussen LEL (2006) Insect pheromones and precursors in female African elephant urine. J Chem Ecol 32:1849–1853

Goodwin TE, Brown PA, Eggert MS, Evola MG, House SJ, Morshedi RG, Weddell ME, Chen CJ, Jackson SR, Aubut Y, Eggert J, Schulte BA, Rasmussen LEL (2007) Use of automated solid phase dynamic extraction (SPDE)/GC-MS and novel macros in the search for African elephant pheromones. In: Hurst J, Beynon R, Wyatt T, Roberts C (eds) Chemical signals in vertebrates 11. Springer, New York, pp 25–35

Greenwood DR, Comeskey D, Hunt MB, Rasmussen LEL (2005) Chirality in elephant pheromones. Nature 438:1097–1098

Hollister-Smith JA, Poole JH, Archie EA, Vance EA, Georgiadis NJ, Moss CJ, Alberts SC (2007) Age, musth and paternity success in wild male African elephants, *Loxodonta africana*. Anim Behav 74:287–296

Jainudeen MR, McKay GM, Eisenberg JG (1972) Observations on musth in the domesticated Asiatic elephant. Mammalia 36:247–261

Lazar J, Rasmussen LEL, Greenwood DR, Bang I-S, Prestwich GD (2004) Elephant albumin: a multi-purpose pheromone shuttle. Chem Biol 11:1093–1100

Lee PC (1986) Early social development among African elephant calves. Natl Geogr Res 2:388–401

Lee PC, Moss CJ (1999) The social context for learning and behavioural development among African elephants. In: Box HO, Gibson KR (eds) Mammalian social learning: comparative and ecological perspectives. Symposia of the Zoological Society of London (No. 72), London, p 102–125

Loizi H, Goodwin TE, Rasmussen LEL, Whitehouse AM, Schulte BA (2009) Sexual dimorphism in the performance of chemosensory investigatory behaviours by African elephants (*Loxodonta africana*). Behaviour 146:373–392

McKay GM (1973) Behavior and ecology of the Asiatic elephant in southeastern Ceylon. Smithsonian Contrib Zool 125:1–113

Merte CE, Goodwin TE, Schulte BA (2010) Male and female developmental differences in chemosensory investigations by African elephants (*Loxodonta africana*) approaching waterholes. Behav Ecol Sociobiol 64:401–408

Meyer JM (2006) Sexually dimorphic social development and female intrasexual chemical signaling of African elephants (*Loxodonta africana*). M.Sc. thesis, Georgia Southern University

Meyer JM, Goodwin TE, Schulte BA (2008) Intrasexual chemical communication and social responses of captive female African elephants, *Loxodonta africana*. Anim Behav 76:163–174

Moss C (1996) Getting to know a population. In: Kangwana K (ed) Studying elephants, African Wildlife Foundation Handbook 7. AWF, Nairobi, Kenya, p 58–74

Nasseri NA (2009) The relationship of herpetofaunal community composition to an elephant (*Loxodonta africana*) modified woodland of Northern Tanzania, and bioassays with African elephants. M.Sc. thesis, Georgia Southern University

Paley RGT, Kerley GIH (1998) The winter diet of elephant in Eastern Cape subtropical thicket, Addo Elephant National Park. Koedoe 41:37–45

Perrin TE, Rasmussen LEL (1994) Chemosensory responses of female Asian elephants (*Elephas maximus*) to cyclohexanone. J Chem Ecol 20:2557–2566

Poole JH (1987) Rutting behavior in African elephants: the phenomenon of musth. Behaviour 102:283–316

Poole JH (1989) Announcing intent: the aggressive state of musth in African elephants. Anim Behav 37:140–152

Poole JH (1999) Signals and assessment in African elephants: evidence from playback experiments. Anim Behav 58:185–194

Rasmussen LEL, Greenwood DR (2003) Frontalin: a chemical message of musth in Asian elephants (*Elephas maximus*). Chem Senses 28:433–446

Rasmussen LEL, Greenwood DR (2005) Reproduction in Asian elephants: precise chemical signaling has behavioural and biochemical foundations. ChemoSense 8:1–4

Rasmussen LEL, Schulte BA (1998) Chemical signals in the reproduction of Asian and African elephants. Anim Reprod Sci 53:19–34

Rasmussen LEL, Wittemyer G (2002) Chemosignaling of musth by individual wild African elephants, *(Loxodonta africana)*: implications for conservation and management. Proc R Soc Lond 269:853–860

Rasmussen LEL, Lee TD, Roelofs WL, Zhang A, Daves GD Jr (1996) Insect pheromones in elephants. Nature 379:684

Rasmussen LEL, Lee TD, Zhang A, Roelofs WL, Daves GD Jr (1997) Purification, identification, concentration and bioactivity of (Z)-7-dodecen-1-yl acetate: sex pheromone of the female Asian elephant, *Elephas maximus*. Chem Senses 22:417–437

Rasmussen LEL, Riddle HS, Krishnamurthy V (2002) Mellifluous matures to malodorous in musth. Nature 415:975–976

Rasmussen LEL, Krishnamurthy V, Sukumar R (2005) Behavioural and chemical confirmation of the preovulatory pheromone, (*Z*)-7-dodecenyl acetate, in wild Asian elephants: its relationship to musth. Behaviour 142:351–396

Schulte BA, Rasmussen LEL (1999) Signal-receiver interplay in the communication of male condition by Asian elephants. Anim Behav 57:1265–1274

Schulte BA, Bagley KR, Groover M, Loizi H, Merte C, Meyer JM, Napora E, Stanley L, Vyas DK, Wollett K, Goodwin TE, Rasmussen LEL (2008) Comparisons of state and likelihood of performing chemosensory event behaviors in two populations of African elephants (*Loxodonta africana*). In: Hurst JL, Beynon RJ, Roberts SC, Wyatt TD (eds) Chemical signals in vertebrates 11. Springer, New York, pp 81–90

Thornton A, Clutton-Brock TH (2011) Social learning and the development of individual and group behaviour in mammal societies. Philos Trans R Soc B 366:978–987

Vyas DK (2006) Sexually dimorphic developmental patterns of chemosensory behaviors in African elephants (*Loxodonta africana*). M.Sc. thesis. Georgia Southern University

Whitehouse AM, Hall-Martin AJ (2000) Elephants in Addo Elephant National Park, South Africa: reconstruction of the population's history. Oryx 34:46–55

Whitehouse AM, Schoeman DS (2003) Ranging behaviour of elephants within a small, fenced area in Addo Elephant National Park, South Africa. Afr Zool 38:95–108

Wrangham RW (1980) An ecological model of female bonded primate groups. Behaviour 75:262–300

Part IV
Odour Profiles and Social Organisation

Chapter 11
Sex Differences in Olfactory Behaviours Reflect the Importance of Scent Marking for Social Integration in Adult Females and Competition Between Reproductively Active Males in the Spotted Hyena

Marion L. East, Nicole Gusset-Burgener, and Heribert Hofer

Abstract We investigated the effect of sex, social status and life-stage on three behaviours (pasting, dry-pasting and pawing) associated with olfactory communication in the spotted hyena (*Crocuta crocuta*). Pasting is a behaviour used to place a fatty, anal scent gland secretion on vegetation. Dry-pasting behaviour resembles pasting except that scent is not deposited, but instead scent pasted on vegetation by other clan members is smeared on the ventral side of a dry-pasting animal. Pawing is vigorous scratching of the ground with one then the other front paw, which deposits scent from inter-digital glands at scratching sites. Our focus was olfactory communication within social groups termed clans. We obtained data from individually known members of three large clans in the Serengeti National Park, in northern Tanzania. Data were collected at clan communal dens, which are important social centres, and elsewhere in the clan territory. Adult female's pasting and dry-pasting rates significantly declined with increasing age which indicated that younger adult females invested more than older females in depositing scent and anointing themselves with scent from other clan members. Females pawed the ground rarely. Pasting and dry-pasting rates of reproductively active males (mostly immigrant and a few natal males) above median social status were significantly higher than rates of those below median status at the clan communal den, but not elsewhere in the clan territory. This may result from a reduced opportunity for males below median status to scent mark at the clan den because of the presence of socially dominant males. Alternatively, a lower-ranking male (i.e. more recent immigrant) may gain fitness benefits from concealing its olfactory identity from den-bound juvenile females, because this would increase the number of females that would consider it

M.L. East (✉) • N. Gusset-Burgener • H. Hofer
Department of Evolutionary Ecology, Leibniz Institute for Zoo and Wildlife Research, Alfred-Kowalke-Street 17, 10315 Berlin, Germany
e-mail: east@izw-berlin.de

M.L. East and M. Dehnhard (eds.), *Chemical Signals in Vertebrates 12*,
DOI 10.1007/978-1-4614-5927-9_11,

a potential mate throughout its tenure in the clan. Males above median status pawed the ground at a higher rate both at the communal den and elsewhere in the territory than males below median status. We conclude that these sex differences reflect the importance of (1) olfactory communication and scent mixing between females during their initial years of social integration into the adult hierarchy; (2) male-male competition in terms of pawing throughout the territory and possibly pasting at the den; (3) social communication and scent mixing between males mostly away from the communal den.

11.1 Introduction

All four extant members of the family Hyaenidae 'paste' (Kruuk 1972) scent from anal scent glands onto vegetation within their territory (Fox 1971; Kruuk 1972; Mills et al. 1980; Richardson 1991). The most social species in this family is the spotted hyena (*Crocuta crocuta*) which forms multi-female, multi-male fission-fusion social groups termed clans in which all natal females plus their offspring are socially dominant over all immigrant males (Kruuk 1972; Frank 1986; East and Hofer 2002). Both sexes produce a fatty secretion in anal glands that open just above the anus which is deposited by dragging vegetation, such as a grass stalk, along their ventral side, between the hind legs and across the protruded scent glands (Fig. 11.1, termed an anal bulge by Kruuk 1972). Pasting has been described in relation to territorial scent marking in the spotted hyena (Gorman and Mills 1984; Mills and Gorman 1987). Recent studies indicate that pasting is also an important component of communication between clan members (Hofer et al. 2001; Burgener et al. 2009).

Spotted hyenas often paste their scent on a stalk with the scent mark of another clan member. This behaviour is termed 'over-pasting' (Burgener et al. 2008, 2009) or overmarking (Theis et al. 2008), and requires the protrusion of the anal gland. When over-pasting there is the strong chance that scent from another clan member may enter the protruded gland. It has been suggested that over-pasting is a remote mechanism to mix anal gland scent between clan members and thus the development of clan-specific olfactory cues (Burgener et al. 2008). Pasting without gland protrusion and thus without the deposition of scent has been termed 'dry-pasting' (Burgener et al. 2008). When spotted hyenas either paste or dry-paste over an existing scent mark, scent from the mark will be transferred to their ventral side, which may be an important component of within-clan olfactory communication, particularly during greeting ceremonies (Burgener et al. 2009).

Pawing the ground is another type of scent marking about which little is known (Kruuk 1972). Adults of both sexes paw the ground which is thought to deposit scent from inter-digital glands on the front feet at pawing sites. Kruuk (1972) reported a male pawing when his attempt to copulate was frustrated by an uncooperative female and also during territory marking, at kills and in the presence of lions, and Szykman et al. (2007) described pawing as a component of male courtship

Fig. 11.1 A spotted hyena pasting scent from its protruded anal scent gland on a grass stalk plus a scent mark on a grass stalk

behaviour. Spotted hyenas can locate clan members they have lost visual contact with by following odour trails at ground level (pers. obs.), which suggests inter-digital gland secretions are individually distinct. In fact, inter-digital gland secretions may be more individually distinct than anal gland secretions because pawing sites appear to be used by individuals (pers. obs.) whereas over-pasting results in a degree of mixing of anal gland scent.

Spotted hyena clans are socially structured into two linear dominance hierarchies: one among adult natal females and the other among reproductively active, mostly immigrant males. The mechanisms by which animals acquire their social positions in these two hierarchies differ. An adult female's social position is mostly determined by the behavioural support provided by its mother and other close female relatives, and as a result adult daughters normally acquire a rank close to and below that held by their mother (Smale et al. 1993; Hofer and East 2003; East et al. 2009). The social position of an immigrant male is mostly determined by its length of tenure in a clan because these males generally observe strict queuing conventions (East and Hofer 2001). Not all reproductively active males are immigrants (Höner et al. 2007), and males that reproduce in their natal clan typically hold a high social position in the male hierarchy when they become reproductively active (East and Hofer 2001).

Adult females and immigrant males of high social status produce long-distance calls (whoops) more frequently than animals subordinate to them in their respective hierarchies (East and Hofer 1991). If the status of an animal is linked to the frequency at which it deposits anal gland scent, high-ranking adult females and immigrant males would be expected to scent mark (paste and paw) more frequently than animals below them in their respective hierarchies. Furthermore, high-ranking animals might also be expected to use competitive counter marking, as shown by urinary marking in rodents (Hurst and Beynon 2004). However, it has been shown that over-pasting behaviour at least by adult female spotted hyenas is not competitive counter marking because scent marks of high-ranking females are preferentially over-pasted by clan females subordinate to them (Burgener et al. 2009). Less is known about over-pasting behaviour by immigrant male spotted hyenas.

Previously we have shown that anal gland scent marks from individually known spotted hyenas in three large clans in the Serengeti National Park in northern Tanzania contain individually distinct chemical profiles, and that behavioural responses to scent marks indicate that spotted hyenas can determine the identity of the clan member which deposited them (Burgener et al. 2009). These findings support the idea that in the fission-fusion society of a spotted hyena clan, scent marks pasted by females near the clan communal den function as persistent and socially important 'calling-cards' that communicate continued clan membership to animals visiting the den (Hofer et al. 2001), even when a female is absent from the clan territory for several days on long-distance foraging trips (Hofer and East 1993a, b).

The chemical profiles of anal gland scent marks of both natal and immigrant members of a clan indicate the presence of group odour cues which probably result from the mixing of scent between clan members (Burgener et al. 2008, 2009). In order to contribute scent to the group odour badge, an individual needs to paste scent within the clan territory at locations where other clan members are likely to find it. Scent from other clan members also has to be acquired by finding and over-pasting scent marks deposited in the clan territory.

Here we test predictions on the functions of pasting, dry-pasting and pawing using the rate and the location of these behaviours. As the clan communal den is an important social centre, particularly for females, olfactory communication of continued clan membership and the mixing of anal gland secretions between females would be more efficiently achieved at the communal den than elsewhere in the territory. If the sole function of pasting at the communal den is to communicate continued membership to the clan, all adult clan members should paste at roughly the same rate. If pasting is also an essential part of the mechanism by which clan members integrate their individual anal gland odours into the group odour of their clan, young adult females and recent (and hence lower-ranking) immigrant males would be expected to have higher pasting and over-pasting rates than long-term, well-established clan members. If scent marking (pasting and pawing) is an important mechanism to display social status in adults of either sex, we would expect scent marking rates to increase as social status increases.

Males below mid-rank are often excluded from venturing close to the communal den area by females and high-ranking males (East and Hofer 1991, 2001; East et al. 1993). This may reduce their ability to deposit scent marks near the communal den and to acquire scent pasted close to the den (Hofer et al. 2001; Burgener et al. 2009). Since young adult females show strong mate choice preference for males that joined the clan after the females were born (Höner et al. 2007), it is likely that lower-ranking, more recent immigrant males would gain fitness benefits by depositing their scent marks in locations where juvenile females are unlikely to encounter them, away from the communal den. By concealing their olfactory identity from juvenile females, recent immigrants would increase the number of clan females that would view them as a potential mate throughout their reproductive careers in the clan. This predicts that lower-ranking males should avoid scent-marking in the vicinity of the communal den where den-bound juvenile females stay until the age of approximately 12 months (Hofer and East 1993b).

Finally, if the anointment of fur with the scent of another clan member by dry-pasting is important during social interactions such as greeting ceremonies, we expect dry-pasting rates to be roughly similar across adult members of the clan.

11.2 Methods

Data were obtained from three clans under long-term observation in the Serengeti National Park, Tanzania, between 2001 and 2011. All clan members were individually known and habituated to observers in a vehicle (Burgener et al. 2009). Details of the study site, clan structure and feeding ecology are described elsewhere (Hofer and East 1993a, b). Methods used to individually identify and sex clan members and assess the age of natal animals are described by East et al. (2003). Animals less than 2 years of age were categorised as juveniles and those of 2 or more years of age as adults (Hofer and East 2003). Natal males that displayed sexual interest in females and exhibited behaviour typical of immigrant males towards adult females were considered reproductively active natal males (East and Hofer 2001; Höner et al. 2007). The term 'reproductively active male' was applied to all immigrant males and reproductively active natal males (Höner et al. 2007). When we refer to males we refer to reproductively active males unless otherwise stated.

The dominance hierarchies of adult females and reproductively active males were determined using submissive responses in dyadic interactions as previously described (East and Hofer 2001; East et al. 2003). To compare animals with the same social position across clans, individuals were assigned a standardised rank evenly distributed between the highest social status (standardised rank +1) and lowest status (standardised rank −1) with median rank set to 0 (Hofer and East 2003).

Data were collected during observation periods at dusk (approximately between 1700 and 1930 h) and dawn (approximately between 0600 and 0930 h) either at communal den sites (i.e. within a radius of 50 m of the den area) or elsewhere in the territory. We categorised an animal that protruded its gland when exhibiting pasting behaviour as pasting (i.e. depositing scent) and when the gland was not protruded as dry-pasting (Burgener et al. 2008). When the position of the animal prevented observation of whether or not the gland was protruded the rate for that observation period was excluded from the analysis. A pawing event was one bout of ground scratching separated by at least 1 min from another bout. The rate at which individuals pasted or pawed the ground was calculated by dividing

Table 11.1 Mean rates of pasting, dry-pasting and pawing h^{-1} of reproductively active males with high (above median standardised rank) or low social status (below median standardised rank) when at the clan communal den or elsewhere in the clan territory.

	Communal den		Elsewhere	
Social status	Above median rank	Below median rank	Above median rank	Below median rank
N	16 males	6 males	24 males	22 males
Pasting rate	1.07±0.31	0.0	1.10±0.27	1.11±0.39
Dry-pasting rate	0.43±0.16	0.0	0.32±0.16	0.13±0.07
Pawing rate	1.09±0.30	0.0	1.10±0.30	0.19±0.11

the total number of marking behaviour events by the duration of the observation period for each individual. Rates of pasting, dry-pasting and pawing were obtained from observations of at least 30 min, from 53 males and 82 females (mean duration of observation period 73.4±3.8 min). The male sample included 52 immigrant males, 1 reproductively active natal male, 29 males above median status and 24 below median status. A mean rate per male of pasting, dry-pasting and pawing was calculated within each dominance category (above and below median standardised rank) at each location (at the clan communal den and elsewhere in the territory, Table 11.1). Mean observation period for males at the communal den was 78.8±9.2 min and elsewhere in the territory 87.7±6.2 min. Fourteen males contributed a mean rate both at the den and elsewhere. Females were between 1 and 14 years of age. For each female, a mean rate of pasting, dry-pasting and pawing was calculated for each year of life. Eight females contributed mean rates in two different age categories. Incidental observations of pawing behaviour were collected from 10 adult females and 62 reproductively active males.

Statistical analyses were conducted using SYSTAT 13.0 (Systat Software Inc., Richmond, USA). Rates are quoted as means±standard error of the mean. We used a General Linear Model (GLM) to simultaneously consider the effect of age and social status on female pasting and dry-pasting rates. Data were not normally distributed and thus were log-transformed.

11.3 Results

11.3.1 Female Rates of Pasting and Dry-Pasting

Only one of eight juvenile females (between 12 months and less than 24 months of age) pasted during the periods they were monitored (mean observation period 58.1±5.9 min), whereas five of these juvenile females dry-pasted. Nine of 14 young adult females (between 24 months and less than 36 months of age) pasted and eight of these 14 females dry-pasted during the period they were monitored (mean observation

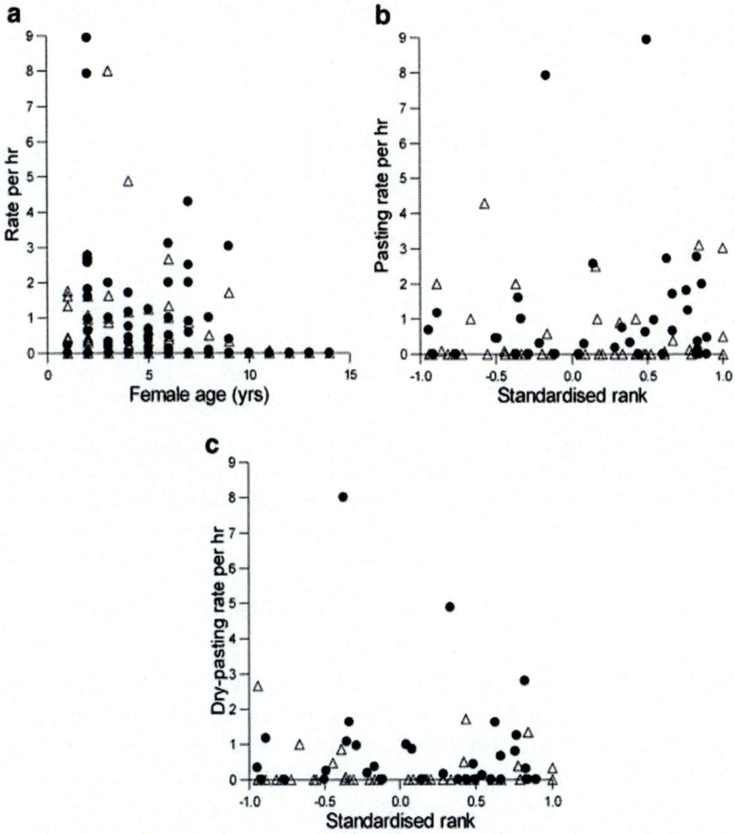

Fig. 11.2 Mean behaviour rates for female spotted hyenas. (**a**) Pasting (*filled circles*) and dry-pasting (*open triangles*) rates of females in relation to age at the clan communal den. (**b**) Mean pasting rate of adult females (≥2 years of age) in relation to their standardised rank (+1: top rank). (**c**) Mean dry-pasting rate of adult females in relation to their standardised rank. Younger adult females between 2 and 5 years of age (*open triangle*), older adult females between 6 and 14 years of age (*filled circle*)

period 60.4±8.2 min), a significant increase in the likelihood of pasting by animals during the transition from being a juvenile to becoming an adult (Chi-square test of independence, $\chi^2=5.506$, $df=1$, $p=0.019$).

The GLM investigating the effect of female age (Fig. 11.2a) and female standardised rank (Fig. 11.2b) on pasting rates ($F_{2\text{–}84}=6.104$, $p=0.003$) revealed a significant decline in pasting with increasing age ($t=-3.245$, $p=0.0017$) but no effect of standardised rank ($t=0.604$, $p=0.55$). Similarly, the effect of these factors on dry-pasting rates ($F_{2\text{–}77}=6.883$, $p=0.002$) also revealed a decline in dry-pasting with increasing age ($t=-3.710$, $p=0.0004$) and no significant effect of standardised rank ($t=-7{,}014$, $p=0.49$, Fig. 11.2c).

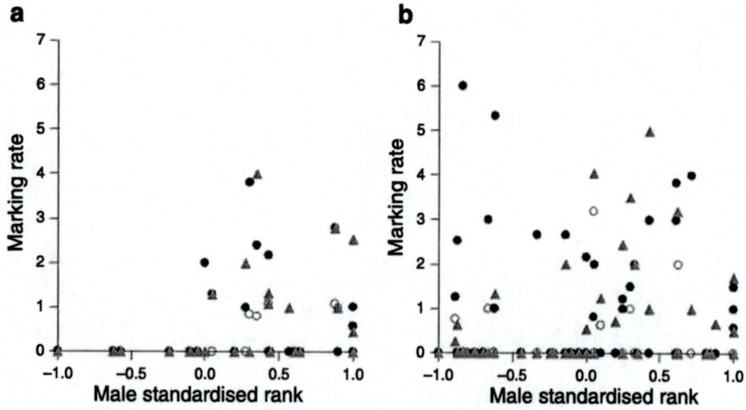

Fig. 11.3 Mean rates of pasting (*filled circles*), dry-pasting (*open circles*) and pawing (*grey triangles*) of reproductively active male spotted hyenas in relation to their standardised rank (+1: top social status). (**a**) Rates at the clan communal den. (**b**) Rates elsewhere in the clan territory

11.3.2 Rates of Marking by Reproductively Active Males

Males above median social status (above standardised rank 0, Fig. 11.3a) pawed the ground at a higher mean rate than males below median status when at the communal den (Mann–Whitney U test, $U=78.0$, $p=0.02$) and when elsewhere in the territory ($U=75.5$, $p=0.004$, Table 11.1). Males above median status also pasted at a higher rate than males below median status at the clan communal den ($U=75.0$, $p=0.025$, Table 11.1). There was no difference in either the rate of pasting or dry-pasting of males above and below median status (Fig. 11.3b) when they were elsewhere in the territory (pasting: $U=300.0$, $p=0.392$; dry-pasting: $U=283.0$, $p=0.516$, Table 11.1).

11.3.3 Pawing Behaviour

We observed 328 bouts of pawing, including 295 bouts (89.9%) by 65 immigrant males (mean 4.5 ± 0.6, range 1–19), 8 bouts (2.4%) by 4 reproductively active natal males (1.8 ± 0.8, range 1–4) and 25 bouts (7.7%) by 14 adult females (1.8 ± 0.4, range 1–4). All females observed pawing the ground held a social position above median status. In many cases (29.3%) pawing was accompanied by pasting, urination or defecation (or a combination of these behaviours). Pawing was observed in the vicinity of the den, at daytime resting sites, by the sides of tracks and roads (particularly at communal scent marking sites) and where clan members fed on large carcasses. A variety of factors were associated with pawing by males (Table 11.2). High-ranking males and females pawed the ground during periods of social instability when their status was challenged and when they challenged the established hierarchy.

Table 11.2 The context in which reproductively active male and adult female spotted hyenas exhibited pawing behaviour.

Reproductively active male	*N*=99 (%)
While shadowing or defending an adult female	62.2
At a communal scent marking area or latrine	15.5
During periods of social instability in the male hierarchy	8.7
Immediately before or after whooping	5.8
After a greeting ceremony	5.8
After aggression from an adult female	1.9
After baiting an adult female	0.9
Adult female	*N*=6
During periods of social instability in the female hierarchy	Yes
While defending food	Yes
After greeting a female	Yes

Pawing by females was too infrequently to warrant calculation of percentages

11.4 Discussion

Our results indicated that adult female pasting and dry-pasting rates were significantly influenced by age but not social status. Juvenile females at the communal den frequently dry-pasted which suggests that anointing their underbellies with scent from other clan members is socially important (Burgener et al. 2008; Theis et al. 2008). Only when females approached adulthood did they start pasting scent by protruding their anal gland. The relatively high pasting rate of younger adult females and the significant decline in adult female pasting rates with age (Fig. 11.2a) indicate a greater need for adult females when young to both advertise their clan membership and to ensure their scent is rapidly incorporated into the clan odour badge (Burgener et al. 2008). By placing scent marks at the communal den, young females probably increase the chance that their scent is taken-up by another clan member, particularly if they place their scent near a previous scent mark of a high-ranking female, which has a higher chance of being over-pasted by another female (Burgener et al. 2009). Adult female dry-pasting rates also significantly declined with increasing age, suggesting that it is more important for young females to anoint their fur with the scent of other clan members than older females. Together, these results indicate that age-related changes in pasting and dry-pasting rates reflect the greater importance of olfactory integration for younger than older adult females.

Adult female spotted hyenas can identify the member of their group which deposited a scent mark and are more likely to respond and respond faster to scent of high-ranking females (Burgener et al. 2009). As scent marks contain information about the identity of the animal which pasted and female clan members are drawn to paste and dry-paste over scent marks of high-ranking animals, it is perhaps not surprising that we found no correlation between female social status and pasting rate. Given that high-ranking females are more often present at the clan den than other females (East and Hofer 1991; Hofer and East 1993b, 2003) it is probable that more scent marks of high-ranking females accumulate around the communal den than those of other females, even if scent marking rates of females do not increase with social status. When animals dry-paste, they anoint their fur with the anal gland scent of other clan members, but do not deposit scent. As adult females preferentially dry-paste over scent marks of high-ranking females (Burgener et al. 2009), it is possible that younger adult females mostly carry the odours of high-ranking females on their under-belly fur.

The rate at which males pasted and dry-pasted at the communal den was influenced by their social status, as males above median status pasted and dry-pasted at the communal den at a higher rate than males below them in the hierarchy. Interestingly, away from the communal den area, male social status had no significant effect on the rate at which males pasted and dry-pasted. Unlike older females, males above median social status with longer tenures than males below median status maintained an overall higher rate of pasting and dry-pasting. Furthermore, unlike younger females, males below median status (with shorter tenures than males above median status) did not invest in pasting and dry-pasting at the communal den to advertise their clan membership or mix their anal gland scent with scent from natal female clan members. Instead, they focused their pasting and dry-pasting behaviour on areas of the territory away from the communal den (Table 11.1, Fig. 11.3a). In contrast, longer-tenured males pasted at similar rates at the communal den as elsewhere (Table 11.1). Together, these results suggest that anal gland scent marking by males is directed at other clan males, rather than clan females, and may also serve to signal to intruding males their 'ownership' of clan females.

Males above median status pawed the ground at a higher rate than males below median status both at the den and away from the den. This is expected, as most cases of pawing by males were in the context of shadowing and defending which are mating tactics rarely used by males below median status (East and Hofer 2001). Pawing in this context may be an olfactory display to a shadowed or defended female, as suggested by Szykman et al. (2007). Our results suggest that pawing in this context may also be used by higher ranking males as an olfactory signal to competing males that they are in attendance. If pawing is an important olfactory display to females, we would expect males to paw the ground more often when near females, regardless of whether the male was shadowing or defending the female. Adult females above median status pawed the ground, but did so rarely. When pawing the ground, or shortly before or afterwards, both males and females also urinate, defecate or paste, thus adding additional olfactory information to the area where they scratch the ground.

Why do males below median status paste and dry-paste at such a low rate at the clan communal den? In part this may be because they are often excluded from the communal den by males above median status. More important may be the strong effect of female mate choice on male behaviour. Immigrant males can have tenures that span many years and thus potentially they could sire offspring with their daughters (East and Hofer 2001). However, the vast majority of female spotted hyenas effectively avoid incest (East et al. 2003; Höner et al. 2007) through the use of simple mate-choice rules, which include a preference for males that become reproductively active in the clan after a female was born (Höner et al. 2007). One likely mechanism which would allow young adult females to discriminate against potential mates present in their clan when the females were cubs would be for them to prefer males they did not encounter when they were young. This would suggest that there may be a trade-off for young males between visiting the communal den to interact with adult females and to gather olfactory information (urine and anal gland scent) placed by females at the communal den, and maintaining a distance to prevent young females learning their odours and thus discriminating against them as potential mates, later in life. Thus, short-tenured males should generally avoid pasting close to the communal den and interacting with cubs, which is what they do, and they should paste away from the communal den, which they also do, and at similar rates to long-tenured males.

11.5 Conclusion

Our study provides further evidence of the importance of olfactory communication among members of fission-fusion spotted hyena clans.

Acknowledgements We thank the Tanzania Commission for Science and Technology for permission to conduct the study, the Tanzania Wildlife Research Institute and Tanzania National Parks for their cooperation and the Leibniz Institute for Zoo and Wildlife Research for funding. We thank Oliver Höner for helpful comments and A. Francis, T. Shabani, D. Thierer and K. Wilhelm for their assistance.

References

Burgener N, East ML, Hofer H, Dehnhard M (2008) Do spotted hyena scent marks code for clan membership? In: Hurst J, Beynon RJ, Roberts SC, Wyatt TD (eds) Chemical communication in vertebrates XI. Springer, Berlin, New York, pp 169–177

Burgener N, Dehnhard M, Hofer H, East ML (2009) Does anal gland scent signal identity in the spotted hyena? Anim Behav 77:707–715

East ML, Hofer H (1991) Loud calling in a female-dominated mammalian society: II. Behavioural contexts and functions of whooping of spotted hyaenas, *Crocuta crocuta*. Anim Behav 42:651–669

East ML, Hofer H (2001) Male spotted hyenas (*Crocuta crocuta*) queue for status in social groups dominated by females. Behav Ecol 12:558–568

East ML, Hofer H (2002) Conflict and co-operation in a female-dominated society: a re-assessment of the "hyper-aggressive" image of spotted hyenas (*Crocuta crocuta*). Adv Study Behav 31:1–30

East ML, Hofer H, Wickler W (1993) The erect 'penis' is a flag of submission in a female-dominated society: Greetings in Serengeti spotted hyenas. Behav Ecol Sociobiol 33:355–370

East ML, Burke T, Wilhelm K, Greig C, Hofer H (2003) Sexual conflict in spotted hyenas: male and female mating tactics and their reproductive outcome with respect to age, social status and tenure. Proc R Soc Lond B270:1247–1254

East ML, Höner OP, Wachter B, Wilhelm K, Burke T, Hofer H (2009) Maternal effects on offspring social status in spotted hyenas. Behav Ecol 20:478–483

Frank LG (1986) Social organization of the spotted hyena (*Crocuta crocuta*). II. Dominance and reproduction. Anim Behav 34:1510–1527

Fox MW (1971) Ontogeny of a social display in *Hyaena hyaena*: anal gland protrusion. J Mammal 52:467–469

Gorman ML, Mills MGL (1984) Scent marking strategies in hyaenas (Mammalia). J Zool Lond 202:535–547

Hofer H, East ML (1993a) The commuting system of Serengeti spotted hyaenas: how a predator copes with migratory prey. I. Social organization. Anim Behav 46:547–557

Hofer H, East ML (1993b) The commuting system of Serengeti spotted hyaenas: how a predator copes with migratory prey. II. Intrusion pressure and commuters' space use. Anim Behav 46:575–589

Hofer H, East ML (2003) Behavioral processes and costs of co-existence in female spotted hyenas: a life history perspective. Evol Ecol 17:315–331

Hofer H, East M, Sämmang I, Dehnhard M (2001) Analysis of volatile compounds in scent-marks of spotted hyenas (*Crocuta crocuta*) and their possible function in olfactory communication. In: Marchelewska-Koj A, Lepri JJ, Müller-Schwarze D (eds) Chemical signals in vertebrates IX. Kluwer Academic/Plenum, New York, pp 141–148

Höner OP, Wachter B, East ML, Streich WJ, Wilhelm K, Burke T, Hofer H (2007) Female mate choice drives the evolution of male-biased dispersal in a social mammal. Nature 448:798–801

Hurst JL, Beynon RJ (2004) Scent wars: the chemobiology of competitive signaling in mice. Bioessays 26:1288–1298

Kruuk H (1972) The spotted hyena. Chicago University Press, Chicago

Mills MGL, Gorman ML (1987) The scent marking behaviour of the spotted hyaena *Crocuta crocuta* in the southern Kalahari. J Zool Lond 212:483–497

Mills MGL, Gorman ML, Mills MEJ (1980) The scent marking behaviour of the brown hyaena, *Hyaena brunnea*. S Afr J Zool 15:240–248

Richardson PRK (1991) Territorial significance of scent marking during the non-mating season in the aardwolf *Proteles cristatus* (Carnivora: Protelidae). Ethology 87:9–27

Smale L, Frank LG, Holekamp KE (1993) Ontogeny of dominance in free-living spotted hyaenas: juvenile rank relations with adult females and immigrant males. Anim Behav 46:467–477

Szykman M, Van Horn RC, Engh AL, Boydston EE, Holekamp KE (2007) Courtship and mating in free living spotted hyenas. Behaviour 144:815–846

Theis KR, Heckla AL, Verge JR, Holekamp KE (2008) The ontogeny of pasting behavior in free-living spotted hyenas, *Crocuta crocuta*. In: Hurst JL, Beynon RJ, Roberts CS, Wyatt TD (eds) Chemical signals in vertebrates XI. Springer, Berlin, New York, pp 179–187

Chapter 12
Is the Perception of Dietary Odour Cues Linked to Sexual Selection in Humans?

Jitka Fialová, S. Craig Roberts, and Jan Havlíček

Abstract A growing body of evidence suggests that various animal species use diet-associated cues to assess the condition of a potential mate. Such cues are not restricted to the visual channel (e.g. brightness of colouration) but could also be perceived by olfaction. Here, we first summarise current knowledge about the role of diet in vertebrate semiochemistry. Consequently, we employ an adaptationist approach to communication to explore whether cues to diet are or were subject to sexual selection in humans. These thoughts are speculative as research in this area is rather fragmental.

12.1 Introduction

The majority of research on human communication has focused on the vocal and visual domains. Comparatively little research has considered chemosignalling in humans. However, an accumulating body of evidence indicates that human odour cues provide information about the producer's sex, genetic compatibility, reproductive and emotional state, health and diet (for recent reviews see Grammer et al. 2005; Havlicek and Roberts 2009; Lenochova and Havlicek 2008). These body odour cues appear to be used for communication and formation of attributions, although sometimes on a subconscious level and/or in form of hedonic assessment.

J. Fialová (✉) • J. Havlíček
Department of Anthropology, Charles University,
Prague, Czech Republic
e-mail: jitka.fialova@fhs.cuni.cz; jan.havlicek@fhs.cuni.cz

S.C. Roberts
Department of Psychology, School of Natural Sciences, University of Stirling,
Stirling, Scotland, UK
e-mail: craig.roberts@stir.ac.uk

M.L. East and M. Dehnhard (eds.), *Chemical Signals in Vertebrates 12*,
DOI 10.1007/978-1-4614-5927-9_12, © Springer Science+Business Media New York 2013

12.2 Adaptationist Approach to Communication

Theoretical approaches to communication vary widely, not only across different scholarly fields, but also within those taking an evolutionary perspective. Here we employ the adaptationist approach elaborated by Maynard Smith and Harper (1995) as it specifically focuses on evolution of biological signalling. According to them, a signal is any act or structure which alters the behaviour of other organisms, which evolved because of that effect, and which is effective because the receiver's response has also evolved. Such behaviours or traits are thought to be often formed through the process of ritualisation, where a behaviour originally had a different function, which has been lost, but receivers started to use it as a cue and under selection the cue evolved into a signal (Scott-Phillips 2008). In this view, a signal is differentiated from a cue: a cue is anything which affects another organism's behaviour but did not develop for that effect (Maynard Smith and Harper 1995). Under this definition there is transfer of one of two types of semantic information (information about an event or a state). The first can carry information about the sender and is often sexually selected ('self-reporting signals', which could, for instance, be represented by displays of physical quality or status, often to the opposite sex), while the second carries information about the environment ('other-reporting signals', e.g. alarm calls and the dances of honey bees) (Maynard Smith and Harper 1995).

If the signal is more costly to produce than the minimum required for effective information transfer, we may refer to cost-added signals. The excess cost evolved because receivers have been more likely to respond appropriately to costly signals. In models of sexual selection it is assumed that if signals reveal the quality of an individual's genes or potential parental care, there is a correlation between signal cost and the quality of the signaller. In mate choice, for example, the correlation exists because high-quality males can afford to produce more costly signals and females benefit from choosing higher quality males. Such signals are likely to reflect a signaller's overall condition, rather than be related to a specific disease or nutritional state (Wyatt 2003). Furthermore, such costly signals are considered to be 'honest' reflections of a signaller's quality, because poor quality cheaters cannot fake them (Maynard Smith and Harper 1995; Zahavi et al. 1999).

12.3 Nutrition-Related Signals

Numerous studies have shown that secondary sexual displays (e.g. brightness of colour) 'honestly' reveal quality via a link between signal characteristics and foraging success or diet (Maynard Smith and Harper 1995; Kirkpatrick 1996). An example is the plumage colouration of male house finches (*Carpodacus mexicanus*), which varies from pale yellow to bright red. Colourful plumage is a result of carotenoid pigmentation. The distribution of carotenoid colouration is inherited, but the intensity of pigmentation is a result of variation in the type and quantity of carotenoids ingested. This is because carotenoids cannot be synthesised *de novo* and must be derived from food. Carotenoid pigmentation therefore represents a condition-dependent trait, where the condition reflects access to carotenoids. In experiments, female house finches prefer males with the most colourful plumage who have consumed food richer in carotenoids (Hill 1990). Pairing success has also been positively linked to carotenoid-based colouration in the American goldfinch (*Carduelis tristis*) (MacDougall and Montgomerie 2003) and golden-collared manakin (*Manacus vitellinus*) (Stein and Uy 2006). Similarly, male guppies (*Poecilia reticulata*) raised on a high-carotenoid diet were more attractive to females than males raised on a carotenoid-free diet (Grether 2000). As in bird and fish species, carotenoid colouration is a perceivable cue to human health, and may also be relevant to mate choice: people with higher dietary intake of carotenoids in fruit and vegetables have increased skin yellowness which improves their assessed attractiveness (Stephen et al. 2009). Moreover, when participants were asked to maximise apparent health of faces in photographs, they preferred increased skin carotenoid colouration (Stephen et al. 2011).

As with visually perceived traits, females also choose between males using odour. Ferkin et al. (1997) examined how varying amounts of protein in the diet (9, 15 or 25%) affect the attractiveness of odour in meadow voles (*Microtus pennsylvanicus*). Both males and females preferred odours from individuals on high-protein diets: they spent more time investigating odours of opposite-sex conspecifics on a 25% protein diet compared to those on a 9% protein diet (Ferkin et al. 1997). Furthermore, when female meadow voles were not fed for 24 h, their odour was less attractive to males compared with odour of females fed *ad libitum*, but this difference disappeared in 48 h after the food-deprived animals were fed (Pierce and Ferkin 2005). Similarly, high-quality food positively influences attractiveness of urine odour in guinea pigs (*Cavia porcellus*) (Beauchamp 1976). In the red-backed salamander (*Plethodon cinereus*), males extract information concerning territorial ownership from chemical signals in faecal pellets. Walls et al. (1989) found that they could distinguish between faecal pellets that were derived from high-quality or low-quality diets. This ability could be advantageous for females because they can quickly assess territory quality by examination of male faecal pellets alone (Walls et al. 1989). Finally, the salience of diet in regulating odour-mediated social interaction can also be illustrated by how it modulates genetically mediated odour cues. Urinary odours of mice from the same strain but fed different diets were more readily discriminated than faeces from two different mouse strains that differed at one MHC locus, but were fed on the same diet (Brown et al. 1996). Also spiny mouse pups (*Acomys cahirinus*) prefer the odour of a heterospecific females fed on the maternal diet compared to a conspecific female on a different diet (Porter and Doane 1977; Kwak et al. 2008).

12.4 Effect of Diet on Human Body Odour

Diet is also a source of environmental variability in human body odour. This statement is supported by evidence from studies of monozygotic twins. For example, trained dogs could discriminate between the odours of monozygotic twin infants fed on different diets and between the odours of twins who differed in genetic relatedness, but were unable to discriminate between the odours of monozygotic twins on the same diet (Hepper 1988). People asked to distinguish between hand odours of two unrelated women on the same diet, monozygotic twins on the same diet, two sisters on the same diet, and between monozygotic twins on a different diet were most successful in distinguishing odours from the two unrelated women and were also able to discriminate between the odours of twins on different diet, but could not discriminate hand odours of twins on the same diet (Wallace 1977).

Body odours of new born babies are used by obstetricians as diagnostic markers for specific metabolic disorders, such as maple syrup syndrome (Monastiri et al. 1997). Unusual odours may also be caused by maternal diet. For example, one misdiagnosis of maple syrup syndrome was subsequently explained by the mother having eaten fenugreek seeds immediately prior to delivery (Yalcin et al. 1999). In other cases, babies that smelled of cumin were attributed to maternal diet before delivery (Hauser et al. 1985).

12.5 Experimental Evidence

The above-reviewed evidence suggests that diet profoundly impacts body odour and could be a cue to the quality of ingested food and perhaps to nutritional status. However, we still know relatively little about how particular dietary components shape human body odour. A first study investigated the effect of red meat consumption on male odour. A sample of 17 adult males was placed on a prescribed "meat" or "non-meat" diet for 2 weeks. During the last 24 h, they wore cotton pads in their armpits and, immediately after these were collected, adult females rated the odours for pleasantness, attractiveness, intensity and masculinity. The same procedure was repeated a month later with the same males in each experimental group placed on the opposite diet. Samples from men on the meat diet were rated as less pleasant, less attractive and more intense than samples from the same individuals on the non-meat diet. This suggests that red meat consumption had a negative effect on male body odour attractiveness (Havlicek and Lenochova 2006). Although it is often assumed that the perception of odours associated with a high meat diet would be more attractive for women as it might point to good hunting skills (in an analogous manner to the protein-rich diet of meadow voles or intense colouration of house finches as described above), the results did not support this assumption. Perhaps the consumption of relatively large amounts of meat is instead suggestive of some sort of metabolic disorder.

Recently, we conducted similarly designed experiments to test whether consumption of garlic has an effect on human axillary odour as it has previously been reported to influence body odour (Borrelli et al. 2007; Amagase et al. 2001) and breast milk and amniotic fluid (Mennella and Beauchamp 1991; Mennella et al. 1995). The consumption of garlic is often prohibited in studies employing body odour sampling (Kuukasjärvi et al. 2004; Lenochova et al. 2009; Havlíček et al. 2005, 2006; Roberts et al. 2011).

To test the effect of garlic on body odour we performed two experiments and used a balanced within-subject experimental design (i.e. we compared hedonic ratings of the samples taken from the same odour donors in the garlic and control conditions). In the first experiment, 10 male odour donors participated (mean age 25.2; range 18–31 years), with 16 male donors in the second (mean age 25.1; range 20–34 years). All received instructions to refrain from eating spicy food, from using soap and perfumes, and to avoid demanding physical activities both on the day before sampling and during the sampling day.

In both experiments, males were randomly divided into two groups (A, B). Males in group A were given bread and cream cheese spread blended with garlic (experiment 1: 6 g and experiment 2: 12 g of garlic, this roughly corresponds to two and four cloves of fresh garlic, respectively), while men in group B were given bread and plain cream cheese spread. The procedure was reversed 1 week later. Cotton pads worn in the armpits for 12 h following ingestion served as stimuli. To avoid odour contamination from external odours, they wore new white 100% cotton T-shirts as a first layer of clothing.

Stimuli were rated by 14 female raters (mean age 24.6; range 20–35 years) in the first experiment and by 40 female raters (mean age 22.5; range 19–32 years) in the second experiment. Raters assessed odour pleasantness, attractiveness, masculinity and intensity on 7-point scales anchored by verbal descriptions (e.g. 1—very unpleasant and 7—very pleasant). Female participants were all using hormonal contraception to avoid fluctuations in olfactory perception across the menstrual cycle (Doty 1981; Navarrete-Palacios et al. 2003). The samples from each donor were presented in pairs and raters were instructed not to use same value within each pair. The ratings were written down immediately after sniffing each stimulus, but the time spent sniffing was not restricted. Each rater assessed both sets (i.e. 20 stimuli in the first experiment, 32 stimuli in the second experiment).

The statistical package STATISTICA 7.0 was used for all testing. As our design was within-subjects, paired *t*-tests were used. The results of the first experiment showed that assessments of attractiveness and pleasantness of the samples taken in the experimental (garlic) condition were higher compared to the control (non-garlic) condition; however these differences were not statistically significant. In the second experiment, the odour of donors in the experimental (garlic) condition was judged as significantly more attractive, more pleasant and less intense than in the control (non-garlic) condition (Fig. 12.1). These preliminary results unexpectedly suggest that garlic consumption positively influences body odour.

There could be several explanations of this effect. Garlic could influence armpit odour indirectly through the well-documented antioxidant properties of garlic: sulphur-containing compounds help protect blood cells and vessels against oxidative stress (Borek 2001) which is caused by highly reactive oxygen molecules, possibly by decreasing production of these molecules and/or their metabolites by axillary glands. Garlic enhances levels of two antioxidant enzymes, catalase and glutathione peroxidase, key enzymes in the cell´s defence mechanism against oxidative stress (Kemper 2000). Garlic has also bactericidal properties against various bacteria that contribute to the production of armpit odour. If the quantity of bacteria

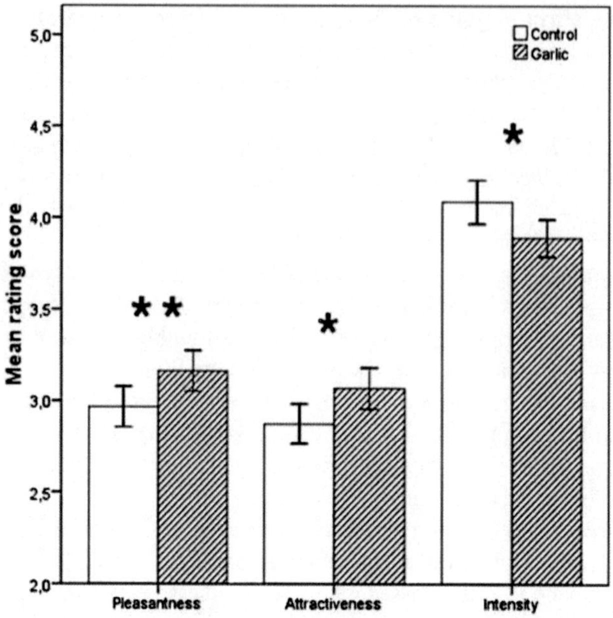

Fig. 12.1 Mean ratings (±SE) of 16 pairs of axillary odours on pleasantness, attractiveness and intensity in the experimental (garlic) condition (*shaded bars*) and control (non garlic) condition (*white bars*) by 40 women. Ratings were on 7-point scale (e.g. 1—very unpleasant and 7—very pleasant). *Asterisks* indicate level of significance in paired *t*-tests. $*p<0.05$ level; $**p<0.01$ level

is reduced, the intensity of armpit odour could be reduced; the intensity of odour is often negatively correlated with pleasantness and attractiveness.

The results on perception of axillary odour after garlic consumption are consistent with the above-reviewed findings on odour-based mate preferences affected by diet in other species. The health benefits of garlic consumption are well established and far-reaching. They include antidiabetic, antithrombotic, antiplatelet, antihypertensive, antioxidative and neuroprotective effects. It is thus plausible that human odour preferences were shaped by sexual selection to be sensitive to odour cues of current metabolic functioning in potential mates. These cues are affected by the amount and quality of food such as garlic digested by the producer. At this point, our knowledge of the mechanisms underlying how garlic consumption shapes axillary odour is highly limited. Therefore, we find it premature to refer to garlic-related odours as quality signals and we conservatively refer to them as cues. To label these as signals, it should be first demonstrated that a particular mechanism has been selected for. Such criteria would, for instance, include condition-dependent production of specific chemical compounds or mixtures of compounds. Furthermore, to avoid deception, production of such compounds should be restricted only to individuals on a high-quality diet.

Finally, all empirical studies testing the effect of diet on human body odour have so far focused on specific dietary compounds. However, perceivers might be more sensitive to cues related to general nutritional status, as has been neatly shown in voles (Ferkin et al. 1997; Pierce and Ferkin 2005). In an analogous study in humans, we might expect people who are starving or on a low nutrition diet to smell less pleasant than those whose food consumption is roughly equal to their caloric expenditure. If this were the case, the mechanism would probably involve catabolic processes which include production of, for instance, acetone, which one can detect in the breath odour of starving individuals or those suffering from diabetes (Laffel 1999).

12.6 Conclusions

We aimed to assess the role of diet in the production of chemical cues involved in communication processes. In particular, we asked whether diet-related odour cues are used in assessments of condition and quality of potential mates. This idea is exemplified in the case of eating garlic, which is known for its antioxidative properties and consumption of which, we have found, increases the pleasantness of axillary body odour. This suggests that perception of odour cues linked to diet and/or nutrition may be subject to sexual selection in humans, as in animals. Having said this, the number of studies focusing on this issue remains very limited and we strongly advocate more empirical work in what we see as a fascinating area of semiochemistry.

Acknowledgements We would like to thank Jindra Havlickova for many helpful comments and proof-reading. JH is supported by Czech Science Foundation grant (GACR P407/10/1303) and Charles University Research Centre (UNCE 204004).

References

Amagase H, Petesch BL, Matsuura H, Kasuga S, Itakura Y (2001) Intake of garlic and its bioactive components. J Nutr 131:955S–962S

Beauchamp GK (1976) Diet influences attractiveness of urine in guinea-pigs. Nature 263:587–588

Borek C (2001) Antioxidant health effects of aged garlic extract. J Nutr 131:1010S–1015S

Borrelli F, Capasso R, Izzo AA (2007) Garlic (*Allium sativum* L.): adverse effects and drug interactions in humans. Mol Nutr Food Res 51:1386–1397

Brown RE, Schellinck HM, West AM (1996) The influence of dietary and genetic cues on the ability of rats to discriminate between the urinary odors of MHC-congenic mice. Physiol Behav 60:365–372

Doty RL (1981) Olfactory communication in humans. Chem Senses 6:351–376

Ferkin MH, Sorokin ES, Johnston RE, Lee CJ (1997) Attractiveness of scents varies with protein content of the diet in meadow voles. Anim Behav 53:133–141

Grammer K, Fink B, Neave N (2005) Human pheromones and sexual attraction. Eur J Obstet Gyn R B 118:135–142

Grether GF (2000) Carotenoid limitation and mate preference evolution: a test of the indicator hypothesis in guppies (*Poecilia reticulata*). Evolution 54:1712–1724

Hauser GJ, Chitayat D, Berns L, Braver D, Muhlbauer B (1985) Peculiar odors in newborns and maternal prenatal ingestion of spicy food. Eur J Pediatr 144:403

Havlicek J, Lenochova P (2006) The effect of meat consumption on body odor attractiveness. Chem Senses 31:747–752

Havlicek J, Roberts SC (2009) MHC-correlated mate choice in humans: a review. Psychoneuroendocrino 34:497–512

Havlicek J, Roberts SC, Flegr J (2005) Women's preference for dominant male odor: effects of menstrual cycle and relationship status. Biol Lett 1:256–259

Havlíček J, Dvořáková R, Bartoš L, Flegr J (2006) Non-advertized does not mean concealed: body odor changes across the human menstrual cycle. Ethology 112:81–90

Hepper PG (1988) The discrimination of human odor by the dog. Perception 17:549–554

Hill GE (1990) Female house finches prefer colourful males—sexual selectioin for a condition-dependent trait. Anim Behav 40:563–572

Kemper KJ (2000) Garlic (*Allium sativum*). The Longwood Herbal Task Force. (http://www.mcp.edu/herbal/default.htm)

Kirkpatrick M (1996) Good genes and direct selection in evolution of mating preferences. Evolution 50:2125–2140

Kuukasjärvi S, Eriksson CJP, Koskela E, Mappes T, Nissinen K, Rantala MJ (2004) Attractiveness of women's body odors over the menstrual cycle: the role of oral contraceptives and receiver sex. Behav Ecol 15:579–584

Kwak J, Willse A, Matsumura K, Opiekun MC, Yi WG, Preti G, Yamazaki K, Beauchamp GK (2008) Genetically-based olfactory signatures persist despite dietary variation. PLoS One 3:e3591

Laffel L (1999) Ketone bodies: a review of physiology, pathophysiology and application of monitoring to diabetes. Diabetes Metab Res 15:412–426

Lenochova P, Havlicek J (2008) Human body odor individuality. In: Hurst JL, Beynon RJ, Roberts SC, Wyatt TD (eds) Chemical signals in vertebrates 11. Springer, New York, pp 189–198

Lenochova P, Roberts SC, Havlicek J (2009) Methods of human body odor sampling: the effect of freezing. Chem Senses 34:127–138

MacDougall AK, Montgomerie R (2003) Assortative mating by carotenoid-based plumage colour: a quality indicator in American goldfinches, *Carduelis tristis*. Naturwissenschaften 90:464–467

Maynard Smith J, Harper DGC (1995) Animal signals: models and terminology. J Theor Biol 177:305–311

Mennella JA, Beauchamp GK (1991) Maternal diet alters the sensory qualities of human milk and the nurslings behavior. Pediatrics 88:737–744

Mennella JA, Johnson A, Staley C, Beauchamp GK (1995) Garlic ingestion by pregnant women alters the odor of amniotic fluid. Chem Senses 20:207–209

Monastiri K, Limame K, Kaabachi N, Kharrat H, Bousnina S, Pousse H, Radhouane M, Gueddiche MN, Snoussi N (1997) Fenugreek odour in maple syrup urine disease. J Inherit Metab Dis 20:614–615

Navarrete-Palacios E, Hudsonm R, Reyes-Guerrero G, Guevara-Guzman R (2003) Lower olfactory threshold during the ovulatory phase of the menstrual cycle. Biol Psychol 63:269–279

Pierce AA, Ferkin MH (2005) Re-feeding and the restoration of odor attractivity, odor preference, and sexual receptivity in food-deprived female meadow voles. Physiol Behav 84:553–561

Porter RH, Doane HM (1977) Dietary-dependent cross-species similarities in maternal chemical cues. Physiol Behav 19:129–131

Roberts SC, Kralevich A, Ferdenzi C, Saxton TK, Jones BC, DeBruine LM, Little AC, Havlicek J (2011) Body odor quality predicts behavioral attractiveness in humans. Arch Sex Behav 40:1111–1117

Scott-Phillips TC (2008) Defining biological communication. J Evol Biol 21:387–395

Stein AC, Uy JAC (2006) Plumage brightness predicts male mating success in the lekking golden-collared manakin, *Manacus vitellinus*. Behav Ecol 17:41–47

Stephen ID, Smith MJL, Stirrat MR, Perrett DI (2009) Facial skin coloration affects perceived health of human faces. Int J Primatol 30:845–857

Stephen ID, Coetzee V, Perrett DI (2011) Carotenoid and melanin pigment coloration affect perceived human health. Evol Hum Behav 32:216–227

Wallace P (1977) Individual discrimination of humans by odor. Physiol Behav 19:577–579

Walls SC, Mathis A, Jaeger RG, Gergits WF (1989) Male salamanders with high-quality diets have feces attractive to females. Anim Behav 38:546–548

Wyatt TD (2003) Pheromones and animal behaviour—communication by smell and taste. Cambridge University Press, Cambridge

Yalcin SS, Tekinalp G, Ozalp I (1999) Peculiar odor of traditional food and maple syrup urine disease. Pediatr Int 41:108–109

Zahavi A, Balaban A, Ely MP (1999) The handicap principle: a missing piece of Darwin's puzzle. Oxford University Press, USA

Chapter 13
Hemiterpenoids and Pyrazines in the Odoriferous Urine of the Maned Wolf (*Chrysocyon brachyurus*)

Thomas E. Goodwin, Nucharin Songsasen, Laura J. Broederdorf, Blake A. Burkert, C. Joi Chen, Stephen R. Jackson, K. Bryan Keplinger, Margaret E. Rountree, Zach J. Waldrip, Margaret E. Weddell, Linda P. Desrochers, William K. Baker Jr., and Johannes P.F.G. Helsper

Abstract Maned wolves (*Chrysocyon brachyurus*) are endemic to South America, monogamous, solitary, and threatened in the wild. Maned wolf urine has a pungent and powerful odour and is used for scent marking. There is evidence to suggest that the presence of a male may be required to initiate oestrus and/or ovulation, thus implying the presence of a primer pheromone. We have employed solid phase dynamic extraction (SPDE)/GC-MS to identify a number of volatile compounds in maned wolf urine. These include sulphur-containing hemiterpenoids which are predominantly responsible for the distinctive urinary odour, hemiterpenoid alcohols which are known bark beetle pheromones, and a variety of pyrazines, some of which are known to be insect pheromones. Hemiterpenoids are most likely biosynthesised via a shunt of the mevalonate pathway, while pyrazines are thought to be products of amino acid metabolism. The abundance of some of these compounds increases as the urine ages, perhaps providing a timed release of putative chemical signals.

T.E. Goodwin (✉) • L.J. Broederdorf • B.A. Burkert • C.J. Chen • S.R. Jackson • K.B. Keplinger • M.E. Rountree • Z.J. Waldrip • M.E. Weddell • L.P. Desrochers
Department of Chemistry, Hendrix College, Conway, AR 72032, USA
e-mail: goodwin@hendrix.edu

N. Songsasen
Department of Reproductive Sciences, Center for Species Survival, Smithsonian Conservation Biology Institute, Front Royal, VA 22630, USA

W.K. Baker Jr.
Little Rock Zoo, Little Rock, AR 72205, USA

Panthera Research, Rockwall, TX 75032, USA

J.P.F.G. Helsper
Business Unit Bioscience, Plant Research International, Wageningen, The Netherlands

M.L. East and M. Dehnhard (eds.), *Chemical Signals in Vertebrates 12*, DOI 10.1007/978-1-4614-5927-9_13,

13.1 Introduction

13.1.1 General Information About Maned Wolves

The maned wolf (*Chrysocyon brachyurus*) has an evolutionary history that dates back six million years (Dietz 1985), is endemic to parts of South America (Argentina, Bolivia, Brazil, and Paraguay), and is the largest canid species on the continent. The maned wolf is a monotypic canid (i.e. the only member of its genus); the scientific name is derived from Greek and means "short-tailed gold dog". The species is unique in morphology, social behaviour, and physiology compared to other wild canids. Maned wolves are monogamous and mates interact primarily during the breeding season (Rodden et al. 2004; Jácomo et al. 2009). Maned wolves are currently on the IUCN Red List as a "Near Threatened" species due to declining numbers caused primarily by habitat loss and human encroachment on their native territory.

13.1.2 Maned Wolf Reproduction

Maned wolves are facultatively monogamous; although they share the same territory, pairs do not live together. The females are monoestrous (one ovulation per year), and although the breeding season may span several months (April to June in South America, and October to February in North America), oestrus (and receptivity) is short, normally between 1 and 10 days (Songsasen et al. 2006). Since maned wolves are generally solitary with home ranges mainly inhabited only by the breeding pair that interact primarily during the short period of female receptivity, territorial urinary scent marking is important for communication (Bestelmeyer 2000; Jácomo et al. 2009).

Recent research indicates that the reproductive physiology of this species is distinct from that of other canids. Most canid species are spontaneous ovulators; females ovulate irrespective of the presence of a male. However, there is evidence to suggest that the presence of a male may be required to initiate oestrus and/or ovulation in the maned wolf (Songsasen et al. 2006). Specifically, the progesterone concentration of singleton females remains at the baseline level throughout the breeding season, suggesting that ovulation does not occur in solitary individuals. Furthermore, when administering maned wolves with a gonadotropin releasing hormone agonist, ovulation occurs in unpaired females housed next to a male but not in singleton individuals housed far away from the opposite sex (Songsasen, unpublished). Therefore, it is likely that a male semiochemical stimulates a luteinising hormone surge and ovulation in the female. Induced ovulation ensures that pairs have the maximum chance of successful reproduction given their solitary behaviour and short reproductive receptivity period (1–10 days/year).

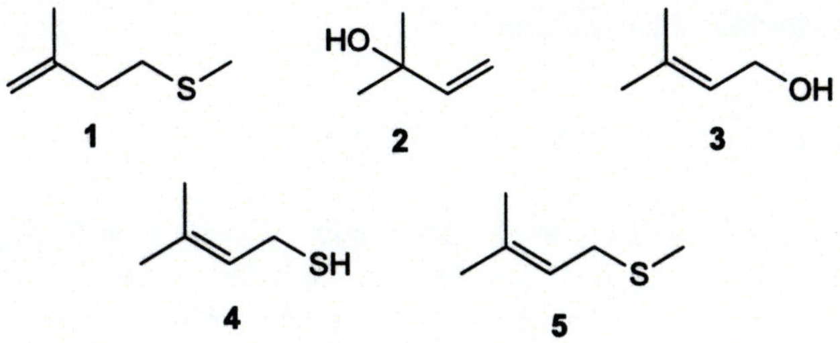

Fig. 13.1 Hemiterpenoids in maned wolf urine

It has been shown that carnivores, including canids, use various scent marking methods (e.g. urination and defecation) to communicate with conspecifics about their territory, hierarchical status, and reproductive state (Albone 1984; Asa et al. 1984; Rodden et al. 1996; Kleiman 2011). We hypothesise that the maned wolf ovulation process is regulated by volatile chemicals excreted in male urine or faeces during scent marking. Because urine scent marking reaches maximum frequencies for both sexes during the proestrous period (Dietz 1984), we initiated a study to identify the putative semiochemicals in maned wolf urine.

13.1.3 Maned Wolf Scent

The most noticeable feature of the maned wolf, apart from its fox-like colour and black mane, is its odoriferous urine. Volatile sulphur-containing compounds are reasonably suspected as contributors to the distinctive urinary odour because: (1) thiols and thioethers are often malodorous; (2) 3-methyl-1-methylthiobut-3-ene (**1**; Fig. 13.1) has been identified in the urine of other canids, namely the red fox, coyote, and gray wolf (Wilson et al. 1978; Raymer et al. 1984, 1986; Schultz et al. 1988); and (3) cystinuria, common in maned wolves (Bovée et al. 1981), is a genetic disorder in which intestinal and renal cystine transport is impaired (Mattoo and Goldfarb 2008), thus cysteine is an abundant sulphur source for biosynthesis. The only reported analysis of volatile organic compounds in maned wolf urine (using solid phase microextraction (SPME)/GC-MS) identified no sulphur-containing compounds (Childs-Sanford 2005). We have previously used automated solid phase dynamic extraction (SPDE)/GC-MS to identify volatile organic compounds in elephant secretions and excretions (Goodwin et al. 2006, 2007, 2012; Goodwin and Schulte 2009) and have now employed this methodology to investigate maned wolf urine. We will report herein the identification of hemiterpenoids, including malodorous sulphur-containing compounds, as well as pungent pyrazines. Several of these compounds are known to be semiochemicals in other species (*vide infra*) and may contribute to scent signalling in maned wolves.

13.2 Materials and Methods

13.2.1 Sample Collection and Preparation

We analysed urine samples from the Smithsonian Conservation Biology Institute and from the Little Rock Zoo. Samples were collected in clean glass containers when maned wolves urinated on clean stainless steel trays in their enclosure, and the urine was kept frozen (–20 or –70°C) until analysis of the volatile organic compounds using automated SPDE/GC-MS. A 1.0–5.0 mL aliquot of thawed urine and a small Teflon®-coated stir bar were sealed in a 20 mL screw-top vial with a threaded, metallic septum cap (silicone/PTFE layered septum; www.autosamplerguys.com). For some samples, 0.3–1.5 g of NaCl (depending on the urine sample size) was added to force more volatiles into the headspace. For experiments designed to assess effects of urine aging and possible microbe-mediated metabolism, glassware and plasticware were autoclaved at 18 psi and 121°C for 20 min. On the first day of thawing, an aliquot of urine was centrifuged at 10–15,000 × *g* for 2 min. The supernatant was passed through a 0.22 μm sterilising filter to remove insolubles, including microbes.

13.2.2 Urine Analysis

Multiple samples were programmed to run automatically using the Combi PAL robot and associated SPDE hardware and software (www.chromtech.de). The SPDE needle was internally coated with activated charcoal (Carboxen®)-polydimethylsiloxane (AC-PDMS). After incubating the stirred sample at 37°C for 15 min, the headspace was sampled for 13 min (200 up-and-down 1 mL strokes of the syringe). Desorption of extracted analytes was at 250°C in the GC inlet. GC/MS analyses were conducted using an Agilent 6890N GC and 5973N Mass Selective Detector. The capillary GC column was an Equity 1 (bonded, polydimethylsiloxane, 60 m × 0.32 mm ID, 1 μm film thickness, Supelco cat. No. 28058U). The GC oven was programmed to hold for 2 min at 35°C, followed by ramping to 180°C at 3.75°C/min where it was held for 5 min before ramping at 20°C/min to a final temperature of 250°C where it was held for 10 min. The GC was run in the splitless mode. The mass spectrometer was programmed to collect 3.09 scans/s over the mass range of 30–500 amu. Unless stated otherwise, compound identities were verified by comparisons of retention times and mass spectra to authentic standards and were confirmed by comparison to the NIST02 mass spectral library (see annotations on the Tables). Data reported herein indicate merely the presence or absence of specific compounds in the urine samples (Table 13.1), or relative abundances (Tables 13.2 and 13.3), and should not be used to infer absolute quantities or concentrations. NMR analysis was carried out with a JEOL ECS-400 spectrometer.

Table 13.1 Hemiterpenoids and pyrazines in 36 samples of male and female maned wolf urine[a]

Compound	Number of samples where present (36 total)
2-Methylbut-3-en-2-ol[b] (**2**)	35
3-Methylbut-2-en-1-ol[b] (**3**)	19
3-Methylbut-2-ene-1-thiol[b] (**4**)	10
3-Methyl-1-methylthiobut-2-ene[c] (**5**)	36
Methylpyrazine[b] (**8**)	12
2,5-Dimethylpyrazine[b] (**9**)	36
Trimethylpyrazine[b] (**10**)	30
2-Ethenyl-6-methylpyrazine[d] (**11**)	28
3-Ethyl-2,5-dimethylpyrazine[d] (**12**)	30
2-Methyl-6-(1-propenyl)pyrazine[d, e] (**13**)	34
2,5-Dimethyl-3-propylpyrazine[d] (**14**)	12

[a]Numbers in parentheses refer to structures in the Figures
[b]Verified by comparison of retention time and mass spectrum to an authentic sample and to the NIST02 mass spectral library
[c]Tentative structural assignment (see Sect. 13.3)
[d]Identified by comparison to the NIST mass spectral library only
[e]Possibly an isomeric 2-methyl,(1-propenyl)pyrazine; *E/Z* isomer undetermined

Table 13.2 Effect of ageing on maned wolf urine samples (one male, one female) at the ambient temperature for 48 h[a]

Compound	Female Day 1[b]	Female Day 3[c]	Male Day 1[b]	Male Day 3[c]
2-Methylbut-3-en-2-ol[d] (**2**)	1.00	2.37	2.96	7.51
3-Methylbut-2-en-1-ol[d] (**3**)	1.00	2.36	2.95	2.58
3-Methyl-1-methylthiobut-2-ene[e] (**5**)	1.78	7.91	1.00	5.46
Methylpyrazine[d] (**8**)	1.14	6.85	1.00	11.9
2,5-Dimethylpyrazine[d] (**9**)	1.18	1.51	1.00	2.10
Trimethylpyrazine[d] (**10**)	1.00	2.13	1.02	4.09
3-Ethyl-2,5-dimethylpyrazine[f] (**12**)	2.94	5.05	1.00	12.3
2-Methyl-3-(1-propenyl)pyrazine[f, g] (**13**)	1.17	1.60	1.00	3.62
2,5-Dimethyl-3-propylpyrazine[f] (**14**)	–	1.15	–	1.00

Ratios are based on peak areas from GC-MS total ion chromatograms
[a]A dash indicates non-detectable or inadequate amount for accurate integration
[b]Day of thawing
[c]Forty-eight hours after thawing
[d]Verified by comparison of retention time and mass spectrum to an authentic sample and to the NIST mass spectral library
[e]Tentative structural assignment (see Sect. 13.3)
[f]Identified by comparison to the NIST mass spectral library only
[g]Possibly an isomeric 2-methyl,(1-propenyl)pyrazine; *E/Z* isomer undetermined

Table 13.3 Effect of ageing, centrifuging, and filtering (CF) on maned wolf urine samples (one male)[a]

	Male	Male	Male	Male
Compound	Day 1 CF[b]	Day 1 untreated[b]	Day 3 CF[c]	Day 3 untreated[c]
2-Methylbut-3-en-2-ol[d] (**2**)	1.00	1.01	4.78	4.94
3-Methyl-1-methylthiobut-2-ene[e] (**5**)	1.00	2.07	3.19	1.67
Methylpyrazine[d] (**8**)	–	–	1.21	1.00
2,5-Dimethylpyrazine[d] (**9**)	1.00	1.18	1.68	2.09
Trimethylpyrazine[d] (**10**)	1.00	1.37	2.54	3.04
2-Ethenyl-6-methylpyrazine[f] (**11**)	1.00	1.16	1.33	1.66
3-Ethyl-2,5-dimethylpyrazine[f] (**12**)	1.00	1.40	4.69	6.06
2-Methyl-3-(1-propenyl)pyrazine[f, g] (**13**)	1.00	1.80	5.50	6.77

Ratios are based on peak areas from GC-MS total ion chromatograms
[a]A dash indicates non-detectable or inadequate amount for accurate integration
[b]Day of thawing
[c]Forty-eight hours after thawing
[d]Verified by comparison of retention time and mass spectrum to an authentic sample and to the NIST mass spectral library
[e]Tentative structural assignment (see Sect. 13.3)
[f]Identified by comparison to the NIST mass spectral library only
[g]Possibly an isomeric 2-methyl,(1-propenyl)pyrazine; *E*/*Z* isomer undetermined

13.2.3 Procurement of Authentic Standard Compounds

Thiol **4** (Fig. 13.1) was prepared as described (Helsper et al. 2006). Sulphide **5** (Fig. 13.1) was prepared as described for the isomeric sulphide **1** (Wilson et al. 1978; Woolhouse and Morgan 1995) using sodium thiomethoxide in dimethylformamide at the ambient temperature for 1 h, but using 3,3-dimethylallyl bromide and not the corresponding tosylate (commercially available reagents). The reaction of sodium methoxide and 3,3-dimethylallyl bromide has also been carried out in water (Moir et al. 1980; Roussel-UCLAF patent application 1966). We found that no solvent is required for this reaction, and it was more convenient to stir equimolar amounts of the two reactants at the ambient temperature for 2 h in an SPDE vial (Sect. 13.2.1) before SPDE/GC-MS analysis as described (Sect. 13.2.2) but in the split mode. The identity of synthetic compound **5** was confirmed by mass spectrometry and by ^{1}H NMR spectroscopy which are consistent with published data (Moir et al. 1980). Unless indicated otherwise, the additional compounds reported in the Figures and Tables are commercially available (Sigma-Aldrich) or were donated (see Acknowledgements).

13.2.4 IACUC Approval

All animal-related activities were approved by the Institutional Animal Care and Use Committee (IACUC) or its equivalent at all participating institutions.

13.3 Results and Discussion

13.3.1 General Overview

Mammalian urine is usually rich in volatile organic compounds (Zlatkis et al. 1973; Schwende et al. 1986; Goodwin et al. 2005; Zhang et al. 2005; delBarco-Trillo et al. 2011). It is not uncommon for compounds to function as chemical signals in more than one animal species (Kelly 1996; Rasmussen 1998; delBarco-Trillo et al. 2011). We initially focused primarily on maned wolf urinary volatiles, namely hemiterpenoids (Fig. 13.1) and pyrazines (Fig. 13.3), which are known semiochemicals and/or particularly odoriferous.

13.3.2 Hemiterpenoids in Maned Wolf Urine

Oxygen- and sulphur-containing hemiterpenoids are found in maned wolf urine (Table 13.1). 2-Methylbut-3-en-2-ol (**2**) and 3-methylbut-2-en-1-ol (**3**) are found in maned wolf urine and are known to be pheromones in bark beetles (Bakke et al. 1977; Stoakley et al. 1978; Lanne et al. 1989; Francke 1991; Seybold and Vanderwel 2003). Alcohol **2** has also been identified as a scent marking component in an antelope (Burger and Pretorius 1988) and a component of the alarm pheromone mixture in a hornet (Vieth et al. 1984), while alcohol **3** has been identified in the glandular secretion of fruit flies (Krohn et al. 1992) and in the anal gland secretion of the mink (Schildknecht et al. 1981).

We have identified 3-methyl-2-butene-1-thiol (**4**) in maned wolf urine. This compound is also known to be a photodegradation product of isohumulones and reaction with a sulphur-containing compound (possibly cysteine) in hops, resulting in what is called "lightstruck" or "skunky" beer (Burns et al. 2001, and references contained therein). Additionally, thiol **4** has been detected in the headspace over roasted coffee (where it is considered desirable; Holscher et al. 1992) and is the origin of a "foxy odor" emanating from flower bulbs of the crown imperial (*Fritillaria imperialis*; Helsper et al. 2006). Thiol **4** has also been identified in the shoulder gland secretions of male flying foxes (Wood et al. 2005), and as an aroma component of the sea fig, a shellfish (Senger-Emonnot et al. 2006).

We expected to find 3-methyl-1-methylthiobut-3-ene (**1**) in maned wolf urine because it is found in the urine of other canids, e.g. the red fox, coyote, and gray wolf (Wilson et al. 1978; Raymer et al. 1984, 1986; Schultz et al. 1988). However, we did not find this compound, but rather what appears from the mass spectrum to be an isomer of **1**. For reasons to be discussed below (Sect. 13.3.3), we tentatively identified this isomer as 3-methyl-1-methylthiobut-2-ene (**5**). This compound is known to occur in hop oil (Moir et al. 1980), as well as in the anal gland secretion of the mink (Schildknecht et al. 1981).

6 7

Fig. 13.2 Hemiterpenoid precursors in maned wolf urine

For sulphide **1** the three most abundant mass spectral ions (*m/z*, % of base peak) are 116 (26, M^+), 61 (100), and 41 (20) (Wilson et al. 1978; Schildknecht et al. 1981; one report lists 116 (27), 68 (15), and 61 (100): Sokolov et al. 1980), while for sulphide **5** they are 116 (96, M^+), 69 (97), and 41 (100) (our mass spectral results from synthetic sulphide **5** are in accord with published data; Moir et al. 1980; Schildknecht et al. 1981). The abundances of the *m/z* 116 and 41 ions are much larger in isomer **5** than in isomer **1**, while the *m/z* 61 ion is much smaller (10% of the base peak). The large *m/z* 61 ion abundance for sulphide **1** represents the carbocation $CH_3SCH_2^+$ that results from loss of the stabilised C_4H_7 allylic radical, whereas the large *m/z* 69 ion abundance for sulphide **5** represents the stabilised $C_5H_9^+$ allylic carbocation that results from loss of the CH_3S radical. Therefore it is quite clear that we have not found the typical canid sulphide hemiterpenoid, 3-methyl-1-methylthiobut-3-ene (**1**) in maned wolf urine. The isomer that we have found is one of the most abundant volatile compounds in maned wolf urine and has the same GC retention time as synthetic sulphide **5**. In all samples of maned wolf urine that we have analysed, the three most abundant mass spectral ions for this compound have been *m/z* 116, 69, and 41; however, the slight variability in their ratios leads us to classify structure **5** as probable but tentative at this point in time.

13.3.3 Hemiterpenoid Biosynthesis

Hemiterpenoids are biosynthesised primarily via the mevalonate pathway (Lanne et al. 1989; Francke 1991; Cane 1999; Francke and Schulz 1999; Seybold and Vanderwel 2003; Berg et al. 2012). The end products of this process are isopentenyl diphosphate (**6**; isopentenyl pyrophosphate) and dimethylallyl diphosphate (**7**), the building blocks of terpenes and steroids (Fig. 13.2).

It is reasonable to suggest that isomeric alcohols **2** and **3** are biosynthesised via an S_N1 reaction of water with the allylic carbocation formed from diphosphate **7**. Alcohol **3** (or more likely, diphosphate **7**) has been hypothesised as the precursor of thiol **4** in coyote urine (Schultz et al. 1988), roasted coffee (Holscher et al. 1992), processed foods (Rizzi 1995), and beer (Gros et al. 2009). Rizzi (1995) reasoned that dimethylallyl ("prenyl") carbocation is added to the sulphur atom of cysteine, and that the resulting sulphonium ion is the precursor of thiol **4**. To support this hypothesis, Rizzi cited a report on the decomposition of *S*-(3-methyl-2-butenyl)-L-cysteine to thiol **4** in water when exposed to sunlight (Obata and Tanaka 1965).

Wilson et al. (1978) suggested that isopentenyl diphosphate (**6**) is the probable biological precursor of sulphide **1**. A report from the same laboratories suggested

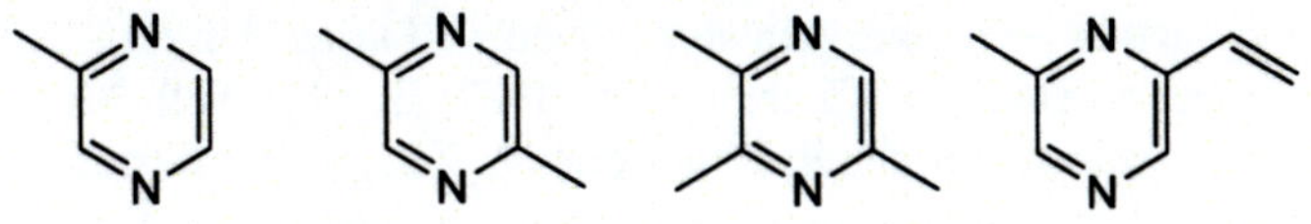

8 9 10 11

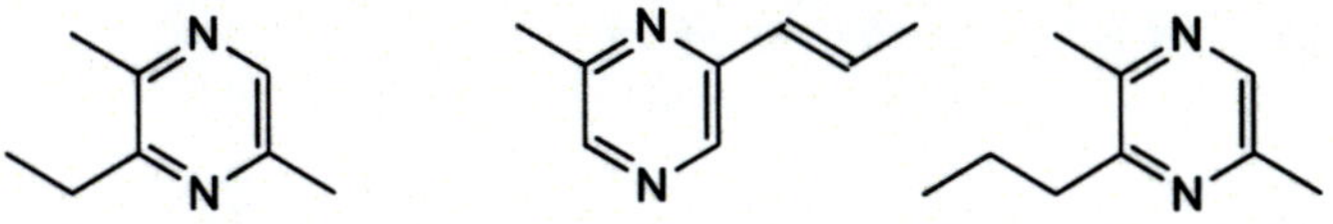

12 13 14

Fig. 13.3 Pyrazines in maned wolf urine

without elaboration that sulphide **1** is derived from compound **6** and methionine via an *S*-methylsulphonium ion intermediate (Jorgensen et al. 1978). Raymer (1984) discussed two hypotheses involving an *S*-methylsulphonium ion intermediate: (a) methionine displaces the diphosphate from compound **6** followed by an unexplained transformation to sulphide **1**; and (b) cysteine displaces the diphosphate from compound **6**, followed by *S*-methylation by *S*-adenosylmethionine (SAM, the major biochemical methyl transfer donor; Berg et al. 2012) to form the sulphonium ion, and subsequent production of sulphide **1** via "an *S*-alkylcysteine splitting enzyme".

There is yet an alternative pathway that can be imagined for the biogenesis of thiol **4** and sulphide **5**. Hydrogen sulphide (H_2S) occurs naturally in mammals and is produced via enzymatic metabolism of cysteine (Teague et al. 2002; Wang 2002). At physiological pH (7.4) and temperature (37°C), H_2S is predominantly converted to its more nucleophilic conjugate base, HS^-, and both are known to react with many intracellular target compounds (Hughes et al. 2009). Among these reactions is the nucleophilic displacement of a phosphate group in the generation of 4-thiouridine in tRNA (Mueller et al. 2001). While not proven, it is reasonable to suggest that H_2S or HS^- could react with diphosphate **7** via a nucleophilic substitution mechanism to produce thiol **4**, subsequent SAM methylation of which would provide sulphide **5**. (Sulphide **1** could be produced from diphosphate **6** via a similar reaction pathway.) Thiol **4**, sulphide **5**, and H_2S could each contribute to the powerful odour of maned wolf urine.

13.3.4 Pyrazines in Maned Wolf Urine

Pyrazines are widespread natural products that are often involved in chemical signalling, particularly among insects (Guilford et al. 1987; Brophy 1989; Woolfson and Rothschild 1990). They are thought to be biosynthesised from amino acids via α-amino ketone or aldehyde intermediates (Rizzi 1988; Dickschat et al. 2010; Nawrath et al. 2010). Childs-Sanford (2005) identified pyrazines **9**, **10**, and **13** (Fig. 13.3) in maned wolf urine. We have found these as well, which along with four additional pyrazines are quite common in maned wolf urine (Table 13.1). Pyrazines **8–12** are insect pheromones (Chuman et al. 1987; Brophy 1989; Morgan 2009). Pyrazines also act as semiochemicals in excretions and secretions from some mammals, including the brown antechinus (**11**; Toftegaards et al. 1999), the house mouse (**9**; Jemiolo and Novotny 1993), and the pine vole (**9**; Boyer et al. 1989). Pyrazines often have a distinctive odour (Woolfson and Rothschild 1990) and may contribute to the scent and scent marking of maned wolves.

13.3.5 Increase in Compound Abundances as Maned Wolf Urine Ages

We have previously shown that a number of urinary compounds from female (Goodwin et al. 2006) and male (Goodwin et al. 2012) African elephants increase in abundance as the urine ages. In some cases this increase can be prevented by centrifuging and filtering to remove microbes, thus implying microbial metabolism of nutrients in the elephant urine as a contributor to this increase (Goodwin et al. 2012; see also Albone et al. 1977; Albone 1984). Our preliminary studies on the aging of maned wolf urine have revealed increases in the abundance of most of the urinary hemiterpenoids and pyrazines after 48 h at ambient temperature (Table 13.2). It is possible that this continued compound production serves as a time release of chemical signals to conspecifics. Unlike our results with elephant urinary components, in most cases centrifuging and filtering maned wolf urine does not prevent the increase in compound abundance after aging (Table 13.3), thus microbial action does not appear to be involved.

13.4 Conclusion

We have identified a number of odoriferous compounds which contribute to the odour of maned wolf urine and may function as semiochemicals to conspecifics. Continued chemical analyses may validate additional urinary compound structures and reveal sexual and seasonal variations in compound abundances. Further studies on the effect of urine aging on compound abundance will continue with a larger data set to probe aspects and causes of the temporal release of putative semiochemicals. These chemical studies combined with extensive behavioural bioassays should clarify the role of chemical signalling among maned wolves, particularly as it relates to reproduction.

Acknowledgements Urine samples were provided by the Center for Species Survival, Smithsonian Conservation Biology Institute, and the Little Rock Zoo. We thank Paul Caster of the Little Rock Zoo for assistance in urine collection. We are grateful for the hemiterpenoid and pyrazine samples that were provided by Joe Brophy, Doug Crump, Peter Schieberle, and Tony Woolhouse.

TEG appreciates financial support provided by the Hendrix College Odyssey Program, Julia Mobley, and John and Laura Byrd. Funding was provided by the U.S. National Science Foundation for the GC-MS (NSF CRUI grant to T.E.G., Bruce Schulte, and the late L.E.L. Rasmussen, Award Nos. 02-17062, -17068 and -16862, respectively) and the NMR spectrometer (NSF MRI grant to Hendrix College, Award No. 1040470).

References

Albone ES (1984) Mammalian semiochemistry. Wiley-Interscience, New York

Albone ES, Gosden PE, Ware GC (1977) Bacteria as a source of chemical signals in mammals. In: Müller-Schwarze D, Mozell MM (eds) Chemical signals in vertebrates. Plenum Press, New York, pp 35–43

Asa CS, Mech LD, Seal US (1984) The use of urine, feces, and anal-gland secretions in scent marking by a captive wolf (*Canis lupus*) pack. Anim Behav 33:1034–1036

Bakke A, Frøyen P, Skattebøl L (1977) Field response to a new pheromonal compound isolated from *Ips typographus*. Naturwissenschaften 64:98–99

Berg JM, Tymoczko JL, Stryer L, Gatto GJ Jr (2012) Biochemistry. W.H. Freeman, New York

Bestelmeyer SV (2000) Solitary, reproductive and parental behaviors of maned wolves (*Chrysocyon brachyurus*). PhD Thesis, Colorado State University, Fort Collins, Colorado, USA

Bovée KC, Bush M, Dietz J, Jezyk P, Segal S (1981) Cystinuria in the maned wolf of South America. Science 212:919–920

Boyer ML, Jemiolo B, Andreolini F, Wiesler D, Novotny M (1989) Urinary volatile profiles of pine vole, *Microtus pinetorum*, and their endocrine dependency. J Chem Ecol 15:649–662

Brophy JJ (1989) Pyrazines obtained from insects: their source, identification, synthesis and function. In: Atta-ur-Rahman (ed) Studies in natural products chemistry, Vol. 5, structure elucidation (Part B). Elsevier, Amsterdam, p 221–293

Burger BV, Pretorius PJ (1988) Mammalian pheromone studies, VI. Compounds from the preorbital gland of the blue duiker, *Cephalophus monticola*. Z Naturforsch 42c:1355–1357

Burns CS, Heyerick A, De Keukeleire D, Forbes MDE (2001) Mechanism for formation of the lightstruck flavor in beer revealed by time-resolved electron paramagnetic resonance. J Chem Eur 7:4553–4561

Cane DE (1999) Isoprenoid biosynthesis: overview. In: Barton D, Nakanishi K, Meth-Cohn O (eds) Comprehensive natural products chemistry, Vol. 2: isoprenoids including carotenoids and steroids. Elsevier, Amsterdam, pp 1–13

Childs-Sanford SE (2005) The captive maned wolf (*Chrysocyon brachyurus*): nutritional considerations with emphasis on management of cystinuria. M.S. Thesis, University of Maryland, College Park, Maryland, USA

Chuman T, Landolt PJ, Heath RR, Tumlinson JH (1987) Isolation, identification, and synthesis of male-produced sex pheromone of papaya fruit fly, *Toxotrypana curvicauda* Gerstaecker (Diptera: Tephritidae). J Chem Ecol 13:1979–1992

DelBarco-Trillo J, Burkert BA, Goodwin TE, Drea CM (2011) Night and day: the comparative study of strepsirrhine primates reveals socioecological and phylogenetic patterns in olfactory signals. J Evol Biol 24:82–98

Dickschat JS, Wickel S, Bolten CJ, Nawrath T, Schulz S, Wittmann C (2010) Pyrazine biosynthesis in *Corynebacterium glutamicum*. Eur J Org Chem 2687–2695

Dietz JM (1984) Ecology and social organization of the maned wolf (*Chrysocyon brachyurus*). Smithsonian Institution Press, Washington, DC

Dietz JM (1985) Chrysocyon brachyurus. Mammal Spec 234:1–4

Francke W (1991) Semiochemicals: mevalogenins in systems of chemical communication. In: Muller PM, Lamparsky D (eds) Perfumes: art, science and technology. Elsevier, London and New York, pp 61–99

Francke W, Schulz S (1999) Pheromones. In: Barton D, Nakanishi K, Meth-Cohn O, Mori K (eds) Comprehensive natural products chemistry, Vol. 8: miscellaneous natural products including marine natural products, pheromones, plant hormones, and aspects of ecology. Elsevier, Amsterdam, pp 197–261

Goodwin TE, Schulte BA (2009) Prospecting for mammalian chemical signals via solventless extraction techniques: an elephantine task. ChemoSense 11:9–15

Goodwin TE, Rasmussen LEL, Schulte BA, Brown PA, Davis BL, Dill WM, Dowdy NC, Hicks AR, Morshedi RG, Mwanza D, Loizi H (2005) Chemical analysis of preovulatory female African elephant urine: a search for putative pheromones. In: Mason RT, LeMaster MP, Müller-Schwarze D (eds) Chemical signals in vertebrates 10. Springer Press, New York, pp 128–139

Goodwin TE, Eggert MS, House SJ, Weddell ME, Schulte BA, Rasmussen LEL (2006) Insect pheromones and precursors in female African elephant urine. J Chem Ecol 32:1849–1853

Goodwin TE, Brown PA, Eggert MS, Evola MG, House SJ, Morshedi RG, Weddell ME, Chen CJ, Jackson SR, Aubut Y, Eggert J, Schulte BA, Rasmussen LEL (2007) Use of automated solid phase dynamic extraction (SPDE)/GC-MS and novel macros in the search for African elephant pheromones. In: Hurst J, Beynon R, Wyatt T, Roberts C (eds) Chemical signals in vertebrates 11. Springer Press, New York, pp 25–35

Goodwin TE, Broederdorf LJ, Burkert BA, Hirwa HH, Mark DB, Waldrip ZJ, Kopper RA, Sutherland MV, Freeman EW, Hollister-Smith JA, Schulte BA (2012) Chemical signals of elephant musth: temporal aspects of microbially-mediated modifications. J Chem Ecol 38:81–87

Gros J, Nizet S, Collin S (2009) Hop allylic alcohols are precursors of sulfur-containing odorants in fresh beer. Acta Hortic 848:273–278

Guilford T, Nicol C, Rothschild M, Moore BP (1987) The biological roles of pyrazines: evidence for a warning odour function. J Linn Soc 31:113–128

Helsper JPFG, Bucking M, Muresan S, Blaas J, Wietsma WA (2006) Identification of the volatile component(s) causing the characteristic foxy odor in various cultivars of *Fritillaria imperialis* L. (Liliaceae). J Agric Food Chem 54:5087–5091

Holscher W, Vitzthum OG, Steinhart H (1992) Prenyl alcohol—source for odorants in roasted coffee. J Agric Food Chem 40:655–658

Hughes MN, Centelles MN, Moore KP (2009) Making and working with hydrogen sulfide, the chemistry and generation of hydrogen sulfide *in vitro* and its measurement *in vivo*: a review. Free Radic Biol Med 47:1346–1353

Jácomo ATDA, Kashivakura CK, Ferro C, Furtado MM, Astete SP, Tôrres NM, Sollmann R, Silveira L (2009) Home range and spatial organization of maned wolves in the Brazilian grasslands. J Mammal 90:150–157

Jemiolo B, Novotny M (1993) The long-term effect of a urinary chemosignal on reproductive fitness in female mice. Biol Reprod 48:926–929

Jorgensen JW, Novotny M, Carmack M, Copland GB, Wilson SR (1978) Chemical scent constituents in the urine of the red fox (*Vulpes vulpes* L.) during the winter season. Science 199:796–798

Kelly DR (1996) When is a butterfly like an elephant? Chem Biol 3:595–602

Kleiman DG (2011) Canid mating system, social behavior, parental care and ontogeny: are they flexible? Behav Genet 41:803–809

Krohn S, Fletcher MT, Kitching W, Moore CJ, Drew RAI, Francke W (1992) Chemistry of fruit flies: glandular secretion of *Bactrocera* (*Polistomimetes*) *visenda* (Hardy). J Chem Ecol 18:2169–2176

Lanne BS, Ivarsson P, Johnsson P, Bergström G, Wassgren A-B (1989) Biosynthesis of 2-methyl-3-buten-2-ol, a pheromone component of *Ips typographus* (Coleoptera: Scolytidae). Insect Biochem 19:163–167

Mattoo A, Goldfarb DS (2008) *Cystinuria* Semin. Nephrology 28:181–191

Moir M, Gallacher IM, Seaton JC, Suggett A (1980) Terpene methyl sulphides in the essential oil of hops. Chem Ind, London, pp 624–625

Morgan ED (2009) Trail pheromones of ants. Physiol Entomol 34:1–17
Mueller EG, Palenchar PM, Buck CJ (2001) The role of the cysteine residues of ThiI in the generation of 4-thiouridine in tRNA. J Biol Chem 276:33588–33595
Nawrath T, Dickschat JS, Kunze B, Schulz S (2010) The biosynthesis of branched dialkylpyrazines in myxobacteria. Chem Biodivers 7:2129–2144
Obata Y, Tanaka H (1965) Studies of the sunlight flavor of beer Part IX. Formation of prenylmercaptan from S-prenyl-L-cysteine by sunlight. Agric Biol Chem 29:104–107
Rasmussen LEL (1998) Chemical communication: an integral part of functional Asian elephant (*Elephas maximus*) society. Ecoscience 5:410–426
Raymer JH (1984) Investigation into the chemical nature of chemo-olfactory communication in the wolf (*Canus lupus*). PhD Thesis, Indiana University, Bloomington, Indiana, USA
Raymer J, Wiesler D, Novotny M, Asa C, Seal US, Mech LD (1984) Volatile constituents of wolf (*Canis lupus*) urine as related to gender and season. Experiencia 40:707–709
Raymer J, Wiesler D, Novotny M, Asa C, Seal US, Mech LD (1986) Chemical scent constituents in urine of wolf (*Canis lupus*) and their dependence on reproductive hormones. J Chem Ecol 12:297–314
Rizzi GP (1988) Formation of pyrazines from acyloin precursors under mild conditions. J Agric Food Chem 36:349–352
Rizzi GP (1995) Formation of sulfur-containing flavor compounds from allylic alcohol precursors. Dev Food Sci 37:289–302
Rodden MD, Sorenson LG, Sherr A, Kleiman DG (1996) Use of behavioral measures to assess reproductive function in maned wolves (*Chrysocyon brachyurus*). Zoo Biol 15:565–585
Rodden M, Rodrigues F, Bestelmeyer S (2004) Maned wolf (*Chrysocyon brachyurus*). In: Sillero-Zubiri C, Hoffman M, Macdonald DE (eds) Canids: foxes, wolves, jackals and dogs. Status survey and conservation action plan. IUCN/SSC Canid Specialist Group, Gland, Switzerland and Cambridge, UK, pp 38–43
Roussel-UCLAF patent application (1966) Cyclic sulfones. Chem Abstr 64:15849h
Schildknecht H, Birkner C, Krauß D (1981) Struktur und wirkung der musteliden-ökomone II. Erweiterte analyse des analbeutelsekretes des nerzes Mustela vison L. Chem Ztg 105:273–286
Schultz TH, Flath RA, Stern DJ, Mon TR, Teranishi R, Kruse SM, Butler B, Howard WE (1988) Coyote estrous urine volatiles. J Chem Ecol 14:701–712
Schwende FJ, Weisler D, Jorgensen JW, Carmack M, Novotny M (1986) Urinary volatile constituents of the house mouse, *Mus musculus*, and their endocrine dependency. J Chem Ecol 12:277–296
Senger-Emonnot P, Rochard S, Pellegrin F, George G, Fernandez X, Lizzani-Cuvelier L (2006) Odour active aroma components of sea fig (*Microcosmus sulcatus*). Food Chem 97:465–471
Seybold SJ, Vanderwel D (2003) Biosynthesis and endocrine regulation of pheromone production in the Coleoptera. In: Bloomquist G, Vogt R (eds) Insect pheromone biochemistry and molecular biology. Elsevier, Amsterdam, pp 137–200
Sokolov VE, Albone ES, Flood PF, Heap PF, Kagan MZ, Vasilieva VS, Roznov VV, Zinkevich EP (1980) Secretion and secretory tissues of the anal sac of the mink, *Mustela vison*. J Chem Ecol 6:805–825
Songsasen N, Rodden M, Brown JL, Wildt DE (2006) Patterns of fecal gonadal hormone metabolites in the maned wolf (*Chrysocyon brachyurus*). Theriogenology 66:1743–1750
Stoakley JT, Bakke A, Renwick JAA, Vité JP (1978) The aggregation pheromone of the larch bark beetle, *Ips cembrae* Heer. J Appl Entomol 86:174–176
Teague B, Asiedu S, Moore PK (2002) The smooth muscle relaxant effect of hydrogen sulphide *in vitro*: evidence for a physiological role to control intestinal contractility. Br J Pharmacol 137:139–145
Toftegaards CL, Moore C, Bradley AJ (1999) Chemical characterization of urinary pheromones in brown antechinus, *Antechinus stuartii*. J Chem Ecol 25:527–535

Vieth HJ, Koeniger N, Maschwitz U (1984) 2-Methyl-3-butene-2-ol, a major component of the alarm pheromone of the hornet *Vespa crabro*. Naturwissenschaften 71:328–329

Wang R (2002) Two's company, three's a crowd: can H_2S be the third endogenous gaseous transmitter? FASEB J 16:1792–1798

Wilson SR, Carmack M, Novotny M, Jorgenson JW, Whitten WK (1978) Δ^3-Isopentyl methyl sulfide. A new terpenoid in the scent mark of the red fox (*Vulpes vulpes*). J Org Chem 43:4675–4676

Wood WF, Walsh A, Seyjagat J, Weldon PJ (2005) Volatile compounds in shoulder gland secretions of male flying foxes, genus *Pteropus* (Pteropodidae, Chiroptera). Z Naturforsch C 60:779–784

Woolfson A, Rothschild M (1990) Speculating about pyrazines. Proc R Soc Lond B 242:113–119

Woolhouse AD, Morgan DR (1995) An evaluation of repellents to suppress browsing by possums. J Chem Ecol 21:1571–1583

Zhang JX, Soini HA, Bruce KE, Wiesler D, Woodley SK, Baum MJ, Novotny MV (2005) Putative chemosignals of the ferret (*Mustela furo*) associated with individual and gender recognition. Chem Senses 30:727–737

Zlatkis A, Bertsch W, Lichtenstein HA, Tishbee A, Shunbo F, Liebich HM, Coscia AM, Fleischer N (1973) Profile of volatile metabolites in urine by gas chromatography–mass spectrometry. Anal Chem 45:763–767

Chapter 14
The Perfume-Body Odour Complex: An Insightful Model for Culture–Gene Coevolution?

Jan Havlíček and S. Craig Roberts

Abstract Olfaction is involved in various human social interactions, ranging from mother-offspring attachment to mate choice; as in other species, such chemical signalling is thought to be shaped by evolution. However, across many human cultures, individuals manipulate their body odour by means of various fragrances and these may significantly affect the outcome of social encounters in a context-specific fashion. Here we employ the framework of dual-inheritance theory, which advocates that cultural practices should be incorporated into the analysis of evolution of human behaviour, to explore cultural means of olfactory signalling such as ethnic and status markers. Further, we review studies showing that perfumes interact with body odour in an individual fashion and that people tend to choose perfumes according to their genetic make-up. This indicates that biologically evolved chemical signalling might operate in concert with cultural human practices. Finally, we propose two scenarios: (1) how culturally based preferences and use of perfume might impact gene frequencies in individual populations, and (2) how evolved cognitive biases might affect selection of scents that are appropriate for body adornments. This, in our view, makes the perfume-body odour complex a potentially insightful model for culture–gene coevolution.

J. Havlíček (✉)
Department of Anthropology, Charles University,
Prague, Czech Republic
e-mail: jan.havlicek@fhs.cuni.cz

S.C. Roberts
Department of Psychology, School of Natural Sciences,
University of Stirling, Stirling, Scotland, UK
e-mail: craig.roberts@stir.ac.uk

M.L. East and M. Dehnhard (eds.), *Chemical Signals in Vertebrates 12*,
DOI 10.1007/978-1-4614-5927-9_14, © Springer Science+Business Media New York 2013

14.1 Biology of Human Semiochemistry

Humans, like other vertebrates, emit from their bodies a variety of aromatic chemical compounds that can be perceived by others and thus become involved in communication processes. Body odours can be categorised as cues of: (1) individuality and relatedness; (2) reproductive state; (3) affective state; (4) individual quality; and (5) genetic complementarity. These phenomena are reviewed elsewhere (e.g. Havlicek et al. 2010; Lenochova and Havlicek 2008; Havlicek and Roberts 2009; Roberts and Havlicek 2012). Here we detail the richness of human semiochemistry in two areas: (1) how affective states may be communicated via body odours, and (2) how genetic complementarity affects body odour preferences in mate choice.

In a classic study, Karl von Frisch (1938) found that skin damage in European minnows causes a flight reaction in conspecifics. The phenomenon later described as an alarm pheromone is present in many other social species, e.g. in black-tailed deer (Müller-Schwarze 1971). Inspired by these findings, researchers have tested whether affective states might also be communicated in humans. Although there is ongoing debate about whether these should be called cues or signals, evidence supports the idea that affective states can influence body odours in a way that can be perceived by others. Axillary odours collected from individuals when they feel socially anxious induce a higher startle reflex and this reaction is specifically pronounced in individuals scoring high in social anxiety (Prehn et al. 2006). The startle reflex is used as an indicator of individual alertness. Similarly, smelling "anxious odours" increases the state of anxiety in others (Prehn-Kristensen et al. 2009). Individuals also tend to use more risky strategies in experimental games after exposure to 'anxious' odours (Haegler et al. 2010), while Adolph et al. (2010) reported that body odours collected under a competitive situation (a badminton match) tend to increase skin conductance in individuals smelling them. Communication of affective states by body odours is a fascinating and fertile area that will yield further exciting results.

Another principal context in which human chemical communication takes place is within mate choice. Studies in several vertebrates show that odour cues provide information about genetic make-up of conspecifics (Bernatchez and Landry 2003). Particular attention has focused on the link between odour and genes in the major histocompatibility complex (MHC). The MHC is extremely variable, with hundreds of different known alleles in some human loci. MHC gene products play a critical role in vertebrate immune functioning by discriminating between self/non-self molecules. Higher MHC-heterozygosity, and thus vigour, in resulting offspring is achieved by individuals often preferring MHC-dissimilar sexual partners (Apanius et al. 1997). This is also true in humans: in a pioneering study, Wedekind et al. (1995) found that women judged as more attractive the axillary odour of men whose MHC was dissimilar to their own. This finding was further supported by several other studies (Wedekind and Füri 1997; Santos et al. 2005; Thornhill et al. 2003; reviewed in Havlicek and Roberts 2009). Interestingly, Wedekind et al. (1995) also found that women using hormonal contraception showed elevated preference for

MHC similarity. Shifts in preferences due to hormonal contraception were confirmed by a more recent study in women tested before and after initiation of oral contraceptive use (Roberts et al. 2008). Although MHC studies sometimes produce inconsistent results, may be context-dependent, and may be modulated by other factors, we concluded that the evidence suggests that mate-related odour preferences are associated with one's genetic make-up (Havlicek and Roberts 2009).

14.2 Universal Use of Fragrances

People can modify their body odour by using fragrances. Estimated sales in the fragrance and flavour industry ($22 billion in 2010: Leffingwell & Associates 2012) show that such tendencies cannot be overlooked. Is fragrance use a relatively recent phenomenon, perhaps linked to the perception of an association between strong body odour and disease, as described by Corbin (1988)? It seems unlikely given that scented oils were used in ancient Mesopotamia and pre-Roman Palestine. In ancient Egypt, fragrances were used to anointed mummies, cult statues, and participants of religious ceremonies, and women used fragrances for secular purposes as depicted on Theban tombs (Stoddart 1990). Ancient societies did not use distillation for perfume production but instead used oils as a base material. These were heated together with specific plants and then mixed with aromatic substances of mostly floral origin (Castel et al. 2009). Depictions on Rekmirekh's tomb, for instance, show perfumers mixing oils with resin and aromatic substances crushed in mortars and then heated in a cauldron to create aromatic oil (Brun 2000). Compared to Egypt, the use of aromatic compounds in Ancient Greece was of a more secular character and commonplace in the Classical period. It flourished in the Hellenistic period, becoming a well-established industry (Brun 2000). Perfumes were also popular amongst the Romans, with Campania as the centre of perfume production. Scented oils were commonly used during bathing, which was a sophisticated and lengthy procedure during which many important business contracts were made and political discussions held. In the part of the baths called the caldarium, aromatic oils were rubbed over the body. Romans also tended to control or forbid perfume usage (as in, for example, the general edict of Rome in 188 BC) as they were seen as seductive and immoral (Stoddart 1990).

Anthropologists report that, in diverse cultures and geographically distant regions, people modify their body odour with various odorous substances. An example is the Ongees, hunter-gatherers of the Andaman Islands (Pandya 1990), who believe each living creature (including humans) emanates smell that can be traced by malevolent spirits that cause disease or kill incautious individuals by eating their smell. To avoid this, Ongees decorate their face and body with red and white clay paints. Red clay is thought to warm the body and increases smell release whereas white paint cools the body and restrict sweating. Clay paint is used by pig hunters to ensure they are not smelled by the animals they aim to kill (Pandya 2007). Ancestral (i.e. benevolent) spirits are believed to identify their descendants by the specific smell each person produces. A combination of visual and olfactory adornment by scented body paints is also performed by the Desana of Upper Amazonia and the inhabitants of the Kiriwina (formerly Trobriand) Islands of New Guinea. Another use of fragrances is illustrated by the Nauruans of the Southern Pacific, who place fragrant herbs on hot stones to produce aromatic steam which is thought to cleanse the body of bad smells and create a lasting perfume (Classen et al. 1994). Altogether, this evidence clearly indicates that olfactory adornment has been an integral part of human cultures in historical societies as well as cross-culturally.

14.3 Social Effects of Perfumes

Some regard the use of fragrances as a form of olfactory jewellery, adding extra cachet but leaving evolved routes of biological odour communication unaffected. Certainly, psychological studies provide relatively robust evidence that fragrances affect behaviour and social processes such as interpersonal assessment. For instance, formally dressed interview candidates are perceived more negatively when they use perfume compared to informally dressed candidates who after perfume usage were perceived more positively (Baron 1981). Fragrance use may also influence perception of masculinity and hence suitability for a managerial position (Sczesny and Stahlberg 2002). Fragrance can also influence perceivers indirectly via an effect on perfume wearers themselves. Roberts et al. (2009) filmed male participants after application of fragrance or a placebo substance. Observers of the videos rated fragrance users as more attractive even though they could not smell the targets, apparently because of reduced self-confidence in those given the placebo. Similarly, Higuchi et al. (2005) found a decrease in nonverbal behaviours labelled as non-symbolic movements (e.g. self-touching) amongst women wearing perfume during a mock interview; again observers rated perfumed individuals as more self-confident. Taken together, these studies convincingly show that perfumes indeed affect the perception of others, although the nature of the effects might depend on the context of the interaction.

14.4 Dual-Inheritance Theory

Although the use of fragrances is cross-cultural and widespread and affects both wearers and perceiver, there is robust evidence that body odour may cue various biological qualities. How might such apparently contradictory evidence be reconciled? Similar issues in other domains of human behaviour have been addressed by the dual-inheritance paradigm which we believe could also be a useful framework for the study of the fragrance-body odour complex. In essence, this theory takes account of the fact that, as well as inheriting genes, people also acquire beliefs, attitudes, and values through social learning (Richerson and Boyd 2006). The capacity for culture is then seen as a distinct complex of human adaptations.

From this perspective, a full understanding of human behaviour includes knowledge of local cultural practices (Henrich and McElreath 2007). It assumes that humans evolved various cognitive adaptations to learn cultural practices and symbolic systems; or as articulated by Henry Plotkin (2007): "Mankind's natural place is in culture, and culture is a part of human biology because it is our biology that gives us the ability to enter into culture". It also embraces formation of socially constructed systems which shape social reality and are based on mutual sharing, such as marriage, money, a police force, and so on. However, the existence of socially constructed beliefs should not be mistaken for arbitrariness of constructs as is currently thought by many social anthropologists. In contrast, theorists of cultural evolution suggest that human cognition related to cultural capacity was (or is) subject to natural selection (Boyd and Richerson 2005). This manifests itself in cognitive biases in social learning processes and the perception of others. It also suggests that various ways of signalling may take on the form of cultural practices. We will first focus on cultural aspects of chemosignalling and subsequently turn to the possible interactions that may exist between cultural and biological means of signalling.

14.5 Cultural Aspects of Olfactory Signalling

14.5.1 Ethnic Group Markers

Humans live in social groups of different size and complexity. Such groups, here labelled as ethnic groups, cultures or communities, can be identified by variation in beliefs, norms, religion, and other cultural practices (Boyd and Richerson 2005). Cultural group markers include language, dress, cuisine, and adornment. Derogation of other ethnic groups can include the attribution of dubious morality and even the status of full human-ness may be restricted to in-group members. In some cultures, body odours and fragrances are implicated in these prejudices: for example, attributing foul smell to people of other cultures due to their "wrong" way of living. The Desana people of Amazonia say that each tribe emits a specific odour, traceable to the territory they inhabit, and while walking through the forest they constantly sniff to see whether people of other tribe have been present (Classen 1992). It is not uncommon to classify "we" and "others" according to their odour, which varies by ethnic origin, or culinary, hygienic, and religious practices (Low 2005, 2009). These examples indicate that body odours and related scents could be salient cues of ethnic markers.

14.5.2 Status Markers

Social hierarchies are found within most human societies and again, fragrances may denote social status. A characteristic feature of status markers is their limited availability; in complex societies, this is frequently expressed through monetary cost (compare this concept with the idea of 'honest signals' used in behavioural ecology). At least before the advent of synthetic perfumes, fragrances were scarce commodities: ingredients were typically specific plant parts (such as rose petals), which were seasonally or climatically restricted. In addition, the technology involved in their extraction, together with the blending art of perfumers, suggests the manufacturing process was restricted to a few individuals or sites. Add to this the relative instability of ancient perfumes, due to high volatility of some of the compounds and tendency of the bases (e.g. olive oil) to go rancid (Castel et al. 2009), and it is clear how high-quality perfumes became a significant part of trading networks (Brun 2000) and were expensive goods, affordable only to the upper classes, while others used scents containing lower quality ingredients. To more limited extent, this remains true today: the most favoured scents, such as aloewood in the United Arab Emirates market (Classen et al. 1994), are also very expensive. Furthermore, both unpleasant body odour and use of "cheap" perfumes is used to characterise the lower classes in modern Western literature (Largey and Watson 1972). Fragrances are thus interpreted according to cultural constructs of a particular culture, but nonetheless provide information which is also biologically relevant, such as group affiliation and social status.

14.6 Interaction Between Biological and Cultural Chemosignalling

So far, we have shown how socially relevant information is communicated either through body odour itself or through cultural practices of artificial fragrance use. Although we can intellectually distinguish between biological and cultural contributions, these cannot be separated in everyday experience because they are mutually interwoven, often in a highly complex manner. Here we present several scenarios for how cultural practices might interact with biologically evolved signals/cues.

Firstly, cultural practices might disrupt evolved preferences. In this scenario, cultural practices follow a different pattern than evolved preferences. For instance, men's judgments of women's facial (Roberts et al. 2004) or odour (Havlí ek et al. 2006) attractiveness varies across the menstrual cycle, peaking around ovulation. These changes are thought to reflect selection on male cognition; namely sensitivity to cues of female fertility. Cultural practices such as facial cosmetics and perfume may disguise or alter such cues, and hence disrupt male assessment of fertility cues (see Roberts et al. 2010).

Secondly, biologically evolved preferences might shape cultural practices. For example, some aspects of cosmetic use might exploit evolved perception/cognitive biases, such as the use of lipstick to enhance redness of lips, and foundation to obscure poor skin health or rouge to simulate it. Perhaps surprisingly, it might also shape an individual's perfume choice. It is frequently assumed that perfumes mask or obliterate underlying body odour, which would in turn lead to disruption of preferences as reviewed above. However, the only empirical evidence supporting this view is that perfume blurs accuracy in identifying the wearer's sex (Schleidt 1980). Furthermore, if this were the sole function of perfume use, it could not explain the huge diversity that exists in individual perfume preferences. Such considerations have led to the idea that individuals might select perfumes which in fact complement their own body odour (Milinski and Wedekind 2001), raising the intriguing possibility that cultural practices might work in conjunction with, rather than in opposition to, evolved preferences.

Recently, we studied the perception of perfume-body odour blends, testing whether perfumes might mask or interact with body odour (Lenochová et al. 2012). We reasoned that if perfumes mask body odour, the same perfume applied to different body odours would result in a set of odours that were perceptually similar and characterised by the perfume itself. In contrast, if the perfume interacts with individual body odours, then it would have positive effects in some individuals and negative effects in others, while odour variability should be maintained or even enhanced. In two studies, male participants applied perfume to one armpit (the other was left untreated). They wore cotton pads in both armpits, and these were subsequently rated by women for their pleasantness, attractiveness, and intensity. Not surprisingly, the perfume-body odour blends were generally judged as more positive than untreated samples, but analysis also showed a significant interaction between individual odour and perfume treatment (perfume had no effect, or adversely affected odour perception, in some individuals) and treatment did not alter variability among samples. Not only do these results support the interaction hypothesis, but they also indicate that people may choose perfumes in an individual fashion to complement their own body odour. Thus, we then tested the prediction that an individual's body odour would be judged more positively when blended with the wearer's preferred perfume compared with one assigned by the experimenters. This turned out to be the case, even though there were no significant differences in the perceived pleasantness between the assigned and the preferred perfumes. Together, our experiments support the idea that fragrances, rather than simply masking body odour, interact with it in a complementary manner. This is consistent with Milinski and Wedekind's (2001) pioneering work which found an association between MHC-genotype and preference for perfume ingredients for a perfume to be worn by oneself (see also Hämmerli et al. 2012). In contrast, there was no link between assessors' MHC and perfume preferences when they selected perfumes for their partners. Anecdotal complaints by customers of perfume shops indicate it is difficult to buy perfume for someone else and perfumes received as a gift often remain unused, supporting the advice of professional perfumers that perfumes must be tried on for fit (Burr 2007). Our guess is that choosing a perfume for a partner may be more difficult than choice for relatives, who share similarities in body odour (Porter et al. 1985).

14.7 Body Odour-Perfume Coevolution

The sections above point to the idea that perfume and body odour create a mixture with an emergent perceptual quality, which is difficult to predict based on how they smell separately. While this interaction is itself fascinating for researchers in human chemical ecology, we believe we can go even further and argue that it might also help us to understand underlying general processes in gene–culture coevolution.

As we have mentioned, individual communities vary considerably in the substances they employ for perfume production (in most of the speculations below we deliberately ignore recent trends such as technological advancement in global transfer of goods and production of synthetic chemicals used in perfumery: these phenomena appeared only very recently and one might not expect their immediate effect on biological evolution which operates on a much longer time scale). The absence of a specific ingredient in the perfumes of a particular community could be due to the following reasons: (1) the source of the odour is unavailable in the area and is not traded from neighbours. For example, we know that aromatic plants were an important commodity in trading networks in Ancient Egypt or Greece, but some of the scents routinely employed in that era in India were rare or absent in Mediterranean cultures. (2) The community is constrained by a technology. Some of the aromas can be extracted only using a specific technology which might not be available for or discovered by the particular community. In ancient Greece, for instance, ethanol distillation was not used and perfumers instead used mechanic extraction or enfleurage (Brun 2000). (3) Particular scents or their source (e.g. a particular plant) are believed to be inappropriate for body adornment. Such beliefs might stem from religious considerations.

Of course, there could be other reasons for a community not using particular ingredients, and the three we have described are not mutually exclusive. However, considering that only some scent ingredients will complement particular body odours (i.e. particular genotypes) and that a particular community employs only a restricted variety of scents for perfuming, it is plausible that some individuals may not be able to select a perfume which complements their body odour and may therefore suffer a social disadvantage. In the long run, the frequency of genotypes of such individuals would decrease in the particular community. This model of course assumes stability of preferences across generations which might not always be the case. However, as noted above, preferences for particular scents are not incidental and they are frequently part of broader beliefs and might thus be relatively stable. Furthermore, especially if there is preferential in-group mating, we might in consequence observe the differences in frequencies of genes linked to body odour as a result of culture-based processes.

In contrast, evolved cognitive biases in odour preferences might affect cultural beliefs about which scents are used for perfuming. Humans inhabit a world of an almost-infinite number of aromas, and in spite of relatively high cross-cultural variability, only a fraction of them are used for body fragrances. Aside from availability or technological constraints, this could be the result of evolved cognitive biases in formation of olfactory hedonics, restricting the pool of 'acceptable' scents (it is unlikely that any culture would use human excrement-resembling odours such as skatole, for example). Currently, most psychologists regard development of odour preferences to be the result of Pavlovian conditioning or associative learning (Herz 2006), but based on the kinds of evidence reviewed above, we find this overly simplistic and we predict that some biologically relevant odours will be conditioned more easily than others. An example of this would underpin the preference for the odour of human milk in either breastfeeding or formula-fed infants, independently of previous experience with such odour (Marlier and Schaal 2005). If cognitive biases were at work, they might in the long run shape cultural practices.

What we hope to have shown here is that use and choice of perfume, a cross-culturally prominent behaviour which at first sight is largely culturally driven, is also strongly influenced by biology. Patterns of perfume use cannot therefore be fully explained if considered solely in the light of either genetic or cultural evolution. Both influences need to be taken into account because the substrate for selection is likely to be neither body odour in isolation, nor the perfume itself, but the blend of the two odour sources. This blend appears to have dynamic and individually variable emergent properties and is amenable to experimental manipulation. On this basis, we believe the perfume-body odour complex could become an insightful model for biology-culture coevolution.

Acknowledgements We would like to thank Jindra Havlíčková and Lenka Nováková for many helpful comments and proof-reading. JH is supported by Czech Science Foundation grant (GACR P407/10/1303) and Charles University Research Centre (UNCE 204004).

References

Adolph D, Schlösser S, Hawighorst M, Pause BM (2010) Chemosensory signals of competition increase the skin conductance response in humans. Physiol Behav 101:666–671

Apanius V, Penn D, Slev PR, Ruff LR, Potts WK (1997) The nature of selection on the major histocompatibility complex. Crit Rev Immunol 17:179–224

Leffingwell & Associates (2012) Flavor & Fragrance Industry Leaders. www.leffingwell.com/top_10.html. Accessed 2 Feb 2012

Baron RA (1981) Olfaction and human social-behavior—effects of a pleasant scent on attraction and social-perception. Pers Soc Psychol Bull 7:611–616

Bernatchez L, Landry C (2003) MHC studies in nonmodel vertebrates: what have we learned about natural selection in 15 years? J Evol Biol 16:363–377

Boyd R, Richerson PJ (2005) The origin and evolution of cultures. Oxford University Press, New York

Brun JP (2000) The production of perfumes in antiquity: the cases of Delos and Paestum. Am J Archaeol 104:277–308

Burr C (2007) The perfect scent: a year inside the perfume industry in Paris and New York. Picador, New York

Castel C, Fernandez X, Filippi JJ, Brun JP (2009) Perfumes in Mediterranean antiquity. Flavour Fragr J 24:326–334

Classen C (1992) The odor of the other—olfactory symbolism and cultural categories. Ethos 20:133–166

Classen C, Howes D, Synnott A (1994) Aroma—the cultural history of smell. Routledge, London and New York

Corbin A (1988) The foul and the fragrant: odor and the French social imagination. Harvard University Press, Cambridge, MA

Haegler K, Zernecke R, Kleemann AM, Albrecht J, Pollatos O, Bruckmann H, Wiesmann M (2010) No fear no risk! Human risk behavior is affected by chemosensory anxiety signals. Neuropsychologia 48:3901–3908

Hämmerli A, Schweisgut C, Kaegi M (2012) Population genetic segmentation of MHC-correlated perfume preferences. Int J Cosmet Sci 34:161–168
Havlicek J, Roberts SC (2009) MHC-correlated mate choice in humans: a review. Psychoneuroendocrino 34:497–512
Havlíček J, Bartoš L, Dvořáková R, Flegr J (2006) Non-advertised does not mean concealed. Body odour changes across the human menstrual cycle. Ethology 112:81–90
Havlicek J, Murray AK, Saxton TK, Roberts SC (2010) Current issues in the study of androstenes in human chemosignalling. In: Litwack G (ed) Pheromones (Vitamins & Hormones) 83. Academic, London, pp 47–81
Henrich J, McElreath R (2007) Dual-inheritance theory: the evolution of human cultural capacities and cultural evolution. In: Dunbar RIM, Barrett L (eds) The Oxford handbook of evolutionary psychology. Oxford University Press, Oxford, pp 555–570
Herz RS (2006) I know what I like: understanding odor preferences. In: Drobnick J (ed) The smell culture reader. Berg, New York, pp 190–203
Higuchi T, Shoji K, Taguchi S, Hatayama T (2005) Improvement of nonverbal behaviour in Japanese female perfume-wearers. Int J Psychol 40:90–99
Largey GP, Watson DR (1972) The sociology of odors. Am J Sociol 77:1021–1034
Lenochova P, Havlicek J (2008) Human body odour individuality. In: Hurst JL, Beynon RJ, Roberts SC, Wyatt TD (eds) Chemical signals in vertebrates 11. Springer, New York, pp 189–198
Lenochová P, Vohnoutová P, Roberts SC, Oberzaucher E, Grammer K, Havlíček J (2012) Psychology of fragrance use: perception of individual odor and perfume blends reveals a mechanism for idiosyncratic effects on fragrance choice. PLoS One 7:e33810
Low KEY (2005) Ruminations on smell as a sociocultural phenomenon. Curr Sociol 53:397–417
Low KEY (2009) Scent and scent-sibilities. Cambridge Scholar Publishing, Newcastle upon Tyne
Marlier L, Schaal B (2005) Human newborns prefer human milk: conspecific milk odor is attractive without postnatal exposure. Child Dev 76:155–168
Milinski M, Wedekind C (2001) Evidence for MHC-correlated perfume preferences in humans. Behav Ecol 12:140–149
Müller-Schwarze D (1971) Pheromones in black-tailed deer (*Odocoileus hemionus columbianus*). Anim Behav 19:141–152
Pandya V (1990) Movement and space: andamanese cartography. Am Ethnol 17:775–797
Pandya V (2007) Time to move: winds and the political economy of space in Andamanese culture. J R Anthropol Instit 13:S91–S104
Plotkin H (2007) The power of culture. In: Dunbar RIM, Barrett L (eds) The Oxford handbook of evolutionary psychology. Oxford University Press, Oxford, pp 11–19
Porter RH, Cernoch JM, Balogh RD (1985) Odor signatures and kin recognition. Physiol Behav 34:445–448
Prehn A, Ohrt A, Sojka B, Ferstl R, Pause BM (2006) Chemosensory anxiety signals augment the startle reflex in humans. Neurosci Lett 394:127–130
Prehn-Kristensen A, Wiesner C, Bergmann TO, Wolff S, Jansen O, Mehdorn HM, Ferstl R, Pause BM (2009) Induction of empathy by the smell of anxiety. PLoS One 4:e5987
Richerson PJ, Boyd R (2006) Not by genes alone: how culture transformed human evolution. The University of Chicago Press, Chicago
Roberts SC, Havlicek J (2012) Evolutionary psychology and perfume design. In: Roberts SC (ed) Applied evolutionary psychology. Oxford University Press, Oxford, pp 330–348
Roberts SC, Havlicek J, Flegr J, Hruskova M, Little AC, Jones BC, Perrett DI, Petrie M (2004) Female facial attractiveness increases during the fertile phase of the menstrual cycle. Proc R Soc Lond B Biol Sci 271:S270–S272
Roberts SC, Gosling LM, Carter V, Petrie M (2008) MHC-correlated odour preferences in humans and the use of oral contraceptives. Proc R Soc Lond B Biol Sci 275:2715–2722
Roberts SC, Little AC, Lyndon A, Roberts J, Havlicek J, Wright RL (2009) Manipulation of body odour alters men's self-confidence and judgements of their visual attractiveness by women. Int J Cosmet Sci 31:47–54

Roberts SC, Miner EJ, Shackelford TK (2010) The future of an applied evolutionary psychology for human partnerships. Rev Gen Psychol 14:318–329

Santos PSC, Schinemann JA, Gabardo J, Bicalho MD (2005) New evidence that the MHC influences odor perception in humans: a study with 58 Southern Brazilian students. Horm Behav 47:384–388

Schleidt M (1980) Personal odor and nonverbal-communication. Ethol Sociobiol 1:225–231

Sczesny S, Stahlberg D (2002) The influence of gender-stereotyped perfumes on leadership attribution. Eur J Soc Psychol 32:815–828

Stoddart DM (1990) The scented ape—the biology and culture of human odour. Cambridge University Press, Cambridge

Thornhill R, Gangestad SW, Miller R, Scheyd G, McCollough JK, Franklin M (2003) Major histocompatibility complex genes, symmetry, and body scent attractiveness in men and women. Behav Ecol 14:668–678

Von Frisch K (1938) Zur Psychologie des Fisch-Schwarmes. Naturwissenschaften 26:601–606

Wedekind C, Füri S (1997) Body odour preferences in men and women: do they aim for specific MHC combinations or simply heterozygosity? Proc R Soc Lond B Biol Sci 264:1471–1479

Wedekind C, Seebeck T, Bettens F, Paepke AJ (1995) MHC-dependent mate preference in humans. Proc R Soc Lond B Biol Sci 260:245–249

Chapter 15
A Historical Perspective on the Identification of Substances in the Territorial Scent Marks of Male Klipspringer Antelope *Oreotragus oreotragus*

Hans E. Hummel

Abstract Antelopes are territorial, hoofed mammals inhabiting the savannahs of eastern and southern Africa. They are richly endowed with skin glands. The male klipspringer antelope, *Oreotragus oreotragus* (Zimmermann 1783) (Ungulata: Artiodactyla), has well-developed antorbital glands that are frequently used to mark prominent features within its territory. In this chapter I discuss my research on scent marking by male klipspringer in the 1960s and compare my results to those obtained more recently by studies that were able to apply more sophisticated analytical techniques. This comparison provides evidence (1) for the outstanding progress of analytics within the course of nearly three decades and (2) for the validity of the early results obtained in 1968.

15.1 Introduction

Early researchers such as Schaffer (1940), Hediger (1949) and Walther (1964, 1965) described the territorial behaviour of antelopes. Male antelopes mark their territories and most likely use specific pheromones (Karlson and Lüscher 1959; Law and Regnier 1971) produced by specific skin glands (Schaffer 1940; Walther 1964, 1965). Except for early reports by Ruzicka (1926a, b, c; see also Wheeler 1977) the chemical nature of territorial scent marks was largely obscure until Brownlee et al. (1969) identified the tarsal gland pheromone of the black-tailed deer, *Odocoileus hemionus columbianus*, as (Z)-6-dodecenyl-γ-lactone. Black-tailed deer and

H.E. Hummel (✉)
Justus-Liebig-University Giessen, Organic Agriculture, Karl-Gloeckner-Str. 21 C, D-35394 Giessen, Germany

University of Illinois Urbana-Champaign, Prairie Research Institute, 1816 S. Oak Street, Champaign, IL 61820, USA
e-mail: Hans.E.Hummel@agrar.uni-giessen.de

M.L. East and M. Dehnhard (eds.), *Chemical Signals in Vertebrates 12*,
DOI 10.1007/978-1-4614-5927-9_15, © Springer Science+Business Media New York 2013

klipspringer, *Oreotragus oreotragus*, are both members of the family Bovidae, subfamily Antilopinae.

Tarsal and antorbital glands both belong to the skin gland ensemble typically involved in territorial marking behaviour (Birch 1974; Müller-Schwarze and Mozell 1977; Müller-Schwarze 1971, 2006; Müller-Schwarze et al. 1974; Mykytowicz 1972, 1973, 1974; Mykytowicz et al. 1976; Stoddard 1974; Wheeler 1977; Wyatt 2003).

In 1965 available analytical techniques included column chromatography, paper- and thin layer chromatography (TLC), infrared (IR), ultraviolet (UV) and visible light (Vis) spectroscopy, and nuclear magnetic resonance (NMR) spectrometers. Gas–liquid chromatography (GLC or GC), the work-horse of modern analytical labs, was just starting to become commercialised. Packed and also open tubular GC columns were available for isothermal analyses, and were used less commonly in the temperature programmed mode. High-pressure liquid chromatography (HPLC) was still about 15 years in the future. Organic mass spectrometry (MS) existed in the laboratories of Biemann (1962) who published the first textbook emphasising organic mass spectrometry, followed by Spiteller (1966). GC-MS systems (Ryhage 1964; Schomburg 1966; Leemans and McCloskey 1967; McFadden 1967) were just on the verge of being commercialised. At this time I started a qualitative analysis of the klipspringer antorbital secretions.

15.2 Materials and Methods

By means of the workup and purification procedure depicted in Fig. 15.1, the antorbital secretion was crudely separated into five major fractions according to solubility in organic solvents, boiling point, molecular mass, steam volatility and chromatographic mobility. These techniques were standard procedures in the mid-1960s and can be found in Stahl (1967) and Organikum (1970). However, micro-techniques for the analysis of pheromones were still mostly lacking, with a few notable exceptions (Hecker 1955; Brownlee and Silverstein 1968). Bioassay procedures were self-developed arrangements (Hummel 1968) adapted according to strategies known to natural product chemists using bioassays as guides for enrichment and finally for isolation of certain components of complex natural mixtures (Hummel and Miller 1984; Müller-Schwarze 2006; Haynes and Millar 1999).

Bioassays were carried out by using artificial pheromone dissemination sources consisting of porous ceramic suction probes ("Filterkerzen" used in soil chemistry), which were inserted into hollow cylinders made from polyamide in such a way that only the tip of the pheromone-loaded probe protruded 10 mm from the tip of the polyamide cylinder. Mounted on a washable and heatable polymer rack, and attached as a unit to the fence of the animal enclosure, this arrangement was sturdy enough to even withstand horn attacks by a frustrated male antelope trying to remove the odour signals perceived to be those left by rivals. After a given observation period, usually 5 min, the treated probe was removed from the hollow cylinder and exchanged for another clean probe as control with minimal disturbance of the experimental animal. Another treated probe could then be secretly inserted into the cylinder. After half a

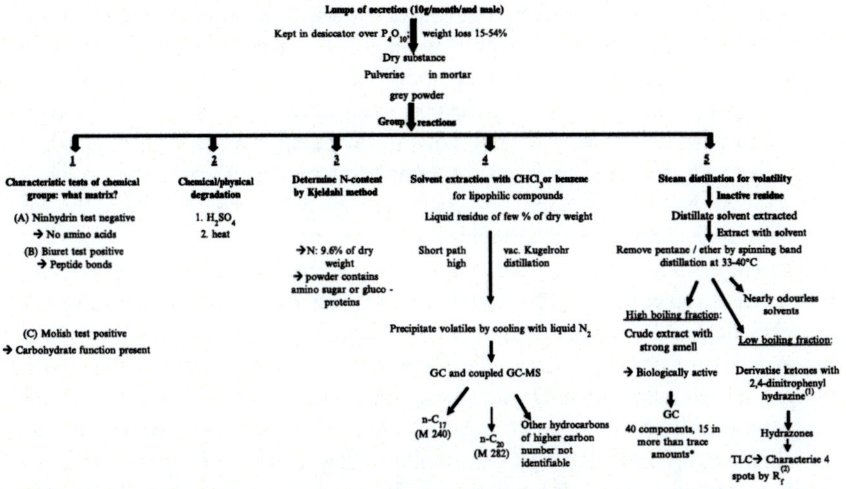

Fig. 15.1 General workup and fractionation scheme for *O. oreotragus* secretion. Steps 1–3 are for preliminary orientation about composition, steps 4 and 5 for characterisation and identification of several compounds. *In these fractions, the *n*-C_{17} and *n*-C_{20} hydrocarbon were positively identified by mass spectrometry and/or comparison of GC retention times of the natural materials with those of reference compounds on three columns with different polarities (ApL, SE-30, DEGS). [(1)]Ariga (1972), [(2)]Stahl (1967)

day, the antelopes were habituated to the presence of the observer. Therefore, a control phase without samples always preceded the test phase.

The need for suitable pheromone research techniques was bothersome, but this problem was later resolved by the papers of Claesson and Silverstein (1977), Silverstein et al. (2005), as well as by monographic reviews of Hummel and Miller (1984), Millar and Haynes (1998), Wyatt (2003) and Hummel and Hecker (2012). Recently, Müller-Schwarze (2006) emphasised the need for specific research techniques in the field of vertebrate chemical ecology.

15.3 Results

A general workup scheme (Fig. 15.1) was developed to characterise the multicomponent mixture of the klipspringer secretion. It consisted of five parallel steps and allowed (1) for crude characterisation of major compound classes, (2) heat treatment and reaction with concentrated sulphuric acid, followed by (3) determination of its nitrogen content (N-content) by the Kjeldahl method, (4) solvent extractability of compounds and (5) determination of the steam volatility of its constituents. The results of step 1 indicated no free amino acids, but the presence of peptide bonds and carbohydrate functionality. Step 2 indicated degradability with sulphuric acid, resulting in the formation of an ammonium sulphate solution from which ammonia was released by treatment with KOH. After quantitative Kjeldahl distillation of the NH_3 (step 3) into a standardised acid solution, a quantitative back titration gave an N-equivalent value from which the N content of the sample was calculated as 9.62±0.12 % of the dry weight, a value distinctly different from the N content (7.26±0.22 %) of the Thomson gazelle secretion. Heating the powdered secretion up to 300 °C, the sample decomposed, without a sharp melting point or liberating diagnostic gases or odours, to yield a black residue. Step (1) to (3), taken together, characterised the matrix as a mucopolysaccharide or glycoprotein which contains and probably released the volatiles with some biologically relevant kinetics.

More important are the volatiles characterised in steps (4) and (5). Step (4) started with solvent extraction of the dry powdered secretion. Evaporation of the solvent left a residue that had a fatty appearance. A short-path, Kugelrohr vacuum distillation of this material yielded a small volatile fraction that was frozen out using a liquid-nitrogen cooled microtrap. *n*-Heptadecane (m/z 240) and *n*-eicosane (m/z 282) were positively identified by GC and GC-MS techniques on an LKB 9000 instrument in the laboratory of Prof. Ryhage, Stockholm, using a series of *n*-hydrocarbons of known composition as reference standards. In addition, some unidentifiable (probably branched) higher hydrocarbons were present in this distillate.

Finally, in step (5) the steam volatility of the secretion was investigated. Steam overheated to 110 °C was bubbled through a slurry of the finely powdered klipspringer secretion for 2 h, after which further fractions of the distillate elicited no biologically relevant reaction in behavioural bioassays with the male klipspringer. The volume of aqueous distillate accumulating within 2 h was extracted three times with small volumes of a highly purified pentane-ether (1:1) mixture. Great care was taken to remove peroxides from the ether by flash chromatography through activated alumina, followed by distillation of the main fraction and discarding material of lower and higher boiling point. The combined extracts were distilled under atmospheric pressure over a spinning band column at 40 °C to remove 98 % of the solvent. The solvent distillate had negligible remaining bioactivity. However, the residual material had a strong smell and was highly active in behavioural bioassays. In undiluted form it elicited horn attacks by the male. GC fractionation of this fraction revealed the presence of 15 major and 25 minor components, among others the already identified *n*-heptadecane and *n*-eicosane. Significantly, a portion of this concentrated fraction reacted with an acidified 2,4-dinitrophenylhydrazine solution according to a procedure described by Ariga (1972) to yield a precipitate of 2,4-dinitrophenylhydrazones. These derivatives, indicative of ketones, were isolated by filtration on a microfilter and also centrifuged in a microtube.

TLC analysis of these 2,4-dinitrophenylhydrazones on normal-phase silica plates with benzene-petroleum ether (3:1) as mobile phase gave a total of five spots, all highly UV active at 254 nm. One strong spot remaining at the start at R_f=0.00 contained the unavoidable polar impurities and was discounted. Four other spots appeared at R_f values of 0.54, 0.65, 0.71 and 0.79. For comparison, a number of commercially available *n*-ketones and *n*-aldehydes were also derivatised using the same procedure. However, no match between standards and unknown components could be found. It is possible that the material remaining at the start was the highly polar, protonated 2,4-dinitrophenylhydrazine.

Results from the bioassays revealed that only steam distillate taken during the first 2 h period was active, suggesting that this was sufficient to remove most active ingredients and thus was a useful method to separate volatile fractions (data not shown). However, organic solvents other than pentane and ether have been found to irritate the males and were avoided.

15.4 Discussion

The steam distillate I obtained from male klipspringer territorial scent marks contained three major fractions: a volatile fraction containing solvents, a low boiling fraction with a distinct camphoraceous, anise-like, fruity smell, containing the ketones that were precipitable as hydrazones, and the esters, and finally, a less volatile fraction containing the hydrocarbons probably serving as fixatives for the volatiles. The involatile, inactive residue of mucopolysaccharides remained in the steam distillation flask.

The ketones present in the antorbital secretion were more recently identified by Burger et al. (1997) using GC-MS. The general properties of the pure, commercially available ketones are listed in Table 15.1. In my experiments in 1968 very small quantities of the respective 2,4-dinitrophenylhydrazones derivatives were available, only sufficient for chromatographic separation by TLC and R_f characterisation.

The fruity smell of the secretion suggested the presence of esters in the territorial marks. However, these constituents could not be identified for the following reasons: (1) Esters do not form easily visible coloured derivatives in one-step reactions. The amount of the ester fraction was too small so that conventional methods for ester identification such as saponification followed by the identification of the resulting carboxylic acids and alcohols, for example by esterification with 4′-nitroazobenzene-4-carboxylic acid (Hecker 1955), were not applicable. (2) Esters were more difficult to handle than ketones due to hydrolytic cleavage by superheated steam. (3) Their GC traces could not be matched with those of hydrocarbon standards and of methyl- and ethylester standards of low-boiling fatty acids that were commercially available at that time. (4) Packed GC columns available from 1965 to 1968 were in no way comparable in performance with the high resolution capillary columns available today.

The results of Burger et al. (1997; Table 15.2) reveal the reason why I was unable to identify both the ketones and the esters. Synthetic standards used for comparison were straight-chain compounds, whereas three of the four esters identified by Burger et al. (1997) were branched. Thus no matching of retention times between standards and territorial marking compounds could be expected. Ethyl propanoate was the only unbranched compound that evaded identification because of its relatively high volatility and thus co-elution with the solvent peak tail. Burger et al. (1997) did not find any hydrocarbons in the fresh antorbital secretion of the klipspringer. It is remarkable that the antorbital secretion of the male klipspringer contains only eight volatile organic compounds, whereas antorbital secretions from related species

Table 15.1 Volatile organic constituents of the antorbital secretion of *Oreotragus oreotragus* with structural information on the ketone and ester fractions

Characteristics	3-Pentanone ($C_5H_{10}O$)	4-Methylpentan-2-one ($C_6H_{12}O$)	4-Methyl-3-hexanone ($C_7H_{14}O$)	5-Methyl-3-hexanone ($C_7H_{14}O$)
Ketones[a]				
Boiling point (°C)	102–103	117–118	~144	144–145
Mol. mass (Dalton)	86	100	114	114
2.4-Dinitrophenyl-hydrazones of the ketones[b]	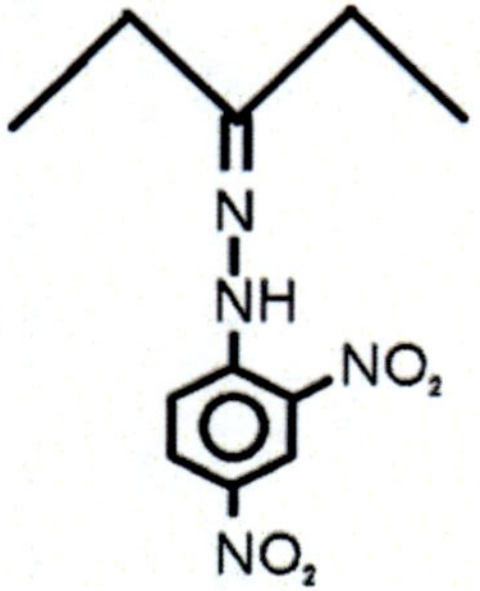 $C_{11}H_{14}N_4O_4$	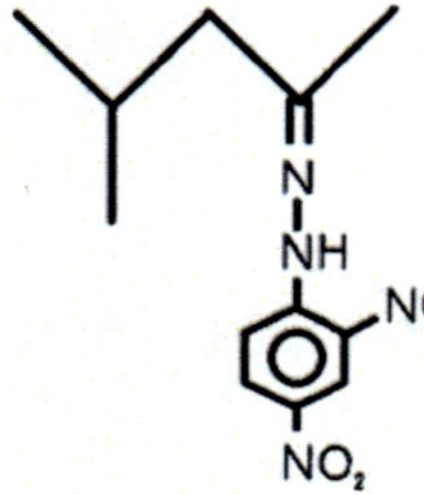 $C_{12}H_{16}N_4O_4$	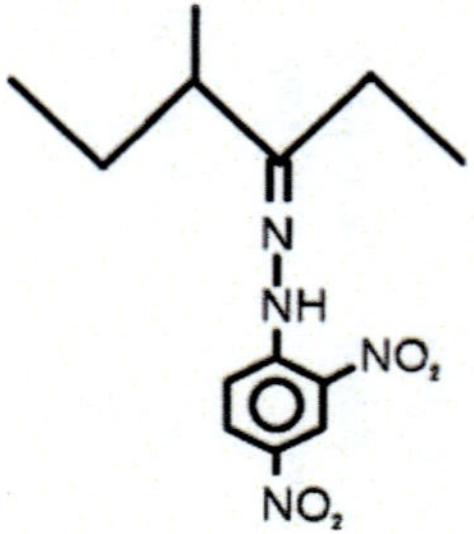 $C_{13}H_{18}N_4O_4$	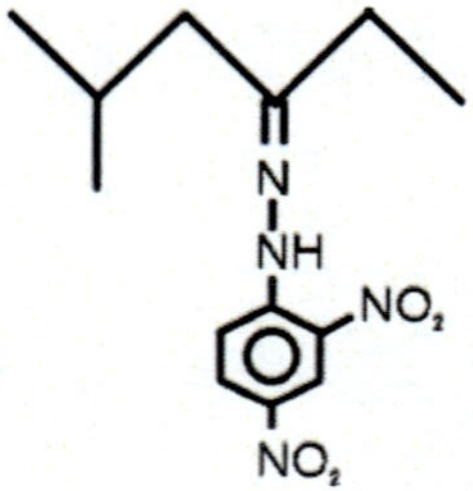 $C_{13}H_{18}N_4O_4$
Mol. mass (Dalton)	M: 266	M: 280	M: 294	M: 294

Characteristics	Ethyl-*n*-propanoate $C_5H_{10}O_2$	Isobutyryl acetate $C_6H_{12}O_2$	Isobutyl-*n*-propanoate $C_7H_{14}O_2$	Ethyl-isovalerate $C_7H_{14}O_2$
Esters				
Boiling point (°C)	99	116–117	137–138	131–133
Mol. mass (Dalton)	102	116	130	130
Structure	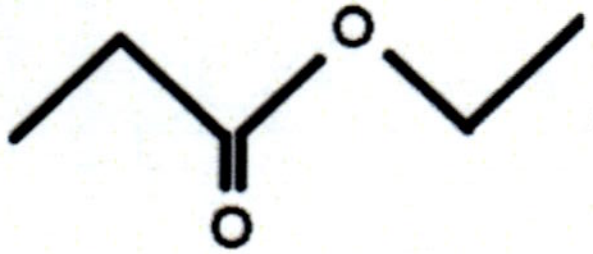	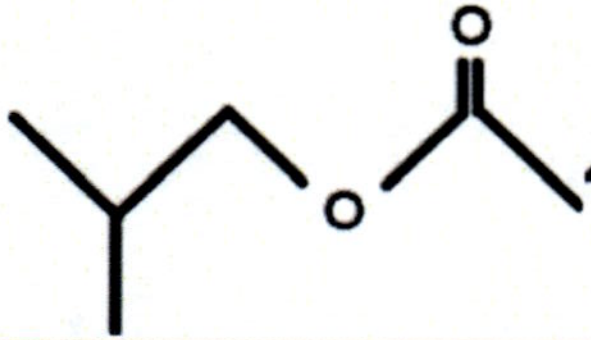	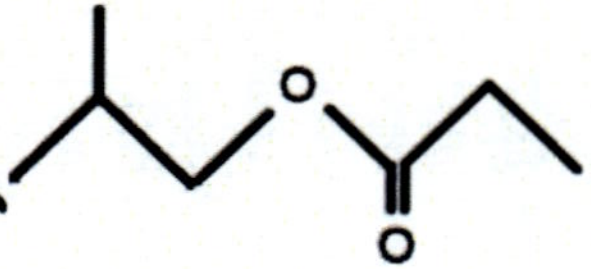	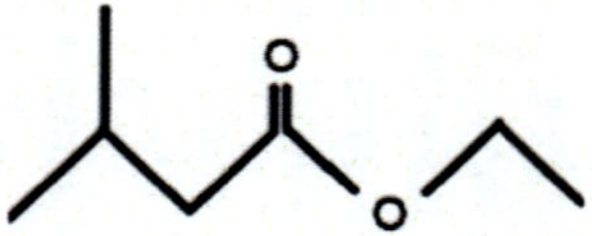

[a]Characterised by Hummel (1968), identified by Burger et al. (1997)
[b]The four spots of the 2.4-dinitrophenyl hydrazones were reddish in colour, and showed strong UV-absorption at 254 nm

Table 15.2 Compounds identified in the fresh antorbital secretion of male klipspringer, *Oreotragus oreotragus* (Burger et al. 1997)

	Compound types
Ketones	3-Pentanone
	4-Methyl-2-pentanone
	5-Methyl-3-hexanone
	4-Methyl-3-hexanone
Esters	Ethyl propanoate
	2-Methylpropyl acetate
	Ethyl 3-methyl-butanoate
	2-Methylpropyl propanoate

Fig. 15.2 Prof. R. M. Silverstein (1916–2007) with one of his juvenile ungulate friends, a black-tailed deer, *Odocoileus hemionus columbianus*

(Burger 2005) contained numerous volatile constituents. However, these eight compounds were exclusively found in klipspringer secretion but not in scents from four related members of the subfamily of territorial Antilopinae and one member of the subfamily of Cephalopinae (Burger 2005).

What is experimentally easy today was a problem in the mid-1960s. Young chemical ecologists might learn from this historical perspective and appreciate the tremendous progress of microanalytics achieved during the last 45 years by authors in this field, among them Professor R. M. Silverstein (1916–2007) (Fig. 15.2) to whose memory this paper is dedicated.

Acknowledgements Prof. P. Karlson as thesis advisor left enough freedom for development of chemical microtechniques and first crude bioassays with *Oreotragus* in the zoos of Frankfurt and Naples. Haarmann und Reimer, Holzminden, kindly provided standards of straight chain ketones and aldehydes. Prof. Dr. B. Grzimek, Zoo Frankfurt, and Director F. Cuneo, Zoo Naples, gave permission to perform behavioural bioassays on territorial marking with their animals. Schwarz Foundation supported the layout and critical completion of this manuscript by Dr. S. Schmid.

References

Ariga N (1972) Thin-layer chromatography of keto acid 2,4-dinitrophenylhydrazones. Anal Biochem 49:436–441

Biemann K (1962) Mass spectrometry, applications to organic chemistry. Mc Graw Hill, New York, p 170

Birch MC (ed) (1974) Pheromones. North Holland Publishing Co, Amsterdam/London. ISBN 07204–71338

Brownlee RG, Silverstein RM (1968) A micro-preparative gas chromatograph and a modified carbon skeleton determinator. Anal Chem 40:2077–2079

Brownlee RG, Silverstein RM, Müller-Schwarze D, Singer AG (1969) Isolation, identification and function of the chief component of the male tarsal scent in black-tailed deer. Nature 221:284–285

Burger BV (2005) Mammalian semiochemicals. In: Schulz S (ed) Topics in current chemistry. Springer, Berlin. ISSN 0340–1022, 240:231–278

Burger BV, Yang T-P, Le Roux M, Brandt WF, Cox AJ, Hart PE (1997) Mammalian exocrine secretions. XI. Constituents of the preorbital secretion of klipspringer, *Oreotragus oreotragus*. J Chem Ecol 23:2383–2400

Claesson A, Silverstein RM (1977) Chemical methodology in the study of mammalian communications. In: Müller-Schwarze D, Mozell MM (eds) Chemical signals in vertebrates. Plenum Press, New York, pp 71–93

Haynes KF, Millar JG (eds) (1999) Methods in chemical ecology. Bioassay methods. Chapman & Hall, Dordrecht

Hecker E (1955) 4′-Nitroazobenzol-4-Carbonsäure-4-chlorid als Reagenz auf Alkohole. Chem Ber 88:1666–1675

Hediger H (1949) Säugetier-Territorien und ihre Markierung. Bijdr Dierkd 28:171

Hummel HE (1968) Beiträge zur Kenntnis der Pheromone. Teil 1 und Teil 2. 124 Seiten. Dissertation, Universität Marburg

Hummel HE, Hecker E (2012) Half a century of pheromones—their contributions to sustainable management of insect pests. Mitteilungen der Deutschen Gesellschaft für allgemeine und angewandte Entomologie 18:451–460

Hummel HE, Miller TA (eds) (1984) Techniques in pheromone research. Springer, New York

Karlson P, Lüscher M (1959) Pheromones. A new term for a class of biologically active substances. Nature 183:55–56

Law JH, Regnier FE (1971) Pheromones. Annu Rev Biochem 40:533–548

Leemans FAJM, McCloskey JA (1967) Combination gas chromatography–mass spectrometry. J Am Oil Chem Soc 44:11–17

McFadden WH (1967) Mass-spectrometric analysis of gas-chromatographic eluents. In: Giddings JC, Keller RA (eds) Advances in chromatography, vol 4. Marcel Dekker, New York, p 265

Millar JG, Haynes KF (eds) (1998) Methods in chemical ecology. Kluwer, Norwell

Müller-Schwarze D (1971) Pheromones in black-tailed deer. Anim Behav 19:141–152

Müller-Schwarze D (2006) Chemical ecology of vertebrates. Cambridge University Press, Cambridge

Müller-Schwarze D, Mozell MM (1977) Chemical signals in vertebrates. Plenum Press, New York, London

Müller-Schwarze D, Müller-Schwarze C, Singer AG, Silverstein RM (1974) Mammalian pheromone: identification of active component in the subauricular scent of the male pronghorn. Science 183:860–862

Mykytowicz R (1972) The behavioral role of mammalian skin glands. Naturwissenschaften 59:133–139

Mykytowicz R (1973) Reproduction of mammals in relation to environmental odours. J Reprod Fertil 19:501–520

Mykytowicz R (1974) Odour in the spacing behaviour of mammals. In: Birch MC (ed) Pheromones. North-Holland, Amsterdam, pp 327–342

Mykytowicz R, Hestermann ER, Gambale S, Dudzinski ML (1976) A comparison of the effectiveness of the odors of rabbits, *Oryctalogus cuniculus*, in enhancing territorial confidence. J Chem Ecol 2:13–24

Organikum (1970) Trennungsverfahren, Wasserdampfdestillation. VEB Deutscher Verlag der Wissenschaften, Berlin, p 60

Ruzicka L (1926a) Zur Kenntnis des Kohlenstoffringes. I. Über die Konstitution des Zibetons. Helv Chim Acta 9:230–248

Ruzicka L (1926b) Zur Kenntnis des Kohlenstoffringes. VII. Über die Konstitution des Muscons. Helv Chim Acta 9:715–729

Ruzicka L (1926c) Zur Kenntnis des Kohlenstoffringes VIII. Weitere Beiträge zur Konstitution des Muscons. Helv Chim Acta 9:1008–1017

Ryhage R (1964) Use of a mass spectrometer as a detector and analyzer for effluents emerging from high temperature gas–liquid chromatography columns. Anal Chem 36:759–764

Schaffer J (1940) Die Hautdrüsenorgane der Säugetiere. Urban und Schwarzenberg, Berlin, Wien

Schomburg G (1966) Combined use of gas chromatography and spectroscopic methods. 4th Wilkens Gas Chromatography Symposium, Manchester, 25 May 1966. Notes

Silverstein RM, Webster FX, Kiemle DJ (2005) Spectrometric identification of organic compounds, 7th edn. Wiley, New York

Spiteller G (1966) Massenspektrometrische Strukturanalyse Organischer Verbindungen. Verlag Chemie GmbH, Weinheim, p 92

Stahl E (ed) (1967) Dünnschicht-Chromatographie—Ein Laboratoriumshandbuch, 2nd edn. Springer, Berlin

Stoddard DM (1974) The role of odor in the social biology of small mammals. In: Birch MC (ed) Pheromones. North-Holland, Amsterdam, pp 297–315

Walther FR (1964) Einige Verhaltensbeobachtungen an Thomsongazellen (*Gazella thomsoni* Günther, 1884) im Ngorongoro-krater. Z Tierpsychol 21:871–890

Walther FR (1965) Verhaltensstudien an der Grantgazelle (*Gazella granti* Brooke, 1872) im Ngorongoro-krater. Z Tierpsychol 22:167–208

Wheeler JW (1977) Properties of compounds used as chemical signals. In: Mozell MM, Müller-Schwarze D (eds) Chemical signals in vertebrates. Plenum Press, New York, pp 61–70

Wyatt TD (2003) Pheromones and animal behaviour. Communication by smell and taste. Cambridge University Press, Cambridge, specifically chapter 5

Chapter 16
Do Urinary Chemosignals Code for Sex, Age, and Season in the Giant Panda, *Ailuropoda melanoleuca*?

Dingzhen Liu, Hong Yuan, Zhipeng Wang, Rongping Wei, Guiquan Zhang, and Lixing Sun

Abstract The urinary volatile compounds of captive giant pandas, *Ailuropoda melanoleuca*, were analysed with GC-MS (gas chromatography and mass spectrometry). Samples were collected from adult males and females during the mating season, as well as from adult and sub-adult males during the non-mating season. We detected more than 96 compounds, 50 of which were common to all individuals. The major volatile constituents were ketones, phenols, alcohols, and fatty acids. No single compound was found to be specific for characterising sex, age, and season. MANOVA analysis on the relative abundance of compounds showed that adult males and females differ in four urinary compounds in the mating season and adult males differ in seven compounds between the mating and the non-mating seasons. We also found significant age effects on the relative abundance of urinary compounds in males in the non-mating season. This result suggests that those compounds might be the potential chemosignals signalling information about male pandas' age, sex, and reproductive state.

D. Liu (✉) • H. Yuan • Z. Wang
Ministry of Education, Key Laboratory of Biodiversity Science and Ecological Engineering & Institute of Ecology, Beijing Normal University, Beijing 100875, China
e-mail: dzliu@bnu.edu.cn

R. Wei • G. Zhang
China Research and Conservation Center for the Giant Panda, Sichuan 623006, China

L. Sun
Department of Biological Sciences, Central Washington University, Ellensburg, WA 98926-7537, USA

M.L. East and M. Dehnhard (eds.), *Chemical Signals in Vertebrates 12*,
DOI 10.1007/978-1-4614-5927-9_16,

16.1 Introduction

Mammalian urine is rich in organic compounds, presumably various secondary metabolic products which may communicate information about species, sex, and genetic makeup (Yamazaki et al. 1979). Evidence from a variety of mammals shows that urine can provide information about individual identity, age, sex, territoriality, dominance, physiological condition, and reproductive status (Albone 1984; Beynon et al. 1999; Brown 1979; Halpin 1986; Ma and Klemm 1997; Miller et al. 1998; Novotny et al. 1999; Zhang et al. 2005). A wide range of urinary chemical compounds including alcohols, aldehydes, ketones, phenols, fatty acids, and esters are involved in chemical signalling (Andersen and Vulpius 1999; Raymer et al. 1984; Wyatt 2003).

The identification of urinary constituents has mainly involved species commonly used for research on chemical communication, such as small rodents (Ma et al. 1999; Novotny et al. 1985) and white-tailed deer (*Odocoileus virginianus*) (Miller et al. 1998). Rare and exotic species such as the tiger (*Panthera tigris*) (Brahmachary and Dutta 1981; Poddar-Sarkar and Brahmachary 1999), Asian elephant (*Elephas maximus*) (Rasmussen et al. 1996b, 1997) and African elephant (*Loxodonta africana*) (Rasmussen et al. 1996a), and lion (*Panthera leo*) (Andersen and Vulpius 1999) have also been the foci of studies on the mechanisms of chemical communication in large mammals. Notably, the finding of the Asian elephant sharing its female sex pheromone (Z)-7-dodecen-1-yl acetate (Z7-12: Ac) with some 140 species of moth (Rasmussen et al. 1996b) has greatly prompted the studies in this area (Wyatt 2003).

Giant pandas (*Ailuropoda melanoleuca*) rely heavily on chemical signals released with anogenital gland secretions (AGS) and urine to coordinate their reproductive and social activities (Schaller et al. 1985). Scent marking using both sources is higher during the mating season than the non-mating season. In the mating season sexually active males and females scent-mark more frequently than sexually inactive ones (Liu et al. 1998), and females often mark with urine, especially around the time of sexual receptivity (Kleiman 1983; Schaller et al. 1985). Captive pandas discriminate between the odours of conspecifics based on AGS and urine (Swaisgood et al. 1999). Males show preference for oestrous over non-oestrous female urine, suggesting that female urine contains information about their oestrous status (Swaisgood et al. 2002). The fact of elevated urinary marking rates during the mating season also suggests that female pandas may use urine to facilitate mating. A previous study of urine constituents based on one panda pair shows a close association between free fatty acids and urinary oestrogen level in the female (Dehnhard et al. 2005). Gas chromatography reveals some differences between sexually active and inactive males in volatile compounds in urine (Liu et al. 2003). Urinary compounds are similar to those in AGS (Dehnhard et al. 2005), and genetically related individuals have similar AGS chemical profiles (Hagey and MacDonald 2003). Even so, little is known about the effects of sex, age, and season on urinary volatiles or how these volatiles are influenced by sex, age, and season.

To facilitate chemical communication between male and female adult pandas during the mating season, animals at Wolong breeding centre, in China, were periodically translocated between their cage and the cage of animals of the opposite sex which increased the rate of successful mating (Zhang et al. 2004). Based on the elevated level of urinary marking in both male and female pandas during the mating season, we hypothesised that urine contains chemosignals that communicate sex and reproductive state. This hypothesis predicts the composition of urine should differ either in quality or quantity (1) between adult males and females in the mating season, (2) in adult males between the mating and the non-mating seasons (Sun and Müller-Schwarze 1998a), and (3) between adult and sub-adult individuals in the non-mating season. We tested these predictions by identifying and analysing urinary compounds using GC-MS (gas chromatography–mass spectrometry).

16.2 Materials and Methods

16.2.1 Animals

We collected urine from a total of 25 giant pandas housed at the Wolong breeding centre in Sichuan Province, China, including six adult males and seven adult females in the mating season, and six adult males and six sub-adult males in the non-mating season. Individuals younger than 5 years of age were classified as sub-adults and animals older than 5 years as adults. During the mating season, adults were housed individually in enclosures with an indoor room (approximately 5.8 × 2.3 m) and an outdoor yard (5.8 × 13.0 m) with grass, climbing apparatuses, and a small pond. Cement walls separated outdoor enclosures. The animals could see, smell, hear, and have some limited physical contact with neighbouring animals of the opposite sexes through a mesh separator (0.5 × 9.0 m) in the wall between enclosures. All animals received the same daily food including two steamed buns, two pots of milk, and a pile of fresh bamboo (Liu et al. 1998). Fresh urinary samples were collected during the non-mating (August 2001, November 2001 and December 2004) and mating (April to June 2002) seasons.

16.2.2 Sample Collection and Solvent Extraction

Urinary samples were collected with clean syringes from the cement floor between 9:00 and 14:00 h and immediately transferred into clean vials (5–10 mL), sealed with Teflon-line lids and stored at –20°C. Due to the lack of enclosures, the centre housed two or three sub-adult pandas together within a common outdoor yard covered with grass during the mating season. As we were not allowed to separate individuals, we only collected urine samples from six sub-adult males during the non-mating season. We collected a total of 50 urine samples with two parallel urine samples from each panda.

All samples were chilled on ice when transported (about 2-h flight) to the laboratory in Beijing Normal University and were then stored at −20°C until analysed in August 2002. Samples from the adult females were collected in December 2004 (non-mating season) and analysed in the spring of 2005; however, the data were excluded from subsequent statistical analyses due to potential sample degradation.

We used ethyl ether as the solvent based on a pilot study, which showed that ether was excellent for the extraction of urinary compounds. We mixed 1 mL of urine with 1 mL of ether at room temperature. This mixture was shaken for 5 min and then centrifuged for 10 min at 3,000 rpm. We used 1 μL from the upper organic phase for GC-MS analysis.

16.2.3 GC-MS Analysis

Volatile compounds in urine were tentatively identified using a Finnigan TRACE2000 GC-MS (Thermo Electron Corporation, Austin TX) instrument with Xcalibur for Windows 2000 (version 1.2) for data acquisition and processing (Thermo Electron Corporation 2002). The GC analysis was performed with a 30 m×0.25 mm id×0.5 μm film thickness DB-5MS fused-silica capillary column (J &W Scientific Inc., Folsom, CA). Helium was used as carrier gas at a flow rate of 1 mL/min. The splitless mode of injection (1 μL) was used at an injector temperature of 250°C. The transfer line was held at 270°C. We set the column temperature initially at 55°C for 8 min, ramped at 15°C/min to 120°C, then ramped at 6°C/min to 180°C, at 8°C/min to 220°C, at 10°C/min to 270°C, and finally at 12°C/min to 300°C. Electron impact (70 eV) was used for ionisation. We applied the National Institute of Standards and Technology (NIST) library for tentative identification of the compounds without running authentic samples for confirmation. Even so, mass spectra were checked and confirmed by an experienced analytical chemist (Professor Helin Luan who ran all the GC-MS analyses). Occasional fused peaks (shoulder peaks) suggesting the presence of two compounds were investigated by comparing both GC retention times and mass spectra. At least one blank was run between consecutive samples to eliminate the possibility of contamination.

16.2.4 Statistical Analyses

For comparison, each peak was standardised by its percentage of the total amount in the sample and was referred to as the relative abundance (Sun and Müller-Schwarze 1998b). The relative abundances of common compounds (see below) in the urine of the 25 pandas from the two seasons were then compared either by sex, age, or breeding seasons. Rare and minor peaks contributing less than 0.1% to the total area were excluded from all subsequent analyses because of unreliable quantification at such low levels. We use multivariate ANOVA (MANOVA) to examine the sexual difference in the mating season (six males vs. seven females), and seasonal differences for adult males (six and six in the mating and the non-mating season, respectively) and age differences for males in the non-mating season (six adults and six sub-adults). To conduct the MANOVA analysis, we selected 11 common compounds as one subset each time as variables (Ronald R. Swaisgood, personal communication). This was repeated for the next 11 compounds until all the compounds were analysed. This resulted in nine MANOVA analyses for the 96 selected compounds. All statistics were conducted using SPSS for Windows (Release 10.0, Standard Version, SPSS Inc.). The level of significance was set at 0.05 a priori for all tests.

16.3 Results

16.3.1 Individual Difference in the Mating Season

A total of 96 different tentatively identified compound (TIC) peaks were detected from 50 samples collected from 25 individuals in the two seasons (Table 16.1). Only 50 compounds were common to all individuals. Phenols, alcohols, esters, ketenes, and alkanes constituted the majority. Other minor volatiles included fatty acids, fatty acid esters, alkenes, and amines.

The chemical composition of urine differed among individuals. The mean number of peaks per individual was 65.96 ± 6.03 SE peaks (range 57–79 peaks). The relative abundance of the compounds varied greatly among individuals. Many compounds, especially those with small molecular weights and high volatility, were absent in the adult males' urine during the non-mating season (Table 16.1).

16.3.2 Sexual Differences in the Urinary Chemical Composition Between Adult Males and Females in the Mating Season

MANOVA analysis revealed remarkable sexual differences in the relative abundance of urinary compounds between adult male and female pandas in the mating season. Adult males had significantly higher urinary concentrations of compounds 10 (retention time—RT = 13.21 min) and 17 (RT = 14.63 min; see Table 16.1) than females (both $P < 0.05$; Fig. 16.1) while lower concentrations of compounds 27 (RT = 15.92 min) and 74 (RT = 24.60 min; $P < 0.05$ and $P < 0.01$, respectively; Fig. 16.1). No significant sex differences were found regarding the other urinary compounds.

Table 16.1 Chemical compounds tentatively identified in panda urine

Peak No.	RT (min)	Tentatively identified compound	MW	No. of individuals having the compound: Mating season: Adult male (n=6)	Mating season: Adult female (n=7)	Non-mating season: Adult male (n=6)	Non-mating season: Sub-adult male (n=6)
1	11.36	1-(1-Methylethoxy)-butane	116	6	7	6	6
2	11.51	3-Hydroxy-butanoic acid ethyl ester	132	5	7	6	5
3	11.80	3,6-Dimethyl-4, 5-octanediol	174	0	3	0	3
4	11.98	2,7-Dimethyl-4, 5-octanediol	174	0	3	1	2
5	12.02	Benzyl alcohol	108	0	0	0	2
6	12.63	4-Methyl-phenol	108	2	4	2	2
7	12.89	2-Ethoxy-butane	102	6	7	6	6
8	13.11	An ether	?	6	7	6	6
9	13.17	An ether	?	6	7	6	6
10	13.21	Hexyl isopropyl ether	144	6	7	6	6
11	13.33	Phenylethyl alcohol	122	0	2	0	4
12	13.86	2,4-Dimethyl-phenol	122	2	0	0	0
13	13.98	(Z)-3-Nonen-1-ol	142	3	3	5	6
14	14.14	Unknown	122	6	7	6	6
15	14.17	2,3-Ddimethyl-phenol	122	2	1	0	0
16	14.62	1-Dodecene	168	1	0	2	0
17	14.63	1-Methylene-1H-indene	128	6	7	6	6
18	14.76	Dodecane	170	6	7	6	6
19	15.01	2, 3-Dihydro-benzofuran	120	6	7	6	6
20	15.17	2-(1-Methylethyl)-phenol	136	1	0	0	0
21	15.22	An ether	?	6	4	4	6
22	15.27	2-(2-Propenyl)-1,3-dioxolane	114	6	5	5	6
23	15.36	2-Ethyl-5-methyl-phenol	136	1	0	0	0
24	15.61	An ether	?	6	7	6	6

25	15.71	An ether	?	6	7	6	6
26	15.78	Nonanoic acid	158	0	3	2	0
27	15.92	An ether	?	6	7	6	6
28	16.06	4-Ethyl-2-methoxy-phenol	152	0	1	0	3
29	16.25	An ether	?	6	7	6	6
30	16.41	Indole	117	0	1	1	5
31	16.49	Tridecane	184	6	5	6	6
32	16.62	2, 5-Diethylphenol	150	1	1	1	2
33	16.73	2-Methoxy-4-vinylphenol	150	4	4	3	3
34	16.77	Unknown	?	6	7	6	6
35	17.04	3-(1-Methylbutoxy)-2-butanol	160	6	7	6	6
36	17.14	*p*-Mentha-1(7),8(10)-dien-9-ol	152	0	0	0	2
37	17.39	3-(1-Methylbutoxy)-2-butanol	160	6	6	6	6
38	17.46	4-Hydroxy-benzaldehyde	122	2	4	4	2
39	17.64	Pentamethyl-benzene	148	4	3	4	1
40	17.65	9-Hydroxy-2-nonanone	158	0	0	0	4
41	17.70	4-Propyl-phenol	136	0	1	2	0
42	18.14	2-*n*-Butyladamantane	192	6	6	6	6
43	18.28	Tetradecane	198	6	7	6	6
44	18.62	1-(4-Hydroxyphenyl)-ethanone	136	0	4	2	3
45	18.91	1-(3-Hydroxyphenyl)-ethanone	136	1	4	5	3
46	18.93	1-(2-Hydroxyphenyl)-ethanone	136	6	5	6	6
47	19.20	Unknown	?	6	7	6	6
48	19.25	2-Benzothiazolecarboxaldehyde	163	6	7	6	6
49	19.29	Unknown	?	6	7	6	6
50	19.37	Ethyl 4-hydroxy-dl-mandelate	196	1	3	1	0
51	19.86	1-(4-Hydroxy-3-methoxyphenyl)-ethanone	166	6	7	6	4

(continued)

Table 16.1 (continued)

Peak No.	RT (min)	Tentatively identified compound	MW	No. of individuals having the compound			
				Mating season		Non-mating season	
				Adult male ($n=6$)	Adult female ($n=7$)	Adult male ($n=6$)	Sub-adult male ($n=6$)
54	20.44	Unknown	?	6	7	6	6
55	20.52	Unknown	?	6	7	6	6
56	20.62	Paroxypropione	150	6	7	6	6
57	21.15	An ether	?	6	6	6	6
58	21.24	2,6-Dimethoxybenzoquinone	168	4	2	1	0
59	21.32	8-Methyloctahydrocoumarin	168	6	6	6	6
60	21.44	An ether	?	6	6	6	6
61	21.64	2′,4′-Dihydroxy-3′-methylpropiophenone	180	2	4	5	3
62	21.86	Unknown	?	6	7	6	6
63	22.00	Hexadecane	226	6	7	6	6
64	22.09	Unknown	?	6	7	6	6
65	22.22	3-Hydroxy-.beta.-damascone	208	6	7	6	6
66	22.32	5-[1-Hydroxyhexyl]-tetrahydrofuran-2-one	186	5	7	2	4
67	22.56	3-Oxo-.alpha.-damascone	206	6	7	4	2
68	22.86	4-(2,6,6-Trimethylcyclohexa-1,3-dienyl) butan-2-one	192	2	4	6	6
69	23.05	Undecyl-cyclopentane	224	6	7	6	1
70	23.07	4-(2,6,6-Trimethyl-1-cyclohexen-1-yl)-3-buten-2-ol	194	0	2	3	6
71	23.26	3,5,5-Trimethyl-4-(3-oxobutyl)-2-cyclohexen-1-one	208	2	7	5	4
72	23.76	Heptadecane	240	6	7	6	4

73	23.80	4-(3-Hydroxybutyl)-3,5,5-trimethyl-2-cyclohexen-1-one	210	0	2	2	5
74	24.60	2,6-Dimethyl-2,6-dodecadiene	194	0	5	2	0
75	24.69	1H-Indole-3-ethanol	161	0	2	0	3
76	25.13	4-Hydroxy-3,5,5-trimethyl-4-(3-oxo-1-butenyl)-2-cyclohexen-1-one	222	6	7	4	6
77	25.25	Anthracene	178	6	7	6	6
78	25.35	Octadecane	254	6	7	6	6
79	25.41	1H-Indole-3-carboxaldehyde	145	1	3	1	0
80	25.45	A alkyl	?	6	7	6	6
81	26.05	3-Tert-butyl-4-hydroxyanisole	180	5	3	5	0
82	26.14	Unknown	?	1	4	4	5
83	26.31	Undecyl-cyclopentane	224	6	7	5	6
84	26.80	Nonadecane	268	6	7	6	6
85	27.13	Hexadecanoic acid methyl ester	270	1	0	2	1
86	27.92	Unknown	?	4	7	6	3
87	28.14	Eicosane	282	6	7	6	6
88	28.40	Isopropyl palmitate	298	6	7	6	4
89	28.99	1-Nonadecene	266	6	7	6	6
90	29.31	Heneicosane	296	6	7	6	6
91	29.58	Octadecanoic acidmethyl ester	298	1	0	2	0
92	29.93	Octadecanoic acid	284	0	2	0	0
93	30.13	Hexadecanamide	255	2	4	4	3
94	30.38	Docosane	310	6	7	6	6
95	31.36	Tricosane	324	6	7	6	6
96	31.92	(Z)-9-Octadecenamide	281	0	4	4	6

Note: Compounds are numbered in the order of increasing GC retention time. Trace components are not included. "MW" represents molecular weight, "?" represents the compound could not be identified

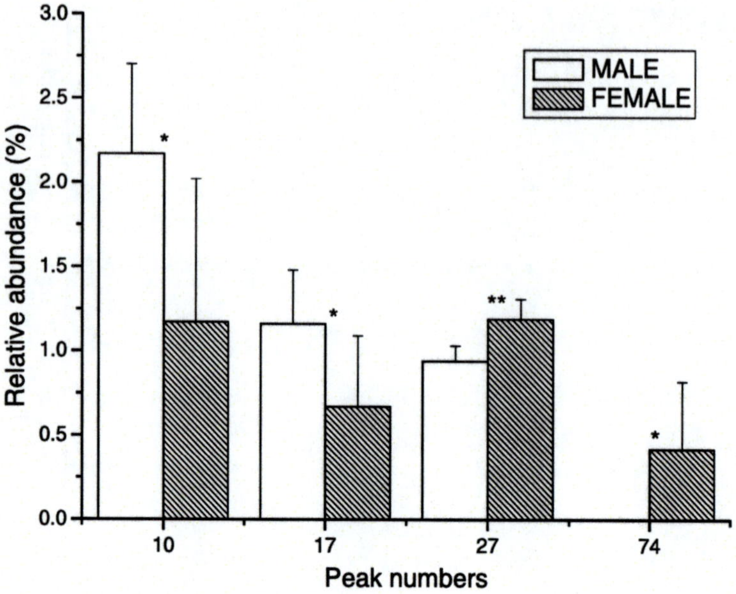

Fig. 16.1 Sexual differences in the relative abundances of urinary compounds for adult males ($n=6$) and females ($n=7$) during the mating season. Data are shown as mean ± SEM. *Single asterisk*: $P<0.05$, *double asterisk*: $P<0.01$

16.3.3 Seasonal Differences in the Relative Abundance of Urinary Compounds in Adult Male Pandas

We found marked seasonal differences in the relative abundance of seven urinary compounds in adult males between the mating and non-mating seasons, with adult males during the mating season having significantly higher concentrations of compounds 7 (RT = 12.89 min), 31 (RT = 16.49 min), 57 (RT = 21.15 min), 64 (RT = 22.09 min), and 87 (RT = 28.14 min) but lower concentrations of compounds 13 (RT = 13.98 min) and 68 (RT = 22.86 min) ($P<0.05$, Fig. 16.2). Figure 16.3 shows gas chromatographs of urine from a male panda in the mating season and non-mating season.

16.3.4 Age Effects on the Relative Abundance of Urinary Compounds During the Non-mating Seasons

We found significant age effects on the relative abundance of urinary compounds when comparing samples from adults and sub-adult male pandas collected during the non-mating seasons. The relative abundances of eight compounds with peak

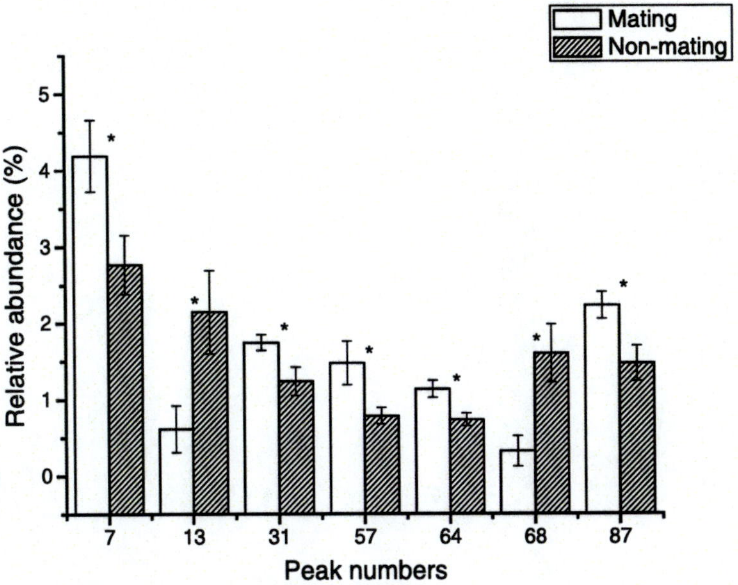

Fig. 16.2 Seasonal differences in the relative abundances of urinary compounds of adult males between the mating season ($n = 6$) and the non-mating season ($n = 6$). Data are shown as mean ± SEM. *Single asterisk*: $P < 0.05$

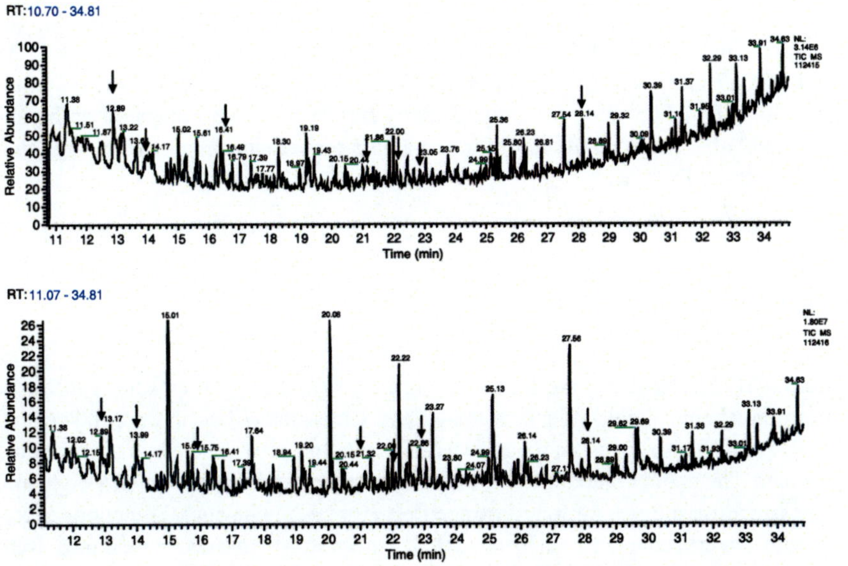

Fig. 16.3 Comparison of the profiles of gas chromatographs of urine from one adult male ("PanPan") during the mating (*upper panel*) and the non-mating season (*lower panel*). *Arrows* show the peaks (compounds) significantly differ in average relative abundance between the mating and the non-mating seasons

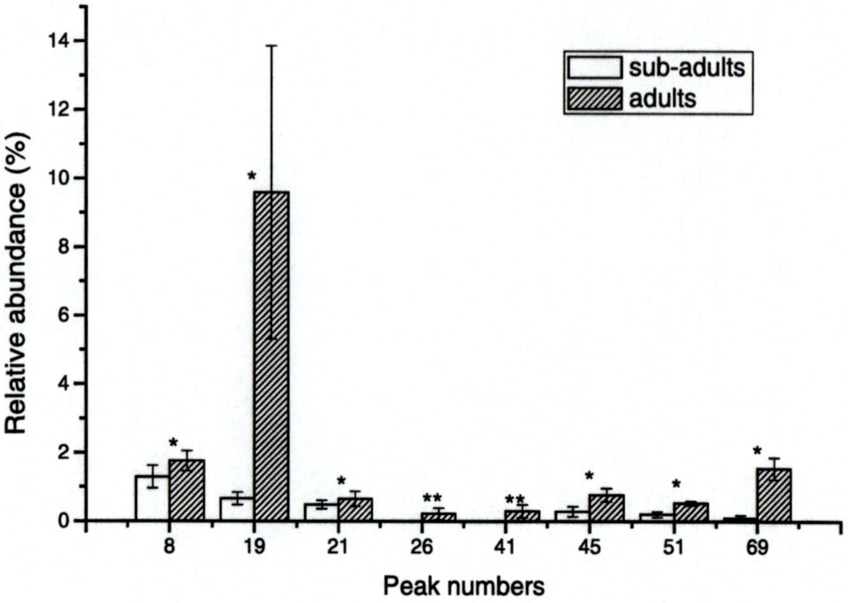

Fig. 16.4 Age differences in the relative abundances of urinary compounds between adult ($n=6$) and sub-adult ($n=6$) males in the non-mating season. Data are shown as mean ± SEM. *Single asterisk*: $P<0.05$, *double asterisk*: $P<0.01$

numbers of 8 (RT = 13.11 min), 19 (RT = 15.01 min), 21 (RT = 15.22 min), 26 (RT = 15.87 min), 41 (RT = 17.70 min), 45 (RT = 18.91 min), 51 (RT = 19.86 min), and 69 (RT = 23.05 min) were significantly higher in adults compared with sub-adults ($P<0.05$ and <0.01, $n=6$ for sub-adults and adults, respectively; Fig. 16.4 and Table 16.1).

16.4 Discussion

Using GC-MS analysis, we found that urinary volatiles were strikingly different from the chemical constituents of anal gland secretions (AGS) in the panda (Yuan et al. 2004). This suggests a functional differentiation between the two chemosignal sources. The major compounds in AGS are steroids with high molecular weight and low volatility, suitable for long-lasting effects for such behavioural functions as territorial advertisement (Yuan et al. 2004). In contrast, urinary compounds were mainly volatiles such as alcohols, phenols, and ketones. These compounds are relatively low in molecular weight and high in volatility and are thus suitable for short-term communication and may be useful in sexual attraction and recognition (Alberts 1992). Our study is among the first to identify chemosignals and their relation to sexual, seasonal, and age differences in the urine of Ursids, where little is known about the role of chemical communication. The significant differences found in the relative abundances of chemical compounds between males and females in the mating season may provide a new lead for further research to identify sex pheromones in the panda. The age difference in the urinary compounds between adult and sub-adult males may provide a simple and useful measurement for determining whether a male is mature and can be used for captive breeding. Finally, the relative amounts of chemicals in urine (this study) and AGS (Yuan et al. 2004) are possibly important in individual recognition.

We found evidence for sexual dimorphism in the relative abundances of urinary chemical compounds in the giant panda during the mating season, even though the differences existed only in four compounds. Similar results have been found in other mammals, such as the wolf (*Canis lupus*) (Raymer et al. 1984), the deer mouse (*Peromyscus maniculatus*) (Ma et al. 1999), and the house mice (*Mus musculus*) (Zhang et al. 2007). Thus, our first prediction was supported. Moreover, our results indicate that the sexual dimorphism was coded by changes in the relative abundances of urinary compounds instead of by their presence or absence (Sun and Müller-Schwarze 1998a, b).

In addition to sex differences, seasonal changes may affect behavioural, physiological, and ecological characteristics in a species. Our results suggest that the relative abundances of chemical compounds in panda urine differed quantitatively between the mating and non-mating seasons in adult males. Similar results have also been found in other mammals such as the lion (Andersen and Vulpius 1999), white-tailed deer (Miller et al. 1998), deer mouse (Ma et al. 1999), and African elephants (Rasmussen and George 2002). These changes might convey information associated with their reproductive state and explain why females pandas show elevated scent-marking and bleating behaviour in response to urine samples from adult males in the mating season compared to the non-mating season (Liu 1996; Xu et al. 2012). Therefore, our second prediction was supported. Our results provide evidence that the relative abundance of urinary compounds might convey reproductive state in male pandas.

Developmental effects have been reported in urinary chemosignals based on behavioural studies on pandas (White et al. 2003, 2004). Here we provide chemical evidence that may explain these behavioural results. We found significant differences in the relative abundance of urinary compounds between adult and sub-adult males during the non-mating season. These might code for information associated with age and/or social status. Obviously, it would be an advantage for sub-adults to detect and identify adult male urine to avoid being attacked. This explains why young pandas try to avoid the urinary marks deposited at a higher position by dominant males (White et al. 2002). Thus, our third prediction was also supported.

Interestingly, all compounds found in males in the non-mating season were also detected in females in the mating season and sub-adult males in the non-mating season. This suggests these compounds might not be sex- or age-specific but species-specific. Comparisons of urinary chemical profiles also showed that no compounds were unique to a certain season and age group. Although some compounds were present in adults or sub-adults, none were found in all individuals in either age group. Hence, coding for these age categories might not depend on particular chemicals but on the combination and ratios of several urinary compounds. This analogue information coding system fits well with the odour-quality model or mosaic signal hypothesis (Johnston 1983; Todrank and Heth 2003).

Despite our findings, there are several pitfalls in our study. We failed to collect urine from sub-adults in the mating season and to run GC-MS analysis on urine from adult females in the non-mating season in the same year that we analysed urine for adult males and females in the mating season. Moreover, the tentative identification of the chemical compounds and lack of behavioural assay also prevented us from making a sound conclusion for this study. These issues need to be addressed in the future. Saying this, our study revealed quantitative differences in the chemical composition of urine in relation to sex, age, and season. It provides some new insights for age discrimination found in behavioural tests (Liu et al. 2005; Swaisgood et al. 1999, 2000). Behavioural assay are required to examine the functions of single compounds or their combinations, especially for those showing seasonal differences and correlation with pandas' ages. In this sense, our results will be valued for providing guidance for focused behavioural tests based on the compounds that are likely to be involved in the recognition of sex, age, season, or genetic relatedness in the panda.

Acknowledgements We are grateful to Professors Helin Luan and Wen Yao for their help in GC-MS analysis. We also thank the staff at the China Conservation and Research Centre for the Giant Panda at Wolong for their assistance in sample collection. This work was supported by grants from the National Natural Science Foundation of China (Grant #30470233, #30230080, #30070107), the Research Fund for the Doctoral Program of Higher Education (Grant #98002717), and the grant from the China Wildlife Conservation Association (Grant #96035) and the International Cooperation Projects of Giant Pandas from State Forestry Administration, P. R. China (Grant #WH0309, #WH0306, #SD0630).

References

Alberts AC (1992) Constraints on the design of chemical communication-systems in terrestrial vertebrates. Am Nat 139:S62–S89

Albone ES (1984) Mammalian semiochemistry: the investigation of chemical signals between mammals. Wiley-Interscience, Wiley & Sons, Chichester, UK, p 360

Andersen KF, Vulpius T (1999) Urinary volatile constituents of the lion, *Panthera leo*. Chem Senses 24:179–189

Beynon RJ, Robertson DHL, Hubbard SJ, Gaskell SJ, Hurst JL (1999) The role of protein binding in chemical communication. In: Johnston RE, Müller-Schwarze D, Sorensen PW (eds) Advances in chemical signals in vertebrates. Kluwer Academic/Plenum Publishers, New York, pp 137–147

Brahmachary RL, Dutta J (1981) On the pheromones of tigers: experiments and theory. Am Nat 118:561–567

Brown RE (1979) Mammalian social odors: a critical review. Adv Stud Behav 10:103–162

Dehnhard M, Hildebrandt T, Knauf T, Ochs A, Ringleb J, Göritz F (2005) Chemical signals in giant panda urine (*Ailuropoda melanoleuca*). In: Mason RT, LeMaster MP, Müller-Schwarze D (eds) Chemical signals in vertebrates 10. Springer US, New York, NY, pp 110–117

Hagey L, MacDonald E (2003) Chemical cues identify gender and individuality in giant pandas (*Ailuropoda melanoleuca*). J Chem Ecol 29:1479–1488

Halpin ZT (1986) Individual odors among mammals: origins and functions. In: Rosenblatt JS, Beer C, Busnel M, Slater PJB (eds) Advances in the study of behavior, vol 16. Academic, New York, pp 39–70

Johnston RE (1983) Chemical signals and reproductive behavior. In: Vandenbergh JG (ed) Pheromones and reproduction in mammals. Academic, New York, pp 3–37

Kleiman DG (1983) Ethology and reproduction of captive giant pandas (*Ailuropoda melanoleuca*). Z Tierpsychol 62:1–46

Liu D (1996) Studies on the behavioral ecology of captive giant pandas (*Ailuropoda melanoleuca*). PhD Dissertation (in Chinese), Beijing Normal University, Beijing

Liu D, Fang J, Sun R, Zhang G, Wei R, Zhang H (1998) Behavioral comparison in individuals of different sexual ability in giant panda (*Ailuropoda melanoleuca*). Acta Zool Sinica (in Chinese) 44:27–34

Liu D, Wang LW, Zhang XT, Wei RP, Zhang GQ, Yun ZH, Sun RY (2003) Urinary volatile profiles of the giant pandas pertaining to gender, season and sexual ability. J Beijing Normal University (Natural Science) (in Chinese) 39:123–130

Liu D, Zhang G, Wei R, Zhang H, Fang J, Sun R (2005) Behavioral responsiveness of captive giant pandas (*Ailuropoda melanoleuca*) to substrate odors from conspecifcs of the opposite sex. In: Mason RT, LeMaster MP, Müller-Schwarze D (eds) Chemical signals in vertebrates 10. Springer, New York, pp 101–109

Ma W, Klemm WR (1997) Variations of equine urinary volatile compounds during the oestrous cycle. Vet Res Commun 21:437–446

Ma W, Wiesler D, Novotny MV (1999) Urinary volatile profiles of the deermouse (*Peromyscus maniculatus*) pertaining to gender and age. J Chem Ecol 25:417–431

Miller KV, Jemiolo B, Gassett JW, Jelinek I, Wiesler D, Novotny M (1998) Putative chemical signals from white-tailed deer (*Odocoileus virginianus*): social and seasonal effects on urinary volatile excretion in males. J Chem Ecol 24:673–683

Novotny M, Harvey S, Jemiolo B, Alberts J (1985) Synthetic pheromones that promote inter-male aggression in mice. Proc Natl Acad Sci USA 82:2059–2061

Novotny MV, Ma W, Zidek L, Daev E (1999) Recent biochemical insights into puberty acceleration, estrus induction, and puberty delay in the house mouse. In: Johnston RE, Müller-Schwarze D, Sorensen PW (eds) Advances in chemical signals in vertebrates. Kluwer Academic/Plenum Publishers, New York, pp 99–116

Poddar-Sarkar M, Brahmachary RL (1999) Can free fatty acids in the tiger pheromone act as an individual fingerprint? Curr Sci 76:141–142

Rasmussen LEL, George W (2002) Chemosignalling of musth by individual wild African elephants (*Loxodonta africana*): implications for conservation and management. Proc R Soc B Biol Sci 269:853–860

Rasmussen LEL, Anthony J, Martin H, Hess DL (1996a) Chemical profiles of male African elephants, *Loxodonta africana*: physiological and ecological implications. J Mammal 77:422–439

Rasmussen LEL, Lee TD, Roelofs WL, Zhang AJ, Daves GD (1996b) Insect pheromone in elephants. Nature 379:684

Rasmussen LEL, Lee TD, Zhang A, Roelofs WL, Daves GD Jr (1997) Purification, identification, concentration and bioactivity of (Z)-7-dodecen-1-yl acetate: sex pheromone of the female Asian elephant, *Elephas maximus*. Chem Senses 22:417–437

Raymer J, Wieler D, Novotny M, Asa C, Seal US, Mech LD (1984) Volatile constituents of wolf (*Canis lupus*) urine as related to gender and season. Experientia 40:707–709

Schaller GB, Hu J, Pan W, Zhu J (1985) The giant pandas of Wolong. University of Chicago Press, Chicago, p 298

Sun L, Müller-Schwarze D (1998a) Anal gland secretion codes for family membership in the beaver. Behav Ecol Sociobiol 44:199–208
Sun L, Müller-Schwarze D (1998b) Anal gland secretion codes for relatedness in the beaver, *Castor canadensis*. Ethology 104:917–927
Swaisgood RR, Lindburg DG, Zhou X (1999) Giant pandas discriminate individual differences in conspecific scent. Anim Behav 57:1045–1053
Swaisgood RR, Lindburg DG, Owen MA (2000) The effects of sex, reproductive condition and context on discrimination of conspecific odours by giant pandas. Anim Behav 60:227–237
Swaisgood RR, Lindburg DG, Zhang H (2002) Discrimination of oestrous status in giant pandas (*Ailuropoda melanoleuca*) via chemical cues in urine. J Zool 257:381–386
Todrank J, Heth G (2003) Odor-genes covariance and genetic relatedness assessments: rethinking odor-based "recognition" mechanisms in rodents. Adv Stud Behav 32:77–130
White AM, Swaisgood RR, Zhang H (2002) The highs and lows of chemical communication in giant pandas (*Ailuropoda melanoleuca*): effect of scent deposition height on signal discrimination. Behav Ecol Sociobiol 51:519–529
White AM, Swaisgood RR, Zhang H (2003) Chemical communication in the giant panda (*Ailuropoda melanoleuca*): the role of age in the signaller and assessor. J Zool 259:171–178
White AM, Swaisgood RR, Zhang H (2004) Urinary chemosignals in giant pandas (*Ailuropoda melanoleuca*): seasonal and developmental effects on signal discrimination. J Zool 264:231–238
Wyatt TD (2003) Pheromones and animal behaviour—communication by smell and taste. Cambridge University Press, London, p 391
Xu M, Wang Z, Liu D, Wei R, Zhang G, Zhang H, Zhou X, Li D (2012) Cross-modal signaling in giant pandas. Chin Sci Bull 57:344–348
Yamazaki K, Yamaguchi M, Baranoski L, Bard J, Boyse EA, Thomas L (1979) Recognition among mice. Evidence from the use of a Y-maze differentially scented by congenic mice of different major histocompatibility types. J Exp Med 150:755–760
Yuan H, Liu D, Sun L, Wei RP, Zhang GQ, Sun RY (2004) Anogenital gland secretions code for sex and age in the giant panda, *Ailuropoda melanoleuca*. Can J Zool 82:1596–1604
Zhang GQ, Swaisgood RR, Zhang HM (2004) Evaluation of behavioral factors influencing reproductive success and failure in captive giant pandas. Zoo Biol 23:15–31
Zhang JX, Soini HA, Bruce KE, Wiesler D, Woodley SK, Baum MJ, Novotny MV (2005) Putative chemosignals of the ferret (*Mustela furo*) associated with individual and gender recognition. Chem Senses 30:1–11
Zhang JX, Rao XP, Sun L, Zhao C, Qin X (2007) Putative chemical signals about sex, individuality and genetic background in the preputial gland and urine of the house mouse (*Mus musculus*). Chem Senses 32:293–303

Chapter 17
On the Body Odour of Wild-Caught Mangrove-Marsh Bengal Tiger of Sundarban

Mousumi Poddar-Sarkar, Sibdas Ray, Subrata Pal Chowdhury, Gopal Samanta, Paramaa Raha, and Ratan L. Brahmachary

Abstract We report the chemical analysis of the body odour collected from three tranquilised wild tigers (*Panthera tigris tigris*), from the Sundarban Biosphere Reserve in India. The electron micrograph of hairs from two tigers has been elucidated. A few saturated, unbranched fatty acid methyl esters like methyl tetradecanoate, methyl pentadecanoate, methyl hexadecanoate, methyl stearate, methyl eicosanoate and methyl eruciate were identified. A few benzenoid compounds were also detected. The possible significance of body odour and hair structure is discussed.

17.1 Introduction

While there are a number of studies on the volatile molecules in the scent marks (synonymous with marking fluid) of the tiger (*Panthera tigris tigris*) that may function as pheromones (Brahmachary and Dutta 1981; Poddar-Sarkar et al. 1991;

M. Poddar-Sarkar (✉)
Department of Botany, University of Calcutta, Kolkata 7000 019, India
e-mail: mpsarkar1@rediffmail.com

S. Ray
Department of Chemistry, University of Calcutta,
Kolkata 700 009, India

S.P. Chowdhury
Department of Forest (Wildlife wing), Government of West Bengal,
Bikash Bhavan, Calcutta 700 091, India

G. Samanta
ABAHC, Alipurdwar I, Jalpaiguri, West Bengal, India

P. Raha
Department of Zoology, Bethune College, Kolkata 700 006, India

R.L. Brahmachary
Embryology Unit, Indian Statistical Unit, Kolkata 700 035, India

M.L. East and M. Dehnhard (eds.), *Chemical Signals in Vertebrates 12*,
DOI 10.1007/978-1-4614-5927-9_17,

Brahmachary et al. 1992; Burger et. al. 2008), the body odour of the tiger has remained unexplored. Tigers are known to rub their head, face and body on tree trunks in the jungle (Thapar 2006), and in captivity (Poddar-Sarkar unpublished) on the walls and other objects (Brahmachary unpublished) and to roll on grasses and herbs (Poddar-Sarkar unpublished). Prey animals flee when they smell the presence of a tiger even though they may not show any signs of alarm when they detect tiger scent smeared shrubs or tree trunks (Brahmachary unpublished).

The odour of the Bengal tiger is sometimes perceptible to human beings (Locke 1954; Sher Jung 1967). We did not detect any body odour on two captive tigers (one subadult male, one adult female) but more than one forest staff of the Sundarban Biosphere Reserve confirmed detecting a strong body odour on tranquilised Sundarban tiger (personal communication). Sankhala (1993) also failed to perceive body odour on captive tigers but registered an unusual odour on five free-ranging tigers. However, one of us (MPS) perceived the specific body odour of a Sundarban wild-caught tiger maintained in a rescue camp. Brander (1923) referred to the nasty odour of wild tigers when shot and skinned.

Between 2008 and 2010 tigers from the core forest of the Sundarban Biosphere Reserve that entered villages were tranquilised for relocation and the attachment of radio-collars. This provided an opportunity to study the chemistry of tiger body odour. Here we describe the chemical compounds we detected in samples from three such tigers and presented the electron microscopic images from the body hair of two of these tigers.

17.2 Methods and Materials

We collected body odour samples from one adult male tiger sampled in 2008 and one female and one male in 2010 that were thought to be adults, by rubbing the upper hind portion of the body (thus avoiding contamination from soil and mud) with 700 mg of cotton-wads soaked with hexane held in a clean pair of forceps. These samples were stored in airtight vials in normal refrigerator at –4°C in field station and then at –20°C in laboratory after transportation until analysed. One microlitre of hexane extract from each tiger was investigated. Extracts were injected separately into the DB5 capillary column for gas chromatography mass spectrometric analysis (GCMS; Varian CP3800, MS 2200) under a temperature programming of between 70 and 200°C with a 5°C rise in temperature per minute while keeping MS ion trap at 200°C, injector port at 260°C and the carrier gas flow rate at 1 mL/min. For confirmation 1 μL aliquot of the second and third sample were injected in a different column (Elite 5 MS) in a Perkin Elmer, Clarus 600C GCMS using a temperature programming of 50°C for 2 min followed by a 10°C rise in temperature per minute to 280°C. Compounds were identified both by comparing their relative retention times with authentic compounds and from NIST mass spectral data base, and more importantly, from the characteristic features of mass fragments. Only those compounds that permitted a clear-cut interpretation were considered. Co-injection was not done because of insufficient remaining solvent extract for further analysis.

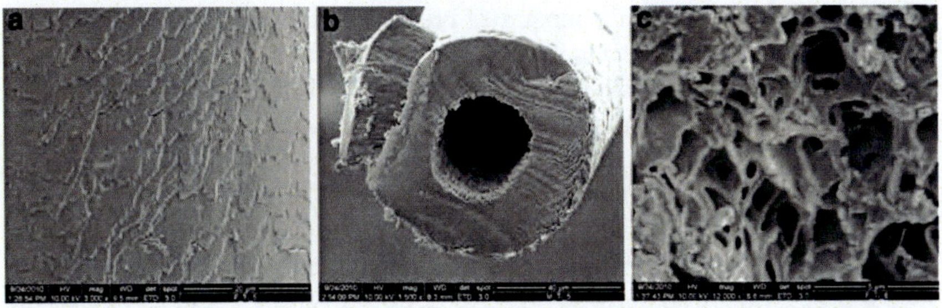

Fig. 17.1 **(a)** Surface characteristic of tiger body hair in SEM. **(b)** Cross section of hairs of tiger showing thick cortical layer. **(c)** Cross section of hairs of tiger showing medulla with tissue anastomosing having channel and holes

Hair samples collected from the rear hind thigh of the tigers sampled in 2010 were gold-coated and subjected to scanning electron microscopy (SEM) (Poddar-Sarkar et al. 2008).

17.3 Results and Discussion

The surface feature of the hair ultrastructure (Fig. 17.1a) shows the usual scaly nature observed in most other mammalian hair. In cross section the cortex region of a hair was thick (Fig. 17.1b) and the lumen was composed of a network of channels and holes (Fig. 17.1c). The SEM cross section of tiger hair revealed a structure that might facilitate the transportation of odour molecules. Results from our chemical analysis of olfactory compounds from the body emanation of one male tigers permitted us to identify five fatty acid methyl esters (FAME) of higher alkanoic acids, namely, methyl pentadecanoate, methyl hexadecanoate, methyl stearate, methyl eicosanoate and methyl eruciate. We identified only three FAME in the results obtained from one tigress, namely, tetradecanoic hexadecanoic and octadecenoic acid methyl ester. In the third male tiger, we detected methyl undecanoate based on the very abundant peak at 169, along with peaks at 111 ($CH_2=CH(CH_2)_4CO^+$), 98($CH_2(CH_2)_4CO^+$), 97 ($CH_2=CH(CH_2)_3CO^+$), 84 ($CH_2(CH_2)_3CO^+$), 83($CH_2=CH(CH_2)_2CO^+$) and a low intensity peak at 200 (M^+) peak. The presence of consecutive radical cations supports this interpretation. In addition, nitrobenzene compound was detected in the first sample, and methylbenzaldehyde and hydroxybenzaldehyde in the third sample.

According to Curran et al. (2005), individual body odour is the amalgamated product influenced by genetic, environmental and internal factors that vary with time and individual condition. Many free fatty acids C2–C9 (Poddar-Sarkar et al. 1991) as well as fatty acids of higher carbon chain C14–C22 (Poddar-Sarkar 1996) were identified from the steam distillate and lipid fraction of tiger marking fluid, respectively. Burger et al. (2008) also identified some branched and unbranched fatty acids and dimethyl esters of dicarboxylic acid in addition to some lactone, alkanol and alkanone from the mixture of urine and probably marking fluid. Some of these fatty acids seem to occur in the body smell in methyl ester form. Dodecanoic acid methyl ester, tetradecanoic acid methyl ester and octadecanoic acid methyl esters for example were identified from axillary sweat sample and human skin emanations (Curran et al. 2005). It is also known that among different chemical classes, esters of fatty acids such as methyl undecanoate, methyl decanoate, methyl nonanoate, methyl octanoate and methyl hexanoate show the strongest connection to human leucocyte antigen phenotype and body odour preferences (Janeš et al. 2010).

Our results from wild tigers strongly suggest that tigers have a series of alkanoic acid methyl esters with higher carbon number and benzene derivatives (having strong odour) in the hair/skin which are responsible for their body odour although hexane admittedly misses certain more polar compounds which are soluble in dichloromethane. The biological function of these odours remains to be clarified but might serve as a 'signature mixture', as a 'pheromone' of mating pairs or as an 'allomone' for prey animals. Furthermore, the prey animals may also perceive this and flee.

Acknowledgements The work was supported partially by University Grants Commission and Council for Scientific and Industrial Research, Govt. of India. We are grateful to PCCF, Dept. of Forest, Govt. of West Bengal for his kind permission and to the staff of the forest department who regularly took their risk to capture the wild, dangerous but gracious creature of the nature.

References

Brahmachary RL, Dutta J (1981) On the pheromones of tigers: experiments and theory. Am Nat 118:561–567

Brahmachary RL, Sarkar MP, Dutta J (1992) Chemical signals in tiger. In: Doty RL, Müller-Schwarze D (eds) Chemical signals in vertebrates, vol 6. Plenum, New York, pp 471–475

Brander AAD (1923) Wild animals in Central India. Edward Arnold, London

Burger BV, Viviers M, Bekker JPI, Roux MLE, Fish N, Fourie WB, Weibchen G (2008) Chemical characterization of territorial marking fluid of male Bengal tiger, *Panthera tigris*. J Chem Ecol 34:659–671

Curran AM, Robin SI, Prada PA, Furton KG (2005) Comparison of the volatile organic compound present in human odour using SPME-GC. J Chem Ecol 31:1613–1625

Janeš D, Klun I, Viden-Jeras B, Jeras M, Kreft S (2010) Influence of MHC on odour perception of 43 chemicals and body odour. Cent Eur J Biol 5(3):324–330

Locke A (1954) Tigers of Trengganu. Museum Press, London

Poddar-Sarkar M, Brahmachary RL, Dutta J (1991) Short chain free fatty acids as putative pheromones in the marking fluid of tiger. J Ind Chem Soc 68:255–256

Poddar-Sarkar M (1996) The fixative lipid of tiger pheromone. J Lipid Med Cell Signal 15:89–101

Poddar-Sarkar M, Chakroborty A, Bhar R, Brahmachary RL (2008) Putative pheromones of lion mane and its ultrastructure. In: Hurst JL, Beynon RJ, Roberts SC, Wyatt TD (eds) Chemical signals in vertebrates, vol 11. Springer, New York, pp 61–67

Sankhala KS (1993) Return of the tiger. Lustre, New Delhi

Jung S (1967) Tryst with tigers. Robert Hale, London (Reprinted Hind Pocket Books, Delhi)

Thapar V (2006) Tiger: the ultimate guide. Oxford University Press, New Delhi

Chapter 18
Ultrastructure and Chemical Composition of Elephant Hair in the Context of Chemical Signals in the Asian Elephant *Elephas maximus*

Paramaa Raha, Mousumi Poddar-Sarkar, Utpal K. Nag, Radhaballabh Bhar, and Ratan L. Brahmachary

Abstract Basic knowledge of chemical signalling in Asian elephants (*Elephas maximus*) involves identification of the female-emitted-male-received sex pheromone (Z)-7-dodecenyl acetate and a series of malodorous compounds together with frontalin from the temporal gland secretion of adult males. We describe the ultrastructure and lipid chemistry of the tail hair of the Asian elephant which may have special significance in relation to the scent flagging behaviour of receptive females of the species. Scanning electron micrographs of both the surface and cross sections of the elephant hairs show a marked deviation from those observed in the hair of other mammalian species. Chemical analysis of tail and head hair reveals the presence of C_8–C_{24} fatty acids that included straight-chain saturated, unsaturated and some branched-chain compounds. The significance of these fatty acids in relation to flagging behaviour is discussed.

P. Raha (✉)
Department of Zoology, Bethune College,
Kolkata 700 006, India
e-mail: paramaa@rediffmail.com

M. Poddar-Sarkar
Department of Botany, University of Calcutta, Kolkata 700 019, India

U.K. Nag
Department of Forest, Government of West Bengal, ADFO Wildlife Division I , Darjeeling, India

R. Bhar
Department of Instrumentation Science, Jadavpur University, Kolkata 700 032, India

R.L. Brahmachary
Embryology Unit, Indian Statistical Unit, Kolkata 700 035, India

M.L. East and M. Dehnhard (eds.), *Chemical Signals in Vertebrates 12*,
DOI 10.1007/978-1-4614-5927-9_18,

18.1 Introduction

In elephants, chemical signals play a vital role in sexual communication and include male-to-female signals (Perrin and Rasmussen 1994; Schulte and Rasmussen 1999a, b) and female-to-male signals (Rasmussen et al. 1996, 1997). This is best illustrated by the identification of a single female-to-male preovulatory pheromone (Z)-7-dodecenyl acetate which helps males to determine the location and reproductive status of receptive females (Rasmussen et al. 1997). In contrast, adult male Asian elephants (*Elephas maximus*) in musth signal to both females and males with malodorous compounds and the bicyclic ketal, frontalin, that are produced in their temporal gland (Rasmussen et al. 2002; Rasmussen and Greenwood 2003). Thus, research during the past decade has focused mainly on the identification of odoriferous compounds contained in elephant emissions like urine and temporal gland secretions that elicit biological responses from conspecifics (Rasmussen and Krishnamurthy 2000).

Gajashastra, more than 2,000 years ago, described the 'scent flagging' behaviour of receptive female Asian elephants (Bandopadhyay and Brahmachary 1999). Krishnan (1972) described scent flagging behaviour as the tail being dragged across the clitoris and subsequently raised up and waved or flicked like a flag. This behaviour may disperse volatile chemical compounds of the vaginal fluid as a signal of forthcoming ovulation. This suggests that tail hair may play a part in transporting the putative chemical molecules from the vagina to the environment. In this chapter we investigate the chemical composition of hair lipids and their possible role in communication. As far as we are aware, this topic in relation to social communication in Asian elephants has not yet been described. Although elephants are generally considered hairless animals, both African and Asian elephants are born with thick hair most of which is shed before the elephant is actually born. The distribution of hair is thickest on the tail tuft while they are sparsely distributed on the head and back.

Epidermal structures such as hair, quills or spines may play a role in storage and transportation of putative semiochemicals (Poddar-Sarkar et al. 2008, 2011). From the assumptions of these previous studies, we have investigated the tail hair lipid chemistry which might play a pheromonal and fixative role in addition to those of vaginal secretion and also the ultrastructure of tail hair which might be a receptacle of pheromone.

18.2 Methods and Materials

The tail hairs (three females and two males) and forehead hairs (two females and one male) of Asian elephant were collected with the help of Forest Departmental authorities. These were then processed for the ultrastructural study and chemical analysis.

18.2.1 Preparation of Samples for Scanning Electron Microscope Images

The hairs were washed thoroughly with distilled water before use. Thin sections were cut and subjected to gold coating. The scanning electron microscope (SEM) images of the dorsal surface and the cross section of the hairs were taken by FEI Quanta 200 (Netherlands) at different magnifications.

18.2.2 Chemical Analysis

Lipid was extracted from the hairs by a modification of the method of Bligh and Dyer (1959). The lower chloroform phase was separated out and collected over anhydrous sodium sulphate and spotted on high performance thin layer chromatography (HP-TLC-Aluminium sheets, E. Merck- 20×20 cm Silica gel $60F_{254}$) plate for fractionation of different lipid classes (Stahl 1965). The different classes of neutral lipid were detected with the help of HPTLC apparatus (Camag, USA) by applying different staining reagents like 0.1% 2,7-dichlorofluorescein in methanol and Liebermann-Büchard reagent (Kates 1972; Huang et al. 1961).

An aliquot of the total lipid in chloroform was derivatised to methyl ester for detailed analysis of fatty acid composition (Poddar-Sarkar 1996). The hexane extracts of the fatty acid methyl ester (FAME) were subjected to gas chromatography mass spectrometry (GC-MS, Varian CP3800, MS 2200) with DB5 capillary column under a temperature programming of 60–230°C with an initial hold of 10 min and then 5°C rise/min keeping the MS ion trap at 200°C, manifold at 40°C and transferline at 260°C. The flow of the carrier gas, helium, was 1 mL/min. The solvent (hexane) used for extraction was subjected to GC-MS as a pre-run prior to injection of the sample. The FAMEs were identified by comparing retention time of authentic compounds (Supelco™ Bellefonte, PA) along with NIST mass spectral data bank and by scrutinising the mass fragment pattern. Solvents of HPLC grade for different experiments were procured from E. Merck, India.

18.3 Results

As opposed to the usual ultrastructural feature of most species of mammalian hair, the surface characteristic of both the tail and forehead hair reveals a non-scaly feature (Fig. 18.1a). The cross-sectional images show a solid matrix formation devoid of any channel or hole-like appearance and with some granular structures embedded on the matrix (Fig. 18.1b). This variation in structure of the elephant hairs from that of the other mammals could be interesting from an evolutionary aspect as elephants are believed to have evolved from a semi-aquatic ancestor (Gaeth et al. 1999).

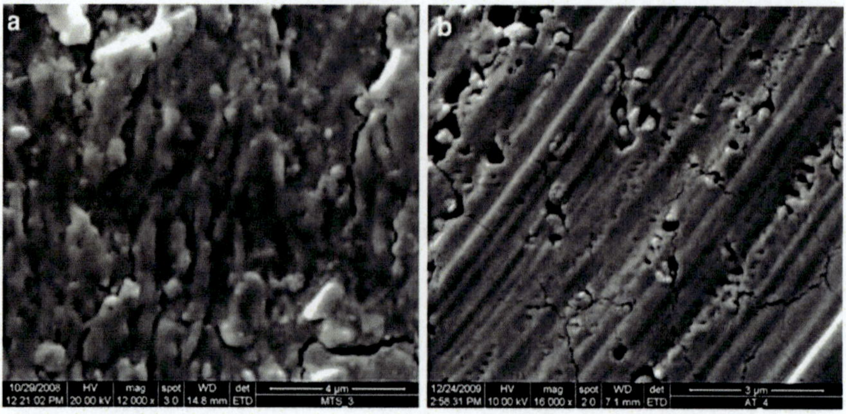

Fig. 18.1 (**a**) Surface image of elephant (*Elephas maximus*) tail hair. (**b**) Cross section of elephant (*Elephas maximus*) tail hair

The HPTLC profile of the neutral lipid fraction of both head and tail hair showed the presence of free sterol, long chain alcohol, free fatty acid, triglyceride, wax- and sterol ester, of which the free fatty acid content is maximal in both cases.

GC-MS analysis of fatty acid methyl ester of the lipid fraction of *E. maximus* revealed the presence of fatty acids ranging from C_8–C_{24} (Table 18.1). The fatty acids were identified using the NIST mass spectral data bank, relative retention time (RRt) values and most importantly by analysing the mass spectra fragmentation pattern. The low-molecular-weight, volatile alkanoic acids like caprylic acid ($C_{8:0}$), capric acid ($C_{10:0}$) and lauric acid ($C_{12:0}$) were detected in both head and tail hair. Fatty acids like myristic acid ($C_{14:0}$), palmitic acid ($C_{16:0}$) stearic acid ($C_{18:0}$) and eicosanoic acid ($C_{20:0}$) along with the comparatively high-molecular-weight fatty acids like docosanoic $C_{22:0}$) and tetracosanoic acid ($C_{24:0}$) were identified in the forehead and tail hairs. Beside these saturated fatty acids some unsaturates like palmitoleic acid ($C_{16:1}$), linoleic acid ($C_{18:2}$), oleic acid ($C_{18:1}$), eicosenoic acid ($C_{20:1}$) and erucic acid ($C_{22:1}$) were also detected in both forehead and tail hair. Few odd carbon fatty acids like pentadecanoic acid ($C_{15:0}$) (Fig. 18.2) and heptadecanoic acid ($C_{17:0}$) along with some branched chain saturated fatty acids were evident from some of the samples (Table 18.1).

18.4 Discussion

Scanning electron microscopic images of the surface view of epidermal structures of most of the mammals show a scaly appearance (Poddar-Sarkar et al. 2008; Church et al. 2009; Poddar-Sarkar et al. 2011; unpublished data from this laboratory on tiger (*Panthera tigris tigris*), leopard (*Panthera pardus*), cat (*Felis catus*), deer (*Axis axis*), gorilla (*Gorilla gorilla*), etc.). In contrast, the hair covering the surface of specialised integumentary glands in some mammals is known to retain material containing

Table 18.1 Summary of the fatty acid profile obtained from elephant (*E. maximus*) hairs

E. maximus		
FAME of total lipid	Forehead hair (adult)	Tail hair (adult)
Even		
$C_{8:0}$	✓	✓
$C_{10:0}$	✓	✓
$C_{12:0}$	✓	✓
$C_{14:0}$	✓	✓
$C_{16:0}$	✓	✓
$C_{18:0}$	✓	✓
$C_{20:0}$	✓	✓
$C_{22:0}$	✓	✓
$C_{24:0}$	✓	✓
$C_{26:0}$	✓	–
Odd		
$C_{15:0}$	–	✓
$C_{17:0}$	✓	✓
Unsaturates		
$C_{16:1}$	✓	✓
$C_{18:2}$	✓	✓
$C_{18:1}$	✓	✓
$C_{20:1}$	✓	✓
$C_{22:1}$	✓	✓
$C_{24:1}$	–	✓
2-Hydroxy hexadecanoic acid	✓	✓
2-Hydroxy octadecanoic acid	✓	–
12-Methyl tridecanoic acid	✓	✓
12-Methyl tetradecanoic acid	–	✓
14-Methyl pentadecanoic acid	✓	✓
17-Methyl octadecanoic acid	✓	✓

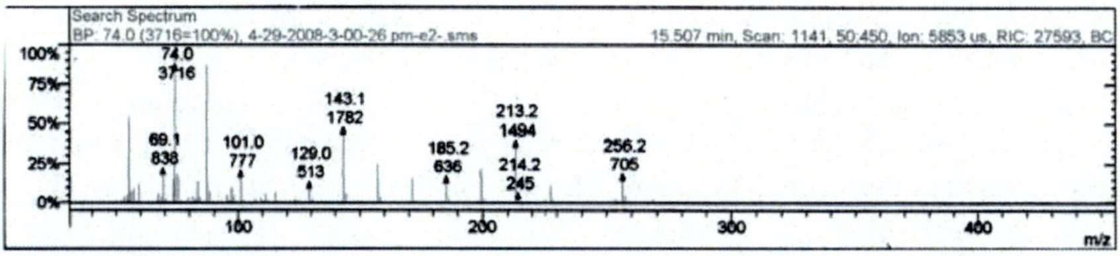

Fig. 18.2 Mass fragmentation pattern of methyl ester of pentadecanoic acid of elephant (*Elephas maximus*) tail hair

olfactory information. The classic examples are the honeycombed surface of the crest hair of the African crested rat (Stoddart 1979), black-tailed deer (Müller-Schwarze et al. 1977), antechinus (Toftegaard and Bradley 1999) and some bat species (Scully et al. 2000). Such hairs are structurally different from hairs on other body regions and are known as 'osmetrichia'. However, our results revealed that the surface features of the forehead and tail hairs of the Asian elephant do not show the two above-mentioned features. Each tail hair consists of a number of finer bristles thus assuming the form of a miniature brush. This special organisation of the tail hair, i.e. a brush (a single hair) within a brush (tail tuft) is better adapted to mopping up vaginal fluid (Brahmachary et al. 2007). The cross-sectional views do not reveal any orientation of channels as observed in other mammalian hairs. Theoretically, the products of various glands are stored in and released through the hair associated with the gland (Albone 1983) thus facilitate transmission of olfactory signals to conspecifics. In elephants, the sebaceous glands with their lipid secretions and body hairs with substrate effects have been reduced during evolution (Rasmussen 1999); this could be a possible explanation for the different ultrastructure of their body hairs.

In sheep wool the presence of wax (lanolin), sterol, fatty acids and aliphatic alcohols was confirmed whereas lipids such as triglycerides were not found (Demanze 1996). In Asian elephant hairs, besides the presence of wax ester and triglycerides, we found an appreciable amount of the free fatty acid and free sterol in HP-TLC plates. In human hair, in addition to waxes, free fatty acids and triglycerides, monoglycerides and diglycerides were also identified (Aitzetmuller and Koch 1978). Triglyceride was detected in lion mane (Poddar-Sarkar et al. 2008) but it was hardly traced in the lower mammalian groups like porcupine and hedgehog (Poddar-Sarkar et al. 2011).

The GC-MS analysis of elephant hair shows a number of fatty acids ranging from C_8–C_{22} of which C_8–C_{12} with high volatility might have a pheromonal role. The occurrence of odd carbon number fatty acid like pentadecanoic acid may be of special interest as animal fats and oils generally have even number of carbon atoms. As in human hair (Florkin and Mason 1962) and in the lower mammals viz., porcupines and hedgehog (Poddar-Sarkar et al. 2011), hexa and octadecanoic acids are the chief fatty acids in elephant hairs. $C_{21:0}$ai fatty acid or 18-methyl-eicosanoic acid has been identified as the principal fatty acid in wool wax (Downing et al. 1960), human hair (Wertz and Downing 1989) and Tasmanian devil fur (Church et al. 2009). We have traced this compound in the tail hair of both male and female elephants with the help of authenticated standard. The evolutionary conservation of the $C_{21:0}$ai fatty acid implies that these lipids play an essential function as structural elements of the lipid-rich envelopes which surround the hair cells (Wertz and Downing 1989).

The lower fatty acids ($C_{8:0}$–$C_{12:0}$) of the tail hairs might act as pheromones while the higher fatty acids presumably may play a role of fixative for the putative volatile pheromones of the vaginal secretion (which itself may have fixative molecules).

Acknowledgements We are grateful to University Grants Commission, Government of India for financial support. We thank the Officer-in-Charge, Bethune College, Kolkata 700006 for giving us infrastructural facility. We acknowledge Dr. Jeff Church, Principal Research Scientist CSIRO, Material Science & Engineering, Australia for his kind donation of the authentic compound $C_{21:0}$ai which helped us a lot for the detection of this compound in elephant hairs. Finally, we are indebted to the PCCF, Department of Forest, Government of West Bengal, India for his kind co-operation.

References

Aitzetmuller K, Koch J (1978) Liquid chromatographic analysis of sebum lipids, and other lipids of medical interest. J Chromatogr 145:195–200
Albone ES (1983) Mammalian semiochemistry. Wiley, Chichester
Bandopadhyay A, Brahmachary RL (1999) Reappraisal of an ancient text on elephants. Curr Sci 77:9–10
Bligh EG, Dyer WJ (1959) A rapid method of total lipid extraction and purification. Can J Biochem Physiol 37:911–917
Brahmachary RL, Biswas S, Bhar R (2007) Is elephant tail hair osmetrichia? Curr Sci 92(9):1199–1200
Church JS, Evans DJ, Woodhead AL (2009) The characterization of Tasmanian devil *Sarcpphilus harrisii*, pelage fibres and their associated lipids. Acta Zoologica 90:380–389
Demanze C (1996) Natural waxes. In: Karleskind A (ed) Oils and fats manual, vol 1. Intercept, NewYork, pp 300–319
Downing DT, Kranz FH, Murray KE (1960) Studies in waxes XIV. An investigation of the aliphatic constituents of hydrolysed wool wax by gas chromatography. Aust J Chem 13:80–94
Florkin M, Mason H (1962) Comparative biochemistry, vol II. Academic, New York
Gaeth AP, Short RV, Renfree MB (1999) The developing renal, reproductive and respiratory systems of the African elephant suggest an aquatic ancestry. Proc Natl Acad Sci USA 96:5555–5558
Huang TC, Chen CP, Wefler V, Raltery A (1961) A stable reagent for the Liebermann-Büchard reaction-application to rapid serum cholesterol determination. Anal Chem 33:1405–1407
Kates M (1972) Separation of lipid mixture. In: Work TS, Work E (eds) Laboratory techniques in biochemistry and molecular biology, vol 3, Part-II. North Holland Publishing Company, Amsterdam, pp 393–469
Krishnan M (1972) An ecological survey of the larger mammals of peninsular India: the Indian elephant (*Elephas maximus*, Linnaeus). J Bombay Nat Hist Soc 69:297–321
Müller-Schwarze D, Volkman NJ, Zemanek KF (1977) Osmetrichia: specialized scent hair in black-tailed deer. J Ultrastruct Res 59:223–230
Perrin TE, Rasmussen LEL (1994) Chemosensory responses of female Asian elephants (*Elephas maximus*) to cyclohexanone. J Chem Ecol 20:2577–2586
Poddar-Sarkar M (1996) The fixative lipid of tiger pheromone. J Lipid Mediat Cell Signal 15:89–101
Poddar-Sarkar M, Chakroborty A, Bhar R, Brahmachary RL (2008) Putative pheromones of lion mane and its ultrastructure. In: Beynon RJ, Roberts SC, Wyatt TD, Hurst JL (eds) Chemical signals in vertebrates, vol 11. Springer, New York, pp 61–67
Poddar-Sarkar M, Raha P, Bhar R, Chakroborty A, Brahmachary RL (2011) Ultrastructure and lipid chemistry of specialized epidermal structure of Indian porcupines and hedgehog. Acta Zool 92:134–140
Rasmussen LEL, Lee TD, Roelofs WL, Zhang A, Daves GD (1996) Insect pheromone in elephants. Nature 379:684
Rasmussen LEL, Lee TD, Zhang A, Roelofs WL, Daves GD Jr (1997) Purification, identification, concentration and bioactivity of (z)-7-dodecen-1-yl acetate:sex pheromone of the female Asian elephant, *Elephas maximus*. Chem Senses 22:417–437
Rasmussen LEL (1999) Evolution of chemical signals in the Asian elephant, *Elephas maximus*: behavioural and ecological influences. J Biosci 24(2):241–251
Rasmussen LEL, Krishnamurthy V (2000) How chemical signals integrate Asian elephant society: the known and the unknown. Zoo Biol 19(5):405–423
Rasmussen LEL, Riddle HS, Krishnamurthy V (2002) Chemical communication: mellifluous matures to malodorous in musth. Nature 415:975–977
Rasmussen LEL, Greenwood DR (2003) Frontalin: a chemical message of musth in Asian elephant (*Elephas maximus*). Chem Senses 28(5):433–446

Schulte BA, Rasmussen LEL (1999a) Musth, sexual selection, testosterone and metabolites. In: Johnston RE, Müller-Schwarze D, Sorenson PW (eds) Advances in chemical communication in vertebrates. Plenum, New York, pp 383–397

Schulte BA, Rasmussen LEL (1999b) Signal-receiver interplay in the communication of male condition by Asian elephants. Anim Behav 57:1265–1274

Scully WMR, Fenton MB, Saleuddin ASM (2000) A histological examination of the holding sacs and glandular scent organs of some bat species (*Emballonuridae*, *Hipposideridae*, *Phyllostomidae*, *Vespertilionidae* and *Molossidae*). Can J Zool 78(4):613–623

Stahl E (1965) Thin layer chromatography—a laboratory handbook. Academic, New York

Stoddart DM (1979) A specialized scent-releasing hair in the crested rat *Lophiomys imhausi*. J Zool 189:553

Toftegaard CL, Bradley AJ (1999) Structure of specialized osmetrichia in the brown antechinus, *Antechinus stuartii* (Marsupialia: Dasuridae). J Zool 248:27–30

Wertz PW, Downing DT (1989) Integral lipids of mammalian hair. Comp Biochem Physiol 92B(4):759–761

Chapter 19
Methyl Ketone Production in Juvenile Red-Sided Garter Snakes

Emily J. Uhrig, Michael P. LeMaster, Deborah I. Lutterschmidt, and Robert T. Mason

Abstract The skin lipids of female red-sided garter snakes contain a series of methyl ketones which function as a sexual attractiveness pheromone. The pheromone is likely regulated by oestrogen. As previous studies have focused mainly on adult females, it is unknown whether juvenile garter snakes produce methyl ketones. The current study was designed to determine whether (1) juvenile female garter snakes are physiologically capable of producing methyl ketones and (2) oestrogen initiates or enhances methyl ketone production in juveniles. Snakes were injected with either oestradiol or an oil vehicle, or received no injection. Subsequent chemical analyses of skin lipid samples clearly demonstrated that juveniles were able to produce methyl ketones regardless of treatment. However, we were unable to demonstrate that oestradiol affected methyl ketone production at this stage in the female life cycle.

19.1 Introduction

Regulation of sex pheromone production and/or expression is important for ensuring that reproductive events are properly synchronised. In many species, pheromone regulation is accomplished via hormonal action which, for vertebrates, often

E.J. Uhrig (✉) • R.T. Mason
Department of Zoology, Oregon State University, Corvallis, OR, USA
e-mail: uhrige@science.oregonstate.edu; masonr@science.oregonstate.edu

M.P. LeMaster
Department of Biology, Western Oregon University, Monmouth, OR, USA
e-mail: lemastm@wou.edu

D.I. Lutterschmidt
Department of Biology, Portland State University, Portland, OR, USA
e-mail: d.lutterschmidt@pdx.edu

M.L. East and M. Dehnhard (eds.), *Chemical Signals in Vertebrates 12*,
DOI 10.1007/978-1-4614-5927-9_19,

involves steroid hormones. Androgens, for example, have been shown to mediate the production of primer pheromones in male mice and goats (Lombardi et al. 1976; Iwata et al. 2000). Similarly, female-produced pheromones in the uropygial glands of mallards and the urine of mice are apparently regulated by oestrogen (Bohnet et al. 1991; Achiraman et al. 2010). In comparison to birds and mammals, relatively fewer studies have investigated hormonal control of pheromone production in reptiles. Despite this paucity of studies, there is evidence to suggest that oestrogen may be involved in regulating pheromone production in skinks (Cooper 1995) and garter snakes (Crews 1976, 1985; Mendonça and Crews 1996; Parker and Mason 2012).

The sexual attractiveness pheromone of the red-sided garter snake, a homologous series of 18 long-chain saturated and monounsaturated methyl ketones, is contained within the skin lipids of females (Mason et al. 1989, 1990; LeMaster and Mason 2002). When detected by males, this pheromone elicits stereotypical courtship behaviour that will not occur in the absence of the pheromone (Mason et al. 1990). Thus, proper regulation of the female sexual attractiveness pheromone is crucial to reproduction in this species. Precisely how production and expression of the pheromone is regulated is unknown although it has been hypothesised that oestrogen may have a role in maintaining the pheromone within female skin lipids (Mendonça and Crews 1996). This hypothesis is supported by a recent study demonstrating that intraseasonal changes in pheromone profile following emergence from winter hibernation are accompanied by a decline in oestrogen levels (Uhrig et al. 2012). Previous work has also shown that oestrogen treatment not only enhances the attractivity of female garter snakes (Crews 1976, 1985), but also induces male expression of female-like pheromone profiles (Parker and Mason 2012).

Crews (1985) demonstrated that juvenile female red-sided garter snakes given oestrogen implants received courtship from adult males and hypothesised that their attractivity was due to oestrogen-stimulated pheromone production. However, the pheromone had not yet been chemically identified and thus no chemical analyses were conducted. Following identification of the pheromone (Mason et al. 1989), studies of garter snake chemical ecology have focused on the methyl ketone profiles of adult females. It has not been confirmed whether the skin lipids of juvenile snakes indeed contain methyl ketones or whether juveniles are even physiologically capable of producing such compounds.

Given the apparent activational role of oestrogen on pheromone production (Parker and Mason 2012), we hypothesised that, if juvenile snakes are able to produce methyl ketones, then oestrogen treatment of juveniles could potentially initiate or enhance production of those compounds. To investigate this hypothesis we chemically analysed the skin lipids of juvenile female red-sided garter snakes subjected to one of three treatments: oestradiol injection, peanut oil injection, or no injection. If oestradiol initiates or enhances methyl ketone production, then we would expect the chemical profiles of the oestradiol-treated group to differ from the control treatments in terms of methyl ketone quantity and/or quality.

19.2 Methods and Materials

19.2.1 Study Animals

Juvenile female red-sided garter snakes (*Thamnophis sirtalis parietalis*, $n = 59$) used in this study were randomly selected from a population born in the laboratory to females collected near Inwood, Manitoba, Canada. All procedures used in this study were approved by the Oregon State University Institutional Animal Care and Use Committee (protocol number: 2661) and complied with guidelines established by the National Institutes of Health Guide for the Care and Use of Laboratory Animals.

19.2.2 Hormone Treatment and Analysis

At approximately two months of age, red-sided garter snakes were randomly assigned to one of three treatment groups: oestradiol, vehicle, or non-injected control. Snakes in the oestradiol-treated group ($n = 20$) were injected with a solution of oestradiol in peanut oil (40 μg oestradiol in 100 μL peanut oil), snakes in the vehicle group ($n = 20$) were injected with peanut oil alone (100 μL) and snakes in the non-injected control group ($n = 19$) received no injection. All injections were intraperitoneal and administered once per day for seven days. On the eighth day, all snakes were euthanised with an overdose of Brevital sodium and a subset of each treatment was randomly selected for blood sampling ($n = 10$ snakes per treatment). To obtain a sufficient volume, the tail tip was severed allowing the blood to drain. Samples were centrifuged to separate the plasma which was analysed using direct radioimmunoassay to determine oestradiol concentration (see Lutterschmidt and Mason 2009 for full description and validation). All hormone concentrations were corrected for individual recovery variation. Limits of detectability were approximately 2.4 pg/ml for oestradiol. All samples were analysed in a single assay with an intra-assay coefficient of variation of 9.2%.

19.2.3 Chemical Analysis

To avoid contamination by bodily fluids, the extraction of skin lipids was limited to snakes from which blood samples were not collected ($n = 10$ oestradiol, $n = 10$ vehicle, $n = 9$ control). Following euthanasia on the same day as the blood sampling, snakes used for chemical analyses were placed individually in equal aliquots of hexane and soaked overnight to extract skin lipids. The resulting extracts were dried under reduced pressure by rotoevaporation (35°C) to eliminate excess solvent, weighed and then resuspended in hexane. In order to isolate the methyl ketones, skin lipid extracts were fractionated using column chromatography (described by LeMaster and Mason 2002). The resulting methyl ketone-containing fractions were analysed using gas chromatography/mass spectrometry (Hewlett Packard 5890 series II gas chromatograph and Hewlett Packard 5971 mass selective detector; see LeMaster and Mason 2002 for full description). An internal standard of known concentration (methyl stearate, 10 μg/mL) was injected into the GC/MS with each sample. Following peak integration, the relative concentrations of individual methyl ketones were determined by comparing the area of each methyl ketone peak to that of the standard. To examine quantitative variation in methyl ketones, the amount of methyl ketones expressed (μg) per unit skin surface (cm^2) was determined. Surface area was determined by multiplying snout-vent length (SVL) by mid-body circumference (Mason et al. 1990).

19.2.4 Statistical Analysis

All statistical analyses were carried out using Jandel SigmaStat software (version 3.11, Systat Software, Inc.) and R (v.1.8-8). Comparisons of SVL, mass, lipid quantity and pheromone quantity were conducted using one-way analysis of variance (ANOVA). Hormone data were analysed by a Kruskal–Wallis test. Fisher's exact tests with Tukey-type multiple comparisons (Zar 1999) and multi-response permutation procedures (MRPP) were used for qualitative pheromone analysis (Uhrig et al. 2012; Parker and Mason 2009).

19.3 Results

There were no differences among treatment groups in the SVL or mass of snakes from which blood samples were taken (SVL: ANOVA, $F=1.111$, $P=0.344$; mass: ANOVA, $F=3.273$, $P=0.053$). Plasma oestradiol levels were undetectable in all snakes in the vehicle and non-injected control groups indicating that levels of the hormone were below 2.4 pg/mL, which is the limit of detectability for our radioimmunoassay. The oestradiol level in the hormone-treated group was significantly higher (Kruskal–Wallis test, $H=27.488$, $P<0.001$) with a mean value of 2.48 ng/mL (±0.85 SE).

Following GC/MS analysis, we detected the presence of at least one methyl ketone compound in 27 of the 29 juvenile snakes from which skin lipids were extracted. The methyl ketones identified in the juveniles included saturated and monounsaturated compounds ranging in weight from 394 to 532 Da. All 18 individual compounds recognised in adult snakes (LeMaster and Mason 2002) were identified across the juvenile samples (Fig. 19.1). The low quantity or absence of methyl ketones in several samples suggested that the fractionation method, originally designed for the large lipid quantities extracted from adult snakes, may have

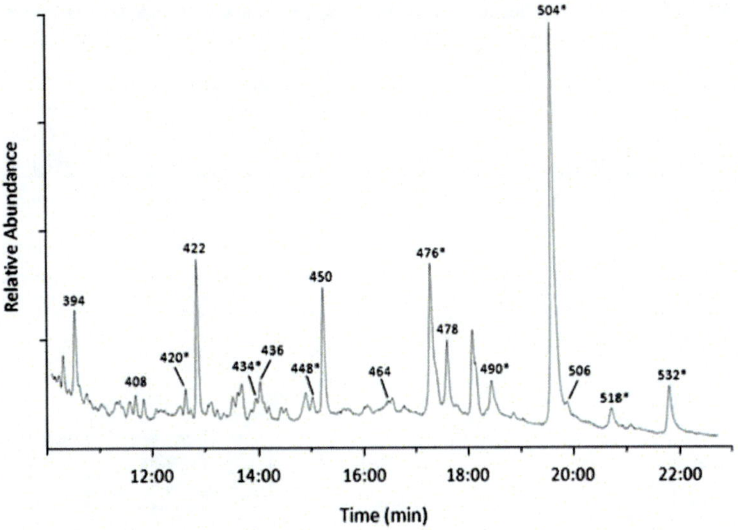

Fig. 19.1 Chemical profile of an individual juvenile female red-sided garter snake from the non-injected control group. There are 16 identifiable methyl ketones present. Numbers above peaks represent molecular weights (Daltons); asterisks denote unsaturated compounds

hindered the detection of low yield compounds. Thus, all statistical analyses have been restricted to snakes with chemical profiles consisting of any nine or more identifiable methyl ketones (control treatment: $n=6$; vehicle treatment: $n=6$; oestradiol treatment: $n=9$). A minimum of nine compounds was chosen as adult females typically express at least nine methyl ketones (LeMaster and Mason 2002, 2003). This restriction was based only on the number of methyl ketones present, not on the identity of the compounds to avoid biasing the results based on the presence or absence of a particular compound.

Neither SVL nor mass differed among treatments for the snakes from which skin lipids were extracted (SVL: ANOVA, $F=0.968$, $P=0.399$; mass: ANOVA, $F=2.141$, $P=0.146$; Table 19.1). The amount of lipids expressed per unit skin surface (μg/cm^2) differed significantly between treatments (ANOVA, $F=12.128$, $P<0.001$; Table 19.1). Holm-Sidak multiple comparisons confirmed that all three treatments were significantly different from one another with the vehicle-treated group having the highest concentration of skin lipids and the control group having the lowest (control versus vehicle: $P<0.001$; control vs. oestradiol: $P=0.026$; vehicle versus oestradiol: $P=0.009$). The amount of methyl ketones expressed per unit skin surface did not differ between treatments (ANOVA, $F=1.015$, $P=0.382$; Table 19.1) nor did the percentage of skin lipids composed of methyl ketones (ANOVA, $F=3.171$, $P=0.066$; Table 19.1).

In terms of pheromone quality, the relative concentrations of individual methyl ketones did not differ among treatment groups (MRPP, $A=0.035$, $P=0.071$; Fig. 19.2). For 17 of the 18 methyl ketones detected, the proportion of snakes

Table 19.1 Mean snout-vent length (SVL), mass, skin lipid and methyl ketone concentrations, and percentage of skin lipids composed of methyl ketones for juvenile female red-sided garter snakes

Treatment	SVL (cm)	Mass (g)	Skin lipids (μg/cm²)	Methyl ketones (μg/cm²)	Methyl ketones (% of total skin lipids)
Oestradiol (n=9)	15.6±0.2	1.5±0.1	695.4±47.7	11.4±2.5	1.7±0.3
Vehicle (n=6)	15.2±0.2	1.3±0.1	916.1±61.9	9.5±2.6	1.1±0.3
Control (n=6)	15.5±0.2	1.3±0.1	513.5±52.8	16.4±4.9	3.2±1.0

Snakes were injected with either oestradiol or a peanut oil vehicle, or received no injection
Values represent mean ± SE

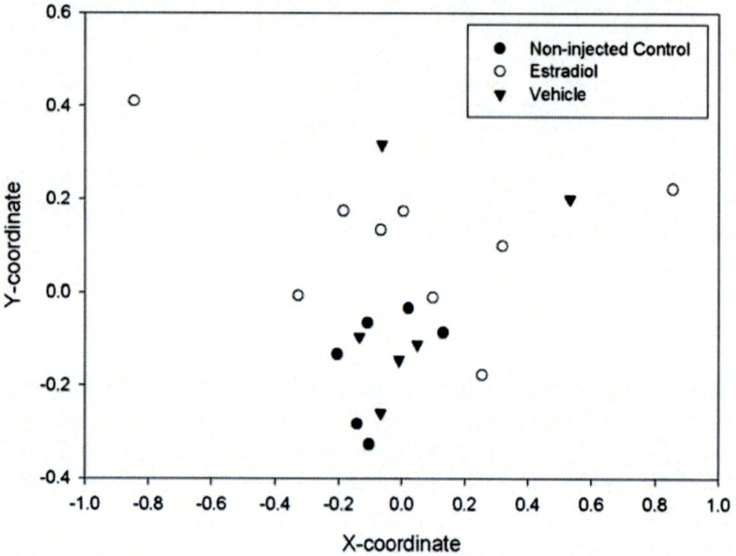

Fig. 19.2 Non-metric multi-dimensional scaling plot of individual pheromone profiles for juvenile female red-sided garter snakes that were injected with either oestradiol or a peanut oil vehicle, or received no injection. Each point represents the profile of an individual snake

expressing a particular compound did not differ between treatments (Fisher's exact tests, $P \geq 0.05$ for all comparisons; Table 19.2). The methyl ketone with a molecular weight of 408 Da was expressed by a significantly greater proportion of snakes in the oestradiol-treated group compared to the vehicle group (Fisher's exact test with Tukey-type multiple comparisons, $P < 0.05$). However, neither the oestradiol group nor the vehicle group differed from the control group in terms of the proportion of snakes expressing the 408 Da methyl ketone ($P > 0.05$ for both comparisons).

Table 19.2 Number of snakes in which each individual methyl ketone was identified

Methyl ketone (molecular weight in Daltons)	Oestradiol ($n=9$)	Vehicle ($n=6$)	Control ($n=6$)	Total ($n=21$)
394	0	0	2	2
408	8	1	4	13
420*	8	6	6	20
422	9	5	6	20
434*	4	3	3	10
436	4	3	6	13
448*	9	6	6	21
450	9	6	6	21
462*	1	3	3	7
464	2	4	5	11
476*	8	6	6	20
478	8	6	6	20
490*	1	3	3	7
492	1	3	3	7
504*	9	6	6	21
506	4	4	6	14
518*	8	6	6	20
532*	9	6	6	21

Snakes were injected with either oestradiol or a peanut oil vehicle, or received no injection
Asterisks denote unsaturated methyl ketones

19.4 Discussion

Our results confirm that juvenile female red-sided garter snakes are capable of producing methyl ketones. Juvenile skin lipids contained methyl ketones ranging in weight from 394 to 532 Da, the same range as the methyl ketones produced by typical adult females (LeMaster and Mason 2002). Although the concentration (μg per cm^2 skin surface) of skin lipids varied significantly between the oestradiol, vehicle and control groups, hormone treatment did not have a measureable effect on methyl ketone concentration. With respect to the quality of methyl ketone profiles, there were no differences in the relative concentrations of individual methyl ketones between groups. With one exception, the proportion of snakes expressing a particular methyl ketone also did not differ among groups.

The presence of methyl ketones in the skin lipids of juvenile female garter snakes likely accounts for the attractiveness of juveniles observed in previous studies (e.g. Crews 1985; Shine et al. 2000). Given the potential costs associated with attracting vigorous male courtship (e.g. increased risk of mortality; Shine et al. 2000), production of a sex pheromone prior to attaining reproductive maturity could be viewed as maladaptive. However, this phenomenon may be beneficial in another aspect of red-sided garter snake life history: females give birth away from the hibernacula and juveniles must travel some distance to locate the den site

(Gregory 1977). Precisely how snakes navigate back to the den is not known, but it has been suggested that they may follow pheromone trails laid down by conspecifics (Costanzo 1989). Thus, juvenile-produced methyl ketones may facilitate the location of den sites.

Based on the results of the radioimmunoassay, we conclude that hormone injections were effective at elevating the oestradiol levels of treated snakes relative to the vehicle and control treatment groups. However, we were unable to detect significant differences among treatments for most variables examined. Given that previous studies have generated strong evidence for the pheromone-regulating role of oestrogen in adult garter snakes (e.g. Mendonça and Crews 1996; Parker and Mason 2012; Uhrig et al. 2012), we suggest that juveniles may have simply not yet developed the biochemical pathways necessary for the hormone to alter methyl ketone expression. If this is the case, then future studies investigating the ontogenetic effects of oestrogen treatment on additional, older age classes may be informative. Alternatively, as sample sizes in the current study were relatively small ($n=6$, 6 and 9 for control, vehicle and oestradiol treatments, respectively), our statistical analyses may have had insufficient power to detect differences among treatments. This lack of statistical power may explain our results regarding expression of the 408 Da methyl ketone.

We recognise that the low sample sizes in this study were exacerbated by our decision to restrict statistical analyses to snakes expressing at least nine methyl ketones and that this restriction may have led to the exclusion of useful data. However, this approach ensured that our analyses were not unduly influenced by individuals from which poor samples were obtained. Our inability to recover satisfactory methyl ketone samples from the lipids of several snakes is likely due to our sampling technique having been originally developed for adult snakes, and our methodology may need to be refined to detect methyl ketones produced in minute quantities.

The concentration of skin lipids per unit skin surface was the only variable that differed significantly among all treatments. However, it is possible that this result was influenced by residual peanut oil that may have remained in the body cavity of some injected animals. Any such residual oil could have been extracted during the lipid extraction process leading to the between-group differences in lipid concentration. This explanation is supported by the finding that the non-injected control group had a lower mean skin lipid concentration than either of the injected treatments. Thus, while the current study clearly demonstrates that juvenile snakes have the biochemical pathways necessary to produce methyl ketones, the role of oestrogen remains unclear. Our treatment-based results may be more easily interpreted in the light of future studies using refined methyl ketone sampling techniques as well as an alternative method of administering oestrogen.

Acknowledgements This research was supported by an HHMI Undergraduate Research Award to E.J.U., a Western Oregon University Faculty Development Award to M.P.L. and an NSF grant (0620125) to R.T.M.

References

Achiraman S, Archunan G, Ponmanickam P, Rameshkumar K, Kannan S, John G (2010) 1-Iodo-2 methylundecane [1I2MU]: an oestrogen-dependent urinary sex pheromone of female mice. Theriogenology 74:345–353

Bohnet S, Rogers L, Sasaki G, Kolattukydy PE (1991) Estradiol induces proliferation of peroxisome-like microbodies and the production of 3-hydroxy fatty acid diesters, the female pheromones, in the uropygial glands of male and female mallards. J Biol Chem 266:9795–9804

Cooper WE (1995) Effects of oestrogen and male head coloration on chemosensory investigation of female cloacal pheromones by male broad-headed skinks (*Eumeces laticeps*). Physiol Behav 58:1221–1225

Costanzo JP (1989) Conspecific scent trailing by garter snakes (*Thamnophis sirtalis*) during autumn: further evidence for use of pheromones in den location. J Chem Ecol 15:2531–2538

Crews D (1976) Hormonal control of male courtship behavior and female attractivity in the garter snake (*Thamnophis sirtalis parietalis*). Horm Behav 7:451–460

Crews D (1985) Effects of early sex steroid hormone treatment on courtship behavior and sexual attractivity in the red-sided garter snake, *Thamnophis sirtalis parietalis*. Physiol Behav 35:569–575

Gregory PT (1977) Life history parameters of the red-sided garter snake (*Thamnophis sirtalis parietalis*) in an extreme environment, the Interlake region of Manitoba. Publications in Zoology, No. 13. National Museum of Canada, Ottawa

Iwata E, Wakabayashi Y, Kakuma Y, Kikusui T, Takeuchi Y, Mori Y (2000) Testosterone dependent primer pheromone production in the sebaceous gland of male goat. Biol Reprod 62:806–810

LeMaster MP, Mason RT (2002) Variation in a female sexual attractiveness pheromone controls male mate choice in garter snakes. J Chem Ecol 28:1269–1285

LeMaster MP, Mason RT (2003) Pheromonally mediated sexual isolation among denning populations of red-sided garter snakes, *Thamnophis sirtalis parietalis*. J Chem Ecol 29:1027–1043

Lombardi JR, Vandenbergh JG, Whitsett JM (1976) Androgen control of the sexual maturation pheromone in house mouse urine. Biol Reprod 15:179–186

Lutterschmidt DI, Mason RT (2009) Endocrine mechanisms mediating temperature-induced reproductive behavior in red-sided garter snakes (*Thamnophis sirtalis parietalis*). J Exp Biol 212:3108–3118

Mason RT, Fales HM, Jones TH, Pannell LK, Chinn JW, Crews D (1989) Sex pheromones in snakes. Science 245:290–293

Mason RT, Jones TH, Fales HM, Pannell LK, Crews D (1990) Characterization, synthesis, and behavioral responses to the sex attractiveness pheromones of red-sided garter snakes (*Thamnophis sirtalis parietalis*). J Chem Ecol 16:2353–2369

Mendonça MT, Crews D (1996) Effects of ovariectomy and oestrogen replacement on attractivity and receptivity in the red-sided garter snake. J Comp Phys A 178:373–381

Parker MR, Mason RT (2012) How to make a sexy snake: oestrogen activation of female sex pheromone in male garter snakes. J Exp Biol 215:723–730

Parker MR, Mason RT (2009) Low temperature dormancy affects the quantity and quality of the female sexual attractiveness pheromone in red-sided garter snakes. J Chem Ecol 35:1234–1241

Shine R, O'Connor D, Mason RT (2000) Sexual conflict in the snake den. Behav Ecol Sociobiol 48:392–401

Uhrig EJ, Lutterschmidt DI, Mason RT, LeMaster MP (2012) Pheromonal mediation of intraseasonal declines in the attractivity of red-sided garter snakes, *Thamnophis sirtalis parietalis*. J Chem Ecol 38:71–80

Zar JH (1999) Biostatistical analysis, 4th edn. Prentice Hall, Upper Saddle River

Part V
Predator Avoidance and Migration

Chapter 20
The Sophistication of Predator Odour Recognition by Minnows

Douglas P. Chivers, Grant E. Brown, and Maud C.O. Ferrari

Abstract Fathead minnows (*Pimephales promelas*), like other prey fishes show considerable sophistication in their ability to recognise the odour signature of their predators. Here we illustrate the way minnows, hatching without the ability to innately recognise predators, acquire subtle information about potential predators through experience. We chronologically summarise the results of key lab-based and field-based studies that have allowed us to untangle the complexity of olfactory risk assessment.

20.1 Introduction

A fundamental prerequisite for prey animals to avoid being captured by predators is being able to identify predators as dangerous (Ferrari et al. 2010). Not surprisingly, many studies have been aimed at addressing what cues prey animals use to distinguish species that pose a threat from those that do not. In aquatic ecosystems, chemosensory cues are paramount in risk assessment. A number of chemical cues are known to mediate predator–prey interactions, some released or mediated by prey: injured conspecific cues (a.k.a., alarm cues), diet cues, and disturbance cues (Ferrari et al. 2010 for a review) and some by predators (predator odour). A wealth

D.P. Chivers (✉)
Department of Biology, University of Saskatchewan,
112 Science Place, Saskatoon, SK S7N 5E2, Canada
e-mail: doug.chivers@usask.ca

G.E. Brown
Department of Biology, Concordia University, 7141 Sherbrooke Street West,
Montreal, QC, H4B 1R6, Canada

M.C.O. Ferrari
Department of Biomedical Sciences, WCVM, University of Saskatchewan,
Saskatoon, SK, S7N 5B4, Canada

M.L. East and M. Dehnhard (eds.), *Chemical Signals in Vertebrates 12*,
DOI 10.1007/978-1-4614-5927-9_20,

of studies have shown that prey can detect and respond to the odour of predators (Ferrari et al. 2010; Kats and Dill 1998). Indeed, the odour signature of predators provides prey animals with a wealth of information upon which to make decisions (Kats and Dill 1998). The purpose of this chapter is to use a model fish system to illustrate the sophistication of predator odour recognition.

Fathead minnows (*Pimephales promelas*) provide an ideal vertebrate model system to study chemosensory assessment of predators. In this paper, we highlight the results of nearly 20 years of research demonstrating the amazing complexity of chemosensory risk assessment of fathead minnows. The lack of examples from other representative systems is not an oversight on our part, but rather illustrates the lack of in-depth studies focusing on this topic. Responses to predator odours are known in other fish taxa (Brown and Chivers 2006; Ferrari et al. 2010), but these studies generally fail to tell us more than the simple fact that prey respond to predator odours. In this chapter, we illustrate the way minnows, hatching without innate predator recognition, acquire incredibly subtle information about potential predators through experience. We chronologically summarise the results of key lab-based and field-based studies that have allowed us to gain more insights into this prey fish's strategy to increase survival.

20.2 Model Species

Fathead minnows are a short-lived (2–3 years), small bodied cyprinid fish (50–70 mm total length) common in many ponds, lakes and streams in central North America. They prefer still waters and hence occupy the shallow margins of lakes where they feed on detritus, aquatic insect larvae and zooplankton (Scott and Crossman 1985). Minnows often inhabit many of the same waterbodies as piscivores, including northern pike, *Esox lucius*. The predation pressure exerted on minnows by different predators often varies with time of year, mainly due to differences in patterns of space use. For example, northern pike hatch in shallow water early in the spring. They begin feeding on invertebrates, but quickly switch to piscivory. As pike grow (>40 cm), they move offshore, preferring to feed on more rewarding prey items, such as larger fish. Hence, as pike grow larger, they become less of a threat to minnows. Throughout their range, minnows are also prey to a host of other piscivores (e.g., bass, walleye, perch, trout) as well as predatory birds (kingfishers, cormorants), snakes, and insects. Little attention has been paid to testing whether the anti-predator strategies of minnows varies with the foraging mode of the predator. This is an area that is ripe for experimentation.

20.3 The Early Days

Mathis et al. (1993) were among the first to test whether minnows exhibit anti-predator behaviour when exposed to the odour of a pike (see also Göz 1941; Magurran 1989). Minnows collected from a lake that contained pike exhibited a very strong anti-predator response (a reduction in movement and an increase in shelter use) when the water from a tank containing a pike was injected into the tank with the minnows. The anti-predator response appeared to be specific to pike cues as the minnows did not respond to water from a tank that held other non-predatory fish (peacock gudgeons, *Tateurndina ocellicauda*). Minnows likewise responded to the sight of a pike presented in an adjacent tank. The original paper by Mathis et al. (1993) provided indirect evidence that learning may be involved in pike recognition by minnows. Minnows collected from a creek in a closed basin that was free of piscivorous fishes showed no anti-predator responses to the pike. In a follow-up study, Chivers and Smith (1994a) demonstrated that minnows collected from a lake with pike recognised pike odour, but minnows from the same population collected as eggs, did not. This again points to the supposition that minnows need experience with pike to recognise pike odour as a threat. Interestingly, wild caught minnows that have been maintained in the laboratory for over 1 year retain their response to pike odour. Thus far we have referred to minnows responding specifically to 'pike odour'. Indeed, Chivers and Smith (1993) demonstrated that the response was mediated through the olfactory system (Fig. 20.1). Minnows that were rendered anosmic with hot Vaseline injected into their nares did not respond to pike odour, while control fish did respond (Chivers and Smith 1993).

20.4 Predator Introduction Studies

Working from the premise that minnows reap enormous fitness benefits from learning to recognise pike odour, Chivers and Smith (1995) introduced ten juvenile pike into a 1 ha pond containing approximately 20,000 minnows. Two weeks later, all of the minnows tested exhibited an anti-predator response to pike odour introduced into their tank, but individuals collected prior to pike stocking did not respond to pike odour. Brown et al. (1997) followed up on this study stocking 39 juvenile pike into a lake containing a population of approximately 78,000 minnows. The minnows were collected from all over the pond 2 days pre-stocking and every 2 days post-stocking. By 8 days post-stocking, minnows began responding to the sight of a pike in an adjacent tank. Interestingly, it took a much shorter time for minnows to learn the odour of the pike. Minnows collected 4 days post-stocking showed significant responses to the pike odour (Fig. 20.2). An intriguing study that remains to be done would be to alter the density of aquatic vegetation and document the speed of acquisition of visual versus chemical information. With a reduction in habitat complexity we should expect to see more emphasis on visual cues over chemical cues (Hartman and Abrahams 2000).

20.5 Untangling the Complexity of Predator Odours

Kusch et al. (2004) were the first to begin to address the complexity of information held in pike odour signatures. They collected odours from pike of different sizes (~25 and 65 cm in length) and tested the response of pike-experienced minnows,

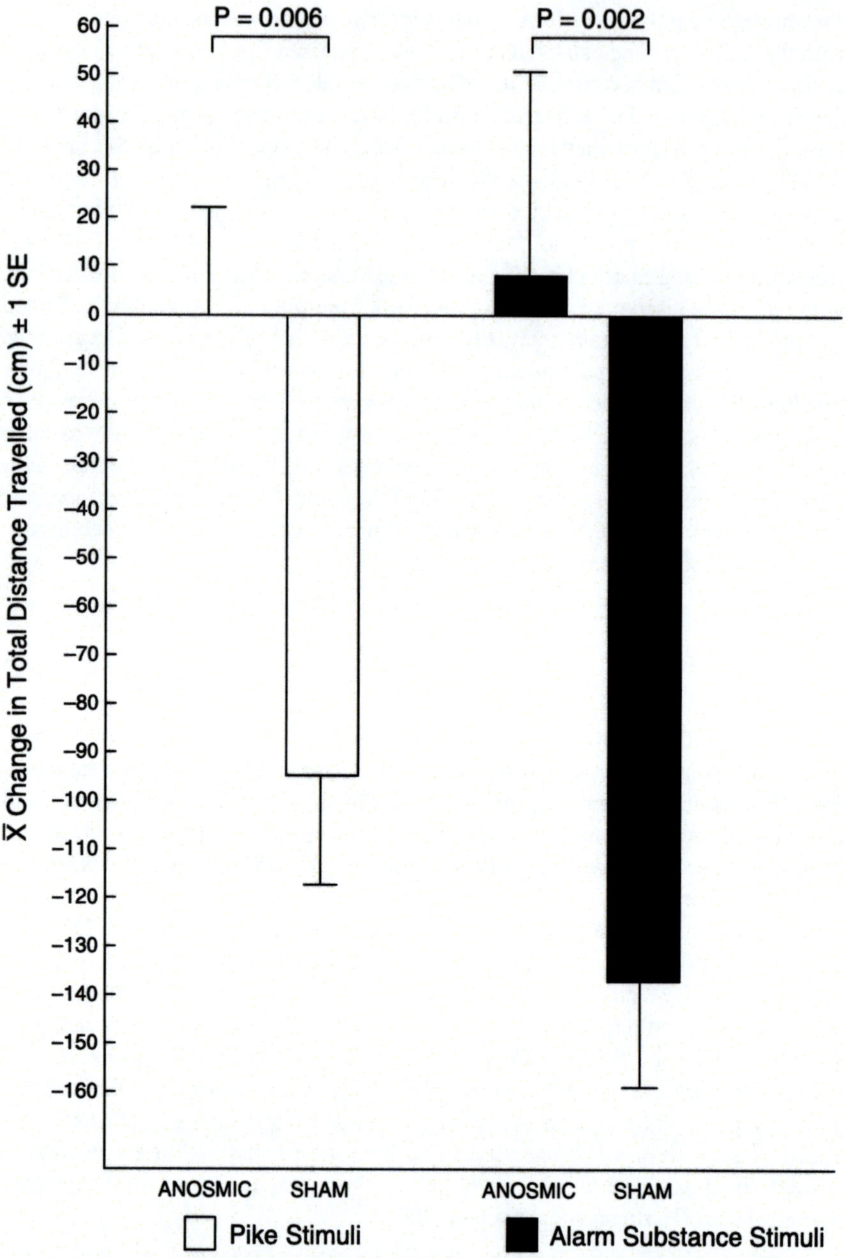

Fig. 20.1 Mean change in distance travelled for anosmic and sham treated minnows following exposure to chemical cues from pike and minnow alarm cues (alarm stimuli). A reduction in activity (distance travelled) is an anti-predator response in minnows. Anosmic fish failed to respond to both pike odour and alarm cues, indicating that both stimuli are detected through the olfactory system (figure taken from Chivers and Smith 1993)

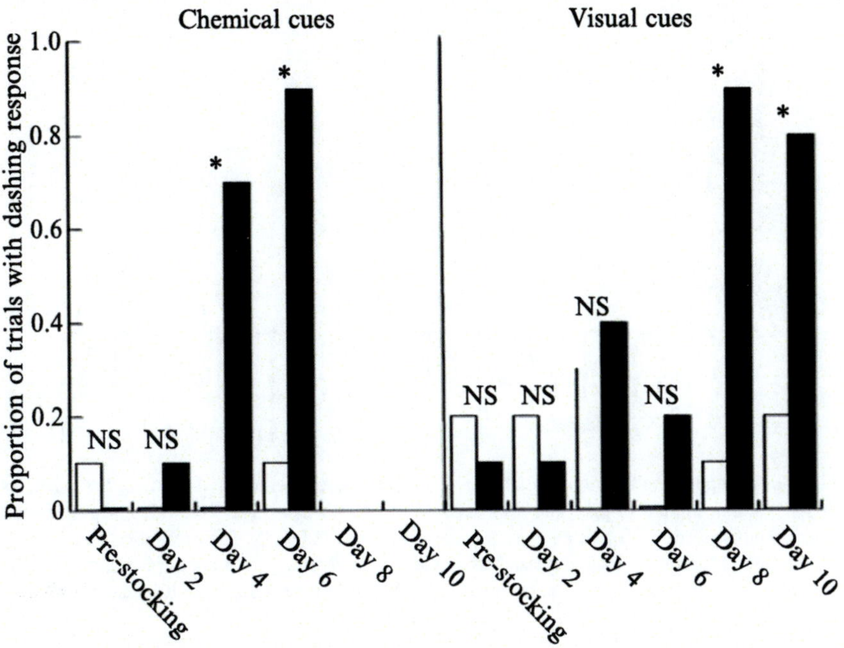

Fig. 20.2 Proportion of minnows exhibiting dashing behaviour (anti-predator behaviour) following exposure to pike odour or the sight of a pike in an adjacent tank (*solid bars*) or control stimuli (water or an empty tank, *open bars*). Minnows were captured 2 days before stocking the pond with 39 juvenile pike and every 2 days following the stocking. Minnows acquired recognition of the odour of the predator faster than the sight of the predator (figure modified from Brown et al. 1997)

collected from a lake containing pike, to varying amounts of pike odour added to their tank. When cues of large pike were introduced to minnow tanks, a relatively low proportion (<25%) of minnows exhibited anti-predator behaviour. In contrast, when the odour of small pike was added to the tanks, a large proportion of minnows responded (up to 90%). Moreover, the minnows exhibited greater responses to greater concentrations of pike odour (Fig. 20.3). At first, this result may seem counter intuitive. Larger pike have a much faster attack speed and should be more dangerous to minnows. However, minnows show a much greater degree of habitat overlap with small pike than large pike, making smaller pike a greater threat to minnows.

The differential behavioural responses to variation in pike odour concentration documented by Kusch et al. (2004) could indicate that minnows are able to assess the relative density of pike in the vicinity or that they are able to assess the relative proximity of the pike. Ferrari et al. (2006) performed a rather complex experiment to tease apart each of these possibilities. They exposed pike-experienced minnows to either 60 mL of pike odour obtained from 12 pike (i.e., 5 mL/pike), 60 mL of pike odour obtained from two pike (i.e., 30 mL/pike) or a control of 60 mL of water. All the pike used in the experiment had a similar size and pike odour was made from

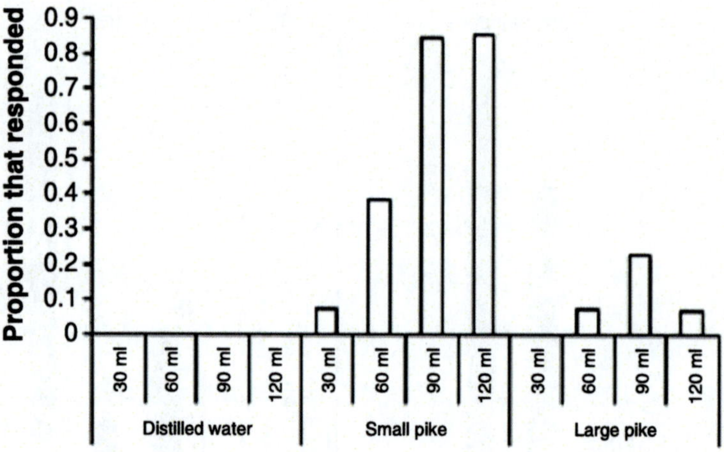

Fig. 20.3 Proportion of minnows that exhibited anti-predator behaviour following the introduction of various amount of pike odour or distilled water into their tank. Pike odour was prepared from pike that were 25 (small pike) or 65 (large pike) cm in length. The proportion of minnows responding to the pike odour increased when cues were obtained from small pike and when the volume of cues introduced into the tank was increased (figure modified from Kusch et al. 2004)

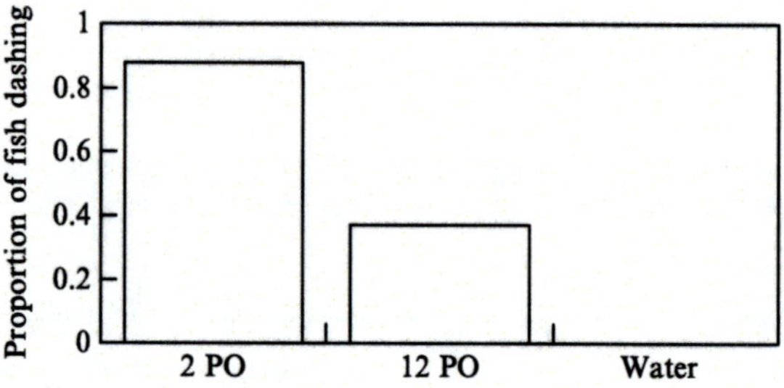

Fig. 20.4 Proportion of minnows exhibiting dashing behaviour following exposure to either 60 mL of pike odour obtained from 12 pike (i.e., 5 mL/pike), 60 mL of pike odour obtained from two pike (i.e., 30 mL/pike) or a control of 60 mL of water. Given that minnows responded with a greater intensity as the per-pike concentration increased while the overall concentration of pike odour remained constant, tells us that minnows can use odours to determine relative proximity to the pike

different individuals for each replicate. Minnows exposed to odours from two pike showed a more intense antipredator response than minnows exposed to odours from 12 pike. This demonstrated that minnows can detect individual pike odour in a mixture of odours from several pike (Fig. 20.4). Put in a broader chemical ecological context, minnows not only recognised pike odour as a kairomone (species-wide chemicals), but they also tease apart signature odours of individual pike (Wyatt 2009). That minnows responded with greater intensity as the per-pike concentration increased while the overall concentration of pike odour remained constant, tells us that minnows can use odours to determine their relative proximity to pike. In a second experiment, Ferrari et al. (2006) exposed pike-experienced minnows to either

5 mL/pike from each of 12 pike, 5 mL/pike from each of two pike or a water control. This time minnows in the 12-pike treatment showed a stronger antipredator response than minnows in the 2-pike treatment, indicating that minnows can assess the relative density of pike using odour signatures.

20.6 Learned Recognition of Predator Odours by Minnows

Given the importance of being able to recognise and respond to predator odours, and that minnows have a sophisticated ability to learn to respond to variations in predator odour, it is not surprising that minnows have become a model system to study mechanisms of learned predator recognition. Indeed, minnows adapt well to laboratory conditions and can be manipulated in field experiments. This is a powerful combination for studies examining predator recognition. Chivers and Smith (1994a, b) demonstrated that minnows can learn to recognise the sight or odour of predators when they are exposed to novel predator cues in conjunction with alarm cues released from injured conspecifics. This mechanism of learned predator recognition is indeed widespread in fishes and other aquatic organisms (Ferrari et al. 2010). Minnows are able to learn not only odours of predators in this way, but they also learn the sight of the predator (Chivers and Smith 1994b) and sounds (Wisenden et al. 2008) that are associated with alarm cues. Minnows learn to recognise predators based on a single pairing of alarm cues and predator cues. Given the survival benefits associated with predator learning (Gazdewich and Chivers 2002), fast effective learning of predators should be highly selected. Minnows also learn the identity of unknown predators when they detect a predator has eaten a diet of minnows. Alarm cues, or their metabolites, leak from the predator to label the predator as dangerous. After minnows learn the identity of a predator, they respond to the predator whether or not its diet contains minnows in its diet (Mathis and Smith 1993). Minnows also learn to recognise predators via social learning, when they watch a conspecific exhibit an anti-predator response in the presence of the predator (Mathis et al. 1996). Such cultural learning may allow the rapid spread of information about the chemical identity of predators. As mentioned above, minnows rapidly learn to recognise the odour of pike stocked into a previously predator free population. It is unlikely that each of the minnows tested would have had direct experience of pike within 2–4 days (Brown et al. 1997).

Many of the early studies on learned recognition of predators asked a relatively simple question. Can minnows learn to recognise predators through a specific mechanism? However, we know that minnows do not just learn to recognise predators, but they learn a considerable amount about the danger level associated with the predator. Ferrari et al. (2005) were the first to address how minnows could adjust the intensity of their responses to predators according to their threat level. They showed that minnows matched the intensity of their learned responses to predator odours with the intensity of risk posed by each predator. For example, when minnows observed a conspecific in an adjacent tank show a mild response to a predator odour, they learned the predator represents only a mild threat. When minnows observed a conspecifics show a strong response to a predator odour, they learned that the predator was a dangerous predator and responded with a higher intensity of response. In a similar way, minnows that were conditioned to recognise the predator with a high concentration of alarm cues learned the predator as highly dangerous while those that were conditioned with a low concentration of alarm cues learned the predator as a mild threat. It is important to remember that all of this learning occurs based on a single conditioning event. What happens if the minnows are given conflicting information? Ferrari and Chivers (2006) showed that minnows update their perception of the danger status of predators. When a conflict arises in the information learned about the risk level posed by a predator, the minnows display a risk-averse strategy, responding to the predator with the highest threat level learned, and also place more emphasis on recent information instead of averaging all of the pieces of information accumulated over several learning events.

One of the most interesting recent developments in predator odour recognition was the finding that minnows could take information that they acquired about known predators to make educated guesses about novel predators (Ferrari et al. 2007; 2008a). This phenomenon is referred to as 'generalisation of predator recognition'. Minnows that were conditioned with alarm cues to recognise the odour of a lake trout displayed a strong respond to lake trout odour, but also generalised their responses to the odour of a novel brook trout (same genus as lake trout) and rainbow trout (same family), but did not generalise to their recognition to distantly related predatory pike or non-predatory suckers (Fig. 20.5). The intensity of antipredator responses to the other species of trout was correlated with the phylogenetic distance to the lake trout; minnows responded with a higher intensity response to brook trout than rainbow trout. In a similar study, Ferrari et al. (2008b) taught fathead minnows to chemically recognise brown trout either as a high level threat or low level threat and then tested the minnows for their responses to brown trout, rainbow trout (closely related predator) or yellow perch (distantly related predator). When the brown trout was considered a high-risk predator, minnows showed an antipredator response to the odour of brown trout and rainbow trout but not to yellow perch. However, when the brown trout was recognised as a low-risk predator, minnows display antipredator responses to brown trout, but not to rainbow trout or yellow perch.

20.7 The Chemistry of Predator Odours

Researchers studying the chemical ecology of predator/prey interactions have been ecologists for the most part. Not surprisingly, we know virtually nothing about the chemistry of predator odours. The demonstration that minnows are able to detect the identity of individual predators in odour mixtures of multiple predators indicates that the odours are certainly complex mixtures of chemicals. Ferrari et al. (2008a) discussed the potential mechanism behind generalisation of predator

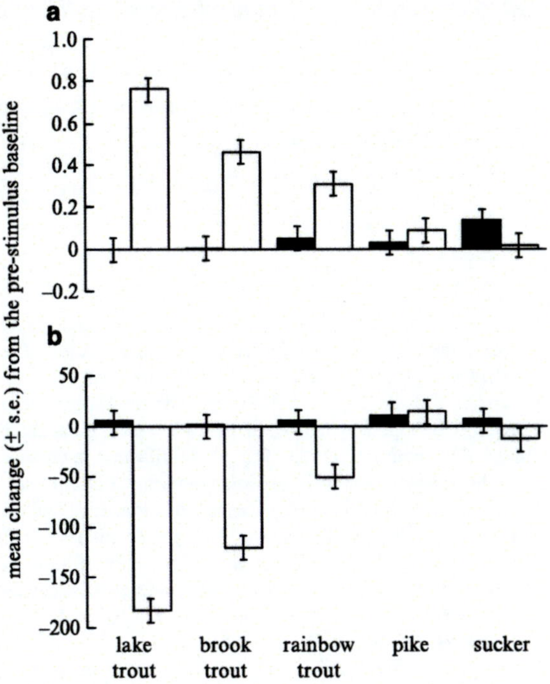

Fig. 20.5 Mean change (post-stimulus minus pre-stimulus) in shoaling index (**a**) and line crosses (**b**) for minnows conditioned to recognise lake trout as a predator and then subsequently exposed to the odour of lake trout and various other fish species. An increase in shoaling index and a reduction in line crosses are both anti-predator responses in minnows. Minnows generalised their recognition of the lake trout to other species of trout, but not to a distantly related predator (pike) or non-predator (sucker) (figure taken from Ferrari et al. 2007)

recognition from different species of trout, arguing that the response patterns may result from small difference in the suite of many molecules that form the trout odour. The addition of a few molecules to a mixture of molecules may be enough to facilitate differentiation. Another possibility is that the same chemicals are present, but in slightly different ratios, allowing for differentiation among species. More effort to identify the suite of chemicals that are responsible for predator odours is in dire need.

20.8 Where to Next?

The past two decades has seen us move ahead in great leaps with regards to our understanding of predator odour recognition in minnows. We started with the basic question of whether minnows could use predator chemicals to recognise predators. We now know that the recognition is amazingly sophisticated. Minnows detect relative risk of predators using diet-based cues released from the predator (Mathis and Smith 1993). They respond to varying concentrations of predator odours (Kusch et al. 2004) and are able to determine the relative density and proximity of predators using odours alone (Ferrari et al. 2006). All of this recognition is a result of learning. Are there other subtle differences in predator odours that minnows can differentiate? We should continue to ask this question, but perhaps more importantly we should start to ask similar questions in other systems. We know that there are a great number of other fish species that recognise predator odours (reviewed by Ferrari et al. 2010), but most of those studies are very simple experiments. Can species X recognize species Y? What we need to know is how much X really knows about Y. We suspect the answer is a great deal.

References

Brown GE, Chivers DP (2006) Learning about danger: chemical alarm cues and predation risk assessment by fishes. In: Brown C, Laland K, Krause J (eds) Fish cognition and behaviour. Blackwell Scientific Press, Oxford, UK, pp 49–69

Brown GE, Chivers DP, Smith RJF (1997) Differential learning rates of chemical versus visual cues of a northern pike by fathead minnows in a natural habitat. Environ Biol Fishes 49:89–96

Chivers DP, Smith RJF (1993) The role of olfaction in chemosensory-based predator recognition in the fathead minnow, *Pimephales promelas*. J Chem Ecol 19:623–633

Chivers DP, Smith RJF (1994a) The role of experience and chemical alarm signaling in predator recognition by fathead minnows, Pimephales promelas. J Fish Biol 44:273–285

Chivers DP, Smith RJF (1994b) Fathead minnows, *Pimephales promelas*, acquire predator recognition when alarm substance is associated with the sight of unfamiliar fish. Anim Behav 48:597–605

Chivers DP, Smith RJF (1995) Free living fathead minnows rapidly learn to recognize pike as predators. J Fish Biol 46:949–954

Ferrari MCO, Chivers DP (2006) Learning threat-sensitive predator avoidance: how do fathead minnows incorporate conflicting information? Anim Behav 71:19–26

Ferrari MCO, Trowell JJ, Brown GE, Chivers DP (2005) The role of learning in the development of threat-sensitive predator avoidance by fathead minnows. Anim Behav 70:777–784

Ferrari MCO, Messier F, Chivers DP (2006) The nose knows: minnows determine predator proximity and density through detection of predator odours. Anim Behav 72:927–932

Ferrari MCO, Gonzalo A, Messier F, Chivers DP (2007) Generalization of learned predator recognition: an experimental test and framework for future studies. Proc Biol Sci 274:1853–1859

Ferrari MCO, Messier F, Chivers DP (2008a) Can prey exhibit threat-sensitive generalization of predator recognition? Extending the Predator Recognition Continuum Hypothesis. Proc Roy Soc B 275:1811–1816

Ferrari MCO, Rive AC, Macnaughton CJ, Brown GE, Chivers DP (2008b) Fixed vs. Random temporal predictability of predation risk: an extension of the risk allocation hypothesis. Ethology 114:238–244

Ferrari MCO, Wisenden BD, Chivers DP (2010) Chemical ecology of predator–prey interactions in aquatic ecosystems: a review and prospectus. Can J Zool 88:698–724

Gazdewich KJ, Chivers DP (2002) Acquired predator recognition by fathead minnows: influence of habitat characteristics on survival. J Chem Ecol 28:439–445

Göz H (1941) Über den Art-und Indivivualgeruch bei Fischen. Z Vergl Physiol 29:1–45

Hartman EJ, Abrahams MV (2000) Sensory compensation and the detection of predators: the interaction between chemical and visual information. Proc Biol Sci 267:571–575

Kats LB, Dill LM (1998) The scent of death: chemosensory assessment of predation risk by prey animals. Ecoscience 5:361–394

Kusch RC, Mirza RS, Chivers DP (2004) Making sense of predator scents: investigating the sophistication of predator assessment abilities of fathead minnows. Behav Ecol Sociobiol 55:551–555

Magurran AE (1989) Acquired recognition of predator odour in the European minnow (*Phoxinus phoxinus*). Ethology 82:216–233

Mathis A, Smith RJF (1993) Fathead minnows (*Pimephales promelas*) learn to recognize northern pike (*Esox lucius*) as predators on the basis of chemical stimuli from minnows in the pike's diet. Anim Behav 46:645–656

Mathis A, Chivers DP, Smith RJF (1993) Population differences in responses of fathead minnows (*Pimephales promelas*) to visual and chemical stimuli from predators. Ethology 93:31–40

Mathis A, Chivers DP, Smith RJF (1996) Cultural transmission of predator recognition in fishes: intraspecific and interspecific learning. Anim Behav 51:185–201

Scott WB, Crossman EJ (1985) Freshwater fishes of Canada. Gordon Soules, Vancouver

Wisenden BD, Pogatshnik J, Gibson D, Bonacci L, Schumacher A, Willett A (2008) Sound the alarm: learned association of predation risk with novel auditory stimuli by fathead minnows (*Pimephales promelas*) and glowlight tetras (*Hemigrammus erythrozonus*) after single simultaneous pairings with conspecific chemical alarm cues. Environ Biol Fishes 81:141–147

Wyatt TD (2009) Fifty years of pheromones. Nature 457:262–263

Chapter 21
Adaptive Responses of Embryonic Amphibians to Predation Risk

Maud C.O. Ferrari and Douglas P. Chivers

Abstract Predation is a pervasive selective force known to shape the behaviour, morphology and life history of prey species. Many prey species, however, are required to learn about potential predatory threats before they can display adaptive responses. When it comes to learning, sooner is better. In this chapter, we present the earliest chemically induced antipredator adaptations known to be displayed by prey species.

21.1 Introduction

Most amphibians have a bipartite life cycle. The adults are terrestrial or semi-terrestrial and often move amongst a mosaic of lakes, ponds and rivers to breed. As a result larvae develop under a range of ecological conditions with considerable variation in resource availability and levels of competition and predation (Orrock et al. 2008; Werner 1986). Not surprisingly, larval amphibians appear to be particularly adapted to deal with such variation, particularly with regard to predation pressure (Ferrari and Chivers 2009a). Many studies have demonstrated that larval amphibians exhibit behavioural responses to risk (Devito et al. 1998; Kats et al. 1988; Kiesecker et al. 1996). Larval salamanders and tadpoles often avoid areas with predators and show reductions in activity or aggregation responses when they detect predators. Larval amphibians also show a remarkable ability to learn to recognise predators from non-predators and recognise the time of day that predators are most

M.C.O. Ferrari (✉)
Department of Biomedical Sciences, WCVM, University of Saskatchewan,
Saskatoon, SK, S7N 5B4, Canada
e-mail: maud.ferrari@usask.ca

D.P. Chivers
Department of Biology, University of Saskatchewan,
112 Science Place, Saskatoon, SK, S7N 5E2, Canada

M.L. East and M. Dehnhard (eds.), *Chemical Signals in Vertebrates 12*,
DOI 10.1007/978-1-4614-5927-9_21,

active (Ferrari et al. 2008, 2010a). In addition to exhibiting behavioural responses to risk, larval amphibians show marked morphological variation in response to varying levels of risk (Relyea 2002; Schoeppner and Relyea 2008). Tadpoles often grow larger, deeper tails when exposed to predators. Such induced responses increase burst speed allowing prey to escape from predators.

In a landmark study, Sih and Moore (1993) demonstrated that embryonic amphibians detect and respond to predation risk cues prior to hatching. They showed that streamside salamander eggs (*Ambystoma barbouri*) that were exposed to odour cues of flatworms delayed hatching. Flatworms are predators of newly hatched larval salamanders. By delaying hatching, the salamanders reach a developmental stage where they are less vulnerable to attacks by flatworms. This chapter spurred on a considerable amount of research aimed at examining how embryonic amphibians recognise and respond to predation risk prior to hatching. Here, we review this expanding literature. In keeping with the theme of this volume, we will consider only those studies where the interactions are mediated by chemosensory information. However, we do note that other researchers, most notably Karen Warkentin, have published some excellent papers examining the ecology of predator-induced hatching in amphibian species that utilise mechanosensory information to mediate risk (Warkentin 1995, 2005).

21.2 Adaptive Variation in the Timing of Hatching

Moore et al. (1996) did a follow-up study examining the adaptive hatching responses in streamside salamanders. The delayed hatching response of salamander embryos to flatworms was mirrored when the embryos were exposed to odours of predatory sunfish, another predator of newly hatched larvae. Again, reaching a developmental stage whereby the prey has an increased probability of escape appears to be highly advantageous.

Larval amphibians are not the only life stage of amphibians that are exposed to predators. Indeed, embryos themselves are often preyed upon by a variety of predators. Consequently, Chivers et al. (2001b) hypothesised that exposure to chemical cues of egg predators (leeches) may likewise alter the timing of hatching. Specifically, the cues may induce early hatching allowing the prey to escape from the imminent threat posed by the leeches. They tested this by exposing Pacific treefrog (*Pseudacris regilla*) embryos to two risk cues, the odour of predatory leeches and alarm cues released from injured eggs, and two non-risk cues, the odour of non-predatory earthworms and distilled water. Both risk cues resulted in the embryos hatching at an earlier developmental stage (Fig. 21.1). In comparison to the controls, the leech cues also induced earlier hatching. Hatching time upon exposure to alarm cues was intermediate between the controls and the leech cues. Chivers et al. (2001b) repeated the experiment using Cascades frog (*Rana cascadae*) embryos. In response to leech cues, the tadpoles again hatched earlier and at an earlier developmental stage than in response to the two risk cues. For those embryos exposed to alarm cues, the timing

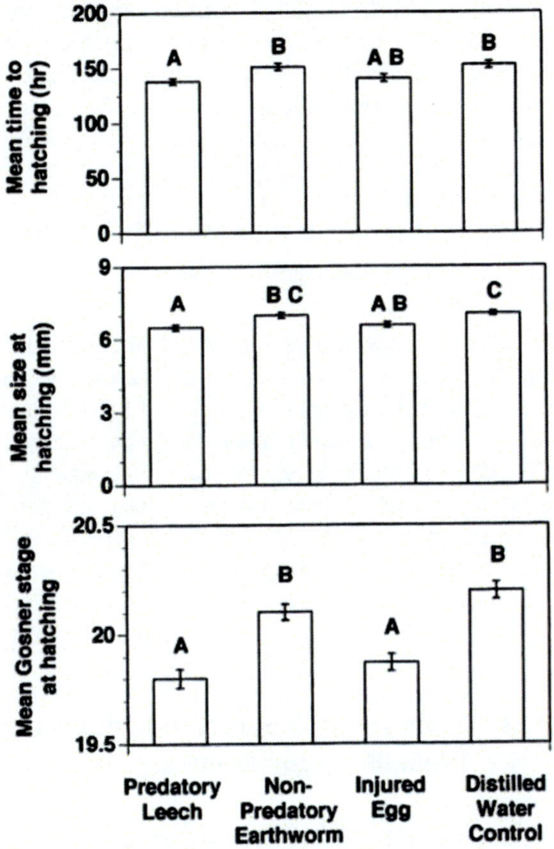

Fig. 21.1 Mean (±SE) time, size and developmental stage at hatching for Pacific treefrog tadpoles exposed to cues from predatory leeches, non-predatory earthworms or nothing (letters indicate differences at alpha = 0.05) (Figure taken from Chivers et al. 2001b)

of hatching was intermediate between the controls and leech treatments. However, exposure to alarm cues resulted in a much shorter time taken for the tadpoles to leave the jelly mass after they hatched. Hatching and leaving the jelly mass at an earlier developmental stage is likely costly; tadpoles that leave early will have a reduced swimming ability but leaving early removes the hatchling from imminent threat of predation.

Taken together, these studies provide strong evidence that two distinctive lineages of amphibians show adaptive plasticity in their hatching ecology; upon exposure to chemical cues indicating risk on tadpoles, embryos appear to delay hatching, while in response to cues that indicate predation on embryos, they show earlier hatching. Following these early studies, a considerable number of researchers have tested this phenomenon in other systems (Capellan and Nicieza 2007; Johnson et al. 2003; Laurila et al. 2001, 2002; Lehman and Campbell 2007; Mandrillon and Saglio 2008; Saenz et al. 2003; Saglio and Mandrillon 2006; Vonesh 2005). Overall, there appears to be strong evidence supporting the considerable hatching plasticity associated with exposure to predation risk. One notable exception was published by

Mandrillon and Saglio (2008), whereby exposure to chemical alarm cues did not alter hatching timing of tadpoles, but nevertheless induced longer, deeper tails in the hatchling tadpoles. Tadpoles with longer deeper tails have a greater burst speed and hence a reduced vulnerability to predators.

21.3 Embryonic Exposure to Risk Influences Post-hatching Behaviour

Based on the convincing evidence that embryonic amphibians recognise predation risk, Mathis et al. (2008) were the first to test whether embryonic exposure to risk also influenced subsequent activity and habitat choice of newly hatched larvae. Ringed salamander (*Ambystoma annulatum*) embryos that were exposed to chemical cues of cannibalistic adult predators subsequently showed reduced activity and greater shelter use behaviour upon hatching compared to embryos exposed to control water or water from non-predatory conspecifics. The idea that embryonic exposure to risk could be a major determinant of microhabitat use of newly hatched individuals is an exciting area of future research.

21.4 Embryonic Exposure to Risk Facilitates Predator Recognition

Mathis et al. (2008) were also the first to consider whether embryonic exposure to risk could influence learning of predators. Many species of aquatic animals, including larval amphibians, do not recognise predators without experience (Chivers and Mirza 2001). They often learn the identity of unknown predators when simultaneously exposed to alarm cues of conspecifics paired with the odour of a novel predator (Mirza et al. 2006). This is a highly effective mechanism of learning, requiring a single conditioning event. Mathis et al. (2008) exposed woodfrog (*Rana sylvatica*) egg clutches to either alarm cues of injured tadpoles paired with the odour of an unknown predatory newt (*Cynops pyrrhogastor*), water paired with newt odour or water alone. The treatment stopped prior to the embryos hatching and the embryos hatched in clean water. Two weeks later, the tadpoles were tested for their response to newt odour. Embryonic exposure to the newt cues facilitated learning. Indeed, those individuals that were exposed, as embryos, to newt odour paired with alarm cues responded to the predator, while those exposed to newt odour paired with water did not, nor did those just exposed to water (Fig. 21.2).

Being able to recognise predators is a fundamental pre-requisite for prey to be able to respond to them. However, prey are not only able to learn to recognise predators from non-predators, but they are also able to learn to distinguish the

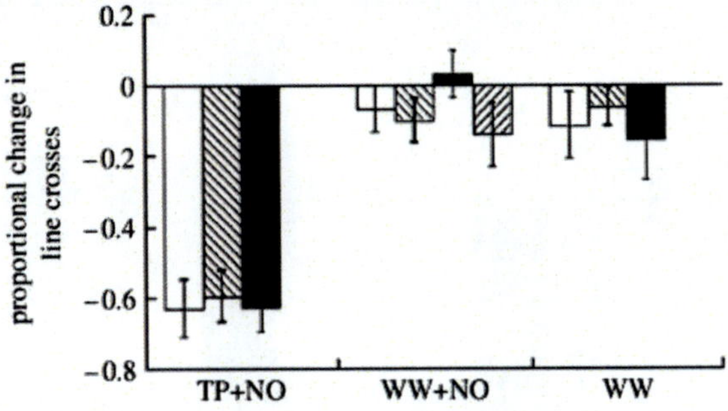

Fig. 21.2 Mean change in activity from pre-stimulus baseline for woodfrog tadpoles exposed to newt odour. The embryos had been exposed to tadpole alarm cues paired with newt odour (TP+NO), well water paired with newt odour (WW+NO) or well water only (WW) in the egg. Different shading patterns represent the responses of tadpoles from difference clutches. A reduction in activity is a typical antipredator response for larval amphibians. Embryos exposed to newt odour in the presence of risk subsequently recognise the newt as dangerous, while those exposed to newt odour in the absence of risk did not (Figure taken from Mathis et al. 2008)

threat level posed by the predator and match their behavioural response to this level of risk. This phenomenon is known as threat-sensitive predator avoidance (Chivers et al. 2001a). Ferrari et al. (2005) were the first to show how such threat-sensitive learning occurs using a model fish system. Minnows (*Pimephales promelas*) exposed to predator odours paired with alarm cues learned to recognise the predator odour as dangerous; as the concentration of alarm cues used in the conditioning increased, the learned behavioural response to the predator odour also increased. Ferrari and Chivers (2010) demonstrated that this same mechanism operates in embryonic learning. Embryos exposed to greater concentrations of alarm cues later showed stronger learned responses to predator odours as tadpoles. One of the most interesting aspects of this study was the suggestion that there was temporal variation in the intensity of responses. The embryos had been treated with alarm cues paired with predator odour between 1,500 and 1,700 h daily. Tadpoles subsequently tested at that time showed stronger responses to the predator odour than tadpoles tested between 1,100 and 1,300 h and those tested between 1,900 and 2,100 h. In a follow-up study, Ferrari et al. (2010a) conditioned woodfrog embryos to recognise tiger salamanders (*Ambystoma tigrinum*) as a predator either in the morning or in the evening, and subsequently exposed the tadpoles to salamander cues both in the morning and the evening. The results provided strong evidence that tadpoles learn the temporal pattern of the risk posed by the predator. Embryos conditioned in the morning subsequently responded to the predator in the morning but not the evening, while those conditioned as embryos in the evening responded in the evening but not the morning.

Some prey lacking innate predator recognition have the ability to display antipredator responses upon their first encounter with those predators if they are

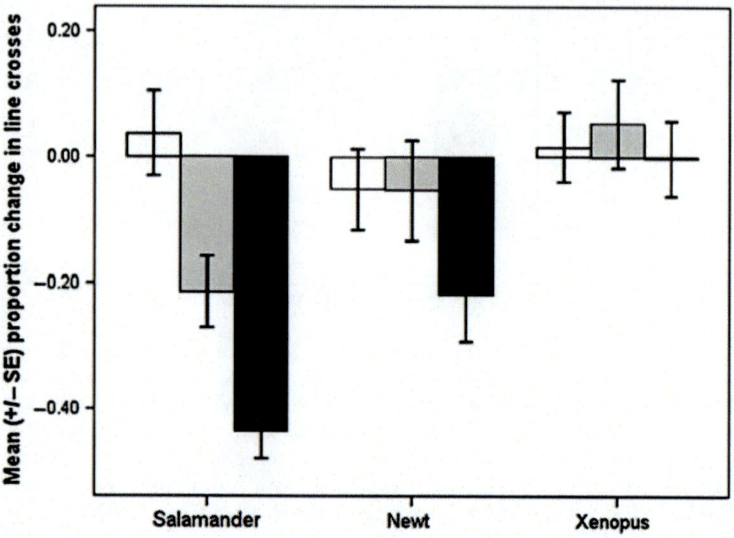

Fig. 21.3 Mean proportion change in line crosses from the pre-stimulus baseline for tadpoles exposed to salamander odour paired with water (*white bars*), a low (*grey bars*) or a high (*black bars*) concentration of injured tadpole cues (alarm cues) as embryos and subsequently tested for their response to the odour of salamander, newt or *Xenopus*. Tadpoles that learned to recognise salamander odour as a threat generalised their response to newt odours when the salamander cue was recognised as a high level threat but not when the salamander was recognised as a low level threat (Figure taken from Ferrari and Chivers 2009b)

similar to predators that the prey has recently learned to recognise (Ferrari et al. 2007). This phenomenon is referred to as generalisation of predator recognition. Ferrari and Chivers (2009b) tested whether embryonic woodfrogs have the ability to learn to generalise their recognition of known predators to closely related novel predators. Similar to previous experiments, embryos exposed to the odour of salamanders paired with alarm cues subsequently displayed fright responses to salamanders. However, the tadpoles also responded to the odour of closely related newts but not to the odour of distantly related African clawed frogs (*Xenopus sp.*) (Fig. 21.3). Tadpoles responded with a greater intensity to the salamanders than to the newts, indicating that a difference between the two odours was perceived by the tadpoles. The odours were close enough to be recognised as threatening, but were yet distinct. Such a pattern of response may result from small difference in the suite of many molecules that form a generalised salamander odour. The addition of a few molecules to a mixture of molecules may be enough to facilitate differentiation between the newt and the tiger salamander. Another possibility is that the same chemicals are present, but these chemicals are in slightly different ratios, allowing for differentiation among species. Clearly, we need more effort to identify the suite of chemicals that are responsible for predator odours.

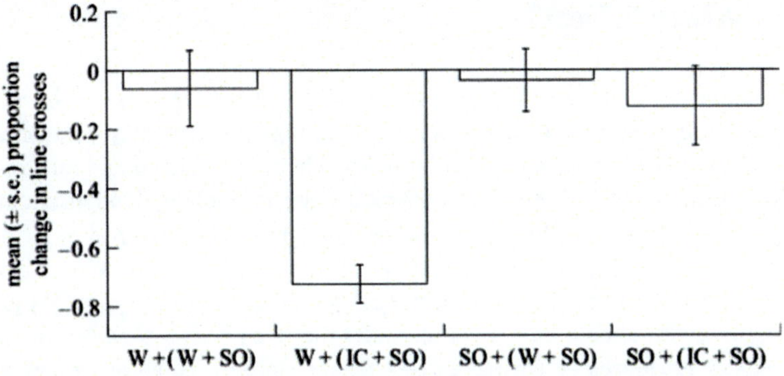

Fig. 21.4 Mean proportion change in line crosses for tadpoles exposed to salamander odour. Woodfrogs were exposed for 5 days to either water (W) or salamander odour (SO) during their embryonic development and later conditioned as tadpoles to learn to recognise salamander as a threat through the pairing of injured conspecific cues and salamander odour (IC+SO) or through a control of water paired with salamander odour (W+SO). A reduction in line crosses is an anti-predator response in tadpoles. Woodfrogs pre-exposed to water learn that salamanders are risky, whereas those exposed to salamanders in the absence of risk do not (Figure taken from Ferrari and Chivers 2011)

21.5 Embryonic Exposure to Safety Facilitates Non-predator Recognition

As discussed earlier, amphibians have very efficient means to recognise predators based on a one-time association between conspecifics alarm cues and cues of unknown predators. The remarkable efficiency of such learning means that recognition mistakes may occur if the tadpoles inadvertently learn that a species is a predator when it is not. Latent inhibition is a mechanism in which prey that are pre-exposed to an unknown species in the absence of negative reinforcement can learn that the unknown animal is likely not a threat (Acquistapace et al. 2003; Ferrari and Chivers 2006). Ferrari and Chivers (2009c) investigated whether woodfrogs would be able to learn to recognise non-predators while in the egg. Woodfrog embryos were exposed to the odour of tiger salamander or a water control daily for 5 days. After hatching, the tadpoles were raised for 2 weeks at which time they undergo a conditioning paradigm, being exposed to conspecific alarm cues paired with salamander odour. When the tadpoles were subsequently tested for their response to salamander odour, the larvae pre-exposed to salamander odour failed to respond to the salamander. However, larvae pre-exposed to water as embryos showed clear evidence of learning (Fig. 21.4). These data provide convincing evidence that embryonic pre-exposure in the absence of risk facilitates non-predator learning. The prey cannot quickly learn to associate risk with the salamander cues because they already recognise them as a safe cue. In a follow-up study, Ferrari and Chivers (2011) showed that woodfrogs that learned as embryos to recognise salamanders as non-predators generalised this response to other closely related species. The tadpoles failed to learn that newts were predators but could easily learn that fish were predators.

21.6 Where to Next?

Amphibian embryos appear to have a remarkable ability to respond to risk prior to hatching. Studies documenting plasticity in hatching time have been conducted in a variety of taxa, both salamanders and tadpoles. Alteration in hatching time is associated with changes in size and developmental stage at hatching. Understanding the long-term effects of this alteration on subsequent growth and survival of larvae is a topic that is ripe for future experimentation. The finding that exposure to risk during embryonic development results in long deeper tails on hatchling tadpoles (Mandrillon and Saglio 2008) should be a good reminder that we need to consider amphibian antipredator responses as an integrated package. Indeed, amphibians can reduce their probability of capture by altering their behaviour, morphology or life history (Ferrari et al. 2010b).

In addition to hatching plasticity, we know that embryonic exposure to risk influences subsequent habitat choice in salamanders (Mathis et al. 2008) and that tadpoles can learn sophisticated information about predators and non-predators during development (Ferrari and Chivers 2009b, 2011). All of the learning studies are restricted to studies of woodfrogs. We clearly need to expand our effort towards testing other amphibian taxa. Studies looking at other vertebrates, particularly fishes, would likely prove fruitful as well.

Nobody writing a review such as ours should pass up the opportunity to highlight that there is an overwhelming lack of understanding about the chemistry of odours that pervades in this field (Chivers and Smith 1998; Ferrari et al. 2010b). We know almost nothing about the identity of predator odours. We hope that researchers will heed our challenge to start identifying the chemical basis of the fascinating responses that are appearing.

References

Acquistapace P, Hazlett BA, Gherardi F (2003) Unsuccessful predation and learning of predator cues by crayfish. J Crustac Biol 23:364–370

Capellan E, Nicieza AG (2007) Trade-offs across life stages: does predator-induced hatching plasticity reduce anuran post-metamorphic performance? Evol Ecol 21:445–458

Chivers DP, Mirza RS (2001) Importance of predator diet cues in responses of larval wood frogs to fish and invertebrate predators. J Chem Ecol 27:45–51

Chivers DP, Smith RJF (1998) Chemical alarm signalling in aquatic predator–prey systems: a review and prospectus. Ecoscience 5:338–352

Chivers DP, Mirza RS, Bryer PJ, Kiesecker JM (2001a) Threat-sensitive predator avoidance by slimy sculpins: understanding the importance of visual versus chemical information. Can J Zool 79:867–873

Chivers DP, Kiesecker JM, Marco A, DeVito J, Anderson MT, Blaustein AR (2001b) Predator-induced life history changes in amphibians: egg predation induces hatching. Oikos 92:135–142

Devito J, Chivers DP, Kiesecker JM, Marco A, Wildy EL, Blaustein AR (1998) The effects of snake predation on metamorphosis of western toads, *Bufo boreas* (Amphibia, Bufonidae). Ethology 104:185–193

Ferrari MCO, Chivers DP (2006) The role of latent inhibition in acquired predator recognition by fathead minnows. Can J Zool 84:505–509

Ferrari MCO, Chivers DP (2009a) Temporal variability, threat sensitivity and conflicting information about the nature of risk: understanding the dynamics of tadpole antipredator behaviour. Anim Behav 78:11–16

Ferrari MCO, Chivers DP (2009b) Sophisticated early life lessons: threat-sensitive generalization of predator recognition by embryonic amphibians. Behav Ecol 20:1295–1298

Ferrari MCO, Chivers DP (2009c) Latent inhibition of predator recognition by embryonic amphibians. Biol Lett 5:160–162

Ferrari MCO, Chivers DP (2010) The ghost of predation future: threat-sensitive and temporal assessment of risk by embryonic woodfrogs. Behav Ecol Sociobiol 64:549–555

Ferrari MCO, Chivers DP (2011) Learning about non-predators and safe places: the forgotten elements of risk assessment. Anim Cogn 14:309–316

Ferrari MCO, Trowell JJ, Brown GE, Chivers DP (2005) The role of learning in the development of threat-sensitive predator avoidance by fathead minnows. Anim Behav 70:777–784

Ferrari MCO, Gonzalo A, Messier F, Chivers DP (2007) Generalization of learned predator recognition: an experimental test and framework for future studies. Proc Roy Soc B 274:1853–1859

Ferrari MCO, Messier F, Chivers DP (2008) Larval amphibians learn to match antipredator response intensity to temporal patterns of risk. Behav Ecol 19:980–983

Ferrari MCO, Manek AK, Chivers DP (2010a) Temporal learning of predation risk by embryonic amphibians. Biol Lett 6:308–310

Ferrari MCO, Wisenden BD, Chivers DP (2010b) Chemical ecology of predator–prey interactions in aquatic ecosystems: a review and prospectus. Can J Zool 88:698–724

Johnson JB, Saenz D, Adams CK, Conner RN (2003) The influence of predator threat on the timing of a life-history switch point: predator-induced hatching in the southern leopard frog (*Rana sphenocephala*). Can J Zool 81:1608–1613

Kats LB, Petranka JW, Sih A (1988) Antipredator defenses and the persistence of amphibian larvae with fishes. Ecology 69:1865–1870

Kiesecker JM, Chivers DP, Blaustein AR (1996) The use of chemical cues in predator recognition by western toad tadpoles. Anim Behav 52:1237–1245

Laurila A, Crochet PA, Merila J (2001) Predation-induced effects on hatchling morphology in the common frog (*Rana temporaria*). Can J Zool 79:926–930

Laurila A, Pakkasmaa S, Crochet PA, Merila J (2002) Predator-induced plasticity in early life history and morphology in two anuran amphibians. Oecologia 132:524–530

Lehman EM, Campbell CD (2007) Developmental window of response to predator chemical cues in rough-skinned newt embryos. Funct Ecol 21:880–885

Mandrillon AL, Saglio P (2008) Developmental windows and origins of the chemical cues mediating hatching responses to injured conspecific eggs in the common frog (*Rana temporaria*). Can J Zool 86:447–455

Mathis A, Ferrari MCO, Windel N, Messier F, Chivers DP (2008) Learning by embryos and the ghost of predation future. Proc Roy Soc B 275:2603–2607

Mirza RS, Ferrari MCO, Kiesecker JM, Chivers DP (2006) Responses of American toad tadpoles to predation cues: behavioural response thresholds, threat-sensitivity and acquired predation recognition. Behaviour 143:877–889

Moore RD, Newton B, Sih A (1996) Delayed hatching as a response of streamside salamander eggs to chemical cues from predatory sunfish. Oikos 77:331–335

Orrock JL, Grabowski JH, Pantel JH, Peacor SD, Peckarsky BL, Sih A, Werner EE (2008) Consumptive and nonconsumptive effects of predators on metacommunities of competing prey. Ecology 89:2426–2435

Relyea RA (2002) Local population differences in phenotypic plasticity: Predator-induced changes in wood frog tadpoles. Ecol Monogr 72:77–93

Saenz D, Johnson JB, Adams CK, Dayton GH (2003) Accelerated hatching of southern leopard frog (*Rana sphenocephala*) eggs in response to the presence of a crayfish (*Procambarus nigrocinctus*) predator. Copeia 2003:646–649

Saglio P, Mandrillon AL (2006) Embryonic experience to predation risk affects tadpoles of the common frog (*Rana temporaria*). Arch Hydrobiol 166:505–523

Schoeppner NM, Relyea RA (2008) Detecting small environmental differences: risk-response curves for predator-induced behavior and morphology. Oecologia 154:743–754

Sih A, Moore RD (1993) Delayed hatching of salamander eggs in response to enhanced larval predation risk. Am Nat 142:947–960

Vonesh JR (2005) Egg predation and predator-induced hatching plasticity in the African reed frog, *Hyperolius spinigularis*. Oikos 110:241–252

Warkentin KM (1995) Adaptive plasticity in hatching age: a response to predation risk trade-offs. Proc Natl Acad Sci U S A 92:3507–3510

Warkentin KM (2005) How do embryos assess risk? Vibrational cues in predator-induced hatching of red-eyed treefrogs. Anim Behav 70:59–71

Werner EE (1986) Amphibian metamorphosis - growth rate, predation risk and the optimal size at transformation. Am Nat 128:319–341

Chapter 22
Response of Lumholtz' Tree-Kangaroos (*Dendrolagus lumholtzi*) to Odours from Native Arboreal and Introduced Terrestrial Predators: A Preliminary Study

Sigrid Heise-Pavlov, Elizabeth Forbes, Clare Andersen, and Michelle Prince

Abstract One main threat to the arboreal Lumholtz' tree-kangaroo (*Dendrolagus lumholtzi*) in north-eastern tropical Australia is fatal attacks by introduced predators such as the domestic dog. When threatened by a predator, tree-kangaroos descend from the canopy and flee on the ground. This preliminary study reports the behaviours of two captive female tree-kangaroos towards odour cues from a native arboreal predator, a python, and an introduced, terrestrial predator, a dog. When exposed to both types of predator odour, the females reduced the duration of their movements while increasing the duration of sniffing. The results indicate that Lumholtz' tree-kangaroos are able to detect predator odours. However, subjects did not show clear differences in their responses to the two different predator odours. Variations in the responses between the two subjects suggest that predator experience may play a role in the behaviour of this species.

22.1 Introduction

Co-evolution of predator–prey relationships results in adaptive behaviours of prey species to counter the hunting strategies of their predators (Sih 1984; Lima and Bednekoff 1999). Anti-predatory behaviour involves the recognition of a predator, predator avoidance and escape from a predator if detected (Apfelbach et al. 2005).

S. Heise-Pavlov (✉)
Center for Rainforest Studies, School for Field Studies, Yungaburra, QLD, Australia
e-mail: sheise-pavlov@fieldstudies.org

E. Forbes
Vassar College, 124 Raymond Avenue, Poughkeepsie, NY 12604, USA
e-mail: forbeselizabeth.s@gmail.com

C. Andersen • M. Prince
Wildlife Management, Wildlife Habitat, Port Douglas, QLD, Australia
e-mail: manager@wildlifehabitat.com.au

M.L. East and M. Dehnhard (eds.), *Chemical Signals in Vertebrates 12*,
DOI 10.1007/978-1-4614-5927-9_22, © Springer Science+Business Media New York 2013

Predator recognition may involve a range of cues (Blumstein et al. 2000), including olfactory cues (Mueller-Schwarze 1972; Kats and Dill 1998). Once a predator is detected, prey species may respond to the presence of predators with increased vigilance, decreased movement, flight and the avoidance of areas where predators are present (Apfelbach et al. 2005; Parsons et al. 2007).

Lumholtz's tree-kangaroo (*Dendrolagus lumholtzi*) is a rare marsupial arboreal folivore endemic to Australia. Tree-kangaroos spend the majority of their time in the forest canopy and, in response to predators, rapidly descend from trees and attempt to flee on the ground using bipedal hopping locomotion (Procter-Gray and Ganslosser 1986). This response may be effective towards arboreal predators, such as pythons and raptors, but proves to be fatal for tree-kangaroos attempting to avoid domestic dogs associated with the European colonisation of Australia. It is estimated that 30 Lumholtz' tree-kangaroos are killed each year from encounters with dogs in their core distribution area, the Atherton Tablelands in north-eastern Queensland (Newell 1999; Kanowski et al. 2001).

In this preliminary study we aim to assess whether Lumholtz' tree-kangaroos are able to detect odours from an arboreal predator, the Amethystine python (*Morelia amethistina*) and from an introduced, terrestrial predator, the domestic dog (*Canis lupus familiaris*), in the same way. Additionally, we want to pre-identify behaviours which reflect the responses of Lumholtz' tree-kangaroos to the presented odours. So far, studies on the responses of marsupials to predatory odours showed that marsupials do not change their feeding behaviour in the presence of predator odour cues, either from native or introduced predators (Russel and Banks 2007; Mella et al. 2011). This suggests that the immediate response of marsupials to predator odours may be expressed in other behavioural patterns, such as a decrease in comfort behaviours, or an increase in discomfort levels (Russel and Banks 2007).

22.2 Methods

Behavioural studies were conducted on two captive tree-kangaroos in April 2011. One subject was an adult, blind, 42-month-old female tree-kangaroo in an 80 m^2 enclosure with a vegetation-covered floor, tree limbs that provided climbing structures and a hut as shelter. The subject was reared in captivity and released in a nearby forest fragment until total blindness confined her to the enclosure. The second subject was an adult, 54-month-old female that was born and reared in captivity at a wildlife sanctuary. Its enclosure measured 27 m^2 with a concrete floor, tree limbs and wooden ledges as climbing structures. We conducted control trials during which no odour cues were placed inside the enclosures and exposure trials during which either an odour from a dog or an amethystine python was placed in the enclosures. Each trial lasted 10 min during which an observer in the enclosure recorded the behaviours of the subjects on a camcord camera (Sony HD Handycam, HDR-XR150E). Python odour consisted of shed python skin, urine and faecal excrement, opportunistically collected from captive pythons. Dog odour sources consisted of urine samples, opportunistically collected by a local veterinarian. All odour samples were stored in sealed plastic containers and kept refrigerated until use. The subjects were exposed to the odour cues by placing open plastic containers, containing the odour source, into the enclosures.

There was a minimum of 15 min of rest in between control and exposure trials. Both subjects were exposed to three repeats of each treatment (dog or python odour cues present). The 42-month-old subject was observed in five control trials, while the 54-month-old subject was observed in four control trials.

Observed behaviours were selected from an ethogram for Lumholtz' tree-kangaroos given in Procter-Gray and Ganslosser (1986). Behaviours included "feeding", "grooming" and "resting" as comfort behaviours; "horizontal movements" along limbs and "climbing up" as locomotor behaviour; "air sniffing" and "food sniffing" as chemosensory behaviours and "alertness", "freezing" and "fleeing" as discomfort behaviour. For each of the behaviours the mean duration was calculated over all trials per animal.

Due to the small sample size in this preliminary study, no statistical analyses were performed.

22.3 Results

Figure 22.1 shows the mean durations of each of the selected behaviours for each animal with respect to the exposure to the odour from a domestic dog and with respect to the odour from a python. For most trials differences in the mean duration of behaviours between control and treatment trials are apparent. Female 1 did not show any "Fleeing" and "Freezing" behaviour in trials. However, this female showed only a very small difference in the duration of "climbing up" and "air sniffing" between control trials and trials with python odour. Female 2 showed no or only very small changes in the duration of "grooming", "climbing up", "air sniffing" and "fleeing" when odours from a dog were presented. In trials with python odour, female 2 did only show small changes in the duration of "resting" and "horizontal movements" between control and treatment trials. While some responses of the two females were opposite, both showed a similar trend in reducing the time engaging in movements and "food sniffing" but increasing time spent "air sniffing" in the presence of dog odour. When presented with the python odour both females reduced the amount of time spending "feeding", but increased time spending "grooming" and "resting". They also showed an increase in time spending "air sniffing" in the presence of a python odour.

22.4 Discussion

The results of this preliminary study suggest that Lumholtz' tree-kangaroos are able to detect odours from predators. Both females showed differences in the duration of most of the observed behaviours between control and exposure trials.

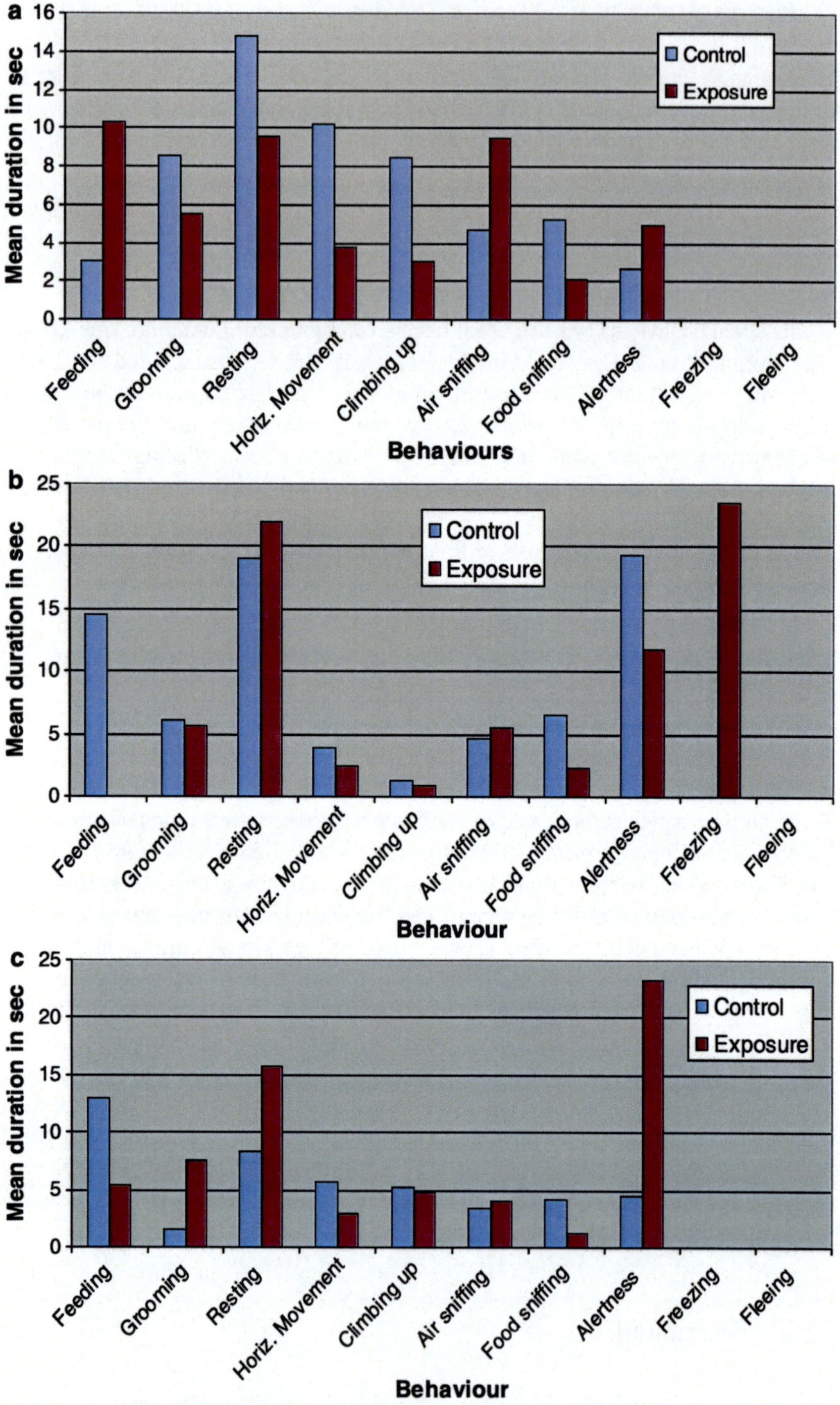

Fig. 22.1 Mean durations of selected behaviours in control and exposure trials of two Lumholtz' tree-kangaroo females. (**a**) Female one: Exposure to the odour from a dog. (**b**) Female two: Exposure to the odour from a dog. (**c**) Female one: Exposure to the odour from a python. (**d**) Female two: Exposure to the odour from a python

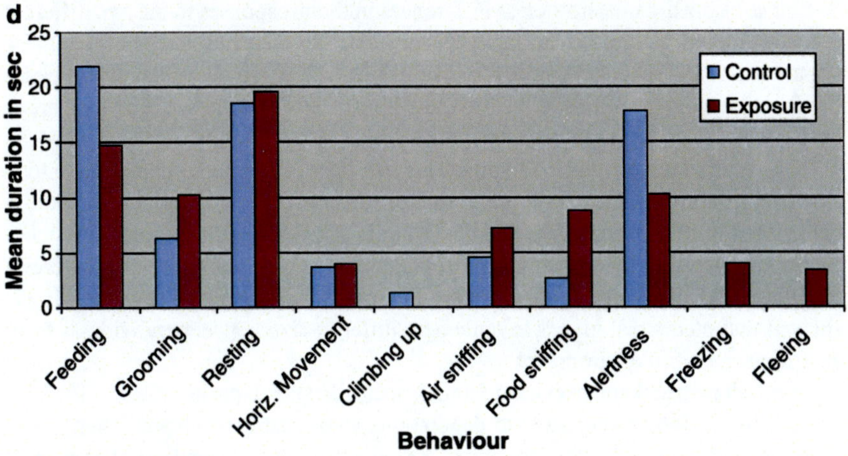

Fig. 22.1 (continued)

The presence of odours from the two predators seemed to release in both subjects a reduction in movements. Anecdotal accounts of Lumholtz's tree-kangaroos suggest that a reduction of locomotor activity might be an anti-predatory strategy as free-ranging individuals have been observed sitting motionless next to an amethystine python (Karen Coombes, personal comm. 2011). Both subjects display an increase in comfort behaviours in the presence of a python odour but different changes in comfort behaviours when presented with odours from a dog. A reduction in locomotor activity and comfort behaviour, and an increase in olfaction have been reported as responses to the presence of predators in some mammalian species (Apfelbach et al. 2005; Ylönen et al. 1992). However, studies on anti-predatory behaviours of marsupials often lack to show a clear response of these species to odours from predators (Ramp et al. 2005). The Southern brown bandicoot (*Isoodon obesulus*) and the Common brushtail possum (*Trichosurus vulpecula*) do not alter foraging behaviour in response to olfactory cues from predators presented in traps (Mella et al. 2011). Similarly, Brown antechinus (*Antechinus stuartii*) do not respond to traps scented with the odour from a native or an introduced predator (Russel and Banks 2007). This may suggest that native Australian marsupials lack the ability to recognise and avoid predator odours (Mella et al. 2011), or that their reactions to predator odours are exhibited in ways which are not recognisable through trapping methods (Russel and Banks 2007). Our results indicate that Lumholtz' tree-kangaroos detect both predatory odours, but that individual differences in some responses exist.

Both females showed different changes in discomfort behaviours between control and exposure trials when presented with different odours. While female 1 displayed increased "alertness" in response to the presented odours, female 2 showed a reduced "alertness". This may be attributable to predator-experiences of the two subjects. While female 1 was attacked by a dog in the past, female 2 had never any encounter with predators. Predator-experiences have been used to train marsupials to recognise introduced predators (McLean et al. 2000) and may also play a role in anti-predatory behaviours of Lumholtz' tree-kangaroos.

Our subjects did not show clear differences in their responses to the two different predator odours. Future studies have to prove whether their initial response to these odours will result in the release of the same anti-predatory strategy, namely descending from the canopy. If this is the case, the high rate of fatalities of tree-kangaroos by domestic dogs can be explained.

The results suggest that tree-kangaroos use olfactory cues to detect predators. However, nothing is known about the role of olfaction in social and anti-predatory behaviours of this species. Marsupials, living in rainforest habitats, may rely more on odour cues than open-habitat marsupials, as the high humidity in rainforests facilitates the distribution of volatiles. For example, Mueller–Schwarze (1987) showed that metatarsal glands in mule deer differ in size between individuals from an arid vs. a humid environment.

Lumholtz's tree-kangaroo is an iconic species for north-eastern Queensland. Its population is likely to continue to decline due to habitat loss, climate change and introduced predators. Further studies of its behaviour and particularly the role that semiochemicals play in its biology may contribute to the development of more efficient conservation strategies.

Acknowledgements We thank Margit Cianelli, a wildlife caretaker, for the access to one of the tree-kangaroo subjects as well as Denise Stanton and Hester Dingle for help in data collection and logistics. The study was done under ethics permit CA 2011/02/487 and the scientific permit No. WISP08961011 provided to Dr. Sigrid Heise-Pavlov.

References

Apfelbach R, Blanchard CD, Blanchard RJ, Hayes RA, McGregor IS (2005) The effects of predator odors in mammalian prey species: a review of field and laboratory studies. Neurosci Biobehav Rev 29:1123–1144

Blumstein DT, Daniel JC, Griffin AS, Evans CS (2000) Insular tammar wallabies (*Macropus eugenii*) respond to visual but not acoustic cues from predators. Behav Ecol 11:528–535

Kanowski J, Felderhof I, Newell G, Parker T, Schmidt C, Stirn B, Wilson R, Winter JW (2001) Community survey of the distribution of Lumholtz's tree-kangaroo on the Atherton Tablelands, north-east Queensland. Pac Cons Biol 7:79–86

Kats LB, Dill LM (1998) The scent of death: chemosensory assessment of predation risk by prey animals. Ecoscience 5:361–394

Lima SL, Bednekoff PA (1999) Temporal variation in danger drives anti-predator behavior: the predation risk allocation hypothesis. Am Nat 153:649–659

McLean IG, Schmitt NT, Jarman PJ, Duncan C, Wynne CDL (2000) Learning for life: training marsupials to recognise introduced predators. Behaviour 137:1361–1376

Mella VSA, Cooper CE, Davies SJJF (2011) Predator odour does not influence trappability of southern brown bandicoots (*Isoodon obesulus*) and common brushtail possums (*Trichosurus vulpecula*). Aust J Zool 58(5):267–272

Mueller-Schwarze D (1972) The responses of young black-tailed deer to predator odors. J Mammal 53:393–394

Mueller-Schwarze D (1987) Evolution of cervid olfactory communication. In: Wemmer CM (ed) Research symposia of the national zoological park: biology and management of Cervidae. Smithonian Institution Press, Washington DC, pp 331–344

Newell GR (1999) Home Range and Habitat Use by Lumholtz's Tree-kangaroo (*Dendrolagus Lumholtzi*) within a Rainforest Fragment in North Queensland. Wildlife Res 26:129–145

Parsons MH, Lamont BB, Kovacs BR, Davies SJJF (2007) Effects of novel and historic predator urines on semi-wild western grey kangaroos. J Wildlife Manage 71:1225–1228

Procter-Gray E, Ganslosser U (1986) The individual behaviors of Lumholtz's tree-kangaroo: repertoire and taxonomic implications. J Mammal 67:343–352

Ramp D, Russel BG, Croft DB (2005) Predator scent induces differing responses in two sympatric macropodids. Aust J Zool 53:73–78

Russel BG, Banks PB (2007) Do Australian small mammals respond to native and introduced predator odours? Aust J Ecol 32:277–286

Sih A (1984) The behavioural response race between predator and prey. Am Nat 123:143–150

Ylönen H, Jedrzejewska B, Jedrzejewski W, Heikkilä J (1992) Antipredatory behaviour of Clethrionomys voles—'David and Goliath' arms race. Ann Zool Fennici 29:207–216

Chapter 23
Voles in Space: Effects of Predator Odour on *Microtus arvalis* Breeding Females

Mateusz Jochym and Stefan Halle

Abstract Voles show an array of adaptations to predation risk. The chance of encountering a specialist, mustelid predator is best minimised by reduced mobility and limited use of space; voles may also shift their activity in time. We inquired whether reproductively active female common voles (*Microtus arvalis*) employ these adaptations in response to olfactory cues of a ferret. We used capture locations and radio-tracking fixes to examine long-term (home range) and short-term (activity range) space use under odour-simulated mustelid predation risk. Range areas were computed with two methods: minimum convex polygon and local convex hull. Using radio-tracking data, we also measured movement distances and distances to the sources of predator odour. The common vole is predominantly diurnal, with activity peaks at dusk and dawn. Using two indexes of circadian activity, we investigated whether females change their temporal behaviour in response to mustelid odours. Non-predator odours served as controls. We found no effect of mustelid odour on home range or activity range size. Additionally, predator odours did not affect movement distances and temporal behaviour. However, females did avoid areas with mustelid odour. We argue that reproductively active females cannot afford spatial and temporal adaptations, as these limit foraging opportunities, and ultimately, female fecundity. Yet, avoidance of ferret odour confirms its significance for breeding female voles.

M. Jochym (✉) • S. Halle
Institute of Ecology, Friedrich Schiller University,
Dornburger Straße 159, D-07743 Jena, Germany
e-mail: mateusz.jochym@gmail.com

M.L. East and M. Dehnhard (eds.), *Chemical Signals in Vertebrates 12*,
DOI 10.1007/978-1-4614-5927-9_23,

23.1 Introduction

Different prey organisms face various levels of predator pressure. Voles are under constant predatory threat (Halle 1993). Vegetation cover provides a refuge from avian predators, but it offers no protection against carnivores, especially mustelids (e.g. weasels, stoats and polecats). Hence, voles are particularly vulnerable to mustelid predation.

Voles show various responses to mustelid predation risk, of which predator avoidance is possibly the most basic. Successful avoidance reduces the chance of an encounter with a predator, which is best achieved by limiting spatial activity. In fact, foraging ranges of voles shrink when predation risk increases, as found in several species including field voles (*Microtus agrestis*; Borowski and Owadowska 2010; Borowski 1998), and southern red-backed voles (*Myodes gapperi*; Anderson 1986). Changes in space use reflect changes in mobility (Madison 1985). In *Microtus* voles, individuals covering longer distances are more likely to be killed by predators, while experimentally reduced risk of predation increases vole mobility (Norrdahl and Korpimäki 1998). Thus, restricted mobility constitutes a basic adaptation to mustelid predation, also demonstrated in field voles and common voles from Orkney Islands (*Microtus arvalis orcadensis*; Gorman 1984).

Adaptations to predation risk may not be equally accessible to both sexes. Mustelid predators prefer female prey (Norrdahl and Korpimäki 1998); females may thus need to go to greater lengths to avoid predation. Furthermore, the response to predation risk may be determined by reproductive status. When reproductively active female voles are challenged with a weasel or its odours, they remain within their ranges, whereas inactive females and males flee (Jedrzejewski and Jedrzejewska 1990).

When conceiving and rearing offspring, breeding females are in a difficult situation. Cushing (1985) demonstrated that female prairie deer mice (*Peromyscus maniculatus bairdii*) in oestrus attract weasels, i.e. predators specialising in rodents of the *Cricetidae* family. Being more vulnerable to predation, female voles have developed a complex adaptation: the predator-induced breeding suppression (Ylönen and Ronkainen 1994). In our earlier work, we verified the occurrence of breeding suppression in common voles (Jochym and Halle 2012), as well as explored the effects of predator odours on breeding females (Jochym and Halle, in review). In this chapter, we investigate whether reproductively active females respond to olfactory cues of predation risk with more conventional adaptations, including a reduction of spatial activity.

Predator evasion may take place not only in space but also in time. Some indication exists that voles can shift their circadian activity to the time of day when the prevailing predator is least active (Eccard et al. 2008). However, this response could depend on many factors, such as the usual modes of activity of prey and predators, and habitat properties (e.g. availability of cover). Our study species, the common vole (*Microtus arvalis*) is predominantly diurnal, with intensive activity at dusk and dawn (Lehmann and Sommersberg 1980). We investigated whether this mode of circadian activity changes in response to mustelid odours. We also used capture location data to examine long-term space use (home ranges) of female common voles exposed to mustelid odour. We used radio-tracking data to verify whether reproductively active females adjust their short-term space use (activity ranges and relocation distances) when risk of predation increases. Lastly, we tested whether reproductively active females avoid mustelid odour (mean distance to the odour source).

23.2 Methods

23.2.1 General Setup

The study was based at the Remderoda field station near Jena, central Germany (50.938°N, 11.529°E). During the breeding seasons of 2007 and 2008, we carried out an intensive capture-mark-recapture (CMR) experiment focusing on the effects of increased predator odours on vole reproduction (Jochym and Halle 2012). Capture location data from this study served to estimate the sizes of long-term home ranges. In addition to the CMR experiment, we also carried out a radio-tracking study on spatial and circadian activity of reproductively active females.

Each year, we monitored six populations of common voles, each enclosed in a separate 50 by 50 m outdoor plot. To allow identification, adult individuals were marked at first capture with passive integrated transponders (PIT, Trovan). We trapped voles using 25 live traps, placed 10 m apart on a regular square grid within each plot. We checked traps twice a day, on 2 days a week. Trapping was carried out from the beginning of the summer until mid-autumn. This period was divided into two phases: during the first phase, termed the monitoring phase, the level of ambient predation risk remained unchanged in all plots; during the second phase, termed the treatment phase, we manipulated the perceived risk of predation with olfactory cues. In 2007, the monitoring phase spanned 12 weeks, whereas in 2008 it spanned 7 weeks. The treatment phase spanned 12 weeks in 2007, and 13 weeks in 2008.

We simulated different levels of predation risk using animal and non-animal odours. The odours, in the form of water extracts, were distributed in equal amounts between 16 permanent odour sources per plot. The odour sources were refilled every 2 days (i.e. a total of three applications per week) throughout the treatment phase. Each odour source was located in the centre of a square demarcated by the four adjacent traps. The arrangement of traps and odour sources was identical in each plot. Each of the six plots was pre-assigned to a level of predation risk; this assignment alternated between the 2 years. In three out of six plots, we simulated high predation risk with faecal odours of captive ferrets (*Mustela putorius furo*). These plots are hereafter referred to as the predator treatment. The remaining three plots served as reference and control for the effect of novelty. Those plots received either faecal odours of captive rabbits (*Oryctolagus cuniculus*), thus constituting the herbivore treatment, or the odour of clean sawdust cage bedding (the neutral treatment).

We assumed that neither of the two control odours would affect the level of predation risk as perceived by the voles. Whenever this assumption was met, we treated the herbivore and neutral treatments collectively as the non-predator treatment.

23.2.2 Capture Location Data

We used the capture records of the CMR experiment to estimate the effect of predator odours on home range size. The minimum number of captures necessary to reliably estimate the home range of *M. arvalis* equals seven to eight (Reichstein 1960). Hence, we limited the calculations of home range area to females captured at least eight times in a given phase, regardless of their reproductive status at capture. To facilitate the computation of home range areas, we added generated data to each individual dataset. The number of added observations was equal to the number of actual observations; the added observations were randomly scattered within ±0.5 m relative of the original barycentre, thus simulating the centre of activity. In case of individuals captured in the same trap more than once, the overlapping coordinates were randomly shifted within ±0.5 m relative of the trap location.

We computed the areas of individual home ranges using two methods: minimum convex polygon (MCP) and adaptive Local Convex Hull (LoCoH; Getz et al. 2007). For a given individual, LoCoH method calculates the area bounded by an isopleth containing 95% of observations (LoCoH 95%); MCP method calculates the area by excluding the outermost 5% of observations (MCP 95%). Despite its poor precision (Burgman and Fox 2003; Börger et al. 2006), we included the MCP method for the sake of comparability with earlier studies. Areas of activity and home ranges were computed using the R package *adehabitat*, version 1.8.3 (Calenge 2006). In calculations of LoCoH 95%, the value of argument *a* was set to 30 for all individuals. Prior to statistical analysis, the areas obtained with both methods were square-root transformed to approximate normal distribution of the data.

23.2.3 Radio-Tracking

The radio-tracking study was carried out in August 2008. It comprised two 48 h sessions. The sessions started at 20:00 CEST. The first session (1–3 August) was conducted in the last 3 days of the monitoring phase—before the beginning of odour application. The second session (15–17 August) was held 2 weeks into the treatment phase—after the odours had been applied six times. This study included only resident, reproductively active females, of which all had given birth at least once. We chose three to four females from each of the five study populations; the sixth, least populated plot did not contain enough females meeting the selection criteria. The females were captured 48–72 h prior to the beginning of each session, equipped with radio-collars

(Pip, Biotrack), and released in their place of capture. We followed 18 females in each session, equally divided between the predator, herbivore and neutral treatments.

The females were located manually at 30-min intervals using Yaesu FT-290RII 2 m receivers (Yaesu Musen Co., Ltd.), fitted with 5 m long handheld detectors. They consisted of a screened coaxial cable with an exposed point antenna, mounted onto a fishing rod. The length of the detectors allowed us to minimise the disturbance of the voles' habitat. Fixes were determined by homing-in (Kenward 1993), and recorded on a Cartesian grid with a resolution of 1 m.

23.2.3.1 Activity Range Area

Areas of the 48 h activity range were calculated using the same methods applied earlier to the capture location data. Adaptive LoCoH 95% argument (Getz et al. 2007) was set individually to the maximum distance between any two given fixes within session. The individual per-session estimates were square-root transformed prior to analysis.

23.2.3.2 Mean Distance to Odour Source

For every female, separately for each session, we identified a 'core area' (a square 2 by 2 m, comprising nine possible locations) containing most of the fixes. We assumed that the core area contained the nest; the fixes falling within the core area were excluded from the calculations. For each of the remaining fixes, we calculated the Euclidean distance to the nearest odour source (DOS). Due to the arrangement of odour sources in an array, the possible values of DOS were limited at 0 (at odour source) and ~7.07 m (at trap, halfway between two diagonally adjacent odour sources). The theoretical mean value of DOS was ~3.85 m.

23.2.3.3 Mean Relocation Distance

We calculated the Euclidean distances between subsequent locations of each female. Fixes falling within a square 2 by 2 m relative to the previous fix were ignored. We assumed that these relocations were either below the resolution of our radio-tracking method, or that they were due to the female's movements in the nest. Relocations greater than ~2.82 m were grouped according to the time of day. The 'time of day' factor contained three levels: day, night and twilight (dusk and dawn). The twilight grouping included relocations nearest to the time of sunrise or sunset, plus the ones occurring within ±30 min; out of 48 individual relocations possible in a diel cycle, the twilight grouping contained up to six. Remaining relocations were attributed to either daytime or nighttime. The length of day differed between the radio-tracking sessions by approximately 1 h, so the number of possible relocations in the daytime or nighttime categories varied accordingly.

23.2.3.4 Crepuscularity and Diurnality Indexes

We used the relocation distances (grouped by time of day) to compute two individual indexes of diel activity: crepuscularity index I_C and diurnality index I_D. Positive values of I_C reflect increased activity at twilight, relatively to the remaining part of the diel cycle; negative values show the opposite (Halle and Lehmann 1987). Diurnality index compares daytime activity with nighttime activity: positive values indicate a relative increase in the former, and vice versa; relocations occurring at twilight are excluded (Halle 1995). The numbers of observations were corrected for each individual to account for missing data.

23.2.4 Statistical Analysis

Statistical analysis was performed in R, version 2.12.1 (R Development Core Team 2010), and employed linear mixed-effects models (LMM) implemented in the *lme4* package, version 0.999375-37 (Bates and Maechler 2010). We included a random intercept for plot ID in models fitted to capture location data. All models based on radio-tracking data included a random intercept for female ID and plot ID. Significance of model terms was estimated using likelihood ratio tests (LR) in the course of backward model selection.

In the analyses, 'phase' factor corresponded to the monitoring and treatment phases of the CMR experiment (capture location data); in the radio-tracking experiment, 'session' factor corresponded to particular sessions. Assignments of the plots to the treatments ('a priori' groupings hereafter) were identified as levels of the 'treatment' factor. As such, this differentiation was only effective when predation risk was manipulated—in the second radio-tracking session, as well as in the treatment phase of the CMR experiment. To test for the effect of predator odours, we estimated the significance of treatment × session interaction (or treatment × phase interaction in case of capture location data). Where additional factors (e.g. year or time of day) were considered, we tested for the significance of applicable three-way interactions.

23.3 Results

23.3.1 Capture Location Data

The total number of female captures was 6155. In the monitoring phase of 2007, the mean number of individual captures was 15.8 (median = 14; maximum = 32), while in the treatment phase it was 18.0 (median = 15.5; maximum = 45). The respective figures for 2008 were 9.9 (median = 9; maximum = 16) and 17.9 (median = 16; maximum = 39). In total, the dataset afforded 359 observations of home range area. As much as 40% of the observations were paired: 67 of 292 females were present in both phases of the experiment. In 2007, as obtained with the MCP 95% method, there were 83 observations in the monitoring phase and 148 observations in the treatment phase. In 2008, the population sizes reached approximately half of those for 2007 (Jochym and Halle 2012); the observations for monitoring and treatment phases were 12 and 82, respectively. Of 359 observations, the LoCoH 95% algorithm failed to compute 34. Of these, 30 occurred in the 2007 treatment phase, and 2 in the monitoring phases of both years.

23.3.1.1 Home Range Area

We did not find significant differences between herbivore and neutral groupings (LR: $\chi^2=9.34$ at 4 df, $p=0.053$) of home range areas calculated with the LoCoH 95% method; hence, we merged them into one non-predator (NP) grouping for further analysis. The following factors were not significant: the year-dependent effect of predator treatment (treatment×phase×year interaction; LR: $\chi^2=0.046$ at 1 df, $p=0.83$), the interaction between the a priori grouping and year (treatment×year; LR: $\chi^2=0.71$ at 1 df, $p=0.40$), the effect of predator treatment (treatment×phase interaction; LR: $\chi^2=1.28$ at 1 df, $p=0.26$) and the phase×year interaction (LR: $\chi^2=2.09$ at 1 df, $p=0.15$). The difference between the phases was marginally non-significant (LR: $\chi^2=3.48$ at 1 df, $p=0.062$). The differences between the predator and non-predator a priori groupings were significant (LR: $\chi^2=4.01$ at 1 df, $p=0.045$). Finally, the year effect was very highly significant (LR: $\chi^2=34.72$ at 1 df, $p<0.001$). In a simplified model containing only the year factor, mean square root of home range area in 2007 was 4.35 ± 0.136 (SE) vs. 5.97 ± 0.282 in 2008.

Areas calculated with the MCP 95% method displayed a similar pattern, but estimates were consistently higher than those obtained with the LoCoH 95% method. Merging of the two non-predator treatments was justified (LR: $\chi^2=7.65$ at 4 df, $p=0.11$). The following factors were not significant: the year-dependent effect of predator treatment (treatment×phase×year interaction; LR: $\chi^2=0.67$ at 1 df, $p=0.41$), the interaction of the a priori grouping and year (treatment×year; LR: $\chi^2=0.19$ at 1 df, $p=0.66$), the effect of predator treatment (treatment×phase interaction; LR: $\chi^2=0.68$ at 1 df, $p=0.41$) and the differences between the a priori groupings (LR: $\chi^2=2.24$ at 1 df, $p=0.13$); the phase×year interaction was marginally non-significant (LR: $\chi^2=3.79$ at 1 df, $p=0.052$). Both phase effect (LR: $\chi^2=12.321$ at 1 df, $p<0.001$) and year effect (LR: $\chi^2=47.006$ at 1 df, $p<0.001$) were very highly significant. As estimated by the simplified model, mean square root of home range area in the 2007 monitoring phase was 7.96 ± 0.393 (SE), compared to 5.86 ± 0.306 in the treatment phase. This pattern was reversed in 2008: the estimates were 9.34 ± 1.260 and 9.69 ± 0.488, respectively.

23.3.2 Radio-Tracking

We started each session with 18 females. During the first session, we lost track of one. Of 17 females followed in the first session, 14 were recaptured for the second session.

We replaced the four missing females with similar ones from respective plots. In the second session, we lost track of two females, both of which were successfully tracked in the first session. The losses during sessions were either due to malfunction or loss of a radio-collar. Additionally, we failed to detect 17.7% of the fixes in the first session (individual range 6–42%) and 2.6% in the second session (individual range 0–8%). Ultimately, the radio-tracking study supplied two datasets: a full one and a paired one. The full dataset contained all radio-tracked females, including those present in one session only. The paired dataset was restricted to the 12 individuals present in both sessions (four females per treatment), allowing pairwise comparisons. We indicate which dataset was used for the calculations.

23.3.2.1 Activity Range Area

We calculated the activity range areas using the full dataset (LoCoH 95%; Fig. 23.1). We found no significant differences between the two non-predator groupings (LR: $\chi^2=5.12$ at 2 df, $p=0.077$), and merged them into one. Predator treatment (treatment × session interaction) had no effect (LR: $\chi^2=0.22$ at 1 df, $p=0.64$). Additionally, there were no differences between the radio-tracking sessions (LR: $\chi^2=1.08$ at 1 df, $p=0.30$) or a priori groupings (LR: $\chi^2=0.035$ at 1 df, $p=0.85$).

Results from the MCP 95% analysis were analogous. There were no significant differences between the two non-predator groupings (LR: $\chi^2=5.95$ at 2 df, $p=0.051$), thus justifying the merging. Again, predator treatment (treatment × session interaction) had no effect (LR: $\chi^2=0.25$ at 1 df, $p=0.62$). Furthermore, there were no differences between the radio-tracking sessions (LR: $\chi^2=0.09$ at 1 df, $p=0.76$) or a priori groupings (LR: $\chi^2=0.07$ at 1 df, $p=0.79$). The MCP 95% estimate (mean square root of activity range area) for the non-predator treatment equalled 9.57 ± 0.606 (SE) in the first session and 9.63 ± 1.209 in the second session. In predator treatment, the respective values equalled 10.17 ± 1.153 and 9.28 ± 0.637.

23.3.2.2 Capture Location Data vs. Radio-Tracking

For 15 females (of which five were present in both phases and sessions), we estimated both the home range (capture locations from 2008) and activity range (radio-tracking). For the MCP 95% method, the individual ratio of 'capture location' area to 'radio-tracking' area fell within the range 0.16–2.48 (mean 1.18), except for one individual, where 'capture location' estimate was almost four times greater than the 'radio-tracking' estimate. The ratios for the LoCoH 95% method ranged within 0.17–1.56 (mean 0.76); here, the exceptionally high ratio was 6.70.

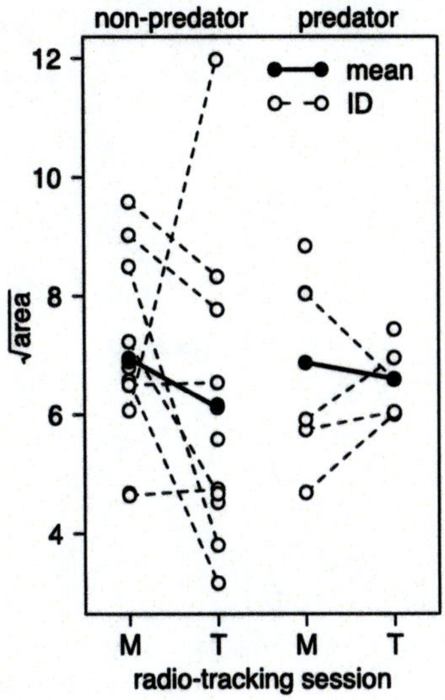

Fig. 23.1 Square-root transformed activity range areas (LoCoH 95%) of reproductively active females, computed from radio-tracking data (*full dataset*). *Connected dots* indicate model estimates for females present in both sessions (*dashed lines*; ID), and of session means (*solid lines*). Significance of model terms is given in the text. Radio-tracking sessions: *M* monitoring; *T* treatment

23.3.2.3 Mean Distance to Odour Source

We calculated the mean distance to the nearest odour source (DOS) using the paired dataset. The proportion of fixes contained within the core area amounted to 41% in the first session and 47% in the second session; these locations were excluded from the analysis. Individual and mean estimates of DOS are shown in Fig. 23.2. As we found no significant differences between the two non-predator groupings (LR: $\chi^2=0.07$ at 2 df, $p=0.97$) these were merged in further analysis. In the second session, DOS was significantly higher in the predator treatment (treatment × session interaction; LR: $\chi^2=9.06$ at 1 df, $p=0.0026$).

23.3.2.4 Mean Relocation Distance

We estimated the relocation distances from the paired dataset. The total number of observations exceeding 2.82 m amounted to 628, obtained over two 48 h sessions from 12 individuals in three groups. In total, 63% of the observations were collected during the day, 12% at twilight and 25% at night. The distinction between herbivore and neutral groupings was retained due to significant differences (LR: $\chi^2=6.46$ at 1 df, $p=0.011$). With regard to the time of day, the effect of the predator treatment was not significant (treatment × session × time of day interaction; LR: $\chi^2=3.44$ at 4 df,

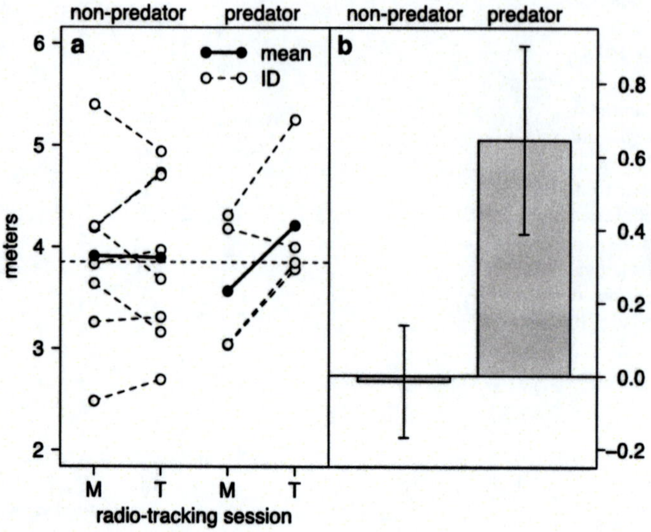

Fig. 23.2 (**a**) Mean distance to the odour source (DOS), calculated with the paired dataset. *Connected dots* indicate model estimates for the individual (*dashed lines*; ID) and mean DOS (*solid lines*). *Horizontal dashed line* marks the theoretical mean value of DOS. Radio-tracking sessions: *M* monitoring; *T* treatment. (**b**) Change in the individual DOS between the radio-tracking sessions. *Bars* indicate the mean; *whiskers* indicate ±1 SE of the mean. Significance of model terms is given in the text

$p=0.49$), as was the sole effect of predator treatment, regardless of time of day (treatment × session interaction; LR: $\chi^2=5.07$ at 2 df, $p=0.079$). The remaining two-way interactions were also non-significant: a priori grouping × time of day (LR: $\chi^2=1.28$ at 4 df, $p=0.87$), as well as session × time of day (LR: $\chi^2=1.63$ at 2 df, $p=0.44$), so was time of day as a simple effect (LR: $\chi^2=1.28$ at 2 df, $p=0.53$) and the effect between the sessions (LR: $\chi^2=0.73$ at 1 df, $p=0.39$). Only the differences between a priori groupings were significant (LR: $\chi^2=6.79$ at 2 df, $p=0.033$). As estimated by the minimum model, the mean relocation distance in the predator grouping was 6.7 ± 0.23 (SE), while it was 5.7 ± 0.23 in the herbivore grouping and 7.2 ± 0.23 in the neutral grouping.

23.3.2.5 Crepuscularity and Diurnality Indexes

Both I_C and I_D were calculated from the paired data. Due to the lack of observations in the second session at twilight, two individuals (one in each of the predator and neutral treatments) were excluded from the calculation of I_C. Estimates of the I_C model are shown in Fig. 23.3a. Merging of the two non-predator groupings was justified (LR: $\chi^2=0.44$ at 2 df, $p=0.80$). The effect of predator

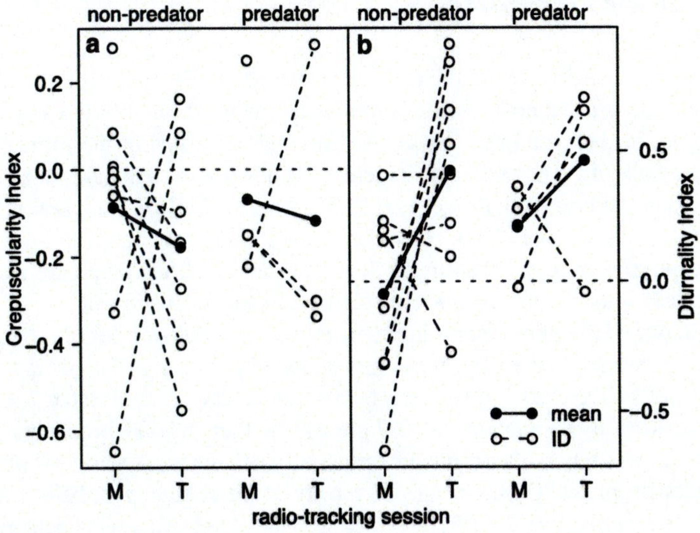

Fig. 23.3 (**a**) Crepuscularity index (*paired dataset*). *Horizontal dashed line* marks equal activity during twilight and the remaining part of the diel cycle. No movements were recorded in two females (one in each treatment) during twilight periods of the second session (*stray dots*). (**b**) Diurnality index (*paired dataset*). *Horizontal dashed line* marks equal activity during day and night. *Connected dots* indicate model estimates for the individual (*dashed lines*; ID) and the mean (*solid lines*). Significance of model terms is given in the text. Radio-tracking sessions: *M* monitoring; *T* treatment

treatment (treatment × session interaction) was not significant (LR: $\chi^2=0.19$ at 1 df, $p=0.67$). Lack of significance was found for both of the simple effects: session (LR: $\chi^2=0.01$ at 1 df, $p=0.92$) and a priori grouping (LR: $\chi^2=0.01$ at 1 df, $p=0.92$). Estimates of the I_D model are shown in Fig. 23.3b. Merging of the two non-predator groupings was justified (LR: $\chi^2=4.43$ at 2 df, $p=0.11$). Again, the effect of predator treatment (treatment × session) was not significant (LR: $\chi^2=0.62$ at 1 df, $p=0.43$), so was the a priori grouping (LR: $\chi^2=1.09$ at 1 df, $p=0.30$). The increase of I_D between sessions was highly significant (LR: $\chi^2=7.48$ at 1 df, $p=0.0063$).

23.4 Discussion

Taken together, our results show that breeding female common voles exposed to predator odour do not alter their spatial and temporal behaviour. They do, however, avoid sources of predator odour, thus confirming its ecological significance.

23.4.1 Home Range Area

The odour of a mustelid predator did not affect either MCP 95% or LoCoH 95% home range size estimates, but both revealed substantial individual variability, a year effect and an effect of the experiment phase. As expected, estimates of the former were higher than those of the latter; the MCP method is known to overestimate home range size (Burgman and Fox 2003; Börger et al. 2006), whereas LoCoH estimates are more accurate (Getz and Wilmers 2004; Huck et al. 2008). Some differences between the a priori groupings were present in the LoCoH 95% dataset. They may have resulted from high individual variability or low resolution of capture location data.

Home range size in microtines is inversely related to population density (Madison 1985). The pattern of home range areas corresponded to changes of population densities revealed by the CMR experiment (Jochym and Halle 2012). Despite identical initial size, the study populations in 2007 grew faster than in 2008, producing much higher densities in the first year; the difference between the years was most obvious in the treatment phase. Thus, changes in density of the studied populations would explain the year effect and the difference between the phases. Changing density also resulted in an unbalanced number of observations. The first year yielded more data and a larger proportion of young individuals. Since newer recruits have smaller home ranges (Mazurkiewicz 1971), numerous individual observations at higher densities may have decreased mean range size. Low to moderate densities in the second year resulted in relatively fewer observations; females with large home ranges may have increased the overall means.

23.4.2 Activity Range Area

Radio-tracking MCP estimates were consistently higher than LoCoH estimates; the same was true with regard to the variability of the estimates. Irrespective of the method, we found no effect of the predator odour on the size of activity ranges in reproductively active females. Lack of difference between sessions showed that the sole manipulation of predation risk also had no effect. Instead, we found considerable variation between individuals under equal conditions. Furthermore, in females with paired observations, their activity range size varied randomly between sessions, indicating that range size was labile (which is in agreement with conclusions of Madison 1985) and was shaped by factors other than mustelid predation risk (e.g. population density, vegetation cover) which were not controlled for in our study.

In Townsend's vole *Microtus townsendii*, a species occupying similar habitat to that of common voles, the spatial reaction to predator odour was only apparent in absence of vegetation cover (Merkens et al. 1991). The activity ranges of voles exposed to predator odours may similarly depend on the availability and structure of cover. We suspect that voles' use of space is reduced in areas with sparse cover compared to areas with good cover. Overall, the vegetation in our plots was dense and without obvious differences between the plots. In areas with thinner vegetation, burrow entrances of some of the females were concentrated under burdock *Arctium* sp. canopy (personal observation), showing that common voles do seek cover from avian predators. We tentatively suggest that, like predator odours, vegetation cover had little or no impact on activity range sizes of radio-tracked females, except for increasing the variability.

We think that the lack of effect of mustelid odours on space use is attributable to the reproductive status of the voles. All radio-tracked females were in breeding condition. Lactating and pregnant females have a very high demand for energy: it exceeds the demand of non-breeding females by 133% and 32%, respectively (Migula 1969). Even though voles in general adapt to mustelid predation risk by reducing spatial activity (Gorman 1984; Borowski 1998; Borowski and Owadowska 2010) and foraging (Bolbroe et al. 2000), this adaptation might not be feasible for breeding females. Reduction of foraging ranges could limit their food intake and consequentially, negatively affect their fitness. Furthermore, it could give more leeway to intraspecific competitors. It seems that such costs of reduced predation risk are too high for the breeding females. This conclusion is supported by the fact that upon exposure to a mustelid predator, breeding females do not abandon their ranges, contrary to voles with a different reproductive status (e.g. inactive individuals or males; Jedrzejewski and Jedrzejewska 1990). With conventional adaptations being too costly, breeding suppression seems the only viable adaptation to predation risk (Koskela and Ylönen 1995; Fuelling and Halle 2004; Jochym and Halle 2012), at least for some female voles.

23.4.3 Capture Location Data vs. Radio-Tracking

For 15 radio-tracked females that also provided enough capture locations, the mean activity range was smaller than the mean home range estimated by MCP; this relation was reversed when the LoCoH method was applied. Individual proportions varied greatly in both instances, but less so for the LoCoH estimates. Madison (1985) predicted that long-term ranges overestimate space use. Home range consists of discrete activity ranges visited over time and enclosed within the polygon marked by the outermost locations. Estimates obtained with MCP probably reflect this fact, whereas LoCoH restricts ranges to where animals were actually trapped. Avoided traps are excluded, often resulting in oblong, ragged or non-convex ranges. Furthermore, spatial resolution of radio-tracking data was 100 times better than the resolution of capture location data. Radio-tracking ranges possibly included areas that would have remained undetected with capture location data, explaining why the mean activity ranges were larger in the LoCoH analysis.

Compared to capture location data, radio-tracking methods acquire information almost instantly. Because individual space use within a population is a dynamic process (Madison 1985), radio-tracking should have a relatively high sensitivity to its changes. However, the behavioural effects of mustelid odours are almost immediate (Eccard et al. 2008), and may also be short-lived, as suggested by Parsons and Bondrup-Nielsen (1996). At some point, habituation to olfactory cues of predation risk may occur. If spatial adaptations of voles to mustelid odours were momentary, neither capture data nor radio-tracking would detect them due to low temporal resolution. Our radio-tracking sessions were separated by 12 days, with six applications of odour during this period. Females that initially perceived an increased risk of predation may have become habituated to mustelid odour. In laboratory rats physiological habituation to ferret odour is claimed to not occur (Masini et al. 2005). Even so, we cannot discount the possibility that in our study the lack of effect of ferret odour was a result of habituation. This could be verified only with additional experiments.

23.4.4 Mean Distance to Odour Source

The difference in mean distance to odour source (DOS) between the monitoring and treatment sessions indicated that ferret odour was repellent. In the predator treatment, the increase of DOS was observed in three out of four females. In the non-predator treatment, the females were not affected by the control odours. Furthermore, the overall mean of DOS approximated the theoretical mean, reflecting a random distribution of females in relation to the source of control odour. Since we had relatively few females in the predator treatment, this result should be confirmed with a larger sample.

Our result is in accord with existing evidence. Gorman (1984) demonstrated that common voles and field voles were repelled by stoat anal gland scent, both in laboratory trials (scented paper sheets) and in the field (soiled traps). Such a response may be species-specific: the distance to odour source was not affected in wild root voles (Borowski 1998). Furthermore, avoidance may occur under laboratory conditions but not in the field, as shown by Parsons and Bondrup-Nielsen (1996) in meadow voles (*Microtus pennsylvanicus*). Despite the general evidence for avoidance of predator odour, the role of voles' reproductive condition remains unclear. Regardless of their reproductive status, bank voles (*Myodes glareolus*) avoided pens visited by a weasel; breeding females responded differently only when weasel entered their vicinity (Jedrzejewski and Jedrzejewska 1990). We suppose that, in contrast to activity ranges, avoidance of predator odour is independent of reproductive status. Further, our results confirm that breeding females avoid ferret odour, but this need not limit their use of space.

23.4.5 Mean Relocation Distance

The distances travelled by females were independent of the time of day (alone) and in interaction with odour treatment. *M. arvalis* is active throughout the diel cycle, with periods of high mobility at dusk and dawn (Lehmann and Sommersberg 1980). Interestingly, we found no indication of this pattern, regardless of other factors. This may be due to the relatively poor temporal resolution of radio-tracking. Given the 30 min fix interval and the 2.82 m distance cut-off, the distribution of resulting movements throughout the sessions was random and relatively sparse. The error was probably greatest during twilight periods as they contained fewest fixes. We also encountered substantial individual variability. It was reflected in the significance of a priori groupings, particularly that each grouping contained only four individuals.

Earlier radio-tracking studies have shown that in *Microtus* voles, mobility is modulated by the risk of predation (Borowski 1998; Norrdahl and Korpimäki 1998; Borowski and Owadowska 2010). However, the role of reproductive status has not yet been determined. Since movement distances and activity ranges are closely related, the arguments provided in the discussion of activity ranges also apply here. We suppose that reproductively active females disregard the information conveyed by predator odour and remain very mobile, searching for food or warding off potential competitors. Given that very mobile voles are more likely to be killed by mustelids (Norrdahl and Korpimäki 1998), the lack of response would make breeding females more prone to predation.

23.4.6 *Crepuscularity and Diurnality Indexes*

The two indexes of circadian activity showed no effect of ferret odours. On average, the females were less active during the twilight phase which is surprising, since peak activity of common voles usually takes place at twilight (Lehmann and Sommersberg 1980). A short twilight period with infrequent fixes that yielded few movement observations may explain our unexpected result. We also found that females were more diurnal than nocturnal, particularly during the treatment session. The prominent shift to daytime activity in some of the radio-tracked females may result from the random nature of recorded movements. However, given better detection of vole locations in the second session (98% vs. 82%), it is likely that the observed increase in daytime activity reflected the actual activity pattern of reproductively active female common voles.

Although voles may, under certain circumstances, respond to predation risk with a shift of the diel activity (e.g. Eccard et al. 2008), it is thought that circadian rhythms are independent of predation risk. Voles face a constantly high risk of predation, from one type of predator or another (Halle 1993). Instead of a flexible circadian rhythm, voles developed a polyphasic pattern, consisting of bouts of foraging activity interlaced with periods of rest. As such, this type of activity is an adaptation to predation risk (Gerkema and Daan 1985; Reynolds and Gorman 1994). Due to high energy demands, it would be impossible for voles to entirely shift activity to a period when predators are inactive. This especially applies to reproductively active females.

23.5 Conclusions

Our results suggest that reproductively active females differ from voles with other reproductive statuses (e.g. adult males or reproductively inactive females) with regard to spatial and temporal adaptations to predation risk. Specifically, we found no evidence that reproductively active females adjusted their long- and short-term space use, or shifted their mode of circadian activity. We suggest that the high energy demand of breeding females prevent such changes in behaviour. If employed, such adaptations would ultimately curtail reproductive output and limit fitness. Still, reproductively active females remain sensitive to predator odours. Avoidance of ferret odour and odour-induced breeding suppression confirms its biological significance.

Since energy demands of non-breeding females are lower than reproductively active females, non-breeding female voles should adopt spatial mobility behaviours likely to decrease predation risk. To verify this, further studies should compare behaviours of reproductively active females with inactive females in response to predator odours. Larger groups and more frequent sampling should be used to account for high individual variability and uncertainty of vole movements.

Acknowledgements The authors would like to thank everyone who helped with the CMR experiment and the radio-tracking of voles, especially Magda Bereza, Daniel Bergelt, Volkmar Haus, Anna Henning, Kerstin Parker and Amelie Zander. MJ was funded with a stipend from the International Max Planck Research School for Chemical Ecology in Jena.

References

Anderson PK (1986) Foraging range in mice and voles: the role of risk. Can J Zool 64:2645–2653

Bates D, Maechler M (2010) lme4: linear mixed-effects models using S4 classes. R package version 0.999375-37. URL:http://CRAN.R-project.org/package=lme4

Bolbroe T, Jeppesen LL, Leirs H (2000) Behavioural response of field voles under mustelid predation risk in the laboratory: more than neophobia. Ann Zool Fenn 37:169–178

Börger L, Franconi N, De Michele G, Gantz A, Meschi F, Manica A, Lovari S, Coulson T (2006) Effects of sampling regime on the mean and variance of home range size estimates. J Anim Ecol 75:1393–1405

Borowski Z (1998) Influence of weasel (*Mustela nivalis* Linnaeus, 1766) odour on spatial behaviour of root voles (*Microtus oeconomus* Pallas, 1776). Can J Zool 76:1799–1804

Borowski Z, Owadowska E (2010) Field vole (*Microtus agrestis*) seasonal spacing behavior: the effect of predation risk by mustelids. Naturwissenschaften 97:487–493

Burgman MA, Fox JC (2003) Bias in species range estimates from minimum convex polygons: implications for conservation and options for improved planning. Anim Conserv 6:19–28

Calenge C (2006) The package "adehabitat" for the R software: a tool for the analysis of space and habitat use by animals. Ecol Model 197:516–519

Cushing BS (1985) Estrous mice and vulnerability to weasel predation. Ecology 66:1976–1978

Eccard J, Pusenius J, Sundell J, Halle S, Ylönen H (2008) Foraging patterns of voles at heterogeneous avian and uniform mustelid predation risk. Oecologia 157:725–734

Fuelling O, Halle S (2004) Breeding suppression in free-ranging grey-sided voles under the influence of predator odour. Oecologia 138:151–159

Gerkema MP, Daan S (1985) Ultradian rhythms in behavior: the case of the common vole (*Microtus arvalis*). Exp Brain Res Suppl 12:11–31

Getz WM, Wilmers CC (2004) A local nearest-neighbor convex-hull construction of home ranges and utilization distributions. Ecography 27:489–505

Getz WM, Fortmann-Roe S, Cross PC, Lyons AJ, Ryan SJ, Wilmers CC (2007) LoCoH: nonparametric kernel methods for constructing home ranges and utilization distributions. PLoS One 2:e207

Gorman ML (1984) The response of prey to stoat (*Mustela erminea*) scent. J Zool 202:419–423

Halle S (1993) Diel pattern of predation risk in microtine rodents. Oikos 68:510–518

Halle S (1995) Diel pattern of locomotor activity in populations of root voles, *Microtus oeconomus*. J Biol Rhythms 10:211–224

Halle S, Lehmann U (1987) Circadian activity patterns, photoperiodic responses and population cycles in voles. Oecologia 71:568–572

Huck M, Davison J, Roper TJ (2008) Comparison of two sampling protocols and four home-range estimators using radio-tracking data from urban badgers *Meles meles*. Wildlife Biol 14:467–477

Jedrzejewski W, Jedrzejewska B (1990) Effect of a predators visit on the spatial distribution of bank voles—experiments with weasels. Can J Zool 68:660–666

Jochym M, Halle S (2012) To breed, or not to breed? Predation risk induces breeding suppression in common voles. Oecologia 170:943–953

Kenward RE (1993) Wildlife radio tagging: equipment, field techniques and data analysis, 3rd edn. Academic, London

Koskela E, Ylönen H (1995) Suppressed breeding in the field vole (*Microtus agrestis*)—an adaptation to cyclically fluctuating predation risk. Behav Ecol 6:311–315

Lehmann U, Sommersberg CW (1980) Activity patterns of the common vole, *Microtus arvalis*—automatic recording of behavior in an enclosure. Oecologia 47:61–75

Madison DM (1985) Activity rhythms and spacing. In: Tamarin RH (ed) Biology of new world microtus. American Society of Mammalogists, Shippensburg, PA, pp 373–419

Masini CV, Sauer S, Campeau S (2005) Ferret odor as a processive stress model in rats: neurochemical, behavioral, and endocrine evidence. Behav Neurosci 119:280–292

Mazurkiewicz M (1971) Shape, size and distribution of home ranges of *Clethrionomys glareolus* (Schreber, 1780). Acta Theriol 16:23–60

Merkens M, Harestad AS, Sullivan TP (1991) Cover and efficacy of predator-based repellents for Townsend vole, *Microtus townsendii*. J Chem Ecol 17:401–412

Migula P (1969) Bioenergetics of pregnancy and lactation in European common vole. Acta Theriol 14:167–179

Norrdahl K, Korpimäki E (1998) Does mobility or sex of voles affect risk of predation by mammalian predators? Ecology 79:226–232

Parsons GJ, Bondrup-Nielsen S (1996) Experimental analysis of behaviour of meadow voles (*Microtus pennsylvanicus*) to odours of the short-tailed weasel (*Mustela erminea*). Ecoscience 3:63–69

R Development Core Team (2010) R: a language and environment for statistical computing. Version 2.12.1. R Foundation for Statistical Computing, Vienna, Austria, ISBN 3-900051-07-0. URL:http://www.R-project.org

Reichstein H (1960) Untersuchungen zum Aktionsraum und zum Revierverhalten der Feldmaus. Z Säugetierkd 25:150–169

Reynolds P, Gorman ML (1994) Seasonal variation in the activity patterns of the Orkney vole *Microtus arvalis orcadensis*. J Zool 233:605–616

Ylönen H, Ronkainen H (1994) Breeding suppression in the bank vole as antipredatory adaptation in a predictable environment. Evol Ecol 8:658–666

Chapter 24
Knowing Your Behaviour: The importance of Behavioural Assays in the Characterisation of Chemical Alarm Cues in Fishes and Amphibians

Reehan S. Mirza, Chantale A. Laraby, and Ashley M. Marcellus

Abstract Predation provides a strong selective pressure that shapes the antipredator responses of aquatic animals. This has led to the evolution of chemical alarm systems where chemicals released from injured prey evoke antipredator responses that enhance survival when encountering predators. The chemical structure of alarm cues is unknown and little work has been conducted on characterising/identifying them. In this study, we use high performance liquid chromatography to separate skin extracts from fathead minnows (*Pimephales promelas*) and larval grey tree frogs (*Hyla vesicolor*) and then deliver the fractions back to test animals to determine which portions of the skin extract leads to antipredator behaviour. In fathead minnows, one fraction of the skin extract evokes antipredator behaviour, but in grey tree frog tadpoles no single fraction of the extract caused antipredator behaviour. These results provide insight into the characterisation of chemical alarm cues and the importance of behavioural assays in this process.

24.1 Introduction

Predation is a strong selective force that shapes antipredator responses in prey animals (Lima and Dill 1990; Chivers and Smith 1998; Ferrari et al. 2010). In aquatic ecosystems many organisms rely on chemical information to assess predation risk (Wisenden 2000) as it may provide more reliable information than other sensory stimuli because it can be detected from a greater distance, may have greater persistence and is available when visual cues are limited, e.g. night time, heavy vegetation (Mirza 2009). Moreover, a large number of chemicals will dissolve in water giving a large number of potential cues to be detected (Kleerekoper 1969; Hara 1975).

R.S. Mirza (✉) • C.A. Laraby • A.M. Marcellus
Department of Biology and Chemistry, Nipissing University,
100 College Drive, North Bay, ON P1B 8L7, Canada
e-mail: reehanm@nipissingu.ca

M.L. East and M. Dehnhard (eds.), *Chemical Signals in Vertebrates 12*,
DOI 10.1007/978-1-4614-5927-9_24, © Springer Science+Business Media New York 2013

Many aquatic organism use chemical alarm cues to assess predation risk (reviewed in Ferrari et al. 2010). Chemical alarm cues fall into four categories and are released at specific points of the predation sequence (Lima and Dill 1990; Chivers and Smith 1998). Chemicals can be released when a predator is detected and categorised as either predator odours or disturbance cues (which may have a nitrogenous component), when a prey animal is captured (injury-released alarm cues) or from the predator after the prey item has been ingested/digested (predator diet cues). Of the four categories, injury-released alarm cues have received the most attention, particularly in fishes.

In response to injury-released alarm cues aquatic prey exhibit a standard suite of antipredator behaviours. The most commonly observed antipredator behaviours are decreased activity, seeking shelter, lack of movement, increased group cohesion and area avoidance (Wisenden 2000). Responses to injury-released alarm cues are believed to be innate. Typically performance of these behaviours provides a survival benefit for receivers (reviewed in Mirza and Chivers 2001). Antipredator behaviours have been documented for several fishes and larval amphibians in response to chemical alarm cues, allowing for a comparison between the two evolutionary related groups.

Fishes have received the majority of attention in the literature particularly in the superorder Ostariophysi (e.g. minnows and carps, catfishes, characins, suckers and gonorynchiformes). Ostariophysans make up approximately 74% of all freshwater fishes and most members populate the prey base in aquatic systems. The fathead minnow, *Pimephales promelas*, is a model species that has demonstrated the complexity and sophistication of how aquatic organisms use chemical alarm cues (reviewed in Mirza 2009; Ferrari et al. 2010). Fathead minnows are small (3–6 cm) fish that are widespread throughout North America and make up a large component of the prey base (Scott and Crossman 1973).

A variety of larval amphibian species also use chemical alarm cues to warn others of potential danger (Chivers and Smith 1998; Summey and Mathis 1998; Rajchard 2006; Mirza et al. 2006; Ferrari et al. 2007). The grey tree frog, *Hyla vesicolor*, is an arboreal dwelling amphibian of North America. Its tadpoles occur in ephemeral summer ponds and have been shown to respond to injury-released chemical alarm cues with typical antipredator behaviour (Schoeppnner and Relyea 2005).

Despite the broad base of knowledge of antipredator responses to injury-released alarm cues, less is known of the chemical constituents of alarm cues. It has been speculated that in Ostariophysan fish species the injury-released alarm chemical may be hypoxanthine-3-N-oxide (H_3NO, Pfeiffer et al. 1985) or purine-N-oxides (PNO, Brown et al. 2000). The N–O functional group appears to be important in the antipredator response because compounds that lacked this functional group do not elicit antipredator responses (Brown et al. 2003). In amphibians little is known regarding the identity of injury-released alarm cues in tadpoles.

In this study, we conducted preliminary work to characterise injury-released alarm cues in larval grey tree frogs and fathead minnows using high performance liquid chromatography (HPLC). We used HPLC to separate the components of both fathead minnow and grey tree frog injury-released alarm cues. We collect the fractions of each chemical constituent and determined which component of the alarm cue contains the active fraction that elicited antipredator behaviour in minnows and tadpoles. We used established behavioural assays to confirm that the potential chemical constituents collected are the injury-released alarm cues used by minnows and tree frog tadpoles.

24.2 Methods

24.2.1 Animal Collection and Maintenance

Fathead minnows and grey tree frog tadpoles were collected from local ponds using minnow traps and dip nets respectively. Fathead minnows were held in 100 L glass aquaria with power filters to maintain water quality. Minnows were maintained on a diet of commercial fish flakes (Nutrafin® Goldfish Food, Hagen, Montréal, Canada). Tree frog tadpoles were held in 30 L plastic tubs filled with dechlorinated water. They were fed boiled lettuce and ground up alfalfa pellets ad libitum. All animals were kept at room temperature (19–22°C) and on a 16 h:8 h light:dark cycle.

24.2.2 Stimuli Preparation

Minnow alarm cue was made from nine donor minnows. Spinal cords were severed followed by pithing in accordance with guidelines set by the Canadian Council on Animal Care. A fillet of skin was removed from each side of the fish and placed in ice-chilled distilled water. A total of 19.64 cm^2 was collected and homogenised in 100 mL of distilled water. The resultant solution was filtered through polywool filter material to remove larger chunks and then diluted by adding distilled water for a total of 400 mL. This gave a concentration of 0.049 cm^2 of skin per 1 mL of fathead minnow skin extract. This was the stock solution of alarm cue and was frozen in 10 mL aliquots at −20°C.

Tree frog tadpoles were euthanised using cervical dislocation in concordance with guidelines set by the Canadian Council on Animal Care. Carcasses were then ground in deionised water with a mortar and pestle. Tree frog tadpoles at Gosner stages 30–36 weighed approximately 0.3 g each. We homogenised ten tadpoles in 50 mL distilled water, filtered through polywool filter floss and then diluted the solution to a final volume of 200 mL with distilled water. This was the tree frog AS (alarm substance) and 100 mL was frozen in 10 mL aliquots at −20°C.

In order to produce the "filtered" and "back-filtered" stimuli, skin extracts were centrifuged for 20 min at 1,000 rpm at 4°C and then filtered through a 0.45 μm filter using a VELP Scientifica JP Recirculating Water Aspirator (Milan, Italy). Anything that passed through the filter was termed "filtered skin extract". The "backfiltered skin extract" then consisted of anything which did not pass through the filter. The filter papers were put in beakers with 20 mL nano water and a sonicator was used to dislodge particles from the filter paper; this formed the backfiltered stimulus. The backfiltered extracts were diluted to 100 mL for both minnow and tree frog AS and then packaged in 10 mL aliquots and frozen at −20°C. A distilled water control was also packaged in 10 mL aliquots and frozen at −20°C.

Based on previous results (Telfer and Mirza, unpub. data), fathead minnow skin extract contained hypoxanthine in the concentration of 0.7 ppm. To create a 0.7 ppm sample of hypoxanthine, a 10 ppm standard solution was created by dissolving 0.001 g of hypoxanthine into 100 mL of distilled water. This 0.7 ppm solution was then aliquoted in 10 mL portions and frozen at −20°C.

24.2.3 HPLC Analysis

HPLC analysis was performed using a Varian PrepStar 218 system (Varian Inc., now part of Agilent Technologies, Mississauga, Ontario, Canada) which is equipped with a dual pump solvent delivery module generating flow rates up to 200 mL/min, a Varian ProStar Autosampler, model 410, a Varian ProStar UV detector, model 325, a Varian ProStar Fraction Collector, model 704, and a Shimadzu CTO-6A column oven (Varian Inc., Mississauga, ON, Canada).

Fathead minnow extracts were collected and separated using a semi-preparative HPLC method. Separation was carried out using a Phenomenex Luna C18 RP column, 100×10 mm, 10 μm (Phenomenex, Torrance, CA, USA). The mobile phase consisted of 25 mM K_2PO_4, pH 6.8 and CH_3CN, (95/5, v/v), with the flow rate set to 4.8 mL/min at 30°C. Each run was done isocratically with a total run-time of 15 min. UV detection was carried out at 254 nm; the injection volume was 1,200 μL.

Tree frog extracts were analysed using a mobile phase consisting of K_2PO_4/CH_3CN (90/10, v/v). A C18 reverse phase analytical column (Macherey-Nagel, 150×4.6 mm, 5 μm pore size) was used. The total run time was 15 min at a flow rate of 0.5 mL/min. Detection was carried out at 254 nm. Tree frog extracts were separated into two fractions based on the UV absorbance profile and the retention times on the semi-preparative column. A C18 reverse phase Macherey-Nagel (125×21 mm, 5 μm pore size) semi-prep column was used. Each run takes 15 min at a flow rate of 4.8 mL/min. Detection was carried out at a wavelength of 254 nm.

The compositions of the present peaks were collected using the Varian ProStar Fraction Collector, model 704. The timed windows were programmed based on chromatograph output, and the fractions collected were used for behavioural assays. Three sets of fractions were collected and used towards behavioural assays. The first collection we suspect is hypoxanthine; the other two collections were classified as 2.1 and 2.2 (Fig. 24.1 top). Tree frog extracts were separated into two fractions based on retention time (Fig. 24.1 bottom). Each fraction collected was a combination of the chemicals in question plus the mobile phase. We also ran Nano water through the HPLC and collected it to use as the control for presence of the mobile

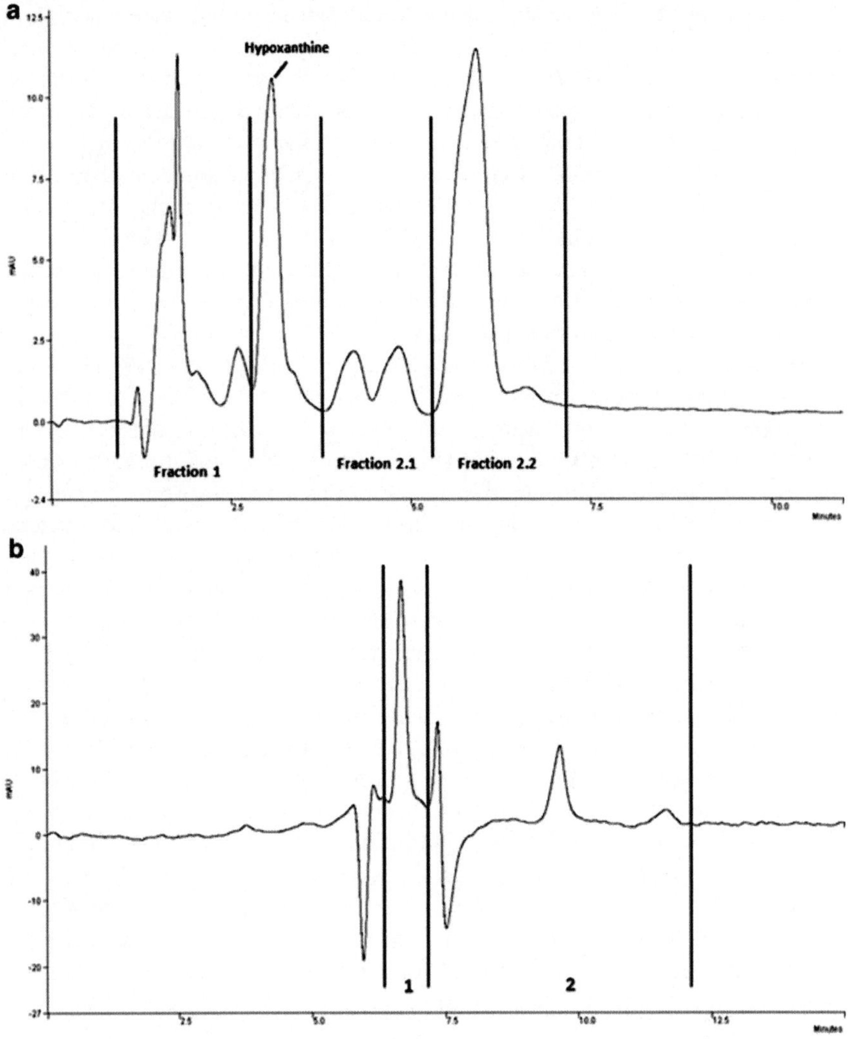

Fig. 24.1 Chromatograms of fathead minnow skin (*top*) and grey tree frog whole body extract (*bottom*) for characterisation of injury-released alarm cues

phase. All stimuli were diluted to a final concentration of 1:100 to provide enough stimuli for the behavioural assays.

24.2.4 Experimental Protocol

To test the antipredator responses of both minnows and tadpoles we used behavioural assays commonly used for each species. Individual fathead minnows were placed into

20 L aquaria filled with room temperature dechlorinated tap water. Each aquaria was equipped with an air stone for aeration, a stimulus injection line and a cover object consisting of one half of PVC pipe. Prior to testing, fish were allowed to acclimate to the test tanks for approximately 18 h. Minnows were fed 1 h prior to testing. We tested a total of 20 minnows for each treatment and minnows were tested only once.

Trials for fathead minnows consisted of an 8 min pre-stimulus observation period, a 1 min stimulus delivery period and an 8 min post-stimulus observation period. Prior to the pre-stimulus period, 60 mL of tank water was removed from the tank via the injection line and discarded. Another 60 mL of tank water was removed and retained. Three millilitres of stimulus was then loaded into the injection line and flushed through with the retained 60 mL of water.

Tadpoles were tested in 3.6 L clear plastic containers (19 cm×19 cm, 10 cm depth). Containers were separated into four quadrants by placing a printed grid underneath the container. Containers were filled with 2.5 L of room temperature (19–22°C) dechlorinated water and a single tadpole was placed in each container and allowed 30 min acclimation time before testing. Tadpoles were observed for 5 min prior to addition of the stimulus. The location of each tadpole was recorded every 15 s using scan sampling. Stimulus was added slowly in the corner of quadrant 3 using a syringe (1 mL was added over 10 s). The location of each tadpole was again observed and recorded every 15 s for 5 min post-stimulus. We tested a total of 20 tadpoles for each treatment and tadpoles were tested only once.

In experiment 1, fathead minnows were exposed to: (1) minnows AS (stock solution), (2) filtered minnows AS, (3) backfiltered minnow AS or (4) distilled water. In experiment 2, minnows were exposed to: (1) filtered minnow AS (diluted 100×), (2) fraction 1 minnow AS, (3) fraction 2 minnow AS, (4) hypoxanthine or (5) control water (run through the HPLC). In experiment 3, minnows were exposed to: (1) filtered minnow AS (diluted 100×), (2) fraction 2.1 minnow AS, (3) fraction 2.2 minnow AS, (4) hypoxanthine or (5) control water (run through the HPLC).

Tree frog tadpoles were exposed to: (1) stock tree frog AS, (2) filtered tree frog AS, (3) backfiltered skin extract or (4) distilled water. In the second experiment, tadpoles were exposed to: (1) stock tree frog AS (diluted 100×), (2) fraction 1 tree frog AS (diluted 100×), (3) fraction 2 tree frog AS (diluted 100×) or (4) distilled water.

24.2.5 Statistical Analysis

For the minnow behavioural assays we recorded the time the minnow spent moving (s), i.e. moving more than one full body length, freezing (motionless for more than 30 s) and dashing (disoriented, unpredictable rapid bursts of swimming). Complete lack of movement, i.e. freezing represents a stronger intensity antipredator response than time spent moving where some movement is still evident but reduced. Changes in time spent moving were calculated (post-stimulus period minus pre-stimulus period) and analysed using a Kruskal-Wallis one-way ANOVA followed by non-parametric pairwise post hoc tests. Dashing and freezing were analysed using the

Likelihood ratio test. Pairwise likelihood ratio tests were used as post hoc tests. For tree frog assays, we recorded line crosses. A line cross was defined by the body of the tadpole moving across the grid line (excluding the tail). The change in number of line crosses (post-minus pre-stimulus) was analysed using a Kruskal-Wallis one-way ANOVA followed by pairwise post hoc tests (Siegel and Castellan 1988).

24.3 Results

24.3.1 Fathead Minnows

Minnows exposed to injury-released alarm cue significantly differed in the time spent under cover ($\chi^2=49.8$, df=3, $P<0.001$). Fathead minnows exposed to the filtered AS spent significantly reduced movement compared to minnows exposed to either back filtered skin or the distilled water control (both $P\leq0.05$, Fig. 24.2). There was no difference in time spent moving by minnows exposed to filtered or stock minnow AS ($P>0.05$; Fig. 24.2). Similarly, there was a significant difference in the proportion of minnows dashing (Likelihood ratio test, $\Lambda=26.6$, df=3, $P<0.001$) and freezing ($\Lambda=33.9$, df=3, $P<0.001$, Fig. 24.2). More minnows dashed in response to stock minnow AS and filtered minnow AS than to backfiltered skin or distilled water ($P<0.001$). There was no difference in dashing or freezing of minnows between stock minnow AS and filtered minnow AS ($P>0.05$: Fig. 24.2).

Significantly more minnows differed in time spent under cover when exposed to minnow AS fractions ($\chi^2=64.2$, df=4, $P<0.001$). Fathead minnows exposed to the filtered AS or fraction 2 moved significantly less than minnows exposed to either fraction 1, hypoxanthine or the distilled water control (all $P\leq0.05$, Fig. 24.3). Minnows exposed to hypoxanthine spent significantly less time swimming than minnows exposed to filtered AS or fraction 2 ($P<0.001$), but significantly more time swimming than minnows exposed to fraction 1 or distilled water ($P<0.001$; Fig. 24.3). There was a significant difference in the proportion of minnows dashing ($\Lambda=35.7$, df=4, $P<0.001$) and freezing ($\Lambda=37.4$, df=4, $P<0.001$) when exposed to the various stimuli. Minnows dashed and froze significantly more when exposed to filtered AS and fraction 2 compared to the other three treatments (all $P<0.001$; Fig. 24.3). There was no significant difference in freezing or dashing between minnows exposed to fraction 2 or filtered AS ($P>0.05$).

There was a significant difference in time spent under cover among the various treatment groups by minnows exposed to chemical stimuli ($\chi^2=64.9$, df=4, $P<0.001$). Fathead minnows exposed to distilled water spent significantly less time under cover than minnows exposed to all other treatments (all $P\leq0.05$, Fig. 24.4). Minnows exposed to fraction 2.1 or filtered AS spent significantly more time under cover than minnows exposed to fraction 2.2 or hypoxanthine (both $P<0.001$). There was no difference in time spent under cover for minnows exposed to fraction 2.1 and filtered AS ($P>0.05$) or fraction 2.2 and hypoxanthine ($P>0.05$). There were

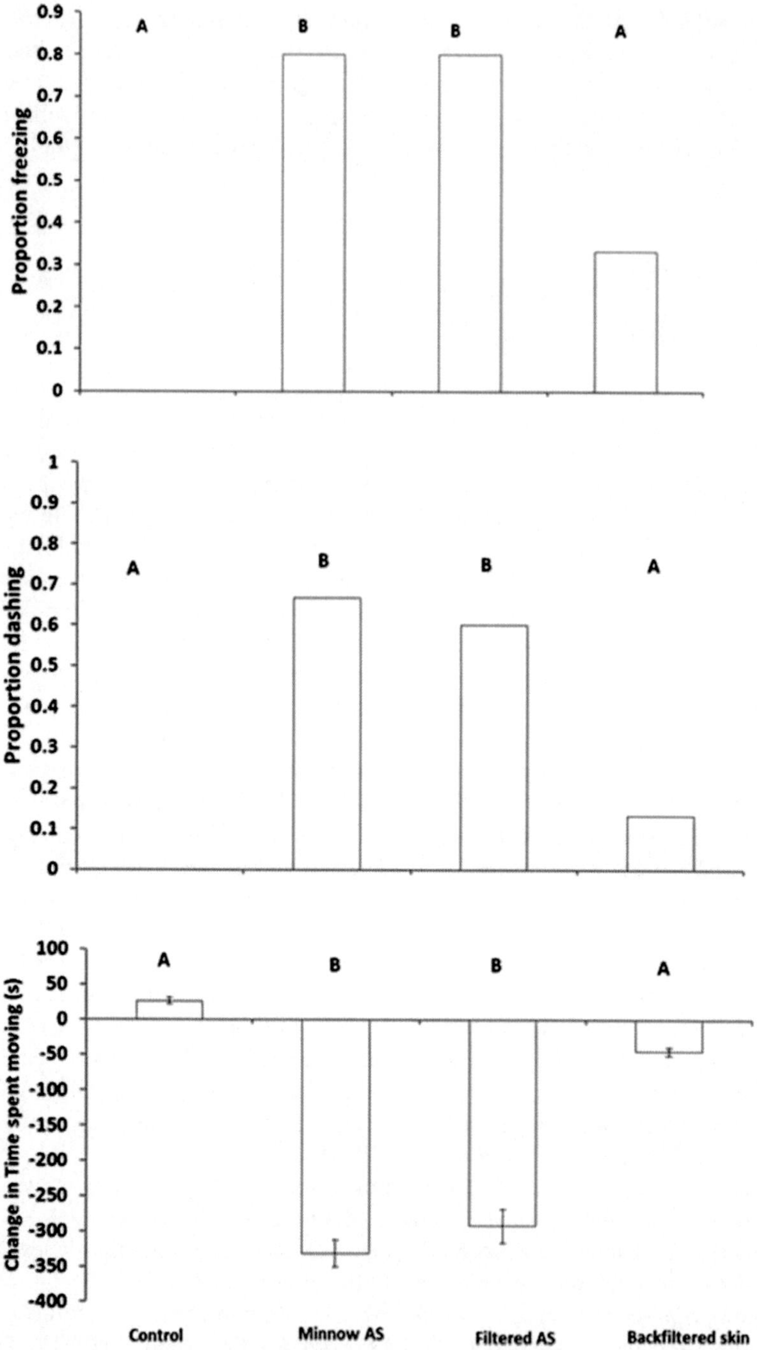

Fig. 24.2 Mean ± SE of change in time spent moving (s) (*bottom*) and proportion of individuals dashing (*middle*) and freezing (*top*) of fathead minnows in response to filtered, backfiltered and non-filtered alarm cues. *Different letters* denote a significant difference at $P \leq 0.05$ (see text for details)

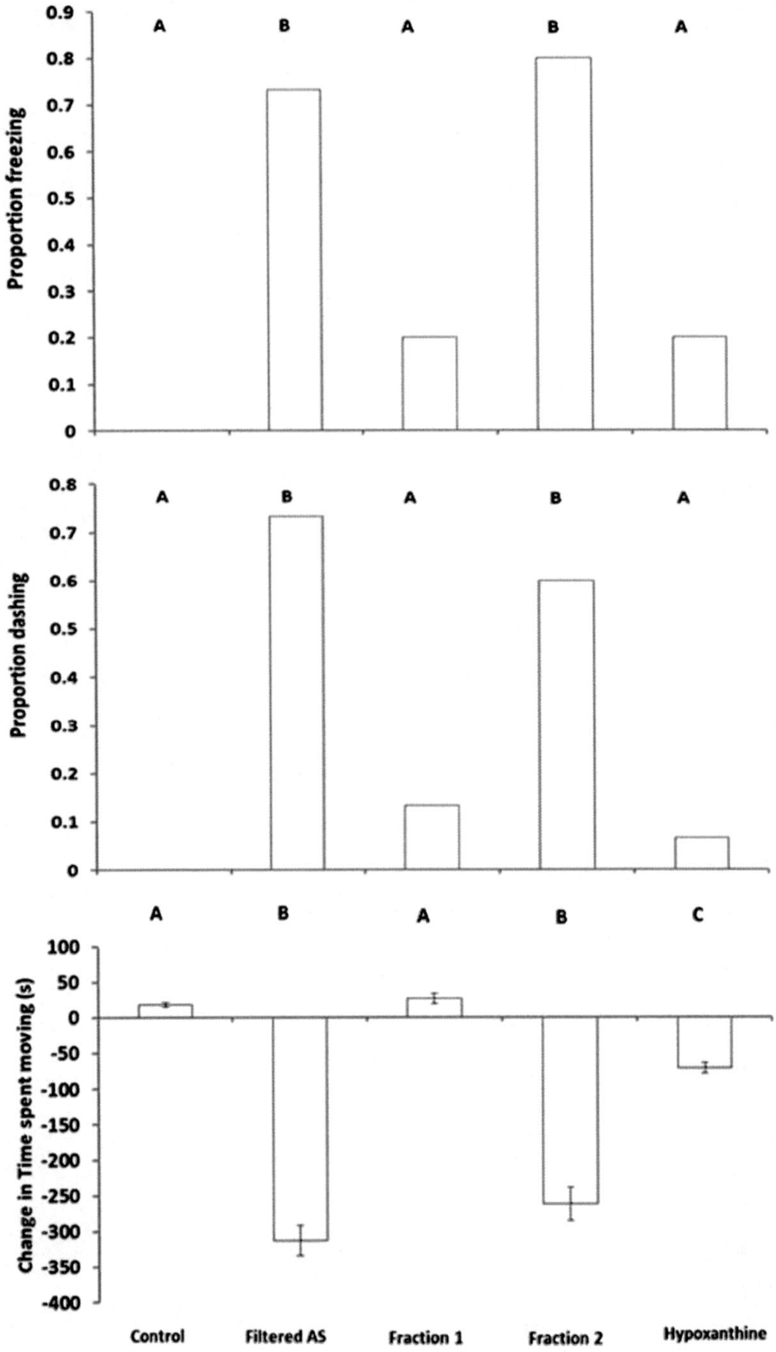

Fig. 24.3 Mean ± SE of change in time spent moving (s) (*bottom*) and proportion of individuals dashing (*middle*) and freezing (*top*) of fathead minnows in response to HPLC fractions of alarm cue. *Different letters* denote a significant difference at $P \leq 0.05$ (see text for details)

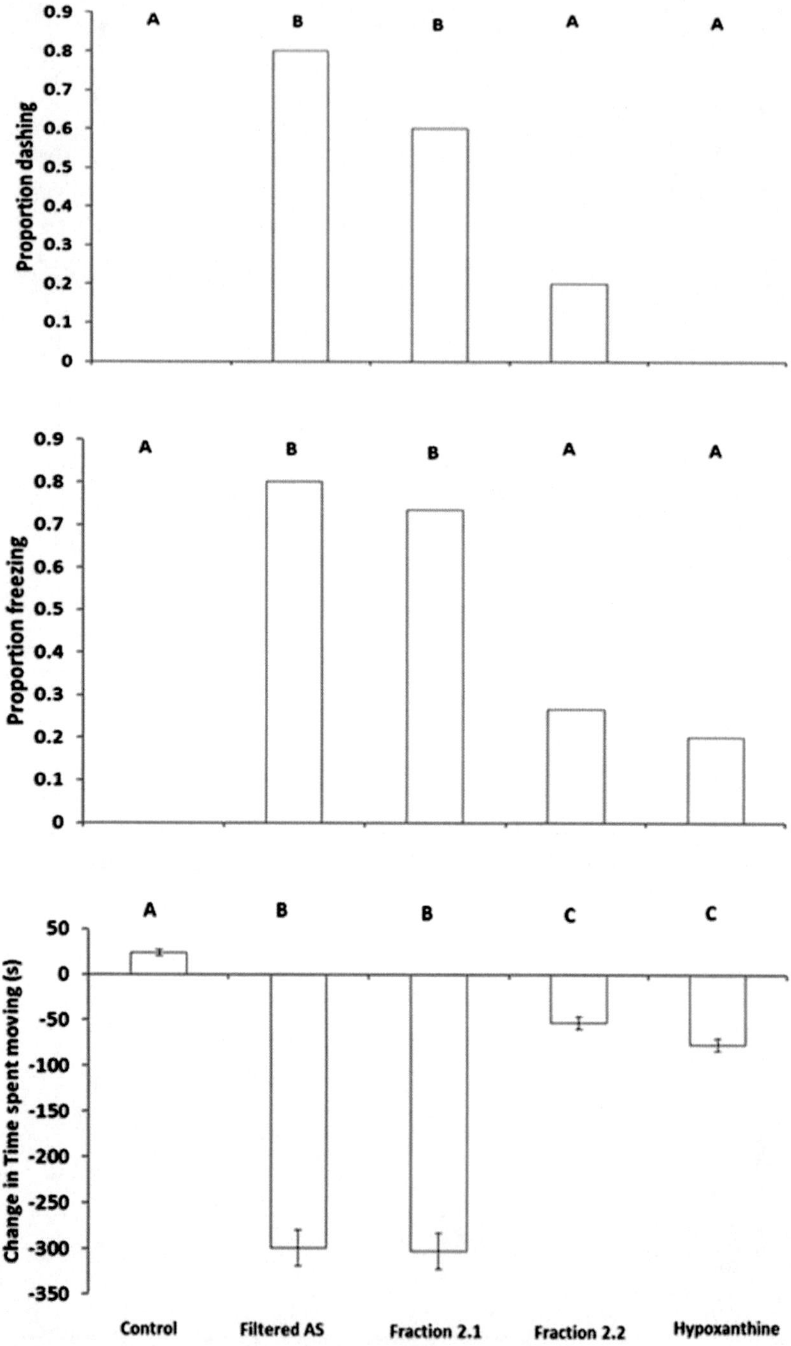

Fig. 24.4 Mean ± SE of change in time spent moving (s) (*bottom*) and proportion of individuals dashing (*middle*) and freezing (*top*) of fathead minnows in response to HPLC fractions 2.1 and 2.2, with hypoxanthine and filtered minnow AS. Different letters denote a significant difference at $P \leq 0.05$ (see text for details)

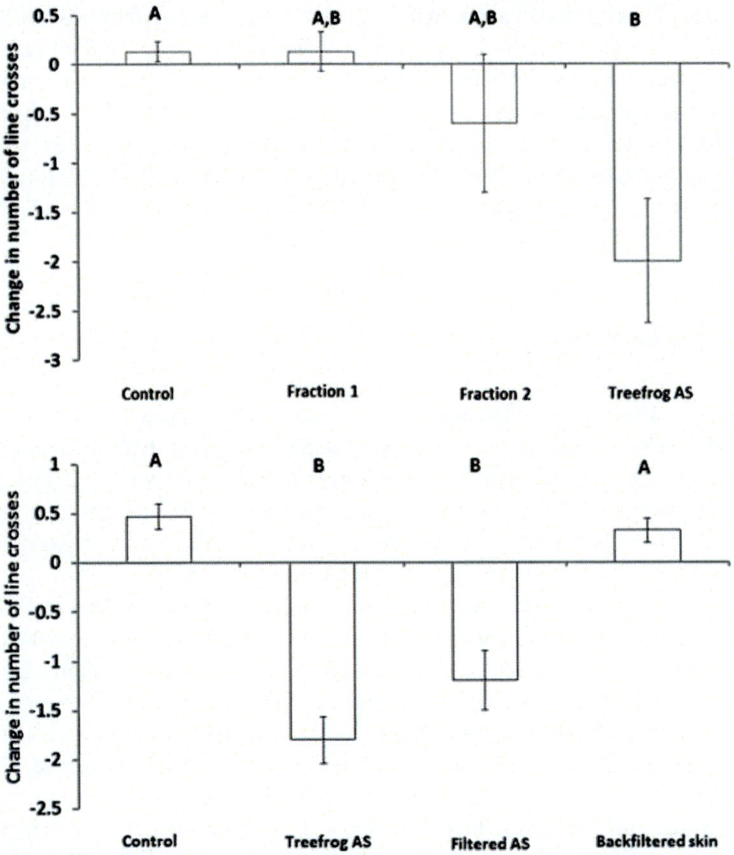

Fig. 24.5 Mean ± SE of change in number of line crosses of grey tree frog tadpoles exposed to filtered and nonfiltered tree frog AS (*bottom*) and HPLC fractions of tree frog AS (*top*). Different letters denote a significant difference at $P \leq 0.05$ (see text for details)

significant differences in dashing ($\Lambda = 43.8$, df = 4, $P < 0.001$) and freezing ($\Lambda = 36.1$, df = 4, $P < 0.001$) among minnows exposed to chemical stimuli. Minnows froze and dashed significantly more if exposed to filtered AS or fraction 2.1 than to the other three stimuli (both $P < 0.001$; Fig. 24.4). Minnows did not differ in freezing or dashing between the fraction 2.1 and filtered groups ($P > 0.05$) or between pairwise comparisons of the other three groups (all $P > 0.05$; Fig. 24.4).

24.3.2 Grey Tree Frogs

There was a significant difference in the number of line crosses in grey tree frog tadpoles among the various treatment groups ($\chi^2 = 39.1$, df = 3, $P < 0.001$). Tadpoles exposed to tree frog AS or filtered AS significantly decreased their movement compared to distilled water and backfiltered skin (all $P < 0.001$; Fig. 24.5). However, there were no significant changes in movement for tree frog tadpoles exposed to tree frog AS vs. filtered AS or backfiltered skin vs. distilled water ($P>0.05$; Fig. 24.5).

There was a significant difference in the number of line crosses in grey tree frog tadpoles among the various treatment groups ($\chi^2=7.87$, df=3, $P=0.049$). Tadpoles exposed to tree frog AS exhibited significantly lower changes in the number of line crosses than tadpoles exposed to distilled water ($P<0.05$; Fig. 24.5). There were no significant differences between fraction 1 and 2 or any other pairs of treatments (all $P>0.05$).

24.4 Discussion

The results of our study indicate that only a portion of the fathead minnow skin (i.e. fraction 2.1) contained chemical injury-released alarm cues, because fraction 2.1 elicited the strongest antipredator response from fathead minnows. The HPLC separation of grey tree frog extract did not result in a clear component eliciting the strongest antipredator behaviour. Instead, it appears that all components of the skin are required for tadpoles to perform antipredator behaviour.

The identity of Ostariophysan alarm cue has been questioned for decades. Pfeiffer et al. (1985) was the first to suggest that H_3NO was the injury-released alarm cue because it caused a decreased feeding response in tetras. This speculation was questionable because the behavioural assay used was weak. Further studies used more rigorous testing methods confirmed antipredator responses to H_3NO in various Ostariophysans (Brown et al. 2000, 2003; Kelly et al. 2006). Brown et al. (2000) found that fathead minnows and finescale dace (*Phoxinus neogaeus*) responded with antipredator behaviour to H_3NO and also pyridine-N-oxide (PNO). Other structurally similar compounds which lacked the nitrogen oxide (N-O) portion did not elicit antipredator behaviour. In a further study, Kelly et al. (2006) exposed glowlight tetras to H_3NO, PNO or mixed ratios of both compounds: 75% H_3NO:25% PNO, 50% H_3NO:50% PNO, 25% H_3NO:75% PNO. They found that as the ratio of H_3NO decreased, tetras increased area use and decreased shoal cohesion indicating no antipredator response. They concluded that the active component of fathead minnow skin was most likely structurally similar with a purine skeleton.

Our results revealed a significantly weaker antipredator response to hypoxanthine than the response to the active fraction of minnow skin, even though it did produced a significantly stronger response than that to distilled water controls. However, previous studies have found that fathead minnows may elicit antipredator responses to hypoxanthine that are of the same intensity as that of fathead minnow skin extract (Brown et al. 2000). This suggests that the response to hypoxanthine can be variable. Hypoxanthine can be produced as a degradation product of nucleotides (Lagoja and Herdewijn 2004) which would explain its presence in epidermal cells, but our results suggest that it is unlikely to be the Ostariophysan injury-released alarm cue. Hypoxanthine-3-N-oxide is also unlikely to be the active component of minnow skin causing fright reactions, because under our

HPLC column conditions, based on the structure of H_3NO it would elute before hypoxanthine (M. Jha, organic chemist, personal communication). Thus both H_3NO and hypoxanthine probably eluted before fraction 2.1 which caused the strongest minnow fright response in our study.

Relatively little is known about the chemical identity of alarm cues in amphibians. Fraker et al. (2009) examined the antipredator behaviour of wood frogs (*Lithobates sylvaticus*) to injury-released alarm cues. The wood frog homogenate was separated into two fractions using HPLC. Wood frog tadpoles did not respond to either fraction alone, but only decreased time spent swimming to a combination of both fractions. Our results did not reveal decreased activity in response to either HPLC fraction of tree frog extract, but showed decreased activity to the whole tree frog homogenate similar to the wood frog study.

One of the keys to successfully determining whether the chemical(s) being tested are chemical alarm cue is having a meaningful behavioural assay. In our minnow study, we have strong confidence in our results. The assay we applied is commonly applied to fishes is highly dependable and provides strong biological significance. Our behavioural results suggest that the active component of minnow alarm cue occurs in fraction 2.1. Our tree frog results failed to produce similarly clear results, perhaps because the intensity of the antipredator response in this species was not strong. Overall tree frog tadpoles reduced activity by only 1–2 line crosses. Antipredator studies in other tadpole species have revealed antipredator responses of stronger intensity than found in our study. Our current assay may require improvements, as the biological response produced seems insufficient. The assay may not be sensitive enough to detect changes in antipredator behaviour that may be present. Moreover, the HPLC analysis produced chromatograms less clean than those produced from minnows.

In conclusion, fathead minnows responded with antipredator behaviour to chemicals from only a portion of their skin. The active components of the skin are most likely not H_3NO or hypoxanthine, but may be structurally similar compounds. Further identification using gas chromatograph/mass spectrometry is needed. Our tree frog results suggest that both the HPLC analysis and behavioural assay require adjustments. To successfully characterise/identify chemical cues, meaningful behavioural assays are required that induce a biologically meaningful response of a sufficient magnitude.

References

Brown GE, Adrian JC Jr, Smyth E, Leet H, Brennan S (2000) Ostariophysan alarm pheromones: laboratory and field tests of the functional significance of nitrogen-oxides. J Chem Ecol 26:139–154

Brown GE, Adrian JC Jr, Naderi NT, Harvey MC, Kelly JM (2003) Nitrogen-oxides elicit antipredator responses in juvenile channel catfish, but not convict cichlids or rainbow trout: conservation of the Ostariophysan alarm pheromone. J Chem Ecol 29:1781–1796

Chivers DP, Smith RJF (1998) Chemical alarm signaling in aquatic predator-prey systems: a review and prospectus. Ecoscience 5:338–352

Ferrari MCO, Messier F, Chivers DP (2007) Degradation of chemical alarm cues under natural conditions: risk assessment by larval wood frogs. Chemoecology 17:263–266

Ferrari MCO, Wisenden BD, Chivers DP (2010) Chemical ecology of predator-prey interactions in aquatic ecosystems: a review and prospectus. Can J Zool 88:698–724

Fraker ME, Hu F, Cuddapah V, McCollum SA, Relyea RA, Hempel J, Denver RJ (2009) Characterization of an alarm pheromone secreted by amphibian tadpoles that induces behavioral inhibition and suppression of the neuroendocrine stress axis. Horm Behav 55:520–529

Hara TJ (1975) Olfaction in fish. Prog Neurobiol 5:273–335

Kelly JM, Adrian JC Jr, Brown GE (2006) Can the ratio of aromatic skeletons explain cross-species responses within evolutionarily conserved Ostariophysan alarm cues?: testing the purine-ratio hypothesis. Chemoecology 16:93–96

Kleerekoper HA (1969) Olfaction in fishes. Indiana University Press, Bloomington

Lagoja IM, Herdewijn P (2004) One-step synthesis of hypoxanthine from glycinamide and diformylurea. Chem Biodivers 1:106–111

Lima SL, Dill LM (1990) Behavioral decisions made under the risk of predation: a review and prospectus. Can J Zool 68:619–640

Mirza RS (2009) The nose knows: chemically mediated antipredator defences in Ostariophysans. In: Giacomo Z, Perrière C, Mathis A, Kapoor BG (eds) Fish defenses: pathogens, parasites and predators, vol 2. Science Publishers, Enfield, pp 291–322

Mirza RS, Chivers DP (2001) Do chemical alarm signals enhance survival of aquatic vertebrates? An analysis of the current research paradigm. In: Marchlewska-Koj A, Lepri JJ, Müller-Schwarze D (eds) Chemical signals in vertebrates, vol 9. Plenum Press, New York, pp 19–26

Mirza RS, Ferrari MCO, Kiesecker JM, Chivers DP (2006) Responses of American toad tadpoles to predation cues: behavioural response thresholds, threat-sensitivity and acquired predation recognition. Behaviour 143:877–889

Pfeiffer W, Riegelbauer G, Meier G, Scheibler B (1985) Effect of hypoxanthine-3(N)- oxide and hypoxanthine-1(N)-oxide on central nervous excitation of the black tetra *Gymnocorymbus ternetzi* (Characidae, Ostariophysi, Pisces) indicated by dorsal light response. J Chem Ecol 11:507–523

Rajchard J (2006) Antipredator pheromones in amphibians: a review. Vet Med 51:409–413

Schoeppnner NM, Relyea RA (2005) Damage, digestion and defence: the roles of alarm cues and kairomones for inducing prey defences. Ecol Lett 8:505–512

Scott WB, Crossman BJ (1973) Freshwater fishes of Canada. Fisheries Research Board of Canada, Ottawa

Siegel S, Castellan NJ Jr (1988) Nonparametric statistics for the behavioral sciences, 2nd edn. McGraw-Hill, New York

Summey MR, Mathis A (1998) Alarm responses to chemical stimuli from damaged conspecifics by larval anurans: tests of three neotropical species. Herpetologica 54:402–408

Wisenden BD (2000) Scents of danger: the evolution of olfactory ornamentation in chemically-mediated predator-prey interactions. In: Espmark Y, Amunsden T, Rosenqvist G (eds) Animal signals: signalling and signal design in animal communication. Tapir Academic Press, Trodenham, Norway, pp 365–386

Chapter 25
Avoiding Predation: The Importance of Chemical and Visual Cues in Poison Frog Reproductive Behaviour

Lisa M. Schulte, Rainer Schulte, and Stefan Lötters

Abstract The detection of biological signals is especially important in predator–prey systems. Anuran amphibians have evolved a remarkable diversity of defense strategies against predators, but the most risk-free is the prevention of a possible danger. This is valid for the protection of offspring as well. The neotropical poison frog *Ranitomeya variabilis* deposits both eggs and tadpoles in phytotelmata. The exploitation of these small pools is advantageous as it lowers the risk of offspring predation compared to larger water bodies. Nonetheless, there are potential predators in these pools as well. We analysed how the parent frogs avoid conspecific cannibalistic tadpoles and damselfly larvae of the species *Microstigma rotundatum*. We compared the use of chemical and visual cues and show that *R. variabilis* avoids conspecific tadpoles for the deposition of its offspring using chemical cues, while visual tadpole models alone were not avoided by the frogs. Damselfly larvae in contrast were avoided when present, but could not be detected by chemical cues alone. We suggest that the invertebrate predators mask their chemical cues, forcing the frogs to use other senses to detect them.

25.1 Introduction

Biological communication is defined as an action on the part of one organism that alters the probability pattern of behaviour on another (Wilson 1970), i.e. sending and receiving honest or manipulative signals (Hart 1996). Animals use a variety of

L.M. Schulte (✉) • S. Lötters
Department of Biogeography, Trier University,
Universitätsring 15, 54286 Trier, Germany
e-mail: s6laschu@uni-trier.de; loetters@uni-trier.de

R. Schulte
Instituto de Investigación Biológica de las Cordilleras Orientales—INIBICO,
Jr. Jose Olaja 500, Tarapoto, Peru
e-mail: info@inibico.org

M.L. East and M. Dehnhard (eds.), *Chemical Signals in Vertebrates 12*,
DOI 10.1007/978-1-4614-5927-9_25,

means to communicate with both conspecifics and heterospecifics (Bradbury and Vehrencamp 1998). Communication between heterospecifics includes predator–prey systems: the danger emanating from the predator organism alters the prey's behaviour, as the behaviour of the prey alters the hunting strategy and the associated success of the predator (Hasson 1991).

The different life-stages of anuran amphibians, i.e. frogs and toads, run the risk of being preyed by a great variety of predators. Toledo et al. (2011) recently found that adult anurans alone showed more than 30 different defensive strategies against potential attacks. The performance of those behaviours is influenced by the stage of predation (i.e. from localisation to digestion) and can be divided into primary and secondary defence strategies. While the former are independent of the presence of a predator, secondary defence strategies are directly elicited by the predator (Edmunds 1974). Examples include the 'unken' reflex (e.g. Haberl and Wilkinson 1997; Williams et al. 2000), puffing up the body (e.g. Duellman and Trueb 1986; Prates et al. 2011), defensive vocalisation (Toledo et al. 2009), biting or different ways of escaping (e.g. McDiarmid and Gorzula 1989).

Primary defence mechanisms can be understood as an avoidance of contact with the predator and can contain both honest and manipulative signals. There are numerous examples of such mechanisms including cryptic colour patterns (Osorio and Srinivasan 1991; Stevens and Merilaita 2009), aposematic colours (e.g. Ruxton et al. 2004; Darst and Cummings 2006) or false eyes (Martins 1989; Toledo and Jared 1995; Lenzi-Mattos et al. 2005), as well as nocturnal activity (Duellman and Trueb 1986) or hiding behaviour (Marchisin and Anderson 1978; Pröhl and Ostrowski 2011). In addition, abilities to recognise honest signals from the predator are considered anuran primary defence mechanisms. These may be conducted via recognition of visual or chemical cues associated with the predator.

With regard to visual predator recognition, frogs and toads tend to avoid all moving objects, larger than themselves (Ewert 1974; Ingle and Hoff 1990; Cooper et al. 2009). More is known about precautions based on chemical predator cues. Especially in aquatic predator–prey systems, chemical cues are the primary vehicle for information about predation risk (Brönmark and Hansson 2000). Accordingly, much is known about tadpoles avoiding either predator cues, cues from predators fed with conspecifics or alarm cues deriving from injured individuals (e.g. Hews 1988; Stauffer and Semlitsch 1993; Eklöv 2000; Ferrari et al. 2007; Smith et al. 2008; Hettyey et al. 2011). Already embryos are able to recognise such cues and either change the time remaining until hatching or their behavioural response to predators after hatching (Saenz et al. 2003; Ferrari and Chivers 2010). Compared to tadpoles and juveniles, there are relatively few studies on the use of chemical cues in adult frogs and toads (Hamer et al. 2011). Avoidance of chemical predator cues by adult anurans is reported by Murray et al. (2004) and Hamer et al. (2011) and indirectly by Wirsing et al. (2005), who noted an avoidance of injured conspecifics by adult frogs.

Besides terrestrial predators such as birds or snakes, aquatic predators represent an indirect danger to frogs and toads as these may prey upon their offspring. It has been shown for several anuran species that they avoid, for tadpole or egg depositions,

Fig. 25.1 A male *Ranitomeya variabilis* carrying one of its larvae to a phytotelm. This species typically uses water-filled leaf axils (phytotelmata) of bromeliad plants. Tadpoles are usually transported one-by-one, but occasionally two or three can be found on the back of the male parent

water containing predators, but it remains mostly unknown how the frogs recognise them (Resetarits and Wilbur 1989; Petranka et al. 1994; Rieger et al. 2004). Some authors suggest that chemical cues may be the main source for predator recognition (Hopey and Petranka 1994; Spieler and Linsenmair 1997; Binckley and Resetarits 2003). This has been demonstrated for predatory fish by Takahashi (2007), showing that chemical cues were avoided for egg deposition by tree frogs. Recently, Schulte et al. (2011) have demonstrated the avoidance of chemical cues of conspecific cannibalistic tadpoles for both egg and tadpole deposition in the Neotropical poison frog *Ranitomeya variabilis* (Dendrobatidae).

To the best of our knowledge, there are no studies on the influence of visual cues of aquatic tadpole or egg predators on parental decisions in anurans. The present study fills this gap and compares the recognition of different chemical and visual tadpole predator cues by *R. variabilis*. Like other poison frogs, this species has evolved a complex reproductive strategy including parental care (Lötters et al. 2007). It deposits its clutches at the water surface of a phytotelm (such as small water bodies in plant axils; Varga 1928) and after hatching the male transports the larvae singly into different phytotelmata (Brown et al. 2008a, b; Schulte et al. 2011; Fig. 25.1). Because of their small size, phytotelmata have nutritive limitations and commonly cannot sustain more than one tadpole at the same time (Lannoo et al. 1987; Brust 1993; Caldwell and de Araújo 1998; McDiamond and Altig 1999; Summers 1999; Lehtinen et al. 2004; Teixeira et al. 2006). As a consequence, tadpoles within these phytotelmata are competitive and often perform cannibalism (Brust 1993; Pramuk and Hiler 1999; Summers 1999; Poelman and Dicke 2007).

Accordingly, parent frogs need to locate unoccupied pools for each tadpole to increase their individual reproductive success.

In pool choice experiments, we showed that wild *R. variabilis* are able to exploit chemical cues produced by conspecific tadpoles and so avoid phytotelmata already occupied (Schulte et al. 2011). Brown et al. (2008a) found evidence that predatory damselfly larvae (Odonata: *Microstigma rotundatum*) were recognised and avoided by *R. variabilis*, when present (both visually and chemically). In the present study, we tested if these invertebrate predators can be recognised by chemical cues alone. Furthermore, we asked whether the recognition of the cannibalistic tadpoles is possible by visual cues alone, using artificial tadpole models. We formulated the following hypotheses: (1) both con- and heterospecific predators are recognised and avoided by *R. variabilis* for egg and tadpole deposition via chemical cues; (2) visual cues do not play a role in this system, as predators are mainly hidden within the phytotelmata.

25.2 Methods and Materials

25.2.1 Study Sites

Field work was conducted in a late-stage secondary rain forest close to the upper Cainarachi River, Región San Martín, Peru. Three study sites were established between 540 and 580 m above sea level on the western side of the river. Minimum distance between study sites was 30 m, thus exceeding the known home range size of *R. variabilis* (Brown et al. 2009) and avoiding pseudo-replication, i.e. repeated measures.

25.2.2 Chemical Cues from Damselfly Larvae

From July to November 2008, we performed pool choice experiments using pairs of plastic cups (200 mL volume, 10 cm height, 7 cm in diameter, made of polypropylene) as artificial phytotelmata (Schulte et al. 2011). These cups were non-transparent and two thirds of the openings were covered to impair visual inspection by *R. variabilis*. The pairwise setup consisted of one cup with 25 mL clean water and one with the same amount of tadpole-treated water. Water was treated by conspecific (i.e. cannibalistic *R. variabilis*) tadpoles collected in the forest nearby. For this purpose, they were housed individually in standard 50 mL plastic tanks and the used water of all tadpoles was taken and mixed for use in the field every other day (for details see Schulte et al. 2011).

From June to August 2010 and April to June 2011, we conducted similar experiments, but with damselfly larvae (Odonata: *M. rotundatum*) as the water treatment.

Brown et al. (2008a) demonstrated that these larvae can be recognised and avoided for tadpole depositions by *R. variabilis* when both visually and chemically present (using pool choice experiments with damselfly larvae with mandibles removed, as described in Wissinger and McGrady 1993). To test the effect of only chemical cues produced by these invertebrate tadpole predators, they were collected from artificial phytotelmata in the field and were kept individually in 50 mL rain water. Water change was conducted every other day and each damselfly larva was fed with one to two mosquito larvae afterwards. The used water was mixed and offered to the frogs in the field, side by side with clean water (as in the 2008 experiments, see above). A water change of both treated and clean water in the field was done every other day, to avoid a potential degradation of damselfly larvae cues and to prevent assimilations of drowned insects and detritus in the cups. At this occasion, cup pairs were controlled for new deposited clutches and tadpoles of *R. variabilis*. Occupied cups were not included in the following water changes. Instead a new pair of cups was fixed on another randomly chosen tree. If there was more than one tadpole deposition in adjacent cup pairs on the same day, just one was chosen randomly for analysis. This was done to minimise pseudoreplication from repeated depositions by the same male, carrying all tadpoles of the same clutch within a short period (Brown et al. 2008a). To ensure that the experimental treatment and no previous depositions had an influence on the deposition choice of the frogs, only single deposition events within each cup pair were recorded.

25.2.3 Visual Cues from Tadpole Models

In 2008, we conducted an experiment similar to Brown et al. (2008a), but using cannibalistic *R. variabilis* tadpoles instead of damselfly larvae as predator stimuli. We gave the frogs the choice between empty pools and pools containing a tadpole within a little cage (to ensure they could not feed on newly deposited tadpoles). In this way, tadpoles were present both visually and chemically.

In 2010, we tested how the frogs reacted if only visual tadpole cues were available. For this purpose, we scattered pairs of artificial phytotelmata within our study plots (as described above), which were transparent and the top completely open. Both cups were filled with 25 mL rain water and in one of them we hung a tadpole model made of heat hardened polymer clay (FIMO®). It was fixed to a transparent fibre hanging down from a wire installed above the water surface. The fragile setup ensured a movement of the model with every vibration of the cup caused by an arriving frog. To exclude the potential influence of chemical cues caused by the clay, we put a flat piece of clay on the bottom of the other cup. Furthermore, we equipped this cup with a horizontal wire, so both cups were exactly the same except for the swimming tadpole model. The cups were checked every other day for egg or tadpole depositions and were cleaned of detritus and drowned insects when necessary.

25.2.4 Statistical Analyses

Egg and tadpole deposition frequencies in each artificial pool type were compared using a G-test (Woolf 1957), which is considered similar to but more exact than a χ^2-test when small sample sizes are used (Sachs 1974; Gotelli and Ellison 2004). To further increase exactness, we adjusted the G-test in the manner of Williams (1976). Our null hypothesis assumed that the frequency of deposition events was random (0.5 in either pool). For comparisons of depositions between the different treatments, contingency tables according to Pearson (1956) were calculated testing the null hypothesis that all distributions were the same. When some of the expected values were too small to use Pearson's χ^2-Test, Fisher's exact test (Fisher 1922) was calculated instead, as suggested by Townend (2002).

25.3 Results

25.3.1 Chemical Cues from Damselfly Larvae

Pooling the results from 2010 to 2011, 36 tadpole and 23 clutch depositions of *R. variabilis* were counted in setups where water treated by damselfly larvae vs. clean water was offered to the frogs. Nineteen tadpoles were deposited in treated and 17 in clean water, while 9 clutches were deposited in clean vs. 14 in treated water. Calculating a G-test with both results, there is no significant avoidance of the damselfly cue, neither for tadpole depositions ($G_1=0.107$, $P>0.05$) nor for clutch depositions ($G_1=1.072$, $P>0.05$).

Comparing the reactions of the frogs to the different chemical signals (from cannibalistic tadpoles vs. predatory damselfly larvae), the difference between them is highly significant for clutch depositions ($\chi^2=20.545$, $P<0.001$) and nearly significant for tadpole depositions ($\chi^2=3.445$, $P=0.06$).

25.3.2 Visual Cues from Tadpole Models

For tadpole depositions, there was no significant preference between empty pools vs. pools containing a tadpole model (18 larvae each in both cup types, $G_1<0.001$, $P=1.00$). For clutch depositions, in total we recorded six depositions only, five in the empty cup and one with the clay model. This was insufficient for calculating a G-test.

For a comparison of our results with those of Brown et al. (2008a) for the presence of damselfly larvae, we used Fisher's exact test to calculate a contingency table, as the quantity of data in Brown et al. (2008a) was too small for Pearson's χ^2-test. The difference between the reactions on purely visual tadpole cues and visual (but also chemical) damselfly cues was significant ($\chi^2=5.250$, $P<0.05$).

25.3.3 Comparisons Between Visual and Chemical Detection

Calculating a contingency table to compare the results of chemical tadpole recognition (Schulte et al. 2011) with those of visual tadpole recognition, the frequencies between both detection strategies differ significantly from each other ($\chi^2=4.383$, $P<0.05$; Fig. 25.2a, b). This suggests that *R. variabilis* is able to recognise cannibalistic tadpoles using only chemical cues but not when only visual cues are given.

Comparisons with experiments by Schulte et al. (2011), where tadpoles in cages (avoided by the frogs) vs. clean water was tested, led to the same conclusion. While the results of the cage experiments did not differ significantly from those experiments using chemical tadpole cues ($\chi^2=0.009$, $P>0.5$), they differed significantly from experiments using tadpole clay models ($\chi^2=3.740$, $P=0.05$). This suggests that in cage experiments chemical cues may have predominated visual cues.

Additionally, tadpole depositions in clean vs. damselfly treated water were compared to the results of Brown et al. (2008a), where damselflies were visually (but also chemically) present. The significant difference (Fisher's exact test: $\chi^2=4.383$, $P<0.05$; Fig. 25.2c, d) revealed that the frogs were able to recognise damselfly larvae only when visual cues were involved but not when chemical cues alone were available.

25.4 Discussion

Our experimental studies demonstrated that adult *R. variabilis* use different cues and accordingly different sensory systems to recognise distinct aquatic predators of their tadpoles. While conspecific cannibalistic tadpoles are detected only by chemical but not visual cues, the presence of predatory damselfly larvae in phytotelmata cannot be detected by chemical cues alone (Fig. 25.3). This contradicts our hypothesis that (1) all predators are recognised chemically by *R. variabilis* and that (2) in this species visual cues do not play a role in the recognition of predators inhabiting phytotelmata.

Though, due to their diurnality, vision is an important sense in poison frogs (e.g. Forester and Wisnieski 1991; Summers et al. 1999; Richards-Zawacki and Cummings 2010), one could assume that chemical recognition of phytotelmata-inhabiting predators is more important than visual recognition. The rationale for this might be that predators in phytotelmata are hidden on the ground and therefore can be easily overseen (authors' unpubl. pers. observation). A possible explanation for the inability to perceive chemical cues from predatory damselfly larvae might be that larval odonate predators mask or break down chemical signals to be cryptic in the phytotelmata. Thus, the invertebrate predator might be involved in an evolutionary predator–prey arms race with *Ranitomeya* species that use the same phytotelmata for breeding (e.g. Brodie and Brodie 1999; Wirsing et al. 2005).

There are several studies documenting the ability of anuran larvae to recognise and avoid chemical cues of predatory odonate larvae (Petranka and Hayes 1998;

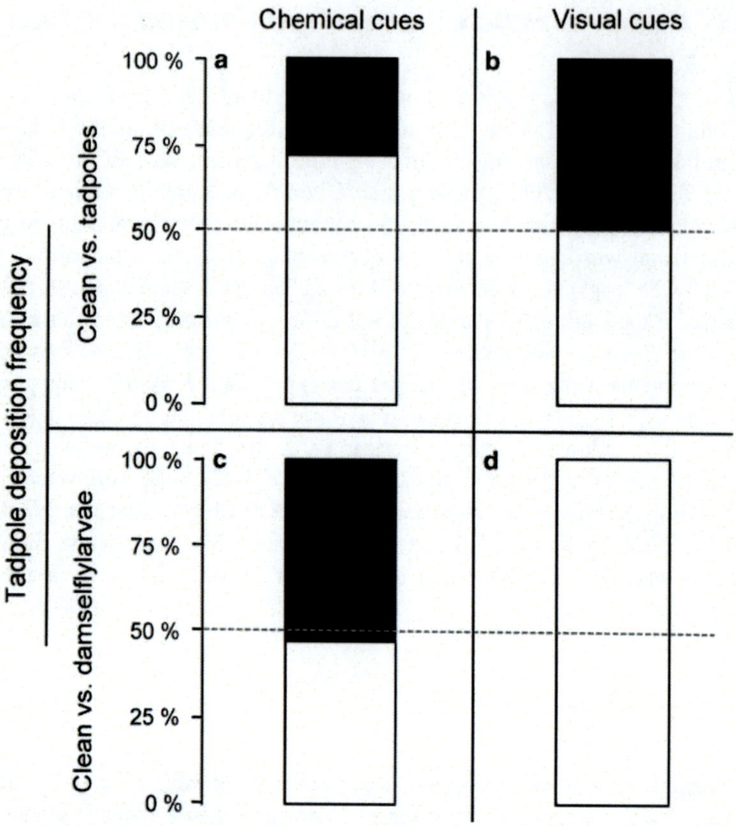

Fig. 25.2 Percentage tadpole deposition distribution of *R. variabilis* in different pool choice experiments. *White* shows depositions in clean water, *black* in treated water. *Top*: water treatment by chemical cues (**a**) vs. visual cues (clay models) of tadpoles (**b**). *Bottom*: water treatment by chemical cues of damselfly larvae (**c**) vs. physical presence of damselfly larvae (**d**)

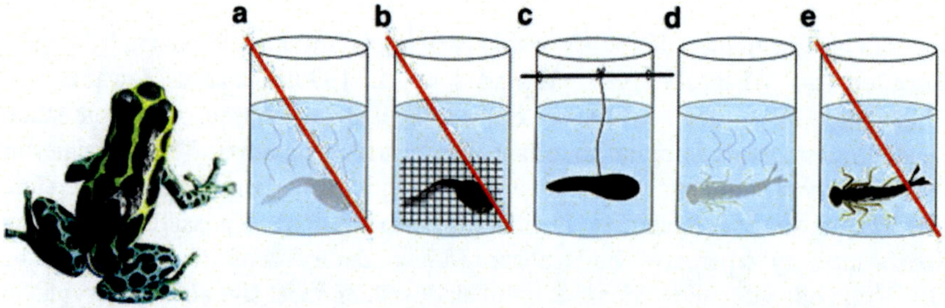

Fig. 25.3 Different cues offered to *R. variabilis* for tadpole depositions. Frogs avoided conspecific cannibalistic tadpoles when only their chemical cues (**a**) or a tadpole individual (in a cage) was present (**b**). Tadpole models (visually recognisable) were not avoided (**c**). Contrary to the tadpole cues, chemical cues of damselfly larvae (*Microstigma rotundatum*) were not avoided (**d**) while visually (and chemically) present damselfly larvae were avoided (**e**)

Takahara et al. 2008). However, some species show only weak or no reactions at all to cues of starved odonate larvae or larvae fed with non-anuran prey compared to those fed with tadpoles before (e.g. Laurila et al. 1997, 1998; Chivers and Mirza 2001; Ferland-Raymond and Murray 2008). Ferland-Raymond et al. (2010) tried to encode the cue avoided by anuran tadpoles and found that dragonfly larvae only produced this cue when previously fed with tadpoles. This might support our suggestion of predator–prey arms races between anurans and odonates. While some predators can be recognised and avoided by their prey, others are able to mask their own cues—but tend to fail after an uptake of alarm cues. However, we never tested if parent *R. variabilis* do react in a different way to *M. rotundatum* after feeding on *R. variabilis* tadpoles. We exclusively fed the invertebrate predators with mosquito larvae in our trials, whilst in the experiments of Brown et al. (2008a) larvae were not fed at all for one week before presented to the frogs (J. L. Brown, pers. comm.).

As we are lacking a comparative test with artificial damselfly larvae we cannot determine if visual cues alone are sufficient to recognise damselflies or if a combination of both visual and chemical cues is essential. Another possible source of predator detection is water vibration as a result of moving damselfly larvae. Despite these remaining questions, our study shows that *R. variabilis* is able to react to different cues, depending on the circumstance and/or the transmitter. Such a situation depending on a switchover between (or combination of) different sensory stimuli is already known in anurans, especially for acoustic and visual cues associated with mate choice or territorial behaviour (e.g. Hödl et al. 2001; Candolin 2003; Narins et al. 2003, 2005; Gomez et al. 2009). A switchover between chemical and acoustic cues is also documented in frogs. Pearl et al. (2000), for instance, indicates that the perception of sexual signals changes according to distance in African dwarf frogs (*Hymenochirus*): while distant males are recognised by females on the basis of calls, close males are chosen by chemical cues. A similar combination of different cues was confirmed by I. Starnberger (pers. comm.): females of African reed frogs (*Hyperolius*) have the option to choose between visual, chemical and acoustical cues to select their mates.

Especially when trying to perceive signals from deadly predators, the ability to recognise the 'right' cue(s) is essential. The avoidance of chemical cues of predators has been demonstrated across a wide range of prey taxa (see Kats and Dill 1998 for review) and is especially important in aquatic systems (Brönmark and Hansson 2000). But if the predator is able to mask such cues, the prey has to adapt by switching to other signals (e.g. from chemical to visual, mechanical or tactile). In our study, we found that chemical communication is an advantageous tool for the detection of some but not all predators of *R. variabilis* and shows how (indirect) prey alters its behaviour depending on the signals available from the predator.

Acknowledgements We thank M. Mayer and E. Rudolf for field assistance, A. Holmes for proofreading our manuscript and O. Finke for the identification of the damselfly larvae. Asociación de Productores de Ranas Venenosas, Progreso (ASPRAVEP) allowed us to use their field station. Research permits were obtained from the Ministry of Agriculture (DGFFS) in Lima, Peru (Authorization No. 0204-2010-AG-DGFFS-DGFFS and 0200-2011-AG-DGFFS-DGFFS). Part of this research was funded with grants from the 'Forschungsfonds' of Trier

University (to S. Lötters), the 'Studienstiftung des deutschen Volkes' (to L. M. Schulte) and the 'Deutsche Forschungsgemeinschaft' to S. Lötters, M. Veith and W. Brack (LO 1681/1-1).

References

Binckley CA, Resetarits WJ Jr (2003) Functional equivalence of none-lethal effects: generalized fish avoidance determines distribution of gray treefrog, *Hyla chrysoscelis*, larvae. Oikos 102:623–629

Bradbury JW, Vehrencamp S (1998) The principles of animal communication. Sinauer Associates, Sunderland

Brodie ED, Brodie ED Jr (1999) Predator-prey arms races. Bioscience 49:557–568

Brönmark C, Hansson LA (2000) Chemical communication in aquatic systems: an introduction. Oikos 88:103–109

Brown JL, Morales V, Summers K (2008a) Divergence in parental care, habitat selection and larval life history between two species of Peruvian poison frogs: an experimental analysis. J Evol Biol 21:1534–1543

Brown JL, Twomey E, Morales V, Summers K (2008b) Phytotelm size in relation to parental care and mating strategies in two species of Peruvian poison frogs. Behaviour 145:1139–1165

Brown JL, Morales V, Summers K (2009) Home range size and location in relation to reproductive resources in poison frogs (Dendrobatidae): a Monte Carlo approach using GIS data. Anim Behav 77:547–554

Brust DG (1993) Maternal brood care by *Dendrobates pumilio*: a frog that feeds its young. J Herpetol 27:96–98

Caldwell JP, de Araújo MC (1998) Cannibalistic interactions resulting from indiscriminate predatory behavior in tadpoles of poison frogs (Anura: Dendrobatidae). Biotropica 30:92–103

Candolin U (2003) The use of multiple cues in mate choice. Biol Rev 78:575–595

Chivers DP, Mirza RS (2001) Importance of predator diet cues in responses of larval wood frogs to fish and invertebrate predators. J Chem Ecol 27:45–51

Cooper WE, Caldwell JP, Vitt LJ (2009) Conspicuousness and vestigial escape behaviour by two dendrobatid frogs, *Dendrobates auratus* and *Oophaga pumilio*. Behaviour 146:325–349

Darst CR, Cummings ME (2006) Predator learning favours mimicry of a less-toxic model in poison frogs. Nature 440:208–211

Duellman WE, Trueb L (1986) Biology of amphibians. Johns Hopkins University Press, New York

Edmunds M (1974) Defence in animals: a survey of anti-predator defences. Longman, New York

Eklöv P (2000) Chemical cues from multiple predator-prey interactions induce changes in behavior and growth of anuran larvae. Oecologia 123:192–199

Ewert JP (1974) The neural basis of visually guided behavior. Sci Am 230:34–42

Ferland-Raymond B, Murray DL (2008) Predator diet and prey adaptive responses: can tadpoles distinguish between predators feeding on congeneric vs. conspecific prey? Can J Zool 86:1329–1336

Ferland-Raymond B, March RE, Metcalfe CD, Murray DL (2010) Prey detection of aquatic predators: assessing the identity of chemical cues eliciting prey behavioral plasticity. Biochem Syst Ecol 38:169–177

Ferrari MCO, Chivers DP (2010) The ghost of predation future: threat-sensitive and temporal assessment of risk by embryonic woodfrogs. Behav Ecol Sociobiol 64:549–555

Ferrari MCO, Messier F, Chivers DP (2007) Degradation of chemical alarm cues under natural conditions: risk assessment by larval woodfrogs. Chemoecology 17:263–266

Fisher RA (1922) On the interpretation of X^2 from contingency tables, and the calculation of P. J Roy Stat Soc Ser A (Stat Soc) 85:87–94

Forester DC, Wisnieski A (1991) The significance of airborne olfactory cues to the recognition of home area by the dart-poison frog *Dendrobates pumilio*. J Herpetol 25:502–504

Gomez D, Richardson C, Lengagne T, Plenet S, Joly P, Léna JP, Théry M (2009) The role of nocturnal vision in mate choice: females prefer conspicuous males in the European tree frog (*Hyla arborea*). Proc R Soc B 276:2351–2358

Gotelli NJ, Ellison AM (2004) A primer of ecological statistics. Sinauer Associates, Sunderland

Haberl W, Wilkinson JW (1997) A note on the Unkenreflex and similar defensive postures in *Rana temporaria* (Anura, Amphibia). Herpetol Bull 61:16–20

Hamer R, Lemckert FL, Banks PB (2011) Adult frogs are sensitive to the predation risks of olfactory communication. Biol Lett 7:361–363

Hart S (1996) The language of animals. Hentry Holt and Company, New York

Hasson O (1991) Pursuit-deterrent signals: communication between prey and predator. Trends Ecol Evol 6:325–329

Hettyey A, Vincze K, Zsarnóczai S, Hoi H, Laurila A (2011) Costs and benefits of defences induced by predators differing in dangerousness. J Evol Biol 24:1007–1019

Hews DK (1988) Alarm response in larval western toads, *Bufo boreas*: release of larval chemicals by a natural predator and its effect on predator capture efficiency. Anim Behav 36:125–133

Hödl W, Amézquita A, Ryan MJ (2001) Visual signaling in anuran amphibians. In: Ryan MJ (ed) Anuran communication. Smithsonian Press, Washington, pp 121–141

Hopey ME, Petranka JW (1994) Restriction of wood frogs to fish-free habitats: how important is adult choice? Copeia 1994:1023–1025

Ingle DJ, Hoff KS (1990) Visually elicited evasive behavior in frogs. Bioscience 40:284–291

Kats LB, Dill LM (1998) The scent of death: chemosensory assessment of predation risk by prey animals. Ecoscience 5:361–394

Lannoo MJ, Townsend DS, Wassersug RJ (1987) Larval life in the leaves: arboreal tadpole types, with special attention to the morphology, ecology, and behavior of the oophagous *Osteopilus brunneaus* (Hylidae) larva. Fieldiana (Zool) 38:1–31

Laurila A, Kujasalo J, Ranta E (1997) Different antipredator behaviour in two anuran tadpoles: effects of predator diet. Behav Ecol Sociobiol 40:329–336

Laurila A, Kujasalo J, Ranta E (1998) Predator-induced changes in life history in two anuran tadpoles: effects of predator diet. Oikos 83:307–317

Lehtinen RM, Lannoo MJ, Wassersug RJ (2004) Phytotelm-breeding anurans: past, present, and future research. Misc Publ Mus Zool Univ Mich 193:1–10

Lenzi-Mattos R, Antoniazzi MM, Haddad CFB, Tambourgi DV, Rodrigues MT, Jared C (2005) The inguinal macroglands of the frog *Physalaemus nattereri* (Leptodactylidae): structure, toxic secretion and relationship with deimatic behaviour. J Zool 266:385–394

Lötters S, Jungfer K-H, Henkel F-W, Schmidt W (2007) Poison frogs: biology, species & captive husbandry. Chimaira, Frankfurt

Marchisin A, Anderson JD (1978) Strategies employed by frogs and toads (Amphibia, Anura) to avoid predation by snakes (Reptilia, Serpentes). J Herpetol 12:151–155

Martins M (1989) Deimatic behavior in *Pleurodema brachyops*. J Herpetol 23:305–307

McDiamond RW, Altig R (1999) Tadpoles: the biology of anuran larvae. University of Chicago Press, Chicago

McDiarmid RW, Gorzula S (1989) Aspects of the reproductive ecology and behavior of the tepui toads, genus *Oreophrynella* (Anura, Bufonidae). Copeia 2:445–451

Murray DL, Roth JD, Wirsing AJ (2004) Predation risk avoidance by terrestrial amphibians: the role of prey experience and vulnerability to native and exotic predators. Ethology 110:635–647

Narins PM, Hödl W, Grabul DS (2003) Bimodal signal requisite for agonistic behavior in a dart-poison frog, *Epipedobates femoralis*. Proc Natl Acad Sci U S A 100:577

Narins PM, Grabul DS, Soma KK, Gaucher P, Hödl W (2005) Cross-modal integration in a dart-poison frog. Proc Natl Acad Sci U S A 102:2425

Osorio D, Srinivasan MV (1991) Camouflage by edge enhancement in animal coloration patterns and its implications for visual mechanisms. Proc R Soc B 244:81–85

Pearl CA, Cervantes M, Chan M, Ho U, Shoji R, Thomas EO (2000) Evidence for a mate-attracting chemosignal in the dwarf African clawed frog *Hymenochirus*. Horm Behav 38:67–74

Pearson K (1956) On the theory of contingency and its relation to association and normal correlation; On the general theory of skew correlation and non-linear regression. In: Pearson ES (Ed) Karl Pearson's early statistical papers. Cambridge University Press, Cambridge, pp 443–475
Petranka J, Hayes L (1998) Chemically mediated avoidance of a predatory odonate (*Anaxjunius*) by American toad (*Bufo americanus*) and wood frog (*Ranasylvatica*) tadpoles. Behav Ecol Sociobiol 42:263–271
Petranka JW, Hopey ME, Jennings BT, Baird SD, Boone SJ (1994) Breeding habitat segregation of wood frogs and American toads: the role of interspecific tadpole predation and adult choice. Copeia 1994:691–697
Poelman EH, Dicke M (2007) Offering offspring as food to cannibals: oviposition strategies of Amazonian poison frogs (*Dendrobates ventrimaculatus*). Evol Ecol 21:215–227
Pramuk JB, Hiler BI (1999) An investigation of obligate oophagy of *Dendrobates pumilio* tadpoles. Herpetol Rev 30:219–220
Prates I, Antoniazzi MM, Sciani JM, Pimenta DC, Toledo LF, Haddad CFB, Jared C (2011) Skin glands, poison and mimicry in dendrobatid and leptodactylid amphibians. J Morphol 273(3):279–90
Pröhl H, Ostrowski T (2011) Behavioural elements reflect phenotypic colour divergence in a poison frog. Evol Ecol 25:1–23
Resetarits WJ Jr, Wilbur HM (1989) Choice of oviposition site by *Hyla chrysoscelis*: role of predators and competitors. Ecology 70:220–228
Richards-Zawacki CL, Cummings ME (2010) Intraspecific reproductive character displacement in a polymorphic poison dart frog, *Dendrobates pumilio*. Evolution 65:259–267
Rieger JF, Binckley CA, Resetarits WJ Jr (2004) Larval performance and oviposition site preference along a predation gradient. Ecology 85:2094–2099
Ruxton GD, Sherratt TN, Speed MP (2004) Avoiding attack: the evolutionary ecology of crypsis, warning signals, and mimicry. Oxford University Press, Oxford
Sachs L (1974) Angewandte Statistik: Planung und Auswertung—Methoden und Modelle. Springer, Berlin
Saenz D, Johnson JB, Adams CK, Dayton GH (2003) Accelerated hatching of southern leopard frog (*Rana sphenocephala*) eggs in response to the presence of a crayfish (*Procambarus nigrocinctus*) predator. Copeia 2003:646–649
Schulte LM, Yeager J, Schulte R, Veith M, Werner P, Beck LA, Lötters S (2011) The smell of success: choice of larval rearing sites by means of chemical cues in a Peruvian poison frog. Anim Behav 81:1147–1154
Smith GR, Burgett AA, Temple KG, Sparks KA, Winter KE (2008) The ability of three species of tadpoles to differentiate among potential fish predators. Ethology 114:701–710
Spieler M, Linsenmair KE (1997) Choice of optimal oviposition sites by *Hoplobatrachus occipitalis* (Anura: Ranidae) in an unpredictable and patchy environment. Oecologia 109:184–199
Stauffer HP, Semlitsch RD (1993) Effects of visual, chemical and tactile cues of fish on the behavioral-responses of tadpoles. Anim Behav 46:355–364
Stevens M, Merilaita S (2009) Animal camouflage: current issues and new perspectives. Philos Trans R Soc Lond Ser B Biol Sci 364:423–427
Summers K (1999) The effects of cannibalism on Amazonian poison frog egg and tadpole deposition and survivorship in *Heliconia* axil pools. Oecologia 119:557–564
Summers K, Symula R, Clough M, Cronin T (1999) Visual mate choice in poison frogs. Proc R Soc Lond Ser B Biol Sci 266:2141–2145
Takahara T, Kohmatsu Y, Yamaoka R (2008) Predator-avoidance behavior in anuran tadpoles: a new bioassay for characterization of water-soluble cues. Hydrobiologia 607:123–130
Takahashi M (2007) Oviposition site selection: pesticide avoidance by gray treefrogs. Environ Toxicol Chem 26:1476–1480
Teixeira RL, Mili PSM, Rödder D (2006) Ecology of anurans inhabiting bromeliads in a saxicolous habitat of southeastern Brazil. Salamandra 42:155
Toledo RC, Jared C (1995) Cutaneous granular glands and amphibian venoms. Comp Biochem Physiol Physiol 111:1–29

Toledo LF, Fernando C, Haddad B (2009) Defensive vocalizations of neotropical anurans. South Am J Herpetol 4:25–42
Toledo LF, Sazima I, Haddad CFB (2011) Behavioural defences of anurans: an overview. Ethol Ecol Evol 23:1–25
Townend J (2002) Practical statistics for environmental and biological scientists. Wiley, Chichester
Varga L (1928) Ein interessanter Biotop der Biocönose von Wasserorganismen. Biol Zentralbl 48:143–162
Williams DA (1976) Improved likelihood ratio tests for complete contingency tables. Biometrika 63:33
Williams CR, Brodie ED Jr, Tyler MJ, Walker SJ (2000) Antipredator mechanisms of Australian frogs. J Herpetol 34:431–443
Wilson EO (1970) Chemical communication within animal species. In: Sondheimer E, Simeone JB (eds) Chemical ecology. Academic, New York, pp 133–155
Wirsing AJ, Roth JD, Murray DL (2005) Can prey use dietary cues to distinguish predators? A test involving three terrestrial amphibians. Herpetologica 61:104–110
Wissinger S, McGrady J (1993) Intraguild predation and competition between larval dragonflies: direct and indirect effects on shared prey. Ecology 74:207–218
Woolf B (1957) The log likelihood ratio test (the G-test); methods and tables for tests of heterogeneity in contingency tables. Ann Hum Genet 21:397–409

Part VI
Chemical Signals in Birds

Chapter 26
Besides Colours and Songs, Odour is the New Black of Avian Communication

Francesco Bonadonna and Jerome Mardon

Abstract Odours are broadly used for individual, sexual and species recognition in vertebrates and may be reliable signals of quality and compatibility. Yet, chemical signals in birds have rarely been investigated. In fact, birds exhibit a wide array of communication mechanisms (e.g. colours and calls) but rarely display obvious olfactory-driven behaviours. This is probably why, despite three decades of physiological and behavioural studies establishing the existence of avian olfactory functions, chemical communication has been essentially ignored.

In spite of the fact that pheromones have never been highlighted in birds, several species produce characteristic scents that may have a social function. For example, odours seem to contribute to the courtship behaviours of ducks and chickens. In crested auklets, a characteristic citrus odour may act as a sexual olfactory ornament broadcasting resistance to ectoparasites. Eventually, it was shown that zebra finches (a passerine bird with a very small olfactory apparatus) display olfactory-driven behaviours.

Petrel seabirds are probably the most striking case-study, since findings relating to many aspects of petrels' ecology including homing, recognition, mate choice and even interspecific competition for nesting sites, provide a comprehensive evidence for avian chemical communication. Some burrowing petrels use the odour of their mate to recognise their own nest and, interestingly, prefer the odour of a conspecific to their own. Using chemical analytical methods, it has been demonstrated that the preen secretions of these birds contain social information including species, sex and identity (i.e. a chemical signature). Results further show that some of this information is still present on the plumage and in the airborne volatiles emitted by birds. Recent results also show that "detective mice" (i.e. biological olfactometers) can identify similarities in individual kin-related petrel odours. Also, one species of

F. Bonadonna (✉) • J. Mardon
Behavioral Ecology Group, CEFE-CNRS,
1919 Route de Mende, 34293 Montpellier Cedex 5, France
e-mail: francesco.bonadonna@cefe.cnrs.fr

M.L. East and M. Dehnhard (eds.), *Chemical Signals in Vertebrates 12*,
DOI 10.1007/978-1-4614-5927-9_26, © Springer Science+Business Media New York 2013

storm petrel has been shown to recognise and avoid kin-related conspecifics in choice experiments.

Together, these results, almost 50 years after the first works on avian olfaction, indicate that chemical signals can contribute, as well as colours, calls and songs, to avian social behaviours; a realisation that has important implications for behavioural processes such as individual recognition and/or mate choice.

26.1 Introduction

A century after the first evidence (and/or speculations) regarding avian olfaction (Benham 1906; Pycraft 1910), the existence of olfactory capabilities in virtually all birds has become generally accepted (Wenzel 2007). The traditional view that birds are anosmic or microsmatic (Soudek 1927; Walter 1943) has indeed been convincingly refuted by anatomical, physiological and behavioural evidence (see Roper 1999 for a review). Despite this the contribution of olfaction to avian behaviours has been largely ignored by ornithologists. This is probably because the general lifestyle of birds seems to emphasise their vocal and visual functions (Wenzel 1973). Furthermore, research on avian olfaction has long been restricted to the responses of a few species (vultures, kiwis, pigeons and procellariiform seabirds) to environmental cues only (Benham 1906; Benvenuti et al. 1977; Grubb 1972; Papi et al. 1974; Shallenberger 1975; Smith and Paselk 1986; Stager 1967; Wenzel 1968). This perspective, traditionally limited to the study of olfactory navigation, both in the context of homing (see Wallraff 2004) and identification of feeding areas (Nevitt 2000, 2008; Nevitt et al. 2006; Nevitt and Bonadonna 2005), has been recently enriched by new evidence from other bird species. For instance, similar to procellariiforms, penguins have been shown to detect feeding areas by smell (Culik 2001; Cunningham et al. 2008; Wright et al. 2011). Land birds, such as passerines and parrots, may also be conditioned to find food near a given odour (Hagelin 2004; Mennerat et al. 2005; Roper 2003). Associating food and odours may therefore be a widespread avian behaviour to locate feeding spots.

Nevertheless, the role and importance of biogenic chemosignals in avian ecology has been essentially overlooked. This chapter reviews and discusses how recent advances in the field of chemosignals in birds, and particularly the results from procellariiform seabirds, should invigorate the field of avian ecology.

26.2 Interspecific Interactions: Predation, Parasitism and Competition

An important step toward an increased awareness of avian olfaction was the discovery that environmental odours may be used by birds to create an aromatic atmosphere around them. Indeed, starlings, *Sturnus vulgaris*, (Clark and Mason 1985; Gwinner 1997;

Gwinner and Berger 2005, 2008; Gwinner et al. 2000) and blue tits, *Cyanistes caeruleus* (Mennerat 2008; Petit et al. 2002), have been observed to incorporate aromatic herbs into their nests, probably for anti-parasitic purposes. This discovery later led to the idea that, as receivers, birds can use their olfactory capabilities in interspecific interaction such as predation and parasitism. For example, blue tits are able to detect predator-related chemical cues and show antipredatory behaviours when exposed to such cues (Amo et al. 2008). Similarly, house finches, *Carpodacus mexicanus*, respond to the odour of mammalian faeces while feeding by reducing the length of feeding bouts (Roth et al. 2008). Heterospecific olfactory cues can also affect parental behaviours as illustrated by a study on dark-eyed juncos, *Junco hyemalis*, (Whittaker et al. 2009) which found that females from a population commonly subjected to brood parasitism by brown-headed cowbirds reduced incubation bout length (i.e. parental investment) following the application of heterospecific preen oils onto their nest.

This last finding is not surprising since many avian taxa are strongly scented (Weldon and Rappole 1997); a characteristic that ornithologists have generally interpreted as a mechanism of chemical defence against predators or ectoparasites (Hagelin and Jones 2007). There are, however, only a few robust examples of predator-deterring avian scents. Some birds from the *Pitohui* or *Ifrita* genera, for instance, emit a sour odour (Dumbacher et al. 1992, 2000) that may deter potential predators such as snakes, raptors and some arboreal marsupials. Importantly, tissues of these birds (in particular feather and skin) contain some batrachotoxins, a potent toxic alkaloid which makes them poisonous. The acrid odour emitted by these birds, whose nature and origin remains unclear, could thus serve a chemical aposematic function (Diamond 1992). Similarly, the foul-scented uropygial secretion of red-billed woodhoopoe, *Phoeniculus purpureus*, that is released upon disturbance (see Hagelin and Jones 2007 for references), is efficient at deterring feline and lizard predators (du Plessis et al., unpublished results in Burger et al. 2004). Another aspect of predation, which can significantly affect the fitness of ground-nesting birds in particular, is nest-depredation by olfactory-searching mammals (Whelan et al. 1994). In that regard, the switch from monoester to diester preen waxes observed in many species of sandpipers during the breeding season has been hypothesised to favour olfactory crypticism at the nest (Reneerkens et al. 2002). Consistent with this hypothesis, mixtures of monoesters are more easily detected by domestic dogs than mixtures of the less volatile diesters (Reneerkens et al. 2005).

Examples of avian chemical defence against ectoparasites are more abundant. A review of this topic, which is beyond the scope of this chapter, can be found in Hagelin and Jones (2007). Interestingly, some of the predator-deterring chemicals mentioned above are also efficient at repelling ectoparasites. This is the case, for example, of *Pitohui* feathers (Dumbacher 1999) and of the secretions of the red-billed woodhoopoe (Law-Brown 2001). A particularly comprehensive case of avian chemical defence against ectoparasites is the study of the tangerine-scented crested auklets, *Aethia cristatella* (Hagelin et al. 2003). Evidence shows that the chemicals present on the feathers of these birds can efficiently deter ectoparasites (Douglas 2008; Hagelin 2008). What is more, the increased intensity of the citrusy scent emitted during the breeding season suggests that it may serve as an olfactory ornament during courtship and mate choice (Hagelin 2007; see also next section). None of the above examples, however, can be considered as *true* chemical communication because none involves the transfer of information that is mutually beneficial for both the emitter and the receiver of the chemosignal (Bradbury and Vehrencamp 1998). In this sense, the results on the existence and importance of a species chemosignal in hypogean petrels (Bonadonna and Mardon 2010; Mardon et al. 2010a) are particularly valuable.

Interspecific variations in the chemical content of procellariiforms' uropygial secretion have been previously studied (Jacob 1976; Jacob and Ziswiler 1982). Jacob's investigations, however, were primarily directed at elucidating and defining chemotaxonomical relationships among birds. He therefore elegantly restricted his interspecific comparison of the complex preen secretions to the analysis of the relative ratio of the different methyl-substituted acids (Fig. 26.1).

Consistent with Jacob's work, our comprehensive analysis of the preen waxes from blue petrels, *Halobaena caerulea*, and Antarctic prions, *Pachyptila desolata*, confirms the existence of a clear species chemosignal in these secretions (Mardon et al. 2010a). What is more, our behavioural results show that blue petrels can discriminate between conspecific and heterospecific odours, and our demographic data indicate that this capability may contribute to nesting decisions (Bonadonna and Mardon 2010). Indeed, several hypogean petrels breed sympatrically on sub-Antarctic islands (Cherel et al. 2002a, b) which expose them to a certain degree of interspecific competition for nesting sites (i.e. burrows). The life-history and breeding ecology of these species should favour a clear broadcasting of burrow ownership so that prospective breeders avoid using heterospecific nests which are more risky. Importantly, this broadcasting of a species-specific signal is beneficial to both the sender (i.e. the original owner of the burrow that needs to find its nest available when it returns to breed) and the receiver (i.e. a prospective breeder that is looking for an empty and safe burrow to complete its breeding attempt) (Bonadonna and Mardon 2010). These results provide therefore an original illustration of interspecific avian chemical communication.

The contribution of a species chemosignal to petrels' interspecific interactions suggests that this signal is more than a physiological side-effect of genetic divergence between species. From an evolutionary perspective, processes driving the evolution of chemosignals involved in species recognition are considered to be a result of selection for pre-mating reproductive isolation, between closely related species, and interspecific competition for communication channels (Johansson and Jones 2007). The possible involvement of chemosignals in the reproductive isolation and speciation events of petrels is entirely speculative at this stage, but offers a fascinating avenue of future phylogenic/chemical research. In contrast, the competitive interactions of closely related, and sympatrically breeding, burrow-nesting petrels offer a realistic context for the evolution of chemical divergence between these species. What is more, this hypothesis could be tested by comparing the chemical distance between several species of procellariiform while accounting for phylogenetic effects. Indeed, a prediction of competition-driven chemical divergence is

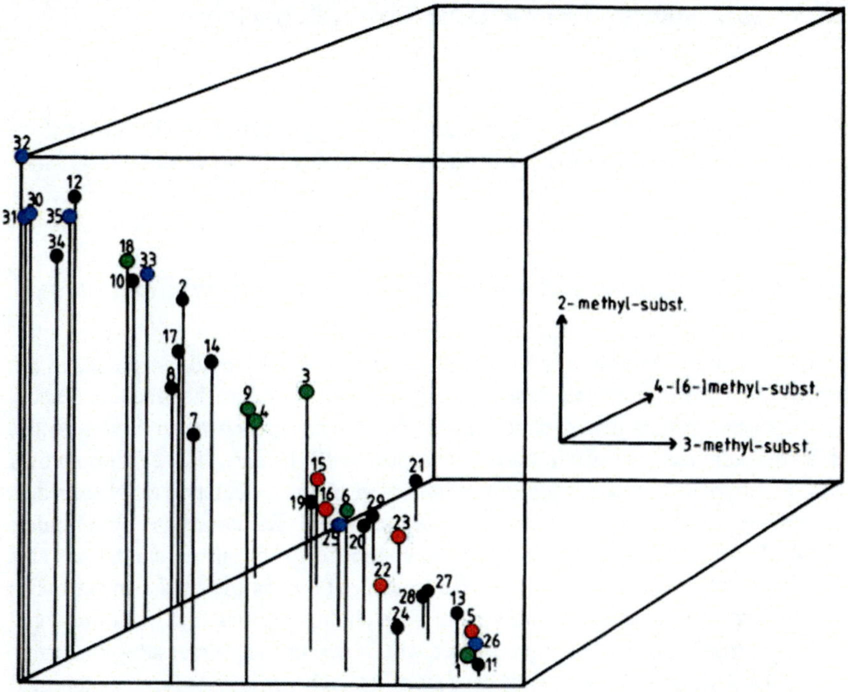

Fig. 26.1 Chemical composition of the wax acids from some procellariiform species. Each species is symbolised by a *circle*: *black*=not nesting on Kerguelen; *blue*=surface-nesting on Kerguelen; *green*=burrow-nesting on Kerguelen; *red*=sympatric burrow-nesting on Ile Verte (Kerguelen). (*1*) *Pelecanoides urinatrix;* (2) *Fregatta grallaria;* (3) *Fregatta tropica;* (4) *Garrodia nereis;* (5) *Oceanites oceanicus;* (6) *Procellaria aequinoctialis;* (7) *Procellaria westlandica;* (8) *Procellaria parkinsoni;* (9) *Procellaria cinerea;* (10) *Puffinus huttoni;* (11) *Puffinus griseus;* (12) *Puffinus assimilis;* (13) *Puffinus tenuirostris;* (14) *Puffinus gavial;* (15) *Halobaena caerulea;* (16) *Pterodroma lessonii;* (17) *Pterodroma inexpectata;* (18) *Pterodroma brevirostris;* (19) *Pterodroma cooki;* (20) *Pachyptila turtur;* (21) *Pachyptila crassirostris;* (22) *Pachyptila desolata;* (23) *Pachyptila belcheri;* (24) *Pachyptila vittata;* (25) *Macronectes giganteus;* (26) *Macronectes halli;* (27) *Thalassoica Antarctica;* (28) *Fulmarus glacialis;* (29) *Fulmarus glacialoides;* (30) *Diomedea chrysostoma;* (31) *Diomedea melanophris;* (32) *Diomedea epomophora;* (33) *Diomedea exulans;* (34) *Diomedea cauta;* (35) *Phoebetria palpebrata.* (adapted from Jacob and Ziswiler 1982)

that chemosignals of sympatric and/or sister-taxa should display particularly dramatic differences (Johansson and Jones 2007). Jacob's chemotaxonomical work on the methyl-substitution of acids contained within procellariiform's uropygial secretions provides, in that regard, a preliminary illustration. Figure 26.1 (adapted from Jacob and Ziswiler 1982) indeed shows that burrowing, and sympatrically breeding, species (red dots) are chemically more divergent than families of surface nesters such as giant petrels (points 25–26, Fig. 26.1) or albatrosses (points 30–35, Fig. 26.1). Further exploration of the relationship between interspecific interactions and the divergence of chemical signals in petrel seabirds may therefore prove particularly informative.

26.3 Intraspecific Interactions: Sexual Behaviours and Recognition

Several recent studies have drawn attention to the contribution of chemosignals to avian intraspecific interactions, with a main emphasis on avian sexual behaviours.

26.3.1 Chemical Communication and Avian Sexual Behaviours

In 1979, a study reported the existence of a chemical sexual dimorphism in the uropygial waxes of domestic ducks (Jacob et al. 1979). During the breeding season, female ducks (but not males) shift their secretions from monoester to diester waxes, under the hormonal control of estradiol (Bohnet et al. 1991). This chemical dimorphism was later hypothetically related to the inhibition of sexual behaviours observed in male ducks whose olfactory nerves had been sectioned (Balthazart and Schoffeniels 1979). A similar contribution of uropygial chemicals to sexual behaviours has also been reported in domestic chickens (Hirao et al. 2009). In this study, intact males courted and copulated significantly more with control females than with uropygial glandectomised females. This preference was, however, not expressed by anosmic males. An olfactory-mediated control of avian sexual behaviours is further supported by the observation that the brain activation of male Japanese quails, *Coturnix japonica*, normally induced by sexual interaction with a female, is significantly affected by olfactory deprivation (Balthazart and Taziaux 2009). Together, these results strongly suggest that the uropygial secretions of some avian taxa contain olfactory cues that favour, possibly through activation of key brain areas, the expression of sexual behaviours such as courtship displays, mounts and copulations.

Interestingly, the latter hypothesis could account for some ambiguous results on the role of olfactory cues in the sexual behaviours of seabirds. The plumage of the highly social crested auklet, for example, emit a strong citrusy odour during the breeding season (Hagelin et al. 2003), when the species exhibit some characteristic courtship behaviours involving essentially an intertwining of necks and the burying of bills in the nape and neck feathers (Jones and Hunter 1993). This has led authors to suggest that the plumage odour of these birds may serve as an olfactory ornament (Hagelin et al. 2003; Jones et al. 2004; Hagelin 2007) and that the so-called ruff-sniff display could constitute a case of alloanointing of ectoparasite-repelling substances between prospective mates (Douglas 2008). According to Hagelin and co-workers, chemicals present on the feathers could act as a sexual trait (on top of an anti-parasite function—see previous section) whereby the most scented individuals are also the most attractive. Yet, field experiments using scented models provided ambiguous results as only male scented models (but not female) were approached more than controls. In contrast with artificial visual ornaments (Jones and Hunter 1993), artificial scents also failed to elicit more stereotyped sexual displays (Jones et al. 2004). However, such results may be expected if, as hypothesised above, sexual olfactory cues in birds act essentially as brain activators and/or in conjunction with other displays (Hagelin and Jones 2007) to facilitate the expression of sexual behaviours.

Similarly, there is currently no evidence that the olfactory sex label identified in petrels (Mardon et al. 2010a, 2011) affects sexual behaviour. Indeed, field experiments have not produced any supportive evidence of olfactory sexual discrimination capabilities, whether in Antarctic prions (Bonadonna et al. 2009) or in blue petrels (Mardon et al. personal communication). Again, such results may be expected if sexual olfactory cues act through a targeted brain activation to elicit specific and context-dependent sexual behaviours. Note that the very presence of this sex label in petrels' secretions supports claims of a behavioural function as such a signal is absent from the secretions of many other bird species (e.g. Burger et al. 2004; Montalti et al. 2005; Piersma et al. 1999).

Finally, it is worth mentioning that most current evidence on the role of olfactory cues on avian sexual behaviours (with the exception of crested auklets—Jones et al. 2004) indicates that the associated signalling is commonly female-biased. In the above examples, male ducks, chickens and Japanese quails all were affected by the deprivation of access to a female signal, both when males olfaction, or female chemical production were manipulated (Balthazart and Schoffeniels 1979; Balthazart and Taziaux 2009; Hirao et al. 2009). In addition, two of the three instances of avian chemical sexual dimorphisms identified to date (ducks and blue petrels) consist of female-specific feature of the chemosignal (Jacob et al. 1979; Mardon et al. 2010a). In budgerigars (*Melopsittacus undulatus*), in contrast, authors claim that sexual dimorphisms consist of male-specific feature (Zhang et al. 2010), but a re-analysis of the data presented suggests, if anything, the presence of a clear female label (Mardon et al. 2010b).

This female bias in sexual chemo-labels may originate from the genetic mechanism of sex determination in birds. Indeed, avian gonosomes work in an opposite pattern to mammals, with males being homogametic (ZZ) while females are heterogametic (ZW) (Fridolfsson and Ellegren 1999). Further work on this particular question should prove particularly valuable.

26.3.2 Chemical Communication and Individual Recognition

Recent research on the homing mechanisms of petrels has proved a fruitful gateway for the study of avian chemical communication as it has revealed the existence of social odours in these birds (Bonadonna 2009). After several authors showed that olfaction is critical for "homing to the nest" (review in Bonadonna et al. 2001), we identified the significant role of a burrow olfactory signature in these processes (Bonadonna et al. 2003, 2004); an olfactory signature, based on the odour of a bird's mate (Bonadonna and Nevitt 2004; Mardon and Bonadonna 2009). While traditionally examined from the perspective of homing mechanisms, these results have drawn attention to the possible use of chemical signals in birds' social behaviours.

Indeed, petrels are to date the only birds shown to possess olfactory capabilities of individual recognition beyond self/non-self discrimination. Following the initial demonstration of self-odour recognition by European storm-petrel chicks, *Hydrobates pelagicus*, (De Leon et al. 2003), three species of burrow-nesting petrels (Antarctic prions, Wilson's storm petrels, *Oceanites oceanicus* and blue petrels) were later shown to recognise not only their own but also the odour of their mate (Bonadonna and Nevitt 2004; Jouventin et al. 2007; Mardon and Bonadonna 2009). The chemical results obtained from blue petrels (Mardon et al. 2010a, 2011), as well as a preliminary study on the feather lipids of Antarctic prions (Bonadonna et al. 2007), show that petrels' individual chemical signatures are secreted through the uropygial waxes and are still present on the plumage of the birds. This work not only provides a robust chemical basis to petrels' recognition behaviours, but it also contributes to elucidate the chemical nature of these chemosignals.

Chemical signatures in the uropygial secretions of blue petrels are not made of individually specific bouquets of compounds. Instead they appear to involve bouquets of relative proportions of as many as 63 compounds present in most individuals (Mardon et al. 2010a). Similar quantitatively coded chemical signatures have been reported in many mammals including mice (Singer et al. 1997), monkeys (Smith et al. 2001), pandas (Hagey and MacDonald 2003), bats (Safi and Kerth 2003) and hyenas (Burgener et al. 2009); in humans, however, individuality appears to be best reflected in the qualitative variations of compounds within the axillary sweat (Penn et al. 2007).

Eventually, birds (and petrels in particular) are well equipped to deal with the level of olfactory complexity suggested by Mardon and co-workers. The structure of the avian neuro-olfactory system is indeed similar to that of most vertebrates (although birds do not have vomeronasal organs—Bertmar 1981). Odour perception and discrimination relies on olfactory sensory neurons distributed over the epithelium of the third nasal chamber (Roper 1999).

26.4 Origin of Avian Chemosignals and Implications

The complex and polymorphic nature of petrels' chemical signatures raises questions about their determinism, a topic which has major implications for avian chemical communication.

26.4.1 Origin of Chemosignals

Volatile signals emitted by birds are essentially limited to the odour of the plumage (but see Hagelin and Jones 2007 for other sources), which is a particular form of olfactory signal. Indeed, this complex volatile emission is not released in a particular behavioural context, or aimed at a particular individual. Instead, it is a durable and passive feature which permanently accompanies its bearer. Furthermore, personal odours, unlike pheromones, rarely trigger a specific response from the receiver (Karlson and Lüscher 1959), nor are they suited for immediate and mutually responsive communication. This is why some authors refer to them as *state* signals, i.e. signals that remain "on" for a prolonged time; as opposed to *event* signals, which are typically short-term manifestations (Hauser 1996).

Existing evidence from chickens (Hirao et al. 2009) and blue petrels (Mardon et al. 2010a, 2011) now clearly indicates that the social signal contained in a bird's odour has an endogenous origin. This is a particularly important finding because it makes these odours appropriate for the transport of subtle physiological and genetic information, such as those typically used by animals for social recognition and quality assessments. Endogenous chemosignals have indeed been found to carry information such as species, group, sex or individual identities (Burgener et al. 2008; Penn et al. 2007; Safi and Kerth 2003; Whittaker et al. 2009), social status (Moore et al. 1997), physiological or health status (Zala and Penn 2004), relatedness (Ables et al. 2007) or even particular genotypes (Reusch et al. 2001).

Animal chemosignals used for recognition are generally considered to be based essentially on genotypic variation. Other pathways may indeed be subjected to environmental and physiological influences, such as ageing, changes in diet or microbial flora (Brennan and Kendrick 2006). The existence of complex chemical signature in petrels (Mardon et al. 2010a, 2011), and their contribution to recognition behaviours, therefore suggests they are genetically determined (at least in part) in birds. Interestingly, a recent experiment exploiting an indirect behavioural method called "biological olfactometer" (Celerier et al. 2010) showed that blue petrels possess a "family" odour that potentially broadcasts kin relatedness (Celerier et al. 2011). The odour of chicks was found indeed to be more similar to their parents' odour than to the odour of unrelated conspecifics, suggesting a genetic basis of this family odour. Recent results from our research group on European storm petrels support this hypothesis. In Y-maze experiments, this species showed a significant avoidance to the odour of related conspecifics (brothers, sisters, parents and grandparents), which could be part of a more general inbreeding avoidance mechanism based on chemosignals (Bonadonna and Sanz-Aguilar 2012). This hypothesis cannot, however, be definitely resolved at this stage and future experiments disentangling the influence of environment and heredity should address this question.

26.4.2 Implications for Mate Choice and Other Behaviours

If identity signals are genetically determined, they should be polymorphic enough to allow a considerable amount of phenotypic variation among individuals. This is why biologists have considered and explored a possible role of the major histocompatibility complex (MHC) in these processes (Johansson and Jones 2007; Penn 2002). The relationship between chemical signals, the MHC and behaviours such as individual recognition and mate choice is rich and complex (Johansson and Jones 2007;

Penn 2002; Tregenza and Wedell 2000). MHC-dependent mating preferences, for example, may provide two different benefits: a better immunocompetence for offspring and/or a way to avoid mating with genetically similar partners (Penn 2002). These processes, although mostly studied in mammals (Brennan and Kendrick 2006), are directly relevant to birds (Zelano and Edwards 2002). Indeed, MHC-dependent mating preferences have been reported for many avian species (Bonneaud et al. 2006; Freeman-Gallant et al. 2003; Gillingham et al. 2009; Johnsen et al. 2000; Richardson et al. 2005) (but see also Ekblom et al. 2004; Westerdahl 2004) although the proximate mechanisms leading to these preferences are still unknown.

Choosing a mate on genetic grounds requires the ability to somehow contrast one's own genetic makeup to that of a potential mate. Current evidence suggests that this task is most achievable through olfactory cues. MHC-dependent preferences based on chemical assessment have indeed been observed in fish and mammals (Penn 2002; Reusch et al. 2001; Wedekind and Furi 1997). What is more, among the many pathways in which MHC can affect sexual traits, its influence on individual odours provides the most direct (and hence reliable) route. To date the processes remain undocumented in birds, essentially because of the limited amount of behavioural and chemical data available on avian chemosignals (Hagelin and Jones 2007). The chemical elucidation of complex chemical signatures in petrel seabirds (Bonadonna et al. 2007; Mardon et al. 2010a, 2011) is an important contribution to the field in this regard. Research investigating the existence of MHC-based mating preferences in petrels is currently being undertaken (Strandh et al. 2011) and a fascinating prospect for avian chemical communication is therefore to manage the integration of genetic and chemical data to unveil the origin of social scents.

26.5 Conclusion

Biogenic chemical substances are a significant aspect of avian biology and have been shown to contribute to various interspecific (e.g. predation, parasitism, competition) and intraspecific (e.g. sexual behaviours, recognition) interactions. Among these, social signals used for communication are essentially limited to the plumage odour, a *state* signal that permanently accompanies the emitter. Avian chemical communication (as it is known today) is therefore principally concerned with the transfer of information supporting neurophysiological activation (as in Japanese quails), social recognition (as in petrels, or in passerines Caspers and Krause 2011) and/or possibly genetically based quality assessments.

The results discussed in this article have contributed to a better understanding of these questions by providing a multidisciplinary and comprehensive investigation of chemical communication in petrel seabirds.

Acknowledgements Our own work presented here was supported by Institut Polaire Français Paul Emile Victor (IPEV, Program no. 354), and by Agence Nationale de la Recherche Française (AMBO ANR-08-BLAN-0117-01 to FB).

References

Ables EM, Kay LM, Mateo JM (2007) Rats assess degree of relatedness from human odors. Physiol Behav 90:726–732
Amo L, Galvan I, Tomas G, Sanz JJ (2008) Predator odour recognition and avoidance in a songbird. Funct Ecol 22:289–293
Balthazart J, Schoffeniels E (1979) Pheromones are involved in the control of sexual behavior in birds. Naturwissenschaften 66:55–56
Balthazart J, Taziaux M (2009) The underestimated role of olfaction in avian reproduction? Behav Brain Res 200:248–259
Benham WB (1906) The olfactory sense in Apteryx. Nature 74:222–223
Benvenuti S, Fiaschi V, Foà A (1977) Homing behaviour of pigeons disturbed by application of an olfactory stimulus. J Comp Physiol A 120:175–179
Bertmar G (1981) Evolution of vomeronasal organs in vertebrates. Evolution 35:359–366
Bohnet S, Rogers L, Sasaki G, Kolattukudy PE (1991) Estradiol induces proliferation of peroxisome-like microbodies and the production of 3-hydroxy fatty acid diesters, the female pheromones, in the uropygial glands of male and female mallards. J Biol Chem 266:9795–9804
Bonadonna F (2009) Olfaction in petrels: from homing to self-odor avoidance. Ann N Y Acad Sci 1170:428–433
Bonadonna F, Mardon J (2010) One house two families: petrel squatters get a sniff of low-cost breeding opportunities. Ethology 116:176–182
Bonadonna F, Nevitt GA (2004) Partner-specific odor recognition in an Antarctic seabird. Science 306:835
Bonadonna F, Sanz-Aguilar A (2012) Kin recognition and inbreeding avoidance in wild birds: the first evidence for individual kin-related odour recognition. Anim Behav 84:509–513
Bonadonna F, Spaggiari J, Weimerskirch H (2001) Could osmotaxis explain the ability of blue petrels to return to their burrows at night? J Exp Biol 204:1485–1489
Bonadonna F, Hesters F, Jouventin P (2003) Scent of a nest: discrimination of own-nest odours in Antarctic prions, *Pachyptila desolata*. Behav Ecol Sociobiol 54:174–178
Bonadonna F, Villafane M, Bajzak C, Jouventin P (2004) Recognition of burrow's "olfactory signature" in blue petrels, *Halobaena caerulea*: an efficient discrimination mechanism in the dark. Anim Behav 67:893–898
Bonadonna F, Miguel E, Grosbois V, Jouventin P, Bessiere JM (2007) Individual-specific odour recognition in birds: an endogenous olfactory signature on petrels' feathers? J Chem Ecol 33:1819–1829
Bonadonna F, Caro S, Brooke M (2009) Olfactory sex recognition investigated in Antarctic prions. PLoS One 4:e4148. doi:10.1371/journal.pone.0004148
Bonneaud C, Chastel O, Federici P, Westerdahl H, Sorci G (2006) Complex MHC-based mate choice in a wild passerine. Proc R Soc Lond B Biol Sci 273:1111–1116
Bradbury JW, Vehrencamp SL (1998) Principles of animal communication. Sinauer Associates, Sunderland, MA
Brennan PA, Kendrick KM (2006) Mammalian social odours: attraction and individual recognition. Philos Trans R Soc Lond B Biol Sci 361:2061–2078
Burgener N, East ML, Hofer H, Dehnhard M (2008) Do spotted hyena scent marks code for clan membership? In: Hurst JL, Beynon RJ, Roberts S, Wyatt TD (eds) Chemical signals in vertebrates, vol XI. Springer, New York, pp 169–177
Burgener N, Dehnhard M, Hofer H, East ML (2009) Does anal gland scent signal identity in the spotted hyaena? Anim Behav 77:707–715
Burger BV, Reiter B, Borzyk O, du Plessis MA (2004) Avian exocrine secretions. I. Chemical characterization of the volatile fraction of the uropygial secretion of the green woodhoopoe, *Phoeniculus purpureus*. J Chem Ecol 30:1603–1611
Caspers BA, Krause ET (2011) Odour-based natal nest recognition in the zebra finch (*Taeniopygia guttata*), a colony-breeding songbird. Biol Lett 7:184–186

Celerier A, Alvergne A, Huchard E, Fejan D, Plard F, Cowlishaw G, Raymond M, Bonadonna F (2010) Detective mice assess relatedness in baboons inquiring into olfactory clues. J Exp Biol 213:1399–1405

Celerier A, Bon C, Malapert A, Palmas P, Bonadonna F (2011) Chemical kin label in seabirds. Biol Lett 7:807–810

Cherel Y, Bocher P, De Broyer C, Hobson KA (2002a) Food and feeding ecology of the sympatric thin-billed *Pachyptila belcheri* and Antarctic *P. desolata* prions at Iles Kerguelen, Southern Indian Ocean. Mar Ecol Prog Ser 228:263–281

Cherel Y, Bocher P, Trouve C, Weimerskirch H (2002b) Diet and feeding ecology of blue petrels *Halobaena caerulea* at Iles Kerguelen, Southern Indian Ocean. Mar Ecol Prog Ser 228:283–299

Clark L, Mason JR (1985) Use the nest material as insecticidal and anti-pathogenic agents by the European starling. Oecologia 67:169–176

Culik B (2001) Finding food in the open ocean: foraging strategies in Humboldt penguins. Zoology 104:327–338

Cunningham GB, Strauss V, Ryan PG (2008) African penguins (*Spheniscus demersus*) can detect dimethyl sulphide, a prey-related odour. J Exp Biol 211:3123–3127

De Leon A, Minguez E, Belliure B (2003) Self-odour recognition in European storm-petrel chicks. Behaviour 140:925–933

Diamond JM (1992) Rubbish birds are poisonous. Nature 360:19–20

Douglas HD (2008) Prenuptial perfume: alloanointing in the social rituals of the crested auklet (*Aethia cristatella*) and the transfer of arthropod deterrents. Naturwissenschaften 95:45–53

Dumbacher JP (1999) Evolution of toxicity in pitohuis: I. Effects of homobatrachotoxin on chewing lice (order Phthiraptera). Auk 116:957–963

Dumbacher JP, Beehler BM, Spande TF, Garraffo HM, Daly JW (1992) Homobatrachotoxin in the genus *Pitohui*: chemical defense in birds? Science 258:799–800

Dumbacher J, Spande TF, Daly JW (2000) Batrachotoxin alkaloids from passerine birds: a second toxic bird genus (*Ifrita kowaldi*) from New Guinea. Proc Natl Acad Sci U S A 97:12970–12975

Ekblom R, Saether SA, Grahn M, Fiske P, Kalas JA, Hoglund J (2004) Major histocompatibility complex variation and mate choice in a lekking bird, the great snipe (*Gallinago media*). Mol Ecol 13:3821–3828

Freeman-Gallant CR, Meguerdichian M, Wheelwright NT, Sollecito SV (2003) Social pairing and female mating fidelity predicted by restriction fragment length polymorphism similarity at the major histocompatibility complex in a songbird. Mol Ecol 12:3077–3083

Fridolfsson AK, Ellegren H (1999) A simple and universal method for molecular sexing of non-ratite birds. J Avian Biol 30:116–121

Gillingham MAF, Richardson DS, Lovlie H, Moynihan A, Worley K, Pizzari T (2009) Cryptic preference for MHC-dissimilar females in male red junglefowl, *Gallus gallus*. Proc R Soc Lond 276:1083–1092

Grubb TC (1972) Smell and foraging in shearwaters and petrels. Nature 237:404–405

Gwinner H (1997) The function of green plants in nests of European starlings (*Sturnus vulgaris*). Behaviour 134:337–351

Gwinner H, Berger S (2005) European starlings: nestling condition, parasites and green nest material during the breeding season. J Ornithol 146:365–371

Gwinner H, Berger S (2008) Starling males select green nest material by olfaction using experience-independent and experience-dependent cues. Anim Behav 75:971–976

Gwinner H, Oltrogge M, Trost L, Nienaber U (2000) Green plants in starling nests: effects on nestlings. Anim Behav 59:301–309

Hagelin JC (2004) Observations on the olfactory ability of the Kakapo *Strigops habroptilus*, the critically endangered parrot of New Zealand. Ibis 146:161–164

Hagelin JC (2007) The citrus-like scent of crested auklets: reviewing the evidence for an avian olfactory ornament. J Ornithol 148:S195–S201

Hagelin JC (2008) New data and new questions for Crested Auklet research. Auk 125:497–498

Hagelin JC, Jones IL (2007) Bird odors and other chemical substances: a defense mechanism or overlooked mode of intraspecific communication? Auk 124:741–761

Hagelin JC, Jones IJ, Rasmussen LEL (2003) A tangerine-scented social odour in a monogamous seabird. Proc R Soc Lond B Biol Sci 270:1323–1329

Hagey L, MacDonald E (2003) Chemical cues identify gender and individuality in giant pandas (*Ailuropoda melanoleuca*). J Chem Ecol 29:1479–1488

Hauser MD (1996) The evolution of communication. MIT Press, Cambridge, MA

Hirao A, Aoyama M, Sugita S (2009) The role of uropygial gland on sexual behavior in domestic chicken *Gallus gallus domesticus*. Behav Process 80:115–120

Jacob J (1976) Chemotaxonomical relationships between penguins and tubenoses. Biochem Syst Ecol 4:215–221

Jacob J, Ziswiler V (1982) The uropygial gland. In: Farner DS, King JR, Parkes KC (eds) Avian biology, vol 6. Academic, New York, pp 199–324

Jacob J, Balthazart J, Schoffeniels E (1979) Sex differences in the chemical composition of uropygial gland waxes in domestic ducks. Biochem Syst Ecol 7:149–153

Johansson BG, Jones TM (2007) The role of chemical communication in mate choice. Biol Rev 82:265–289

Johnsen A, Andersen V, Sunding C, Lifjeld JT (2000) Female bluethroats enhance offspring immunocompetence through extra-pair copulations. Nature 406:296–299

Jones IL, Hunter FM (1993) Mutual sexual selection in a monogamous seabird. Nature 362:238–239

Jones IL, Hagelin JC, Major HL, Rasmussen LEL (2004) An experimental field study of the function of Crested Auklet feather odor. Condor 106:71–78

Jouventin P, Mouret V, Bonadonna F (2007) Wilson's storm petrels *Oceanites oceanicus* recognise the olfactory signature of their mate. Ethology 113:1228–1232

Karlson P, Lüscher M (1959) 'Pheromones': a new term for a class of biologically active substances. Nature 183:155–156

Law-Brown J (2001) Chemical defence in the Red-billed Woodhoopoe *Phoneiculus purpureus*. MSc Thesis, University of Cape Town, Rondebosch, South Africa

Mardon J, Bonadonna F (2009) Atypical homing or self-odour avoidance? Blue petrels (*Halobaena caerulea*) are attracted to their mate's odour but avoid their own. Behav Ecol Sociobiol 63:537–542

Mardon J, Saunders SM, Anderson MJ, Couchoux C, Bonadonna F (2010a) Species, gender, and identity: cracking petrels' sociochemical code. Chem Senses 35:309–321

Mardon J, Saunders SM, Bonadonna F (2010b) Comments on recent work by Zhang & colleagues: 'Uropygial gland-secreted alkanols contribute to olfactory sex signals in budgerigars'. Chem Senses 36:3–4

Mardon J, Saunders SM, Bonadonna F (2011) From preen secretions to plumage: the chemical trajectory of petrels' social scent. J Avian Biol 42:29–38

Mennerat A (2008) Blue tits (*Cyanistes caeruleus*) respond to an experimental change in the aromatic plant odour composition of their nest. Behav Process 79:189–191

Mennerat A, Bonadonna F, Perret P, Lambrechts MM (2005) Olfactory conditioning experiments in a food-searching passerine bird in semi-natural and natural conditions. Behav Process 70:264–270

Montalti D, Gutierrez AM, Reboredo G, Salibian A (2005) The chemical composition of the uropygial gland secretion of rock dove Columba livia. Comp Biochem Physiol A Mol Integrat Physiol 140:275–279

Moore PJ, Reagan-Wallin NL, Haynes KF, Moore AJ (1997) Odour conveys status on cockroaches. Nature 389:25–26

Nevitt GA (2000) Olfactory foraging by Antarctic procellariiform seabirds: life at high Reynolds numbers. Biol Bull (Woods Hole) 198:245–253

Nevitt GA (2008) Sensory ecology on the high seas: the odor world of the procellariiform seabirds. J Exp Biol 211:1706–1713

Nevitt GA, Bonadonna F (2005) Sensitivity to dimethyl sulphide suggests a mechanism for olfactory navigation by seabirds. Biol Lett 1:303–305

Nevitt GA, Bergstrom DM, Bonadonna F (2006) The potential roles of ammonium as signal molecule for procellariiform seabirds. Mar Ecol Prog Ser 315:271–277

Papi F, Ioalè P, Fiaschi V, Benvenuti S, Baldaccini NE (1974) Olfactory navigation of pigeons: the effect of treatment with odorous air currents. J Comp Physiol A 94:187–193

Penn DJ (2002) The scent of genetic compatibility: sexual selection and the major histocompatibility complex. Ethology 108:1–21

Penn DJ, Oberzaucher E, Grammer K, Fischer G, Soini HA, Wiesler D, Novotny MV, Dixon SJ, Xu Y, Brereton RG (2007) Individual and gender fingerprints in human body odour. J R Soc Interface 4:331–340

Petit C, Hossaert-McKey M, Perret P, Blondel J, Lambrechts MM (2002) Blue tits use selected plants and olfaction to maintain an aromatic environment for nestlings. Ecol Lett 5:585–589

Piersma T, Dekker M, Sinninghe Damsté JS (1999) An avian equivalent of make-up? Ecol Lett 2:201–203

Pycraft WP (1910) A history of birds. Methuen and Co., London

Reneerkens J, Piersma T, Damste JS (2002) Sandpipers (Scolopacidae) switch from monoester to diester preen waxes during courtship and incubation, but why? Proc R Soc Lond 269:2135–2139

Reneerkens J, Piersma T, Sinninghe Damsté JS (2005) Switch to diester preen waxes may reduce avian nest predation by mammalian predators using olfactory cues. J Exp Biol 208:4199–4202

Reusch TBH, Häberli MA, Aeschlimann PB, Milinski M (2001) Female sticklebacks count alleles in a strategy of sexual selection explaining MHC polymorphism. Nature 414:300–302

Richardson DS, Komdeur J, Burke T, von Schantz T (2005) MHC-based patterns of social and extra-pair mate choice in the Seychelles warbler. Proc R Soc Lond B Biol Sci 272:759–767

Roper TJ (1999) Olfaction in birds. Adv Study Behav 28:247–332

Roper TJ (2003) Olfactory discrimination in yellow-backed Chattering Lories *Lorius garrulus flavopalliatus*: first demonstration of olfaction in Psittaciformes. Ibis 145:689–691

Roth TC, Cox JG, Lima SL (2008) Can foraging birds assess predation risk by scent? Anim Behav 76:2021–2027

Safi K, Kerth G (2003) Secretions of the interaural gland contain information about individuality and colony membership in the Bechstein's bat. Anim Behav 65:363–369

Shallenberger RJ (1975) Olfactory use in the wedge-tailed shearwater (*Puffinus pacificus*) on Manana is (Hawaii). In: Denton A, Coughlan JP (eds) Olfaction and taste, vol V. Academic, London, pp 355–359

Singer AG, Beauchamp GK, Yamazaki K (1997) Volatile signals of the major histocompatibility complex in male mouse urine. Proc Natl Acad Sci U S A 94:2210–2214

Smith SA, Paselk RA (1986) Olfactory sensitivity of the Turkey Vulture (*Cathartes avra*) to three carrion-associated. Auk 103:586–592

Smith TE, Tomlinson AJ, Mlotkiewicz JA, Abbott DH (2001) Female marmoset monkeys (*Callithrix jacchus*) can be identified from the chemical composition of their scent marks. Chem Senses 26:449–458

Soudek S (1929). The sense of smell in birds. In: Proceedings of X[e] Congres International de Zoology, Budapest, pp 755–765

Stager KE (1967) Avian olfaction. Am Zool 7:415–419

Strandh M, Lannefors M, Bonadonna F, Westerdahl H (2011) Characterization of MHC class I and II genes in a subantarctic seabird, the Blue Petrel, *Halobaena caerulea* (Procellariiformes). Immunogenetics 63:653–666

Tregenza T, Wedell N (2000) Genetic compatibility, mate choice and patterns of parentage: invited review. Mol Ecol 9:1013–1027

Wallraff HG (2004) Avian olfactory navigation: its empirical foundation and conceptual state. Anim Behav 67:189–204

Walter WE (1943) Some experiments on the sense of smell in birds. Arch Neerl Physiol Homme Anim 27:1–72

Wedekind C, Furi S (1997) Body odour preferences in men and women: do they aim for specific MHC combinations or simply heterozygosity? Proc R Soc Lond B Biol Sci 264:1471–1479

Weldon PJ, Rappole JH (1997) A survey of birds odorous or unpalatable to humans: possible indications of chemical defense. J Chem Ecol 23:2609–2633

Wenzel BM (1968) The olfactory prowess of the kiwi. Nature 220:1133–1134

Wenzel B (1973) Chemoreception. In: Farner DS, King JR (eds) Avian biology, vol 3. Academic, New York, pp 389–415

Wenzel BM (2007) Avian olfaction: then and now. J Ornithol 148:S191–S194

Westerdahl H (2004) No evidence of an MHC-based female mating preference in great reed warblers. Mol Ecol 13:2465–2470

Whelan CJ, Dilger ML, Robson D, Hallyn N, Dilger S (1994) Effects of olfactory cues on artificial-nest experiments. Auk 111:945–952

Whittaker DJ, Reichard DG, Dapper AL, Ketterson ED (2009) Behavioral responses of nesting female dark-eyed juncos *Junco hyemalis* to hetero- and conspecific passerine preen oils. J Avian Biol 40:579–583

Wright KLB, Pichegru L, Ryan PG (2011) Penguins are attracted to dimethyl sulphide at sea. J Exp Biol 214:2509–2511

Zala SM, Penn DJ (2004) Abnormal behaviours induced by chemical pollution: a review of the evidence and new challenges. Anim Behav 68:649–664

Zelano B, Edwards SE (2002) An MHC component to kin recognition and mate choice in birds: predictions, progress, and prospects. Am Nat 160:225–238

Zhang JX, Wei W, Zhang JH, Yang WH (2010) Uropygial gland-secreted alkanols contribute to olfactory sex signals in budgerigars. Chem Senses 35:375–382

Chapter 27
Intraspecific Olfactory Communication in Zebra Finches (*Taeniopygia guttata*): Potential Information Apart from Visual and Acoustic Cues

Barbara A. Caspers and E. Tobias Krause

Abstract In this chapter we review evidence for olfactory communication in the zebra finch (*Taeniopygia guttata*). We discuss studies that indicate that zebra finches are capable of olfaction and use olfactory cues for social communication and kin recognition. This finding establishes a so far completely neglected mechanism of kin recognition in songbirds.

27.1 The Sense of Smell in Passerines

Passerines or songbirds, especially males, are primarily known for their acoustic abilities as they use song to attract females and/or to deter male competitors (e.g. Catchpole and Slater 2008; Schmidt et al. 2008; Sprau et al. 2010; Naguib and Mennill 2010). Intraspecific communication in many songbird species is also achieved through visual signals, such as colourful plumage traits (e.g. Hill and Montgomerie 1994; Ohlsson et al. 2002; Hill and McGraw 2006) or nest constructions that attract females with their exaggerated architecture and decoration (e.g. Collias 1964; Borgia 1985, 1995; Brouwer and Komdeur 2004). The idea that olfactory signals might also play an important role in intraspecific communication in songbirds thus seems to be unlikely at first sight, for several reasons. First, because songbirds were for a long time thought to be anosmic, due to their small olfactory bulb (Bang and Cobb 1968). For example, the mean relative olfactory bulb size of 29% for species within the *Procellariiformes*, a bird taxon known for good olfactory

B.A. Caspers (✉) • E.T. Krause
Department of Animal Behaviour, University of Bielefeld,
PO Box 100131, 33501 Bielefeld, Germany
e-mail: barbara.caspers@uni-bielefeld.de; tobias.krause@uni-bielefeld.de

M.L. East and M. Dehnhard (eds.), *Chemical Signals in Vertebrates 12*,
DOI 10.1007/978-1-4614-5927-9_27, © Springer Science+Business Media New York 2013

Fig. 27.1 Visual plumage traits may represent a basis for species (**a** vs. **b**), sex (**a**—male *left*, female *right*) and probably also individual recognition (individual differences between individual in **a**, **b**), as examples from a dimorphic species (**a**) the zebra finch (*Taeniopygia guttata*) and (**b**) a monomorphic species the diamond firetail (*Stagonopleura guttata*)

capacities (e.g. Nevitt et al. 1995; Bonadonna and Nevitt 2004; Nevitt and Bonadonna 2005; Bonadonna et al. 2007; Mardon et al. 2010), is about three times bigger than the average relative olfactory bulb size of 9.7% in the *Passeriformes* (Bang and Cobb 1968). Second, as visual and acoustic signals are known to play such significant roles in bird behaviours such as in the recognition of species (e.g. Emlen 1972; Price 1998; Uy et al. 2009; Fig. 27.1), sex (e.g. Saetre and Slagsvold 1992; Fig. 27.1) and individual identity (Brooks and Falls 1975; Levrero et al. 2009; Jacot et al. 2010), it could be questioned which additional information might be communicated by olfactory signals because passerines, such as other birds, do not show any clear olfactory-driven behaviour.

In this chapter we review evidence that olfactory signals are nevertheless likely to play an additional and important role in social communication of songbirds (e.g. Roper 1999; Zelano and Edwards 2002; Hagelin and Jones 2007) and discuss experimental evidence of olfaction in the zebra finch (*Taeniopygia guttata*).

27.2 Olfaction in Passerines

Within the last decade, evidence that songbirds indeed have a sense of smell has grown. Birds are thought to make use of their sense of smell for predator avoidance (e.g. Amo et al. 2008, 2011), to locate food (e.g. Kelly and Marples 2004; Mennerat et al. 2005), during nest construction (e.g. Mennerat 2008; Gwinner and Berger 2008) and for orientation (Holland et al. 2009; Caspers and Krause 2011). Furthermore, although the olfactory bulb is relatively small, songbirds have a large number of olfactory receptor genes (Steiger et al. 2008, 2009; Warren et al. 2010), indicating that the sense of smell might be more sophisticated and important than traditionally assumed.

Chemical analysis of preen oil secretion of a north American songbird species, the dark-eyed junco (*Junco Hyemalis*), revealed that these birds indeed possess body odours that might encode information about sex, species, individual identity and population affiliation (Whittaker et al. 2010), and perhaps also about condition (Whittaker et al. 2011). Thus the chemical structure of these signals could potentially be used in mate recognition. Nevertheless studies showing that olfactory cues are used for social communication in songbirds are still rare and absent in many passerine species. In contrast to this lack of knowledge in passerines, experimental evidence exists that some other bird species indeed make use of the available body odour differences for sex (e.g. Mardon et al. 2010; Zhang et al. 2010; but see Mardon et al. 2011) and even mate recognition/choice (e.g. Bonadonna and Nevitt 2004; Leclaire et al. 2012). Only recently it has been detected in blue petrels (*Halobaena caerulea*) that closely related individuals have a more similar odour profile than unrelated individuals (Celerier et al. 2011). Thus olfactory cues might, as it has been shown for mammals, fish, amphibians and insects (e.g. Hepper 1991), provide information about kinship. Humboldt penguins (*Spheniscus humboldti*) have been shown to distinguish between kin and non-kin based on olfactory cues alone (Coffin et al. 2011). Is it possible that songbirds might also use olfactory cues for kin recognition?

27.2.1 Mechanisms of Kin Recognition in Songbirds

Inbreeding, i.e. mating with a closely related conspecific, leads to a reduction in genetic variation. This reduction of genetic variation usually leads to a decrease in heterozygosity and consequently to the possibility that inbred offspring suffer from deleterious and/or recessive traits. To avoid inbreeding depression, individuals should therefore preferably mate with non-related individuals. Furthermore, recognition of kin might not only prevent inbreeding, but also is thought to have promoted the evolution of cooperation (Hamilton 1964). By helping closely related individuals to raise their offspring, individuals increase the own inclusive fitness. However, in both cases individuals have to be able to recognise kin. Indeed, there is certain evidence that birds are able to recognise kin. Female zebra finches, for example, adjust their reproductive effort in response to the genetic similarity of their unfamiliar mates (Arct et al. 2010). Peacock (*Pavo cristatus*) males form leks with close-related males that they have not been associated with before (Petrie et al. 1999). These findings lead to the question which underlying mechanism allows individuals to distinguish between kin and non-kin?

Fig. 27.2 Flow pattern of potential errors in kin recognition

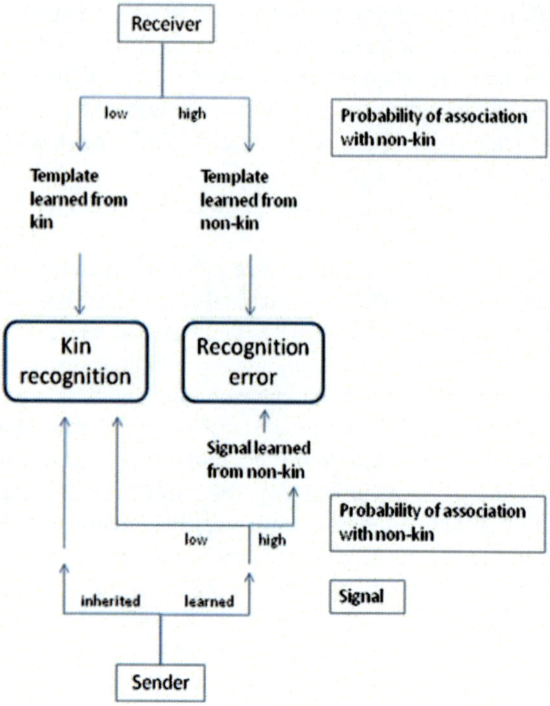

One mechanism of kin recognition is associative learning of acoustic and/or visual cues. Long-tailed tit nestlings learn the call of their family members and are able to recognise kin based on these calls (Sharp et al. 2005). Since the recognition signal is not or only partially genetically based, the possibility of mismatch is higher because there is the possibility that not only the receiver but also the sender learned the signal from non-kin (Fig. 27.2). Male songs, for example, are learned from the father or another tutor and it has been shown that male songs are very similar to the tutor, no matter if it is the father or an unrelated male (Eales 1985; Slater et al. 1988).

Zebra finches are probably one of the best studied songbird species in terms of imprinting and early learning (e.g. Immelmann 1959, 1961, 1968, 1970; Immelmann et al. 1978), and in this species, associative learning of visual and acoustic cues does not explain how individuals recognise kin. Female zebra finches are sexually imprinted on acoustic and visual cues of their fathers during the nestling period or shortly after. In mate choice tests, females showed a strong preference for males that looked (Oetting and Bischof 1996; Witte and Caspers 2006) and sang (Clayton 1988; Riebel 2000) like their fathers do (Fig. 27.3). This suggests that females are likely to preferentially mate with unfamiliar brothers from an earlier brood or cousins (i.e. males which learned their song from the same grandparental pathway). It is likely that female zebra finches do meet unfamiliar relatives from earlier broods

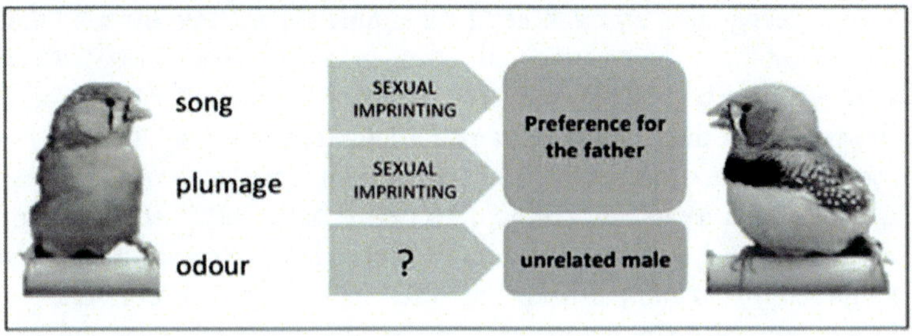

Fig. 27.3 Scheme of the use of visual, acoustic and olfactory cues in mate choice

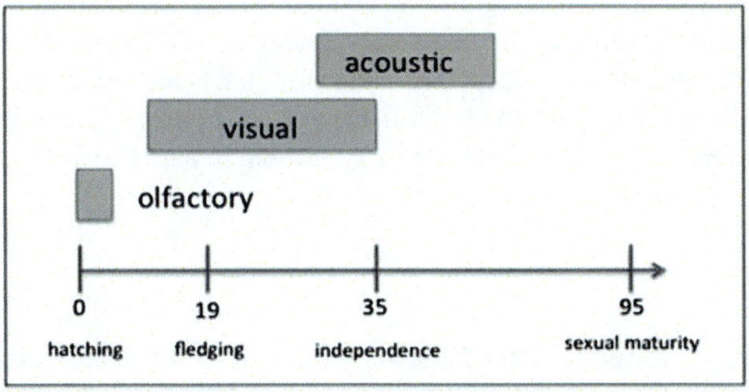

Fig. 27.4 The time frame of the sensitive phases, in which templates are learned, for olfactory, visual and acoustic cues. Olfactory cues are probably learned within the first hours after hatching visual cues between day 10 and 35 (Oetting and Bischof 1996; Witte and Caspers 2006) and acoustic cues between day 30 and 60 (Clayton 1988; Riebel 2000)

in their natal colony as there is no sex-biased dispersal in this species, and despite high annual mortality, about 25% of the breeding birds are natal to the colony (Zann 1996).

The use of visual and acoustic cues for kin recognition also appears error-prone, because contact with non-kin is possible during the sensitive period when sexual imprinting occurs (Fig. 27.4). The sensitive phase, during which visual cues are learned, is between day 10 and 35 (Oetting and Bischof 1996; Fig. 27.4) whereas song learning continues after fledging between day 30 and 60 post-hatching (Clayton 1988; Riebel 2000; Brainard and Doupe 2002; Fig. 27.4). As visual and especially acoustic cues are learned after juveniles have left the nest, this increases the possibility that they learn these cues from non-kin (Figs. 27.3 and 27.4). So if association with the father is necessary to find a potential mate that is similar to the father but not too similar (Bateson 1982) using solely visual and acoustic cues might lead to mistakes in kin recognition. Additionally, associative learning of paternal cues might lead to recognition errors in nests containing chicks sired by more than one male. The ratio of extra pair young varies tremendously in songbirds, with species such as the zebra finch having a relatively low rate of about 2% (Griffith et al. 2010), to species with more than 70% of extra pair young (Griffith et al. 2002).

A recent study found that zebra finches adjust their reproductive effort in response to the genetic similarity to their partner (Arct et al. 2010) such that pairs composed of unfamiliar kin had a lower probability to reproduce successfully compared to pairs composed of random partners. These results suggest that zebra finches have the ability to recognise kin, but the exact underlying mechanism was unknown. In many other vertebrate species (e.g. Waldmann 1991; Penn 2002; Levy et al. 2004; Holmes and Mateo 2007) and even in some non-passerine bird species (e.g. Bonadonna and Nevitt 2004; Bonadonna et al. 2007; Mardon et al. 2010), olfactory cues have been found to be crucially involved in social communication, including kin recognition. Recently zebra finches were found to use olfactory cues for orientation and discrimination between natal and non-natal nest odours (Caspers and Krause 2011). It seems possible that olfactory cues might represent an additional information channel to zebra finches as for example in parent–offspring communication, or kin recognition.

27.3 First Evidence for Olfactory Social Communication

Parent offspring communication is the earliest social communication in the life of a zebra finch. In altricial birds, chicks are reared in a nest, making it difficult to disentangle cues used for nest and/or offspring recognition. To identify the own nest, parents rely on visual cues (e.g. Peek et al. 1972; Trillmich 1976; Hughes et al. 1995), or on acoustic cues, like the begging calls of the own chicks (e.g. Levrero et al. 2009). In order to test whether olfactory cues might also bear an informative value in nest and/or offspring recognition, we performed odour choice tests with breeding zebra finches. In these tests we gave males and females the choice between (1) unused coconut fibres and nest material from their own nest, (2) unused coconut fibre and nest material from a conspecific nest containing nestlings of the same age as their brood, and (3) nest material of their own nest and nest material of a conspecific nest, simultaneously (Krause and Caspers 2012). Males as well as females were tested once during the nestling phase of their offspring and once after the offspring had fledged. Males did not show any preference in all choice situations whereas females preferred the odour of their own nest during the nestling period of their offspring, when given the choice between their own nest and unused coconut fibre. Females also avoided the nest odour of a conspecific nest, when given the choice between unused coconut fibre and nest material of a conspecific nest. But females did not show any preference when given both nest odours simultaneously. After the offspring have fledged, females did not show any preferences. We assume that the behavioural differences are based on motivational differences (Krause and

Caspers 2012). These results revealed that adult zebra finch females gather additional information about their social environment by olfactory cues.

27.3.1 *Olfactory Kin Recognition*

In another experiment (Krause et al. 2012) we investigated whether zebra finch fledglings were able to recognise the nest from closely related individuals by odour alone. Therefore we fostered juveniles shortly after hatchling (either day 1 or 4, mean 2 days ± 1 day SD) into a brood of unrelated juveniles of the same age. After the juveniles fledged, we tested the foster chick and one unrelated nest mate for their preferences towards the odour of the nest in which both chicks were raised and the odour of the nest the foster chick came from.

Interestingly the preference differed significantly between fostered chicks and non-foster chicks, with foster chicks showing a significant preference for the odour of the nest from which they originated. There was no difference in the preference between chicks fostered on their first day of life and chicks fostered on their fourth day of life. In contrast chicks reared in their natal nest showed a preference for their own nest odour in relation to the ratio of full sibs to the unrelated foster chick (i.e. the number of full sibs) in the nest. The more full sibs the non-foster chick had in the nest the higher the preference for the nest was. These results indicate that zebra finches are able to use odour to distinguish kin from non-kin (Krause et al. 2012). However we still do not know whether hatchlings learn the familial odour during their first hours of life or before hatching.

27.3.2 *Potential Benefits of Olfactory Social Communication*

In contrast to visual and acoustic cues, olfactory cues can be learned post-hatching, within the first hours of life (e.g. Cunningham and Nevitt 2011; Caspers and Krause 2011; Krause et al. 2012), as studies on domestic chickens have shown (e.g. Sneddon et al. 1998; Bertin et al. 2010). In altricial birds, such as zebra finches, juveniles are exclusively associated with their parents and siblings during the first days of life, thus the potential of kin recognition errors is minimised (Fig. 27.2).

27.4 Conclusion and Future Perspectives

Within the last few years our understanding of olfactory communication in songbirds has increased and it seems that birds in general and songbirds in particular use olfactory cues not only for orientation and foraging, but also for social communication.

Reconsidering the potential importance of olfactory communication in zebra finches and maybe in songbirds in general is an important issue. The next steps in the research on the sense of scent in songbirds should be conducted to find out in which social (and of course also non-social) context these signals are involved. Further it is important to increase our knowledge about the components and the composition of the signal itself. This will very likely lead to many new insights and probably will help to understand former findings better. We believe that understanding social communication of animals should involve all sensory modes and should be interactively studied to get a comprehensive understanding of the ecology, evolution and life-history of the animals.

Olfactory communication in birds, with respect to the undisputable importance of visual and acoustic communication, might provide important insight and a new understanding of certain avian behaviours.

Acknowledgements We are grateful to Marion East and an anonymous referee for helpful comments on an earlier draft. This study was partly funded by a grant from the University of Bielefeld (Young researchers fund to BC).

References

Amo L, Galvan I, Tomas G, Sanz JJ (2008) Predator odour recognition and avoidance in a songbird. Funct Ecol 22:289–293

Amo L, Visser ME, van Oers K (2011) Smelling out predators is innate in birds. Ardea 99:177–184

Arct A, Rutkowska J, Martyka R, Drobniak SM, Cichon M (2010) Kin recognition and adjustment of reproductive effort in zebra finches. Biol Lett 6:762–764

Bang BG, Cobb S (1968) Size of olfactory bulb in 108 species of birds. Auk 85:55–61

Bateson P (1982) Preferences for cousins in Japanese quail. Nature 295:236–237

Bertin A, Calandreau L, Arnould C, Nowak R, Levy F, Noirot V, Bouvarel I, Leterrier C (2010) In ovo olfactory experience influences post-hatch feeding behaviour in young chickens. Ethology 116:1027–1037

Bonadonna F, Nevitt GA (2004) Partner-specific odor recognition in an antarctic seabird. Science 306:835

Bonadonna F, Miguel E, Grosbois V, Jouventin P, Bessiere JM (2007) Individual odor recognition in birds: an endogenous olfactory signature on petrels' feathers? J Chem Ecol 33:1819–1829

Borgia G (1985) Bower quality, number of decorations and mating success of male bowerbirds (*Ptilonorhynchus-violaceus*)—an experimental analysis. Anim Behav 33:266–271

Borgia G (1995) Complex male display and female choice in the spotted bowerbird—specialized functions for different bower decorations. Anim Behav 49:1291–1301

Brainard MS, Doupe AJ (2002) What songbirds teach us about learning. Nature 417:351–358

Brooks RJ, Falls JB (1975) Individual recognition by song in white-throated sparrows. I. Discrimination of songs of neighbors and strangers. Can J Zool 53:879–888

Brouwer L, Komdeur J (2004) Green nesting material has a function in mate attraction in the European starling. Anim Behav 67:539–548

Caspers BA, Krause ET (2011) Odour-based natal nest recognition in the zebra finch (*Taeniopygia guttata*), a colony-breeding songbird. Biol Lett 7:184–186

Catchpole CK, Slater PJB (2008) Bird song—biological themes and variations, 2nd edn. Cambridge University Press, Cambridge

Celerier A, Bon C, Malapert A, Palmas P, Bonadonna F (2011) Chemical kin label in seabirds. Biol Lett 7:807–810
Clayton NS (1988) Song tutor choice in zebra finches and Bengalese finches—the relative importance of visual and vocal cues. Behaviour 104:281–299
Coffin HR, Watters JV, Mateo JM (2011) Odor-based recognition of familiar and related conspecifics: a first test conducted on captive Humboldt penguins (*Spheniscus humboldti*). PLoS One 6:e25002
Collias NE (1964) The evolution of nests and nest-building in birds. Am Zool 4:175–190
Cunningham GB, Nevitt GA (2011) Evidence for olfactory learning in procellariiorm seabird chicks. J Avian Biol 4:85–88
Eales LA (1985) Song learning in zebra finches: some effects of song model availability on what is learnt and when. Anim Behav 33:1293–1300
Emlen ST (1972) An experimental analysis of the parameters of birds song eliciting species recognition. Behaviour 41:130–171
Griffith SC, Owens IPF, Thuman KA (2002) Extra pair paternity in birds: a review of interspecific variation and adaptive function. Mol Ecol 11:2195–2212
Griffith SC, Holleley CE, Mariette MM, Pryke SR, Svedin N (2010) Low level of extrapair parentage in wild zebra finches. Anim Behav 79:261–264
Gwinner H, Berger S (2008) Starling males select green nest material by olfaction using experience-independent and experience-dependent cues. Anim Behav 75:971–976
Hagelin JC, Jones IL (2007) Bird odors and other chemical substances: a defense mechanism or overlooked mode of intraspecific communication? Auk 124:741–761
Hamilton WD (1964) Genetical evolution of social behavior I. J Theor Biol 7:1–16
Hepper PG (1991) Kin recognition. Cambridge University Press, Cambridge
Hill GE, McGraw KJ (2006) Bird coloration—function and evolution. Harvard University Press, Cambridge
Hill GE, Montgomerie R (1994) Plumage color signals nutritional condition in the house finch. Proc R Soc Lond B 258:47–52
Holland RA, Thorup K, Gagliardo A, Bisson IA, Knecht E, Mizrahi D, Wikelski M (2009) Testing the role of sensory systems in the migratory heading of a songbird. J Exp Biol 212:4065–4071
Holmes WG, Mateo JM (2007) Kin recognition in rodents: issues and evidence. In: Wolff JO, Sherman PW (eds) Rodent societies—an ecological & evolutionary perspective. University of Chicago Press, Chicago
Hughes BO, Petherick JC, Brown MF, Waddington D (1995) Visual recognition of key nest site stimuli by laying hens in cages. Appl Anim Behav Sci 42:271–281
Immelmann K (1959) Experimentelle Untersuchungen über die biologische Bedeutung artspezifischer Merkmale beim Zebrafinken (*Taeniopygia castanotis* Gould). Zool Jahrb Syst 86:437–592
Immelmann K (1961) Biologische Bedeutung optischer und akustischer Merkmale bei Prachtfinken (Aves: Spermestidae). Verh Dtsch Zool Ges 1961:369–374
Immelmann K (1968) Zur biologischen Bedeutung des Estrildidengesanges. J Ornithol 109:284–299
Immelmann K (1970) Zur ökologischen Bedeutung prägunsbedingter Isolationsmechanismen. Verh Dtsch Zool Ges 1970:304–314
Immelmann K, Kalberlah HH, Rausch R, Stahnke A (1978) Sexuelle Prägung als möglicher Faktor innerartlicher Isolation beim Zebrafinken. J Ornithol 119:197–212
Jacot A, Reers H, Forstmeier W (2010) Individual recognition and potential recognition errors in parent-offspring communication. Behav Ecol Sociobiol 64:1515–1525
Kelly DJ, Marples NM (2004) The effects of novel odour and colour cues on food acceptance by the zebra finch, *Taeniopygia guttata*. Anim Behav 68:1049–1054
Krause ET, Caspers BA (2012) Are olfactory cues involved in nest recognition in two social species of estrildid finches? PLoS One 7:e36615
Krause ET, Krüger O, Kohlmeier P, Caspers BA (2012) Olfactory kin recognition in a song bird. Biol Lett 8:327–329

Leclaire S, Merkling T, Raynaud C, Mulard H, Bessiere JM, Lhuillier E, Hatch SA, Danchin E (2012) Semiochemical compounds of preen secretion reflect genetic make-up in a seabird species. Proc R Soc Lond B 279:1185–1193

Levrero F, Durand L, Vignal C, Blanc A, Mathevon N (2009) Begging calls support offspring individual identity and recognition by zebra finch parents. C R Biol 332:579–589

Levy F, Keller M, Poindron P (2004) Olfactory regulation of maternal behavior in mammals. Horm Behav 46:284–302

Mardon J, Saunders SM, Anderson MJ, Couchoux C, Bonadonna F (2010) Species, gender, and identity: cracking petrels' sociochemical code. Chem Senses 35:309–321

Mardon J, Saunders SM, Bonadonna F (2011) Comments on recent work by Zhang and colleagues: "Uropygial gland-secreted alkanols contribute to olfactory sex signals in budgerigars". Chem Senses 36:3–4

Mennerat A (2008) Blue tits (*Cyanistes caeruleus*) respond to an experimental change in the aromatic plant odour composition of their nest. Behav Process 79:189–191

Mennerat A, Bonadonna F, Perret P, Lambrechts MM (2005) Olfactory conditioning experiments in a food-searching passerine bird in semi-natural conditions. Behav Process 70:264–270

Naguib M, Mennill DJ (2010) The signal value of bird song: empirical evidence suggests song overlapping is a signal. Anim Behav 80:e11–e15

Nevitt GA, Bonadonna F (2005) Sensitivity to dimethyl sulphide suggests a mechanism for olfactory navigation by seabirds. Biol Lett 1:303–305

Nevitt GA, Veit RR, Kareiva P (1995) Dimethyl sulfide as a foraging cue for Antarctic Procellariiform seabirds. Nature 376:680–682

Oetting S, Bischof HJ (1996) Sexual imprinting in female zebra finches: changes in preferences as an effect of adult experience. Behaviour 133:387–397

Ohlsson T, Smith H, Raberg L, Hasselquist D (2002) Pheasant sexual ornaments reflect nutritional conditions during early growth. Proc R Soc Lond B 269:21–27

Peek FW, Franks E, Case D (1972) Recognition of nest, eggs, nest site, and young in female red-winged blackbirds. Wilson Bull 84:243–249

Penn DJ (2002) The scent of genetic compatibility: sexual selection and the major histocompatibility complex. Ethology 108:1–21

Petrie M, Krupa A, Burke T (1999) Peacocks lek with relatives even in the absence of social and environmental cues. Nature 401:155–157

Price T (1998) Sexual selection and natural selection in bird speciation. Phil Trans R Soc Lond B 353:251–260

Riebel K (2000) Early exposure leads to repeatable preferences for male song in female zebra finches. Proc R Soc Lond B 267:2553–2558

Roper TJ (1999) Olfaction in birds. Adv Stud Behav 28:247–332

Saetre GP, Slagsvold T (1992) Evidence for sex recognition from plumage colour by the pied flycatcher, *Ficedula hypoleuca*. Anim Behav 44:293–299

Schmidt R, Kunc HP, Amrhein V, Naguib M (2008) Aggressive responses to broadband trills are related to subsequent mating success in nightingales. Behav Ecol 19:635–641

Sharp S, McGowan A, Wood MJ, Hatchwell BJ (2005) Learned kin recognition cues in a social bird. Nature 434:1127–1130

Slater PJB, Eales LA, Clayton NS (1988) Song learning in zebra finches: progress and prospects. Adv Stud Behav 18:1–34

Sneddon H, Hadden R, Hepper PG (1998) Chemosensory learning in the chicken embryo. Physiol Behav 64:133–139

Sprau P, Roth T, Schmidt R, Amrhein V, Naguib M (2010) Communication across territory boundaries: distance dependent responses in nightingales. Behav Ecol 21:1011–1017

Steiger SS, Fidler AE, Valcu M, Kempenaers B (2008) Avian olfactory receptor gene repertoires: evidence for a well-developed sense of smell in birds? Proc R Soc Lond B 275:2309–2317

Steiger SS, Kuryshev VY, Stensmyr MC, Kempenaers B, Mueller JC (2009) A comparison of reptilian and avian olfactory receptor gene repertoires: species-specific expansion of group gamma genes in birds. BMC Genomics 10:446

Trillmich F (1976) Recognition of individual nesting box in budgerigars, *Melopsittacus-Undulatus* Shaw (Aves, Psittacidae). Z Tierpsychol 42:1–11

Uy JAC, Moyle RG, Filardi CE, Cheviron ZA (2009) Difference in plumage color used in species recognition between incipient species is linked to a single amino acid substitute in the melanocortin-1 receptor. Am Nat 174:244–254

Waldmann B (1991) Kin recognition in amphibians. In: Hepper PG (ed) Kin recognition. Cambridge University Press, Cambridge

Warren WC, Clayton DF, Ellegren H, Arnold AP, Hillier LW, Künstner A et al (2010) The genome of a songbird. Nature 464:757–762

Whittaker DJ, Soini HA, Atwell JW, Hollars C, Novotny MV, Ketterson ED (2010) Songbird chemosignals: volatile compounds in preen gland secretions vary among individuals, sexes, and populations. Behav Ecol 21:608–614

Whittaker DJ, Richmond KM, Miller AK, Kiley R, Burns CB, Atwell JW, Ketterson ED (2011) Intraspecific preen oil odor preferences in dark-eyed juncos (*Junco hyemalis*). Behav Ecol 22:1256–1263

Witte K, Caspers B (2006) Sexual imprinting on a novel blue ornament in zebra finches. Behaviour 143:969–991

Zann RA (1996) The Zebra Finch: a synthesis of field and laboratory studies. UK, Oxford University Press, Oxford

Zelano B, Edwards SV (2002) An MHC component to kin recognition and mate choice in birds: predictions, progress, and prospects. Am Nat 160:S225–S237

Zhang JX, Wei W, Zhang JH, Yang WH (2010) Uropygial gland-secreted alkanols contribute to olfactory sex signals in budgerigars. Chem Senses 35:375–382

Chapter 28
Male European Starlings Use Odorous Herbs as Nest Material to Attract Females and Benefit Nestlings

Helga Gwinner

Abstract Male European starlings (*Sturnus vulgaris*) use odorous herbs as green nest material. They show these plants to females to catch their attention before they incorporate them into their nests. Nestlings reared in nests containing herbs carry fewer bacteria, have higher haematocrit levels and more basophile leucocytes, and have a higher fledging weight than those reared in nests without herbs. The incubation temperature is higher in nests with herbs than in herbless nests. Females are attracted by herb nests and herb nests provide energy-saving benefits to incubating females. In choice experiments, starlings employed olfaction to select nest herbs, using innate information and experience acquired as nestlings to identify odours. Measures of head space air in starling nest boxes revealed an increase of volatile substances during the nestlings phase. Manganese-enhanced resonance imaging of the olfactory bulb of starling showed that the neural correlates of olfactory sensitivity change seasonally. Starlings can only discriminate odours during the reproductive season. The association between odour perception and reproductive behaviour suggests that increased testosterone production, related to the increasing photoperiod in spring, may guide these seasonal changes. The size of the olfactory bulb but not its ability to discriminate was affected by testosterone implants in the non-reproductive season.

28.1 Introduction

The pioneering work of Bang, Cobb, and Wenzel (Bang and Cobb 1968; Wenzel 1971, 1973) on anatomical, neuro-anatomical, and physiological aspects of avian olfaction was followed by many studies on chemical signals used by birds in orientation, attachment to familiar environments, food finding, avoidance of

H. Gwinner (✉)
Max-Planck-Institute for Ornithology, Eberhard-Gwinner-Straße, 82319 Seewiesen, Germany
e-mail: hgwinner@orn.mpg.de

M.L. East and M. Dehnhard (eds.), *Chemical Signals in Vertebrates 12*,
DOI 10.1007/978-1-4614-5927-9_28,

predators and noxious substances, mating, and nest building (reviews by Roper 1999; Hagelin 2007; Balthazart and Taziaux 2009).

The incorporation of fresh plants in nests was observed in various bird species, including the woodchat shrike (*Lanius senator*) (Naumann 1838), purple martin (*Progne subis*) (Johnstone and Hardy 1962), Indian house sparrow (*Passer domesticus*) (Sengupta 1981), South American monk parakeet (*Myiopsitta monachus*) (Bucher 1988), American wood stork (*Mycteria americana*) (Rodgers et al. 1988), blue tit (*Parus caeruleus*) (Petit et al. 2002), Cape sparrow (*Passer melanurus*) (Milton and Dean 1998), buzzards (*Buteo buteo*) (Roulin et al. 1997), and European starlings (*Sturnus vulgaris*) (Wimberger 1984; Clark and Mason 1985).

During the evolutionary arms race between plants, their parasites and the herbivores that consume them, some plants have evolved chemical weapons against attack. Birds that incorporate these plants into their nests may have an adaptive advantage because the chemical compounds the plants contain could control the parasite load in bird nests (the "nest protection hypothesis": Wimberger 1984; Clark and Mason 1985, 1988). Experimental tests of this hypothesis are rare and contradictory. Starling in North America prefer the foliage of wild carrots as green nest material and when these are added experimentally to nests they inhibit the development of parasitic mites, and in a laboratory test bacterial growth (Clark and Mason 1985, 1988). Three other studies on starlings, failed to demonstrate an effect of nest herbs on the red fowl mite (*Dermanyssus gallinae*) and the northern fowl mite (*Ornithonyssus silviarum*) the main parasites in starling nests (Fauth et al. 1991; Gwinner et al. 2000; Brower and Komdeur 2004).

To investigate the effect of herbs on various aspects of starling life history, we exchanged natural starling nests for experimental nests which contained herbs (=herb nests) or not (=herbless nests) over several years in a starling nest box colony in Southern Germany (48°N, 11°E).

28.2 Green Nest Material in Starling Nests: Composition and Selection of Herbs and Their Effect on Females and Nestlings

28.2.1 Herbs and Bacteria

Bacteria on starling nestlings were quantified by applying agar pads to their belly skin. Pads were incubated before bacteria colonies were counted as an estimation of a nest's bacterial load. For details, see Gwinner and Berger (2005). The number of bacteria colonies increased during the season with the largest increase occurring in herbless nests (Fig. 28.1). In May, similar numbers of bacteria colonies occurred in herb and herbless nests (*t*-test: $t=-0.16$, $n=34$, $P=0.9$). In June, the number of bacteria colonies was significantly reduced in herb nests ($t=-2.4$, $P=0.02$, Fig. 28.1; Gwinner and Berger 2005). Similar plant effects on bacteria load were observed in nestling blue tits (Mennerat et al. 2009a).

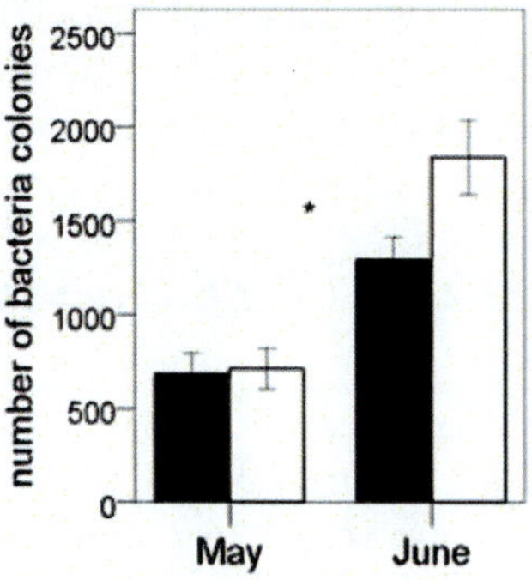

Fig. 28.1 Bacteria load (numbers of bacteria colonies) in experimental May and June nests with herbs ($n = 25$, *black bars*) and without herbs ($n = 22$, *white bars*)

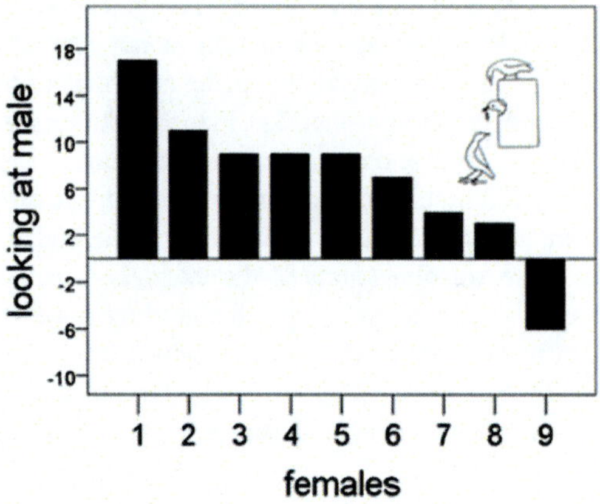

Fig. 28.2 The number of looks nine females gave to a mounted male carrying a herb leaf in his bill minus the looks she gave to a male without a leaf (Gwinner, unpublished data)

28.2.2 *Herbs and Female Attraction*

Other studies investigated the effect of green plants on mate attraction (the "courtship hypothesis": Pinxten and Eens 1990; Fauth et al. 1991; Gwinner 1997). In starlings, males are the main nest builders. They use small twigs and dry grass as nest material and also fresh plants, which they show to prospecting females in an eye catching manner. Most herbal nest material is presented and deposited in the nest at pair formation. Males cease collecting herbal nesting material at the onset of laying (Gwinner 1997; Brower and Komdeur 2004). Male spotless starlings (*Sturnus unicolor*) also collect green nest material and display it to females (Veiga et al. 2006). In an aviary experiment I found that female starlings paid more attention to a mounted starling male carrying a green herb leaf in his bill than to a male without (Fig. 28.2). The presentation of plants is courtship behaviour, but whether plant odours attract females independently from a male's display was not investigated. Experimental manipulations have shown that female starlings prefer nests to which greenery was added compared to nests where it was removed (Brower and Komdeur 2004) and tended to spend more time in nest boxes with aromatic herbs than in boxes without (Lambert 1997). In a dark nest hole where herbs are interwoven into the nest, it is possible that olfactory cues from herbs influence a female choice or rejection of a nest and its constructor.

28.2.3 *Volatile Compounds in Nest Herbs*

A survey conducted on my starling population revealed that volatile plant species were selected more often than non-volatile plant species (*t*-test: $t = -5.0$, $P = 0.0001$). The five volatile plant species incorporated most often were: milfoil (*Achilea millefolia*), bugle (*Ajuga reptans*), hogweed (*Heracleum spondylum*), elder (*Sambucus niger*), and goutweed (*Aegopodium podagraria*).

As males stop the incorporation of plants with laying and the herb fragments loose their volatile substances with time (about a factor 10 in 9 days in an uninhabited nest), to what extent are breeding females and their nestlings exposed to volatile compounds? In cooperation with the Max-Planck Institute for Chemical Ecology, Jena, the head space air in nest boxes was measured during incubation and in the same boxes with 9-day-old nestlings. The five most prominent substances determined in starling nests were: limonene, pinene, sabinene, myrcene, and transcarene. Unexpectedly, the amount of these volatiles (sums of the integrals calculated from chromatograms) increased from the incubation stage to the nestling stage; probably due to increasing humidity and movements of the nestlings which break the plants open (paired samples *t*-test: $n = 27$, $t = 4.24$, df $= 26$, $P = 0.0001$). The amount of volatiles was higher in herb than in herbless nests during both stages (*t*-test: incubation: $t = 2.2$, df $= 44$, $P = 0.035$, nestling stage: $t = 4.2$, df $= 49$, $P = 0.0001$). Nestlings inhale these compounds and may absorb them through their skin. Thus volatiles could directly affect nestling condition as proposed in the "drug hypothesis" (Gwinner et al. 2000).

28.2.4 *Herbs and Fledging Mass*

In herb nests, body mass of chicks at fledging was heavier (*t*-test: $n = 68/65$, $t = 2.7$, $P = 0.008$; Fig. 28.3) and more likely to return to the colony the following year: from 23 yearlings seen in the colony, 15 (65 %) were from herb nests, and 8 (35 %) from herbless nests ($\chi^2 = 4$, $P = 0.05$). Four birds from herb nests bred as yearlings, an exception in starlings, but none from herbless nests (Gwinner et al. 2000).

28.2.5 *Herbs and Red Blood Cells*

Nestlings in herb nests had higher haematocrit levels (Gwinner et al. 2000) and higher red blood cell counts probably indicate good health and a higher aerobic capacity has been shown to increase post-fledging survival in blue tits (Mennerat et al. 2009b).

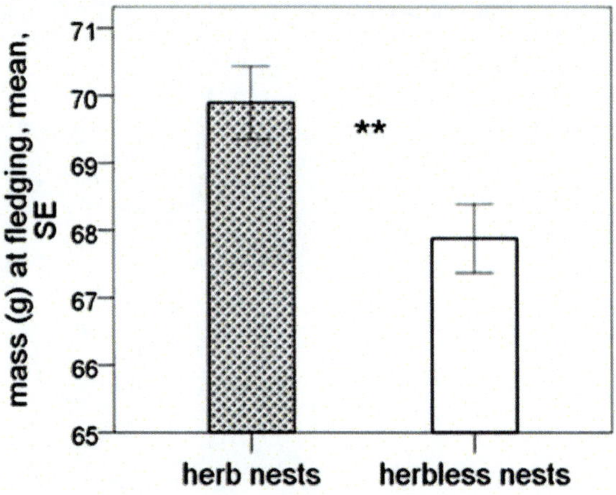

Fig. 28.3 Fledging mass of nestlings from herb nests ($n=68$, *patterned bars*) and herbless nests ($n=65$, *white bars*)

28.2.6 Herbs and Immunity

An element of the immune system, the basophil leucocytes, were increased in the blood of nestlings from herb nests compared with nestlings from grass nests (t=test: $n=13/12$, $t=3.07$, $P=0.006$). In birds, basophils increase the capacity to cope with stress induced by malnutrition or severe climatic conditions (Maxwell and Robertson 1995) and play a role in parasite defence (Roitt et al. 1996). In human medicine herbs such as Echinacea and Bryonia are thought to activate effector mechanisms of the defence system (Wagner and Wiesenauer 2003). Likewise, nest herbs could act as immuno-stimulants, thereby allowing nestlings to cope better or faster with stressful conditions or parasite infection ("the drug hypothesis" Gwinner et al. 2000). Support for the drug hypothesis comes from a recent study on blue tits showing an effect of herbs on nestling feather development, body size, and haematocrit, which was unrelated to parasite load (Mennerat et al. 2009b).

28.2.7 Herbs and Nest Temperature

I investigated whether green nest material affected nest temperature through insulation or composing processes and/or the female's incubation behaviour in experimental herb and herbless nests. Temperatures were higher in herb nests than herbless nests (mixed model: $n=33$, incubation day: $F=111.4$, $P=0.0001$, nest type: $F=29.8$, $P=0.0001$, unpublished data, Fig. 28.4). This difference was caused by differences in incubation behaviour in herb and herbless nests. In herb nests, females sat more frequently without moving on their eggs and thus lost less nest heat and therefore saved energy. Their incubation period was shorter and their offspring hatched

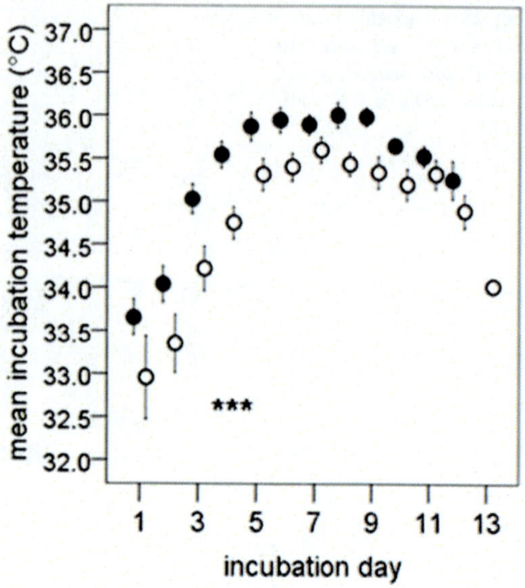

Fig. 28.4 The temperature in herb nests ($n=18$, *black symbols*) and herbless nests ($n=14$, *white symbols*) during the incubation period

earlier (data not shown). Nest temperature affects female and nestling fitness. Warmer nests allow the female to either incubate in shorter spells and give her the opportunity to invest more in self-maintenance or pass the saved energy into longer incubation spells, thus to her offspring, which in general grow faster or hatch larger and earlier in warmer nests (see Dawson et al. 2005; Hepp et al. 2006; Olson et al. 2006; Ardia et al. 2009).

28.2.8 The Choice of Nest Plants

Starlings have the capacity to discriminate odours (Clark and Mason 1987), but how do they acquire this ability? Do they learn from the odours they encounter in the nests in which they are reared or do they use an innate odour pattern for their choice? To answer these questions I exchanged natural starling nests with nests experimentally scented with milfoil oil (milfoil is the mostly preferred nest plant of starlings) or with vanilla (controls). Nestlings were removed shortly before fledging from their nest and were kept in aviaries overwinter. The following spring they were submitted to odour choice experiments in a test aviary with a nest box and a female. Males had to choose leaves from a plate scented with milfoil, vanilla, or unscented, which they used in courtship and/or as nest material. The choices for both behaviours were similar; here data for courtship behaviour are shown in Fig. 28.5. The choice behaviour did not differ in males which grew up in vanilla or in milfoil nests, when milfoil was available: both groups preferred milfoil-scented leaves (M) over vanilla (V) or not scented leaves (N) (One-sample test: M vs. V: $t32=3.5$, $P=0.001$; M vs. N, $t29=3.5$,

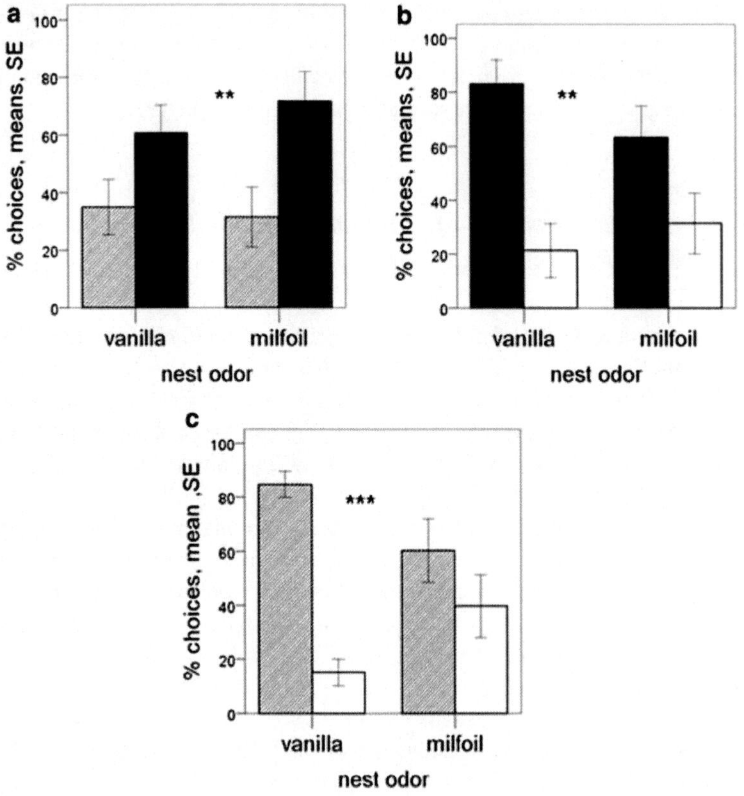

Fig. 28.5 **(a–c)** The effect of the nest odour a male starling was reared in on its choice of artificially scented plants it displayed to potential mates when adult. (**a**) choices of 18 vanilla- and 15 milfoil-experienced males for milfoil- or vanilla-scented leaves, (**b**) choices of 16 vanilla- and 14 milfoil-experienced males for milfoil-scented or unscented leaves, (**c**) choices of 14 vanilla- and 13 milfoil-experienced males for vanilla-scented or unscented leaves. Choices for milfoil-scented plants = *black*, for vanilla-scented plants = patterned, for unscented plants = *white*

$P = 0.002$). However, when they had to choose between vanilla and no scent, males reared in vanilla-scented nests preferred vanilla odour, but milfoil-experienced males did not discriminate between vanilla or unscented plants (One-sample *t*-test: vanilla males: $t12 = 7.1$, $P = 0.0001$, milfoil males: $t10 = 0.88$, $P = 0.4$).

Given that the test plants looked similar but smelled differently, presumably plant choice was based on chemical odours. Since vanilla-experienced birds preferred plants with milfoil odour, which they had not smelled as nestlings, their choice was probably innate. Odours encountered as nestlings may guide decisions, because vanilla-trained birds preferred plants with vanilla odour over plants with no scent, while the milfoil-trained birds did not discriminate between these options. Together, the results of this experiment suggest that male starlings can remember the odour of the nest in which they were reared, but do not necessarily base their selection of herbs for the nests they build as adults on the odour they encountered as nestlings. Male preference for particular odours may be based on the design of the olfactory system, shaped by evolution through the selection advantage of (fitter) offspring from fathers who provide their nests with aromatic plants. All in all, in the starling's olfactory system both experience and an "innate fore-knowledge" influence the behavioural outcome, as suggested by Marler for the auditory system (Marler 1999).

28.2.9 Neural Substrate of Olfactory Sensitivity: The Olfactory Bulb (OB)

Clark and Smeraski (1990) demonstrated in cardiac conditioned starlings that their ability to smell is seasonal. The observation that starling males present herbs especially during pair formation and that they use the sense of smell to identify volatile nest plants lead us to the hypothesis that the volume of the olfactory bulb (OB) may increase with the reproductive season similar to the volume changes of the song nuclei of songbirds.

In cooperation with Geert De Groof from the bio-imaging lab in Antwerp, we used repeated in vivo manganese-enhanced resonance imaging to quantify the seasonal changes in the anatomy and activity of the OB. In each season 14 starling males were measured twice, once with a milfoil scent as stimulus, once without any scent. The olfactory bulb was only able to discriminate milfoil odour from background odours during the reproductive period. There was no seasonal change in the size of the olfactory bulb. By experimentally elevating testosterone in half of the males during the non-breeding season, we showed that the OB volume was increased compared to controls ($P<0.05$). However, the ability of the OB to discriminate odours was not re-established through testosterone (for details see De Groof et al. 2010).

28.3 Conclusion

Volatile substances in the headspace air of starling nestlings have bactericidal, skin soothing, appetising, catalysing, and sleep-inducing properties (Steinegger and Hänsel 1988). Their impact could be the reason for the qualitative differences of nestlings reared in nests with herbs or herbless nests. The most important fitness-related effects of herbal nests are higher fledging weight and probably higher survival to adulthood. Females are attracted to males displaying plants during courtship. Therefore both natural and sexual selection may have designed the olfactory system to discriminate among plants. Whether females assess male or nest quality on the "perfume" they encounter has still to be discovered.

Acknowledgement I thank all the people who were involved in this study: In the field L. Trost, M. Oltrogge, S. Karaki, U. Nienaber, J. Schuller, W. Jensen, E. Koch, S. Gwinner, and S. Habersetzer helped. S. Berger contributed to the study on herbs and bacteria and to the odour choice experiments. F. Grimm and S. Schlemmer from the Veterinarian University, Munich, gave haematological advice and did the white blood cell counts. I. Baldwin from the MPI for Chemical

Ecology, Jena, provided his knowledge and the opportunity to investigate starling head space air, which was analysed by B. Krock. A.V.D. Linden from the Bio imaging lab, Antwerp, and S. Steiger cooperated in the olfactory bulb study, performed by G. De Groof. B. Helm aided us with statistical competence. Ebo Gwinner gave us numerous good advices and discussions and the opportunity to perform this study over years. After his death, 2004, B. Kempenaers helped us to continue with words and deeds. I am very grateful to all of them.

References

Ardia DR, Perez JH, Chad EK, Voss MA, Clotfelter ED (2009) Temperature and life history: experimental heating leads female tree swallows to modulate egg temperature and incubation behaviour. J Anim Ecol 78:4–13

Balthazart J, Taziaux M (2009) The underestimated role of olfaction in avian reproduction? Behav Brain Res 200:248–259

Bang BG, Cobb S (1968) The size of the olfactory bulb in 108 species of birds. Auk 85:55–61

Brower L, Komdeur J (2004) Green nesting material has a function in mate attraction in the European starling. Anim Behav 67:539–548

Bucher EH (1988) Do birds use biological control against nest parasites? Parasitol Today 4:1–3

Clark L, Mason JR (1985) Use of nest material as insecticidal and anti-pathogenic agents by the European starling. Oecologia 67:169–176

Clark L, Mason JR (1987) Olfactory discrimination of plant volatiles by the European starling. Anim Behav 35:227–235

Clark L, Mason JR (1988) Effect of biologically active plants used as nest material and the derived benefit to starling nestlings. Oecologia 77:174–180

Clark L, Smeraski CA (1990) Seasonal shifts in odour acuity by starlings. J Exp Zool 255:22–29

Dawson D, Lawrie C, Brien L (2005) The importance of microclimatic variation in determining size, growth and survival of avian offspring: experimental evidence from a cavity nesting passerine. Oecologia 144:499–507

De Groof G, Gwinner H, Steiger S, Kempenaers B, Van der Linden A (2010) Neural correlates of behavioural olfactory sensitivity changes seasonally in European starlings. PLoS One 5/12:e14337

Fauth PT, Krementz DG, Hines JE (1991) Ectoparasitism and the role of green nesting material in the European starling. Oecologia 88:22–29

Gwinner H (1997) The function of green plants in nests of European starlings (*Sturnus vulgaris*). Behaviour 134:337–351

Gwinner H, Berger S (2005) European starlings: nestling condition, parasites and green nest material during the breeding season. J Ornithol 146:365–371

Gwinner H, Oltrogge M, Trost L, Nienaber U (2000) Green plants in starling nests: effects on nestlings. Anim Behav 59:301–309

Hagelin J (2007) Odors and chemical signaling. In: Jamieson BG (ed) Reproductive behaviour and phylogeny of aves, vol 6B. Science Publishers, Enfield, NH, pp 76–119

Hepp GR, Kennamer RA, Johnson MH (2006) Maternal effects in Wood Ducks: incubation temperature influences incubation period and neonate phenotype. Funct Ecol 20:307–314

Johnstone RF, Hardy JW (1962) Behavior of the purple martin. Wilson Bull 74:243–262

Lambert K (1997) Eintrag von grünem Nistmaterial beim Star, *Sturnus vulgaris*. Diplomarbeit, Universität Würzburg

Marler P (1999) On innateness: are sparrow songs 'learned' or 'innate'. In: Hauser MD, Konishi M (eds) The design of animal communication. MIT Press, Cambridge, MA, pp 293–318

Maxwell MH, Robertson GW (1995) The avian basophilic leucocyte. A review. Worlds Poultry Sci J 51:307–325

Mennerat A, Mirleau P, Blondel J, Perret P, Lambrechts MM, Heeb P (2009a) Aromatic plants in nests of the blue tit *Cyanistes caeruleus* protect chicks from bacteria. Oecologia 161:849–855

Mennerat A, Perret P, Bourgault P, Blondel J, Gimenez O, Thomas DW, Heeb P, Lambrechts MM (2009b) Aromatic plants in nests of blue tits: positive effects on nestlings. Anim Behav 77:569–574

Milton SJ, Dean RJ (1998) The use of green aromatic plants in Karoo bird nests. Ostrich 69:329

Naumann JF (1838) Anhaltische Gartenzeitung

Olson CR, Vleck CM, Vleck D (2006) Periodic cooling of bird eggs reduces embryonic growth efficiency in chick Zebrafinches. Physiol Biochem Zool 79:927–936

Petit C, Hossaert-McKey M, Perret P, Blondel J, Lambrechts MM (2002) Blue tits use selected plants and olfaction to maintain an aromatic environment for nestlings. Ecol Lett 5:585–589

Pinxten R, Eens M (1990) Polygyny in the European starling: effect on female reproductive success. Anim Behav 40:1035–1047

Rodgers JA Jr, Wenner AS, Schwikert ST (1988) The use and function of green nestmaterial by wood storks. Wilson Bull 1003:411–423

Roitt IM, Brostoff J, Male DK (1996) Immunology. Mosby, London

Roper TJ (1999) Olfaction in birds. In: Slater PJ, Rosenblatt JS, Snowdon CT, Roper TJ (eds) Advances in the study of behaviour, vol 28. Academic, Boston, MA, pp 247–332

Roulin A, Jeanmonod J, Blanc T (1997) Green plant material on common buzzard's (*Buteo buteo*) nests during the rearing of chicks. Alauda 65:251–257

Sengupta S (1981) Adaptive significance of the use of Margosa leaves in nests of house sparrows *Passer domesticus*. Emu 81:114–115

Steinegger E, Hänsel R (1988) Lehrbuch der Pharmakognosie und Phytopharmazie. Springer, Berlin

Veiga JP, Polo V, Vinuela J (2006) Nest green plants as a male status signal and courtship display in the spotless starling. Ethology 112:196–204

Wagner H, Wiesenauer M (2003) Immunomodulatory agents from plants. Birkhäuser Verlag, Basel

Wenzel BM (1971) Olfaction in birds. In: Beidler LM (ed) Handbook of sensory physiology IV, 1. Olfaction. Springer, Berlin, pp 432–448

Wenzel BM (1973) Chemoreception. In: Farner DS, King JS (eds) Avian biology 3. Academic, New York, pp 389–415

Wimberger PH (1984) The use of green plant material in bird nests to avoid ectoparasites. Auk 101:615–618

Chapter 29
Embryonic Domestic Chickens Can Detect Compounds in an Avian Chemosignal Before Breathing Air

Julie C. Hagelin, Jacqueline C. Simonet, and Tyler R. Lyson

Abstract We know almost nothing about the chemosensory experiences of birds as they develop within an egg's fluid-filled environment. Given well-established literature on chemical detection of mammals in utero, we explored whether domestic chickens (*Gallus gallus domesticus*) exhibit a similar ability to detect or learn about chemical stimuli prior to breathing air. We incubated 18 eggs from embryonic day (E)9–18 in scented air containing Z-4-decenal and octanal, two key components of a citrusy-scented avian social odour (from Crested Auklets [*Aethia cristatella*]; Proc R Soc Lond 270:1323–1329, 2003). Control eggs were not exposed to scent. Behavioural responses of embryos were quantified by opening the shell and exposing embryos to three different test scents (Crested Auklet, wintergreen [novel scent], and water [unscented control]). Embryos in both the odour and control treatments reacted more to auklet odour than to water consistent with odour detection; the pattern was less clear for wintergreen. Furthermore, odour-treated embryos gave a lower net response (e.g. less kicking, body shifting) when tested with auklet odour, compared to controls. A reduced response is consistent with odour familiarity. We conclude that embryos modified their behaviour after experiencing air-borne compounds that were transmitted into the egg's fluid environment. Emerging data on

J.C. Hagelin (✉)
Department of Biology, Swarthmore College, Swarthmore, PA, USA

Institute of Arctic Biology, University of Alaska, Fairbanks, AK, USA
e-mail: julie.hagelin@alaska.edu

J.C. Simonet
Department of Biology, Swarthmore College, Swarthmore, PA, USA

Perelman School of Medicine, University of Pennsylvania, Philadelphia, PA, USA

T.R. Lyson
Department of Biology, Swarthmore College, Swarthmore, PA, USA

Department of Geology and Geophysics, Yale University, New Haven, CT, USA

M.L. East and M. Dehnhard (eds.), *Chemical Signals in Vertebrates 12*,
DOI 10.1007/978-1-4614-5927-9_29,

avian body odour, nest scent, and maternal-mediated dietary compounds promise to reveal new insights into the role of early chemosensory exposure during bird development.

29.1 Introduction

Embryos of many vertebrate groups are capable of detecting and learning chemical information prior to birth (e.g. humans: Schaal et al. 2000; Mennella et al. 2001; other mammals: Hepper 1988; Bilko et al. 1994) or hatching (e.g. amphibians: Mathis et al. 2008; birds: Porter and Picard 1998; Bertin et al. 2012). Early exposure to chemical stimuli can cause neuroanatomical change, which is currently thought to alter chemosensory perception in a way that adaptively shapes behavioural responses later in life (Todrank et al. 2011). Early chemosensory experiences have the potential to modify key behaviours, such as mate choice decisions, food preferences, or predator avoidance (Todrank et al. 2011).

29.1.1 A Case for Early Exposure of Avian Embryos to Chemical Stimuli

Birds present an interesting situation to study developmental consequences of early odour exposure (Tolhurst and Vince 1976), particularly given the opportunity for volatile odorants to diffuse through the shell (e.g. Rahn et al. 1979). A growing number of avian species are known to emit social chemosignals (e.g. Mardon et al. 2010, 2011; Whittaker et al. 2011; reviewed in Hagelin and Jones 2007; Caro and Balthazart 2010), while others adaptively incorporate scented nest materials into nests (Petit et al. 2002; Gwinner and Berger 2008). The possibility also exists for compounds within the egg itself to relate to maternal diet (Bertin et al. 2012). Hence, chemosensory experiences in ovo may be more germane to avian behavioural ecology than previously recognised (Hagelin 2007).

29.1.2 Two Developmental Opportunities for In Ovo Exposure

Domestic chickens (*Gallus gallus domesticus*), the avian model of development, exhibit two very different opportunities for embryonic exposure to chemical stimuli: (1) before and (2) after air-breathing occurs within the egg. The latter has received the most experimental focus. Tolhurst and Vince's (1976) pioneering work showed that embryos respond to odorants ca. 2 days before hatching, when young pierce the egg's air sac and begin breathing on embryonic day (E)19. Interestingly, some avian chemosensory cells may function prior to E19. Olfactory receptors, for example, are functional 6 days earlier (on E13; Lalloué et al. 2003). At this stage developing chicks are surrounded by amniotic fluid, in a situation that is broadly similar to mammals in utero. Like mammals, avian embryos swallow fluid frequently. Swallowing may facilitate movement of compounds through fluid-filled nasal passages, even though entry to the nares is blocked with tissue (Sneddon et al. 1998).

Since Tolhurst and Vince (1976), many studies have advanced our understanding of avian exposure to chemosensory stimuli in ovo. For example, the process: (1) occurs when compounds are transmitted through air or applied directly to the shell (Sneddon et al. 1998), (2) varies relative to the concentration of the odour stimulus (Bertin et al. 2010), (3) shows a more marked effect during latter stages of development (Gomez and Celli 2008; Bertin et al. 2012), (4) can involve naturally occurring scents, such as nest materials (Gwinner and Berger 2008) and food odours (Burne and Rogers 1999), (5) occurs in multiple species, ranging from domestic chickens (e.g. Porter and Picard 1998) to those with a notable plumage scent and developed olfactory anatomy (e.g. Cunningham and Nevitt 2011).

In all but one of the aforementioned studies (Bertin et al. 2012), embryos were exposed to chemical stimuli during *both* the fluid-filled and breathing stages of development. Consequently, we know almost nothing about the ability of birds to detect chemicals in the egg's aquatic environment. Nor do we have a detailed understanding of how this kind of early chemosensory exposure can adaptively modify avian behaviour.

29.1.3 Purpose and Goals of Investigation

The purpose of our study was to investigate chemosensory responses of avian embryos during only the fluid-filled stage of development (through E18). Our primary goals were to quantify whether chicken embryos could: (1) respond to and (2) modify behaviours based on volatile chemicals transmitted into egg fluids from incubator air. In our tests we used specific compounds previously identified in the plumage odour of an avian chemosignal (Hagelin et al. 2003). We also devised a new test protocol, in order to assess behaviour of avian embryos at an early developmental stage.

29.2 Methods

Our study came about because an undergraduate developmental laboratory had an "excess" of fertile chicken eggs aged E9 (breed: white leghorn [Babcock 300 strain], CBT Farms, Chestertown, MD, USA). All eggs were slated for termination prior to hatching, in compliance with Institutional Animal Care and Use Committee at Swarthmore College. However, we saw an opportunity to experimentally test the chemosensory responses of avian embryos that had not yet begun to breathe air. We divided eggs into two groups ($n = 18$) and incubated each for 10 days (E9–18) in Humidaire™20 incubators (37 °C, 60 % relative humidity). One incubator contained an experimental odour treatment and the other served as the control. The odour treatment exposed embryos for 10 days to two key compounds (octanal and Z-4 decenal) previously identified in the citrusy social odour of Crested Auklets (*Aethia cristatella*; Hagelin et al. 2003).

The two Crested Auklet odour compounds used in our study are convenient and germane to studies of avian development for several reasons. First, both are commercially available, unlike many avian chemical substances (see Campagna et al. 2011). Second, detection in our study species is likely, as chickens produce one compound (octanal; Bernier et al. 2008), and a chemically related version of the second (decanal; Bernier et al. 2008), which is also known from auklets (Douglas et al. 2001). Third, unlike other avian semiochemicals, synthetic mixtures of Crested Auklet compounds have been experimentally tested and implicated as an avian social chemosignal (Hagelin et al. 2003; Jones et al. 2004; Douglas 2007). Finally, both compounds are quite volatile, and hence likely candidates for transmission into eggs.

Our goal was to explore whether embryos exhibited a marked response to experimentally elevated levels of scent (similar to Tolhurst and Vince 1976; Sneddon et al. 1998). We chose to leave less-concentrated "ecologically-relevant" levels of odour for future study. On E9 we added 4.92 mL of synthetic auklet compounds (3.63 mL octanal and 1.26 mL Z-4 decenal; ratio from Hagelin et al. 2003) to 12 cotton balls in a glass Petri dish at the bottom of the incubator. An identical set-up with unscented cotton balls was added to the control incubator. We estimate the volume of auklet scent was roughly similar to the plumage of ~500–750 adult Crested Auklets. Our estimate comes from: (1) plumage weight of an adult is ~11.93 g ($n=5$; Hagelin 2007), (2) concentrations vary on different feather surfaces: 108–590 μg octanal and 59–393 μg Z-4 decenal/g feathers ($n=4$; Douglas 2007), and (3) the density of the two compounds: 0.843 g/mL and 0.864 g/mL, respectively. After 5 days we replenished half of the odour mixture (2.46 mL), as scent depleted noticeably over time. Treatment and control incubators were housed in separate rooms, as the volatile compounds generated a notable scent.

29.2.1 Odour Testing

On E18 we opened each egg at the rounded end to create a ~1.5 cm diameter opening that exposed the embryo (Fig. 29.1). Shell removal first revealed the membrane of the egg's internal airsac (Fig. 29.1b). We verified that each embryo had not yet pierced the membrane, which established that air-breathing had not yet begun. The membrane was removed with forceps, and we carefully poured off some excess fluid to reveal the chick's face and body in preparation for odour tests (Fig. 29.1c).

Our protocol for odour testing was adapted from Porter et al. (1999), in which scent contained in the headspace of a test bottle is gently puffed near the nares/mouth and responses recorded. We expected birds to detect these stimuli, given that rat foetuses removed from amniotic fluid of the womb respond readily to scented gas

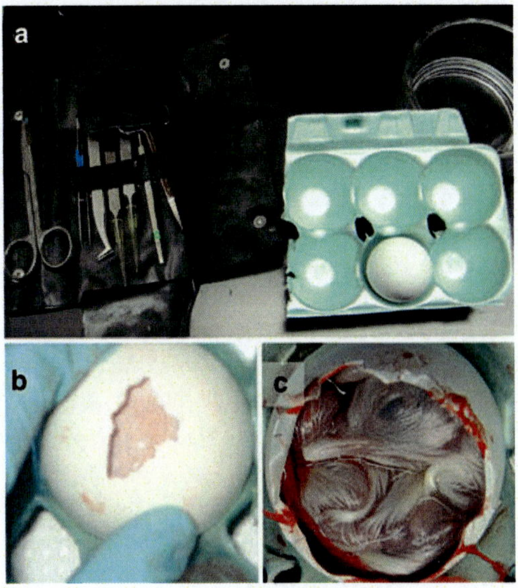

Fig. 29.1 (**a**) E18 egg under laboratory hood with instruments for opening shell. (**b**) Intact internal membrane of air sac verifies that breathing has not yet begun. (**c**) Opened egg with exposed embryo ready for odour testing. Scent was "puffed" within a few mm of the nares using a 250 mL Nalgene™ wash bottle containing the test odorant

delivered into the mouth (Smotherman and Robinson 1990). We tested each embryo to three scents: (1) synthetic auklet odour, (2) wintergreen (novel stimulus), and (3) a distilled water control. We chose wintergreen essential oil (~98 % methyl salicilate), as chicks detect this stimulus and show an aversion, when it is unfamiliar (Jetz et al. 2001). Like wintergreen, other mint-like stimuli also cause marked responses in both embryonic rats and young chicks (Smotherman and Robinson 1990; Porter et al. 1999). Distilled water served as the control, as it was used to maintain humidity in incubators.

We first administered a scent, then recorded embryo responses over a 1-min observation period. Embryos experienced 15 gentle puffs (~10 s) from a 250 mL Nalgene™ test bottle oriented within a few millimetres of the nares. Each bottle contained 4.91 mL of liquid (the same volume added to the odour-treated incubator) and cotton balls. A single observer blind to both the egg's treatment and contents of the odour bottle scored embryo behaviours (described below) during each 1-min observation period. Tests occurred under a laboratory fume hood to keep the work area scent-free.

On test day we alternated taking eggs from the control and experimental incubator for testing, so as to avoid any temporal confounds. We also used three sets of test bottles in a round-robin fashion. This allowed sufficient time for odours to accumulate in the headspace between uses (~1 use per 30 min). To ameliorate any effect of presentation order, each bottle set was given a pre-assigned sequence: Order 1: water, wintergreen, auklet; Order 2: wintergreen, auklet, water; Order 3: auklet, water, wintergreen. Each egg was assigned to a bottle set in advance, such that each set was administered evenly to eggs in each incubator treatment ($n=6$ eggs tested per bottle set per treatment).

29.2.2 Embryo Behaviour

Table 29.1 defines stereotypical behaviours identified in a pilot study of ($n = 12$) E18 embryos. Behavioural responses to odours at E18 show a general similarity to those reported at E20 (Tolhurst and Vince 1976). However, embryos notably lacked head shaking and vocalisations, both of which presumably coincide with air-breathing.

Table 29.1 groups behaviours into two broad categories indicative of response vigour (high-intensity and low-intensity responses). Categories enabled us to combine and compare behavioural scores, as embryos rarely exhibited all behaviours in every trial. During a typical 1-min trial of a single odour bottle, an embryo responded with the greatest number and intensity of behaviours during the first ~30 s (during and shortly after scent delivery). Responses then gradually diminished and stopped over the rest of the trial.

29.2.3 Statistical Analysis

We summed behaviours in the high- and low-intensity categories (Table 29.1) to create two scores for each of the three odour tests run on each embryo. We next analysed each egg treatment group separately (water control and auklet odour). We used Wilcoxon Matched Pairs Signed-Rank tests to determine whether embryos responded more to auklet or to wintergreen relative to the unscented water control. We also used signed-rank tests to determine whether birds discriminated between (i.e. reacted differently to) auklet and wintergreen test scents. Finally, we calculated an embryo's net response to each odour (odour minus control [water] scores) and employed Wilcoxon Rank-Sum tests to compare between treatment groups. All P-values presented in this study are two-tailed.

Analyses were run using JMP™ statistical software. Two embryos (one from each incubator) were excluded from all analyses. One did not develop fully and another was completely unreactive and presumed dead.

29.3 Results

29.3.1 Odour Detection and Discrimination

Embryos in both treatment groups exhibited a greater number of high-intensity behaviours when tested with auklet scent than the water control (Fig. 29.2; $n = 17$, Control treatment: Signed-Rank = 52.0, $P = 0.010$; Odour treatment: Signed-Rank = 30.5, $P = 0.012$). Control embryos specifically showed more kicking ($n = 14$, Signed-Rank = 45.5, $P = 0.002$), and a tendency for more head rolling ($n = 17$,

Table 29.1 Behaviours of E18 embryos scored during odour testing. Actions were largely stereotyped and readily fell into one of two categories indicative of response intensity (high or low)

Response intensity	Behaviour	Description	Sketch
High	Kicking	Rapid or jerky leg movement down into the egg or out of the shell opening	
	Body twisting	Side-to-side squirming or wriggling movement sometimes associated with kicking	
	Head rolling	Rotary-like head movements, in any direction, or from one side to another	
Low	Body shifting	Low-magnitude back and body movement, small shakes, or squirms	
	Beak movement	Opening and closing beak	
	Blinking	Opening and closing eyes	

Signed-Rank=33.5, $P=0.06$) and body twisting ($n=17$, Signed-Rank=35.5, $P=0.07$). Odour-treated embryos, by contrast, exhibited only increased head rolling ($n=13$, Signed-Rank=28.0, $P=0.03$). We detected no pattern for low-intensity responses ($P \geq 0.30$).

High-intensity responses to wintergreen were difficult to detect in both treatments (Fig. 29.2, $P \geq 0.16$). Only control embryos gave more low-intensity responses to wintergreen than to water ($n = 17$, Signed-Rank = 44.0, $P = 0.010$; Odour treatment: Signed-Rank = –2.0, $P = 0.56$). This manifested itself as more body shifting ($n = 17$, Signed-Rank = 24.5, $P = 0.03$). Finally, we detected no strong evidence for odour discrimination between auklet scent and wintergreen in either

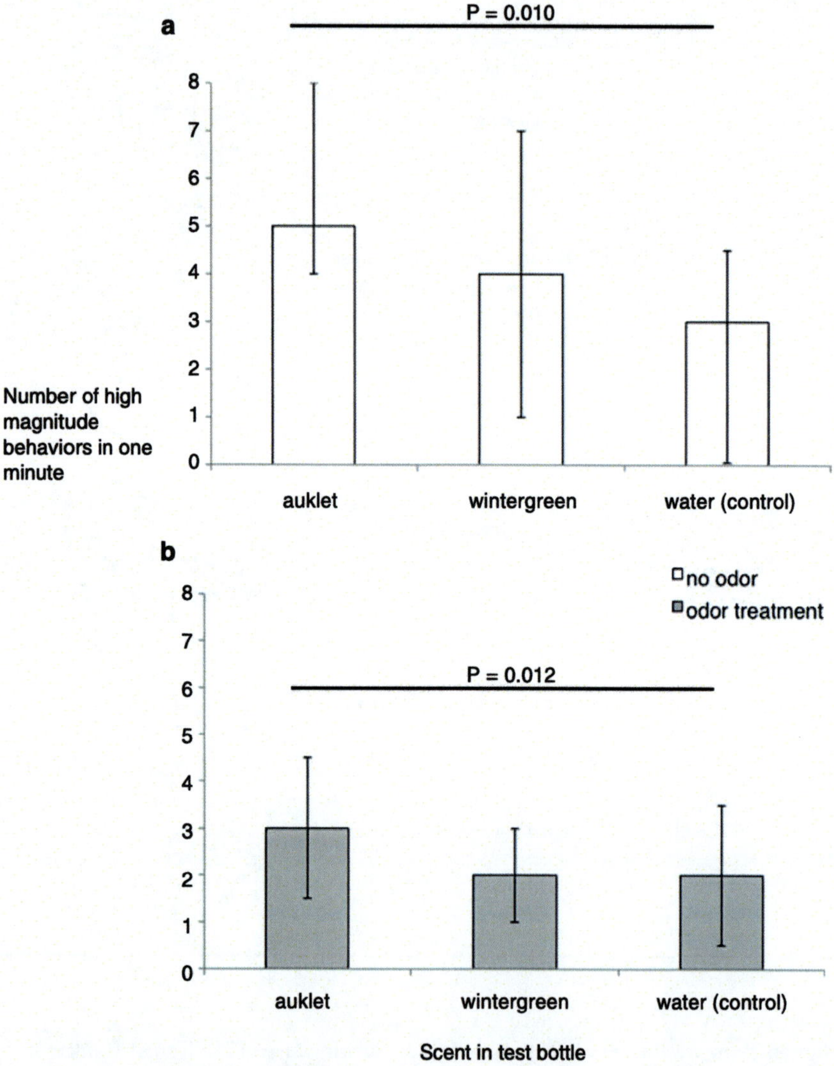

Fig. 29.2 (**a**, **b**) Domestic chicken embryos (aged E18) responded more to synthetic auklet odour than to an unscented water control, regardless of whether eggs had been incubated in: (**a**) control conditions (no odour), or (**b**) odoured conditions (with synthetic auklet scent). $n = 17$ eggs were tested in each treatment. Data represent medians and 25–75 % IQRs. *P*-values are two-tailed

treatment group when we considered high-intensity (Fig. 29.2, $P \geq 0.16$) or low-intensity responses ($P \geq 0.77$).

29.3.2 *Evidence for Behavioural Modification Following Odour Treatment*

During tests of auklet scent, odour-treated embryos gave relatively fewer high-intensity responses than controls (Fig. 29.3; df = 1, $\chi^2 = 4.50$, $P = 0.034$). Specifically, odour-treated embryos showed a reduced frequency of kicking (df = 1, $\chi^2 = 7.59$, $P = 0.006$) and body shifting (df = 1, $\chi^2 = 3.70$, $P = 0.054$). Low-intensity responses to auklet odour showed no treatment effect ($P = 0.47$).

During tests of wintergreen, high-intensity responses did not differ by treatment (Fig. 29.3; $P = 0.27$). However, as indicated above (see 29.3.1), a low-intensity response to wintergreen (body shifting) was noted in control birds. Low-intensity behaviour also tended to be more frequent in controls than in odour-treated birds (df = 1, $\chi^2 = 3.50$, $P = 0.06$).

29.4 Discussion

29.4.1 *Birds Can Detect Chemosensory Stimuli While Surrounded in Amniotic Fluid*

Our study provides two pieces of evidence consistent with the notion that domestic chicken embryos can sense chemical cues in ovo, prior to breathing air. First, regardless of odour treatment, embryos reacted more vigourously to Crested Auklet odour than to water (Fig. 29.2). This indicates that embryos could readily detect one or more compounds in a known avian chemosignal (Hagelin et al. 2003) compared to an unscented control.

Second, odour-treated embryos gave a lower net response to auklet scent than did controls (Fig. 29.3). The behavioural modification is consistent with embryos having prior experience with airborne odorants that entered the egg's aqueous environment (Bertin et al. 2012). Hence it is suggestive of a birds' ability to alter its behaviour following exposure to compounds in an avian chemosignal (Hagelin et al. 2003), even before the onset of breathing. Our data on embryo behaviour independently verifies the conclusions of Bertin et al. (2012), who recently altered the post-hatch feeding activity of domestic chicks by exposing eggs to odour at an early developmental stage (E13–16). Embryo responses are also consistent with physiological studies, in which olfactory neurons (at E18) respond in a manner comparable to other vertebrates (Jung et al. 2005). Domestic chicks at E18 also show some histological evidence consistent with a critical period, during which behavioural associations may form (Gomez and Celli 2008).

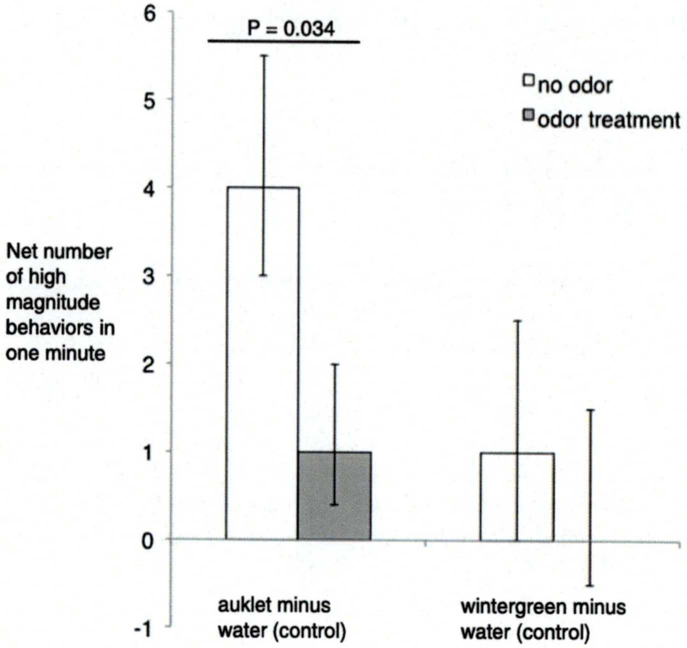

Fig. 29.3 Odour-treated avian embryos (aged E18) exhibited a reduced net number of high-intensity responses during tests of auklet odour, compared to those reared in the unscented (control) condition. The odour treatment exposed eggs to synthetic auklet scent for 10 days. $n = 17$ eggs were tested in both groups. Data represent medians and 25–75 % IQR. *P*-value is two-tailed

29.4.2 Transmission and Detection Mechanisms

Tolhurst and Vince (1976) noted that breakdown of tissue in the nares (at ~E19–20) was essential for scent detection in air-breathing embryos. Yet, birds in our study (age E18) and in Bertin et al. (2012; age E13–16) managed to detect chemical stimuli inside the egg, even though the nares were presumably blocked with tissue. How embryos detected scent during odour tests is unclear; rat foetuses can detect odoured gas administered intra-orally (Smotherman and Robinson 1990). However, the fact that both embryonic (this study) and adult birds (Hagelin et al. 2003; Jones et al. 2004) respond to airborne compounds of a known avian chemosignal is noteworthy. The pattern implicates a common mechanistic link between an embryo's in ovo experience and post-natal development of olfactory-directed behaviour (cf. Smotherman and Robinson 1990).

Unlike our experimental tests at E18, avian embryos under natural conditions are suspended in amniotic fluid. The following mechanisms are relevant to transmission and detection of airborne stimuli within an egg's aqueous environment. First, gas movement through the shell provides a means for volatile compounds to enter amniotic fluid. Interestingly gas uptake increases dramatically between E10 and E14 (Rahn et al. 1979), which overlaps with the early onset of olfactory receptor function (on E13; Lalloué et al. 2003; see also Sect. 29.4.3). Second, swallowing movements of embryos may transport compounds dissolved in amniotic fluid past chemically sensitive tissues, such as olfactory epithelium (Sneddon et al. 1998). Finally, it is also possible that the circulatory system may carry chemical stimuli directly to chemoreceptors (Maruniak et al. 1983).

29.4.3 Role of the Chemical Senses

We do not know the relative contribution of the three chemosenses (olfaction, trigeminal system, taste) to the embryonic responses we observed. However, our data and those of Bertin et al. (2012) suggest chemical stimuli can affect birds at an earlier developmental stage than is generally recognised. Avian olfactory receptor neurons begin to function at E13 (Lalloué et al. 2003). Peripheral projections of the trigeminal nerve are also "well-established" by the end of the first week of incubation (Noden 1980), which coincides with beak sensitivity to tactile stimulation (Hamburger and Narayanan 1969). To our knowledge, however, it is presently unknown when avian trigeminal receptors begin responding to chemical stimuli. Compared to the other chemosenses, avian taste buds mature relatively late in development. Pre-courser cells of taste buds are visible in basal epithelium at E17, exhibit some mature features by E19, and complete development by hatching (Ganchrow and Ganchrow 1987).

29.4.4 Behavioural Modification Following Early Chemosensory Exposure

The reduced response to auklet scent that we observed in odour-treated embryos (Fig. 29.3) is similar to behavioural patterns reported for chicks post-hatch. Chicks are more at ease (i.e. less neophobic) when they sense familiar odours (Jones et al. 2002), including those experienced in ovo (Porter and Picard 1998; Sneddon et al. 1998; Bertin et al. 2010, 2012). The behavioural modification is similar to the generalised reduction in response that occurs in a wide variety of animals following adaptation to an odour stimulus (Dalton 2000).

Conversely, control embryos that experienced auklet odour for the first time exhibited a heightened frequency of high-intensity responses (Fig. 29.3). Behaviours included increased kicking, body twisting, and head rolling. These responses differ somewhat from head shaking and beak clapping, two behaviours generally attributed to scent aversion in older embryos (e.g. Tolhurst and Vince 1976) and chicks (e.g. Ganchrow et al. 1990; Burne and Rogers 1999) that have begun to breathe air.

Two patterns in our data require further investigation. First, embryos were capable of detecting auklet odour during scent testing, but showed only minimal responses to wintergreen (Fig. 29.2). This was surprising, given that newly hatched chicks readily respond to this scent (Jetz et al. 2001). Second, we noted a tendency for control birds to be somewhat more reactive than odour-treated birds, regardless of test scent (Fig. 29.2; see also Sect. 29.3.1). Our analyses accounted for the pattern, as we compared net response data between treatment groups (Fig. 29.3). Nonetheless, we cannot exclude the possibility that our odour treatment may have reduced sensory perception (e.g. Panhuber et al. 1987) in a manner beyond the expected scent-specific decline in sensitivity (Dalton 2000).

29.4.5 Early Chemosensory Exposure and Behavioural Modification in Natural Avian Systems

We agree with Tolhurst and Vince (1976: 778), who noted that natural conditions of avian incubation potentially provide a chemically rich environment and play a role in the "embryo-nest-parent relationship". Chemical detection in utero is well known in mammals (e.g. Hepper 1988; Bilko et al. 1994), where early exposure to maternal dietary metabolites or personal body odour can alter neuroanatomy and affect postnatal preferences (Todrank et al. 2011). There is a growing literature in avian behavioural ecology on the effect of nest odour (e.g. Petit et al. 2002; Gwinner and Berger 2008; Caspers and Krause 2010) and body odour (e.g. Bonadonna and Nevitt 2004; Mardon et al. 2010, 2011; Whittaker et al. 2011) on responses of adult birds and young. Even mate choice in domestic chickens (Hirao et al. 2009; Hirao 2011), quail (Balthazart and Taziaux 2009), and ducks (Balthazart and Schoffeniels 1979) has been linked to the chemosenses.

Bertin et al. (2012) note the intriguing possibility that bird embryos are also exposed to maternal dietary compounds, given that hens fed experimental foods produced eggs bearing the same scent (e.g. Lawlor et al. 2010). The concentration of compounds experienced in ovo is also important, as it affects whether chicks approach or avoid a stimulus (Bertin et al. 2010). Likewise, developmental timing of odour exposure can impact the strength of a bird's response (Bertin et al. 2012).

We are clearly just beginning to learn basic principles of behavioural "pre-programming" (Bertin et al. 2012) that can occur in birds as a result of chemosensory experiences in ovo. A rich literature of in utero studies of mammals can facilitate comparisons cross-taxa (e.g. Schaal and Orgeur 1992; Todrank et al. 2011). Future investigations of birds that use "ecologically-relevant" compounds, presented in natural concentrations and at appropriate times in development, promise to reveal exciting new insight into mechanisms and adaptive behavioural outcomes of early odour exposure.

Acknowledgements This paper would not have been possible without J. Noveral who helped design and carry out the experiment. J. Cebra-Thomas and S. Gilbert provided fertile eggs and access to incubators. F. Bonadonna kindly provided helpful comments on an earlier draft of this manuscript. The Animal Behaviour class at Swarthmore College also gave constructive criticism during data analysis and project synthesis.

References

Balthazart J, Schoffeniels E (1979) Pheromones are involved in the control of sexual behaviour in birds. Naturwissenschaften 66:55–56

Balthazart J, Taziaux M (2009) The underestimated role of olfaction in avian reproduction. Behav Brain Res 200:248–259

Bernier UR, Allan SA, Quinn BP, Kline DL, Barnard DR, Clark GG (2008) Volatile compounds from the integument of White Leghorn chickens (*Gallus gallus domesticus* L.): candidate attractants of ornithophilic mosquito species. J Sep Sci 31:1092–1099

Bertin A, Calandreau L, Arnould C, Nowak R, Lèvy F, Noirot V, Bouvarela I, Leterrier C (2010) In ovo olfactory experience influences post-hatch feeding behaviour in young chickens. Ethology 116:1027–1037

Bertin A, Calandreau L, Arnould C, Lèvy F (2012) The developmental stage of chicken embryos modulates impact of in ovo olfactory stimulation on food preferences. Chem Senses 37(3): 253–261. doi:10.1093/chemse/bjr101

Bilko A, Altbacker V, Hudson R (1994) Transmission of food preference in the rabbit—the means of information-transfer. Physiol Behav 56:907–912

Bonadonna F, Nevitt GA (2004) Partner-specific odor recognition in an Antarctic seabird. Science 306:835

Burne TH, Rogers LJ (1999) Changes in olfactory responsiveness by the domestic chick after early exposure to odorants. Anim Behav 58:329–336

Campagna S, Mardon J, Celerier A, Bonadonna F (2011) Potential semiochemical molecules from birds: a practical and comprehensive compilation of the last 20 years studies. Chem Senses 37:3–25

Caro SP, Balthazart J (2010) Pheromones in birds: myth or reality? J Comp Physiol A Neuroethol Sens Neural Behav Physiol 196:751–766

Caspers BA, Krause ET (2010) Odour-based natal nest recognition in the zebra finch (*Taeniopygia guttata*), a colony-breeding songbird. Biol Lett 7:184–186

Cunningham GB, Nevitt GA (2011) Evidence for olfactory learning in procellariiform seabird chicks. J Avian Biol 42:85–88

Dalton P (2000) Psychophysical and behavioral characteristics of olfactory adaptation. Chem Senses 25:487–492

Douglas HD III (2007) Prenuptial perfume: alloanointing in the social rituals of the crested auklet (*Aethia cristatella*) and the transfer of arthropod deterrents. Naturwissenschaften 95:45–53

Douglas HD III, Co JE, Jones TH, Conner WE (2001) Heteropteran chemical repellents identified in the citrus odor of a seabird (Crested auklet: *Aethia cristatella*). Naturwissenschaften 88:330–332

Ganchrow JR, Ganchrow D (1987) Taste bud development in chickens (*Gallus gallus domesticus*). Anat Rec 218:88–93

Ganchrow JR, Steiner JE, Bartana A (1990) Behavioral reactions to gustatory stimuli in young chicks (*Gallus gallus domesticus*). Dev Psychobiol 23:103–117

Gomez G, Celli A (2008) The peripheral olfactory system of the domestic chicken: physiology and development. Brain Res Bull 76:208–216

Gwinner H, Berger S (2008) Starling males select green nest material by olfaction using experience-independent and experience dependent cues. Anim Behav 75:971–976

Hagelin JC (2007) Odors and chemical signaling. In: Jamieson BGM (ed) Reproductive behavior and phylogeny of aves, vol 6B. Science Publishers, Enfield, NH, pp 76–119

Hagelin JC, Jones IL (2007) Bird odors and other chemical substances: a defense mechanism or overlooked mode of intraspecific communication? Auk 124:741–761

Hagelin JC, Jones IL, Rasmussen LEL (2003) A tangerine-scented social odour in monogamous seabird. Proc R Soc Lond 270:1323–1329

Hamburger V, Narayanan CH (1969) Effects of the deafferentation of the trigeminal area on the motility of the chick embryo. J Exp Zool 170:411–426

Hepper PG (1988) Adaptive fetal learning: prenatal exposure to garlic affects postnatal preferences. Anim Behav 36:935–936

Hirao A (2011) The possible role of the uropygial gland on mate choice in domestic chicken. Internat J Zool. doi:10.1155/2011/860805

Hirao A, Aoyama M, Sugita S (2009) The role of uropygial gland on sexual behavior in domestic chicken (*Gallus gallus domesticus*). Behav Process 80:115–120

Jetz W, Rowe C, Guilford T (2001) Non-warning odors trigger innate color aversions—as long as they are novel. Behav Ecol 12:134–139

Jones RB, Facchin L, McCorquodale C (2002) Social dispersal by domestic chicks in a novel environment: reassuring properties of a familiar odourant. Anim Behav 63:659–666

Jones IL, Hagelin JC, Major HL, Rasmussen LEL (2004) An experimental field study of the function of crested auklet feather odor. Condor 106:71–78

Jung Y, Wirkus E, Amendola D, Gomez G (2005) Characteristics of odorant elicited calcium fluxes in acutely-isolated chick olfactory neurons. J Comp Physiol A Neuroethol Sens Neural Behav Physiol 191:511–520

Lalloué FL, Ayer-Le-Lièvre CS, Sicard G (2003) Analysis of the functional maturation of olfactory neurons in chicks before and after birth. Chem Senses 28:729–737

Lawlor JB, Gaudette N, Dickson T, House JD (2010) Fatty acid profile and sensory characteristics of table eggs from laying hens fed diets containing microencapsulated fish oil. Anim Feed Sci Technol 156:97–103

Mardon J, Saunders SM, Anderson MJ, Couchoux C, Bonadonna F (2010) Species, gender and identity: cracking petrels' sociochemical code. Chem Senses 35:309–321

Mardon J, Saunders SM, Bonadonna F (2011) From preen secretions to plumage: the chemical trajectory of petrels' social scent. J Avian Biol 42:29–38

Maruniak JA, Silver WL, Moulton DG (1983) Olfactory receptors respond to blood-borne odorants. Brain Res 265:312–316

Mathis A, Ferrari MCO, Windel N, Messier F, Chivers DP (2008) Learning by embryos and the ghost of predation future. Proc R Soc B 275:2603–2607

Mennella JA, Jagnow CP, Beauchamp GK (2001) Prenatal and postnatal flavor learning by human infants. Pediatrics 107:e88

Noden DM (1980) Somatotopic organisation of the embryonic chick trigeminal neuron. J Comp Neurol 190:429–444

Panhuber H, Mackay-Sim A, Laing DG (1987) Prolonged odor exposure causes severe cell shrinkage in adult rat olfactory bulb. Dev Brain Res 31:307–311

Petit C, Hossaert-McKey M, Perret P, Blondel J, Lambrechts MM (2002) Blue tits use selected plants and olfaction to maintain an aromatic environment for nestlings. Ecol Lett 5:585–589

Porter RH, Picard M (1998) Effects of early odor exposure in domestic chicks. Reprod Nutr Dev 38:441–448

Porter RH, Hepper PG, Bouchot C, Picard M (1999) A simple method for testing odor detection and discrimination in chicks. Physiol Behav 67:459–462

Rahn H, Paganelli A, Paganelli CV (1979) How bird eggs breathe. Sci Am 240:46–55

Schaal B, Orgeur P (1992) Olfaction in utero: can the rodent model be generalized? Q J Exp Psychol B 44:245–278

Schaal B, Marlier L, Soussignan R (2000) Human foetuses learn odours from their pregnant mother's diet. Chem Senses 25:729–737

Smotherman WP, Robinson SR (1990) Rat fetuses respond to chemical stimuli in gas phase. Physiol Behav 47:863–868

Sneddon H, Hadden R, Hepper PG (1998) Chemosensory learning in the chicken embryo. Physiol Behav 64:133–139

Todrank J, Heth G, Restrepo D (2011) Effects of in utero odorant exposure on neuroanatomical development of the olfactory bulb and odour preferences. Proc R Soc B 278:1949–1955

Tolhurst BE, Vince MA (1976) Sensitivity to odours in the embryo of the domestic fowl. Anim Behav 24:772–779

Whittaker DJ, Richmond KM, Miller AK, Kiley R, Burns CB, Atwell JW, Ketterson ED (2011) Intraspecific preen oil preferences in dark-eyed juncos (*Junco hyemalis*). Behav Ecol 22:1256–1263

Part VII
Reproduction in Domestic Animals

Chapter 30
Measurement of 16-Androstenes (5α-Androst-16-en-3-One, 5α-Androst-16-en-3α-ol, 5α-Androst-16-en-3β-ol) in Saliva of German Landrace and Göttingen Minipig Boars

Martin Dehnhard, Hartmut Rohrmann, and Johannes Kauffold

Abstract Two C16-unsaturated steroids (16-androstenes), 5a-androst-16-en-3-one with a urine-like odour and 5a-androst-16-en-3a-ol with a musk-like odour are present in salivary glands and saliva of mature male pigs. Both 5a-androstenol and 5a-androstenone have a pheromonal function in the pig. They are exhaled by sexually excited boars and stimulate oestrous behaviour in females. Concentrations of 2 ng/ml 5a-androst-16-en-3-one occur in saliva. In European wild and domestic boars there is seasonal variation in gonadal function with plasma testosterone increasing from ~0.5 to ~10.0 ng/ml during the mating season under short-day photoperiod. Anecdotal evidence suggests that Göttingen minipig boars (GMB) are more potent in stimulating females during oestrus than boars commonly used in swine production, which might be due in part to differences in salivary 16-androstene content. We analysed 16-androstene in saliva of GMB and German landrace boars (GLB). A total of 10 GMB and 16 GLB were sampled either in December (GMB) or February (GLB), and saliva (0.5 to 2.0 ml) was obtained during semen collection using a collection systems designed for use in humans. Salivary 5a-androst-16-en-3-one, 5a-androst-16-en-3a-ol and -3b-ol were determined using GC-MS analyses with external standard curves. The MS acquisition was performed in SIM by monitoring the ions m/z 272 for 5a-androst-16-en-3-one, m/z 274 for 5a-androst-16-en-3a-ol/-3b-ol and m/z 202 for 5a-androst-3-one (internal standard). A calibration line prepared by adding various amounts of the analytes and a

M. Dehnhard (✉)
Department of Reproduction Biology, Leibniz Institute for Zoo and Wildlife Research, Alfred-Kowalke Strasse 17, 10315, Berlin, Germany
e-mail: dehnhard@izw-berlin.de

H. Rohrmann
Oberhäger Str. 7, 18182 Rövershagen, Germany

J. Kauffold
Large Animal Clinic for Theriogenology and Ambulatory Services, University of Leipzig, An den Tierkliniken 29, 04103 Leipzig, Germany

M.L. East and M. Dehnhard (eds.), *Chemical Signals in Vertebrates 12*,
DOI 10.1007/978-1-4614-5927-9_30, © Springer Science+Business Media New York 2013

constant amount of 5a-androst-3-one was used for quantification. In addition, salivary testosterone was determined using an in-house enzyme immunoassay (EIA). Mean concentrations of 5a-androst-16-en-3a-ol were 0.18 ± 0.13 and 9.20 ± 8.57 μg/ml, respectively, in GLB and GMB. 5a-androst-16-en-3b-ol concentrations were 0.02 ± 0.01 and 0.42 ± 0.37 μg/ml, respectively. The urine-like compound 5a-androst-16-en-3-one was 0.04 ± 0.03 and 1.70 ± 1.42 μg/ml, respectively. Our results show that both odorous substances were about 50 times higher in GMB than in GLB which indicates a difference in salivary 16-androstene in these two pig breeds. The fact that salivary testosterone was 3.05 ± 3.32 ng/ml in GLB versus 223.7 ± 266.3 ng/ml in GMB supports an effect of breed. However, a seasonal effect cannot be ruled out since samples from GMB were obtained in the middle of the breeding season when sexual activity was highest, while GLB were collected at the end of the breeding season.

30.1 Introduction

Intraspecies communication carried by chemical substances plays a very important role in the control of reproduction as well as in social and sexual interactions. A group of chemicals known as male signalling and priming pheromones represent potent sexual stimuli. When a mature boar is aroused by the presence of an oestrous sow or an unfamiliar boar, the submaxillary salivary glands deliver large amounts of saliva containing odorous 16-androstene steroids. Salivation is facilitated by chewing and champing movements of the boar's mouth delivering copious amounts of frothy saliva. The purpose of the excessive salivation is the release of large amounts of odoriferous 16-androstene steroids into the environment. These odorous steroids, in particular 5α-androst-16-en-3α-ol (3α-androstenol) and 5α-androst-16-en-3-one (5α-androstenone), act as signalling pheromones by facilitating the mating response in oestrous pigs (Melrose et al. 1971; Perry et al. 1980; Reed et al. 1974) and probably indicate their virility status to other boars (Booth and Baldwin 1980). The main compound of the pheromone is 5α-androst-16-ene-3-one, a steroid with 19 carbons, which has a urine-like odour (Patterson 1968). The synthetic substance induces a rigid stance, called the lordosis reflex, in the sow when she is in oestrus. Therefore boar scent sprays consisting of synthetic analogues are applied to determine the best moment for mating or artificial insemination in sows.

There is also evidence suggesting that 16-androstenes do have primer pheromone roles in the 'boar effect' on puberty acceleration in female pigs (Booth 1984; Kirkwood et al. 1983). The presence of a boar can accelerate puberty in gilts by about 30 days (Brooks and Cole 1970) and the sensory cues involved in the boar's effect on puberty were confirmed to be olfactory (Kirkwood et al. 1981). Boar pheromone synthesis primarily proceeds in the boar's testis (Booth 1982). 16-Androstenes are secreted into the blood and transported to the submaxillary salivary glands where they are bound to a specific binding protein called 'pheromaxein' (Booth 1984). It is assumed that pheromaxein is vital for the transportation of the pheromonal steroids in the aqueous medium of saliva, since these steroids are not very hydrophilic, generating very high concentrations in submaxillary glands and saliva of these animals (Booth 1980).

As a result of their fat solubility, high levels of 16-androstenes can accumulate in adipose tissue and are one of the principal causes why pork from boars is usually inedible for humans (Bonneau and Desmoulin 1982; Booth 1982). Since both the signalling and primer pheromone effects of 16-androstenes are mediated by the free steroid (Melrose et al. 1971; Reed et al. 1974), this suggests that the pheromonal steroids dissociate from their binding protein in voided saliva and become volatile as air-borne pheromones. Indeed, preliminary observations on the relative amounts of free and bound pheromone in boar saliva have shown that about 21% of the pheromones are in the free form in freshly voided saliva at 37°C (Booth 1987).

There are discrepancies with respect to the magnitude of boar's salivary 16-androstene concentrations. Levels of 2 ng/mL 5α-androst-16-en-3-one have been reported in saliva assayed by radioimmunoassay (Grinwich et al. 1988). However, it was also found that salivary androstenol concentrations were 10–20 times higher than that of androstenone and that boar saliva may contain up to 50 μg/mL androstenol assayed by gas chromatography (Booth 1982; Claus 1979). In European wild and domestic boars, there is seasonal variation in gonadal function with testosterone increasing from ~0.5 to ~10.0 ng/mL in blood plasma in a wild boar during the mating season under a short-day photoperiod (Weiler et al. 1996). Furthermore, the change in the predominance of individual 16-androstenes that was found in the testis during development of boars reflects the proposed biosynthetic pathway for these compounds in the testis (Brophy and Gower 1972).

During the last decade, the presence of a humoral pathway was demonstrated which transferred the boar pheromone from the nasal cavity into the bloodstream. The pheromone can then pass locally from the perihypophyseal vascular complex into the arterial blood supplying the brain and pituitary gland (Stefanczyk-Krzymowska et al. 2000). Later, Kaminski et al. (2006) verified that 5α-androst-16-en-3-ol acts as a positive neurosteroid in tests on human embryonic kidney 293 cells, and on mouse cerebellar granule cells and slices, and thus may exert physiological effects in the brain and pituitary gland (Kaminski et al. 2006).

One would assume that the olfactory system of pigs is highly sensitive for the detection of its pheromone. Surprisingly, domestic pigs' detection of androstenone was found to be ~1,000× less sensitive than towards geraniol (Dorries et al. 1995), an odour without porcine biological significance. In addition a sexual dimorphism in sensitivity was demonstrated with female pigs detecting androstenone with five times more sensitivity than males (Dorries et al. 1995).

In European wild but also domestic boars, there is seasonal variation in gonadal function. In a wild boar, testosterone in blood plasma increased from ~0.5 to ~10.0 ng/mL during the mating season (Weiler et al. 1996). The fact that high concentrations of 16-androstenes appear in testis of boars and the relationship between testicular 16-androstene and testosterone formation in young pigs suggests that 16-androstenes are preferentially synthesised in the boar testis (Claus et al. 1971). This has been confirmed also by in vitro studies (Ahmad and Gower 1968). However, androgens like testosterone or 5-androsten-3β,17β-diol are probably not precursors of the pheromonal 16-androstene steroids (Booth 1980).

There is anecdotal evidence that Göttingen minipig boars (GMBs) are more potent in stimulating females at oestrus than boars commonly used in swine production. This might be due, at least in part, to differences in salivary 16-androstene content. However, measurements of 16-androstenes in saliva of minipig boars have not been carried out so far. Therefore it was the aim of this study to analyse the concentrations of 16-androstenes in saliva of GMBs and German landrace boars (GLBs) and to compare 16-androstene concentrations with those from salivary testosterone reflecting gonadal activity.

30.2 Methods and Materials

This study was conducted to analyse 16-androstene in saliva of GMB and of GLB. A total of 10 GMB and 16 GLB were sampled either in December (GMB) or February (GLB), and saliva (0.5–2.0 mL) was obtained during semen collection using collection systems designated for use in humans (Salivetten, Sarstedt, Germany).

Saliva (0.5 mL) was diluted with distilled water (1:1 dilution), mixed with 10 μl of a methanolic solution containing 100 ng of 5α-androst-3-one (internal standard) and 2 × 2 mL extracted with *n*-hexane for 30 min in screw-capped vials with horizontally shaking. Following extraction the samples were centrifuged at 3,000 rpm for 20 min to enhance phase separation. After freezing the vial at −80°C, the upper *n*-hexane layer was poured into glass tubes with a conical end. The first extract was evaporated at 22°C to ~100 μl under a stream of nitrogen before the *n*-hexane layer of the second extraction was added. The combined *n*-hexane phases were now carefully evaporated to a volume of about 2 μl that were injected into the gas chromatograph for GC–MS analysis.

The GC-MS determinations were conducted with a Shimadzu GCMS-QP 5050. The MS acquisition was performed in SIM by monitoring the ions *m/z* 272 for 5α-androst-16-en-3-one, *m/z* 274 for 5α-androst-16-en-3α-ol/-3β-ol and *m/z* 202 for 5α-androst-3-one (internal standard). 5α-androst-3-one was used as an internal standard added to saliva prior analyses (100 ng) and monitored at *m/z* 202.

The samples were analysed using a 50 m FS-Supreme-5 ms capillary column (5% phenylpolysilphenylensiloxane), 0.25 mm i.d. and 0.25 μm film thickness (CS, Langerwehe, Germany). Ultrapure helium was used as carrier gas, with a column head pressure setting of 41 kPa. Injector temperature was 300°C; the transfer line temperature was maintained at 250°C and ionising voltage was at 1.4 kV. Splitless injection mode was used into an injection port provided with an inner silanised glass liner containing silanised glass wool. The purge valve was turned on 15 min after injection, with a split flow of 9 mL/min during the GC run. The GC oven was kept at 50°C for 2 min, ramped at 15°C/min to 105°C, ramped at 10°C/min to 175°C, ramped at 4°C/min to 300°C and held for 5 min. The mass spectra were identified by computer MS library research and compared with those of the authentic standards.

For quantification, a calibration line was prepared by adding various amounts of 5α-androst-16-en-3-one and 5α-androst-16-en-3α-ol and a constant amount of 100 ng 5α-androst-3-one to pooled GLB saliva that had previously been shown to contain no detectable amounts of 16-androstenes. Salivary testosterone was determined using an in-house enzyme immunoassay (EIA). The antibody was raised in rabbits immunised against 11-OH-testosterone-HS-BSA. The cross-reactivities of the antibody to different androgens were as follows: 4-androsten-17β-ol-3-one (testosterone)—100%, 5α-androstan-17β-ol-3-one (dihydrotestosterone)—13.6%, 5α-Androst-2-en-17β-ol—1.5%, and <0.1% for 5α-androstan-3β-ol-17-one (epi-androsterone), 5α-androstan-3α-ol-17-one (androsterone) and 5α-androst-2-en-17-one. Testosterone-3-CMO-peroxidase was used as enzyme conjugate.

30.3 Results

Figure 30.1 shows chromatograms of a saliva sample to which 1 μg/mL 5α-androst-16-en-3β-ol and -3-one were added (1a), one sample from a GMB (1b) and one from a GLB (1c) containing the indicated amounts of 16-androstenes. Both standards 5α-androst-16-en-3β-ol and -3-one were well separated corresponding to retention times of 31.04 and 31.80 min. This is also valid for 5α-androst-16-en-3α-ol and -3-one (31.20 and 31.80 min, respectively) in saliva samples from both the GLB and the GMB. However, a baseline separation of the two isomers, 5α-androst-16-en-3α-ol and 3β-ol, occurring simultaneously in samples from both boar breeds was impossible on this column (31.04 and 31.20 min, respectively). Therefore manual integrations were performed to calculate 5α-androst-16-en-3β-ol concentrations based on the *m/z* 274 ion signal intensity, probably leading to some inaccuracies in case of the minor compound 5α-androst-16-en-3α-ol. The four substances that became detectable in GMP boars at later retention times between 32.50 and 34.00 min (1b) probably correspond to unknown 16-en steroids.

16-Androstene concentrations in both breeds showed huge differences between the individuals. In case of the urine-like odorous substance 5α-androst-16-en-3-one mainly causing the boar taint, concentrations varied from 0 to 0.24 μg/mL in GL and from 0.27 to 4.23 μg/mL in GMBs (Table 30.1). The same is obvious from the musk-like substance 5α-androst-16-en-3α-ol that has been measured in ranges from 0.02 to 0.84 μg/mL in GL and from 1.44 to 23.30 μg/mL in GMBs. Altogether, mean 5α-androst-16-en-3α-ol and -16-en-3-one concentrations were 50 and 42, respectively, higher in GMP compared to GLB boars (Table 30.1). Mean concentrations of 5α-androst-16-en-3α-ol GL and GM boars were 0.18 ± 0.13 and 9.20 ± 8.57 μg/mL, respectively, those of 5α-androst-16-en-3β-ol were 0.02 ± 0.01 and 0.42 ± 0.37 μg/mL, respectively. In contrast, 5α-androst-16-en-3β-ol levels measured in minor concentrations in both breeds were also higher (~20 times) in GMBs.

Summarising both, GLB and GMP boars, salivary concentrations of the three 16-androstenes were highly correlated to testosterone ($P < 0.001$, $n = 23$), confirming

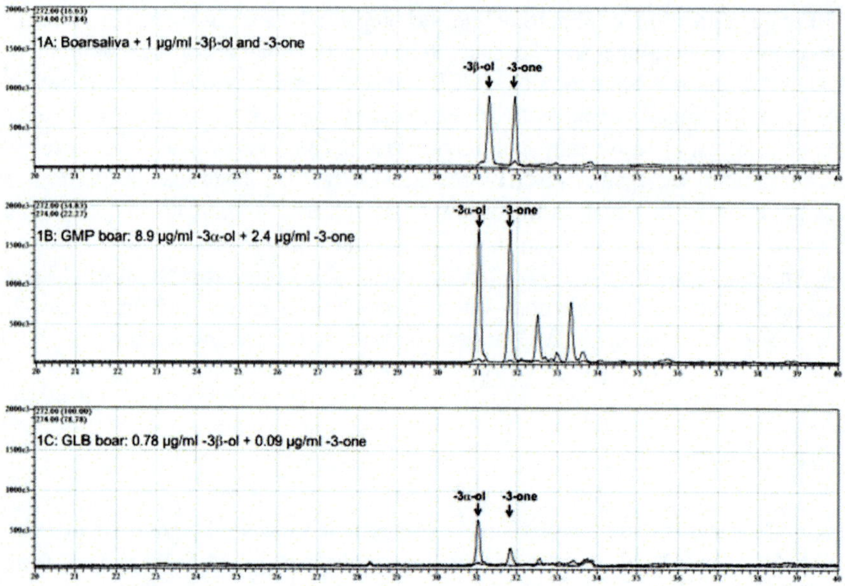

Fig. 30.1 Chromatograms of a saliva sample from a German landrace boar (GLB) (previously shown to contain no measurable endogenous concentrations) to which 1 µg/mL 5α-androst-16-en-3β-ol and -3-one were added (**a**), one sample from a Göttingen minipig boar (GMB) (**b**) and one from a GLB (**c**) containing the indicated amounts of 16-androstenes

their gonadal synthesis. However, when each breed was considered separately, 5α-androst-16-en-3-one in GLB and 5α-androst-16-en-3β-ol concentrations in GMB did not correlate significantly with corresponding testosterone ($P=0.076$; $n=14$ and $P=0.087$; $n=9$, respectively).

30.4 Discussion

The testis of the male pig produces, in addition to testosterone, large quantities of other C_{19} steroids, among them the pheromonal 16-unsaturated C_{19} steroids (16-androstenes). These musk- and urine-like smelling steroids are synthesised by a route independent of androgen production in boar testis (Gower 1972), where they occur in concentrations considerably higher than those of testosterone (Booth 1975). Our analyses in saliva samples collected from GMBs during the mating season revealed large amounts of 16-androstenes and testosterone. The relative concentrations of 16-androstenes were 5α-androst-16-en-3α-ol > 5α-androst-16-en-3-one > 5α-androst-16-en-3β-ol and were similar within both boar breeds which coincides with data published previously (Katkov et al. 1972).

The major salivary glands of mammals are represented by three pairs of organs that cooperate functionally to produce saliva for the oral cavity: the submandibular,

Table 30.1 Mean ± SD and range of 16-androstene and testosterone concentrations in saliva samples from German Landrace (GLB) and Göttingen Minipig boars (GMB) measured by gas chromatography–mass spectrometry (GC-MS) and enzyme immunoassay (EIA), respectively

	5α-A-16-en-3α-ol (μg/mL)		5α-A-16-en-3β-ol (μg/mL)		5α-A-16-en-3-one (μg/mL)		Testosterone (ng/mL)	
Breed	Mean ± SD	Range	Mean ± SD	Range	Mean ± SD	Range	Mean ± SD	Range
GLB	0.18 ± 0.13	0.02–0.84	0.02 ± 0.01	0.00–0.13	0.04 ± 0.03	0.00–0.24	3.05 ± 3.32	0.47–11.18
GMB	9.20 ± 8.57	1.44–23.30	0.42 ± 0.37	0.04–1.28	1.70 ± 1.42	0.27–4.23	223.7 ± 266.3	9.60–929.50

sublingual and parotid glands. They produce a signature secretion that complements the secretions from the other glands (Denny et al. 1997). The ratio of 3α-androstenol and 5α-androstenone in submaxillary gland tissue is presumably individually maintained by a 3α-hydroxysteroid dehydrogenase requiring NADH and NADPH (Katkov et al. 1972).

Concentrations of the urine-like odorous substance 5α-androst-16-en-3-one, mainly causing the boar taint, varied from 0.27 to 4.23 μg/mL in GMBs boars whereas the musk-like substance 5α-androst-16-en-3α-ol reached concentrations between 1.44 and 23.30 μg/mL. These concentrations appear to be extraordinarily high; however, similar concentrations we obtained in earlier studies using GC analyses (Gower 1972). High concentrations of 16-androstenes up to 43 μg/g of tissue were also measured in salivary glands (Babol et al. 1996) and in fatty and parotid tissue of boars reaching 7.5 and 11.4 μg/g tissue, respectively (Claus et al. 1971). With respect to blood plasma concentrations below 40 ng/mL (Sinclair and Squires 2005), both tissues seem to have a storage function.

Mean 5α-androst-16-en-3α-ol and -16-en-3-one concentrations were 50 and 42 times higher, respectively, in GM compared to GLBs. These data might suggest a breed difference in salivary 16-androstene concentrations. However, a seasonal effect cannot be ruled out since samples from GMB were obtained in the middle of the breeding season when sexual activity was highest, while GLB were collected at the end of the breeding season. A distinct seasonal course of serum testosterone has been described in a wild boar under a light reverse program increasing from ~0.2 to ~10 ng/mL during the simulated breeding season (Weiler et al. 1996) and has been confirmed in another study with boars living in mountains (1.5 vs. 7.4 ng/mL) (Macchi et al. 2010). A similar but less pronounced effect of season was observed for GLBs, with lowest testosterone concentrations in May (1.46 ng/mL) and highest in November (4.13 ng/mL; Weiler et al. 1996). Contrary to this is a comparison of serum testosterone concentrations between Duroc and Yorkshire boars, revealing that testosterone concentrations were higher in spring than in summer, autumn and winter (Park and Yi 2002). Another research group, however, was unable to demonstrate a seasonal effect in Duroc, Meishan, Fengjing and Minzhu boars (Borg et al. 1993).

Based on our results and data from the literature, we conclude that: (1) the hormonal concentrations found in saliva of boars are representative of the serum concentrations as described for humans (Sannikka et al. 1983; Wang et al. 1981) and bottlenose dolphins (Hogg et al. 2005); (2) the distinct difference between salivary testosterone concentrations of GMBs in December and those of GLBs in February is most likely not due to a seasonal effect (Borg et al. 1993; Macchi et al. 2010; Park and Yi 2002; Weiler et al. 1996); (3) there is a highly correlated testicular synthesis of 16-androstenes and testosterone in boars as shown from our results and from those of others (Andresen 1976; Claus et al. 1971; Trudeau et al. 1988) and (4) only moderate fluctuations of 16-androstene concentrations in blood with season (month, $P<0.10$, (Trudeau et al. 1988)).

We assume that the extreme differences in salivary 16-androstenes between mating (December) and non-mating season (February) obtained from our inter-breed comparison indicate a clear breed-related difference (as hypothesised); however, a seasonal interference cannot be excluded. This also fits with measures of 16-androstenes in adipose tissue/saliva indicating only moderate seasonal differences in domestic landrace boars.

References

Ahmad N, Gower DB (1968) The biosynthesis of some androst-16-enes from C21 and C19 steroids in boar testicular and adrenal tissue. Biochem J 108:233–241

Andresen O (1976) Concentrations of fat and plasma 5-alpha-androstenone and plasma testosterone in boars selected for rate of body-weight gain and thickness of back fat during growth, sexual-maturation and after mating. J Reprod Fertil 48:51–59

Babol J, Squires EJ, Bonneau M (1996) Factors regulating the concentrations of 16-androstene steroids in submaxillary salivary glands of pigs. J Anim Sci 74:413–419

Bonneau M, Desmoulin B (1982) Boar taint and possible utilization of boar meat. Ann Zootech 31:319

Booth WD (1975) Changes with age in the occurrence of C19 steroids in the testis and submaxillary gland of the boar. J Reprod Fertil 42:459–472

Booth WD (1980) A study of some major testicular steroids in the pig in relation to their effect on the development of male characteristics in the prepubertally castrated boar. J Reprod Fertil 59:155–162

Booth WD (1982) Metabolism of C19 steroids *in vitro* by the submaxillary salivary gland of mature and immature domestic pigs. J Endocrinol 93:91–97

Booth WD (1984) Sexual dimorphism involving steroidal pheromones and their binding protein in the submaxillary salivary gland of the Göttingen miniature pig. J Endocrinol 100:195–202

Booth WD (1987) Factors affecting the pheromone composition of voided boar saliva. J Reprod Fertil 81:427–431

Booth WD, Baldwin BA (1980) Lack of effect on sexual behaviour or the development of testicular function after removal of olfactory bulbs in prepubertal boars. J Reprod Fertil 58:173–182

Borg KE, Lunstra DD, Christenson RK (1993) Semen characteristics, testicular size, and reproductive hormone concentrations in mature Duroc, Meishan, Fengjing, and Minzhu boars. Biol Reprod 49:515–521

Brooks PH, Cole DJ (1970) The effect of the presence of a boar on the attainment of puberty in gilts. J Reprod Fertil 23:435–440

Brophy PJ, Gower DB (1972) The metabolism *in vitro* of 16-unsaturated C 19 keto steroids in boar testis. Biochem J 127:23P–24P

Claus R (1979) Mammalian pheromones with special reference to the boar taint steroid and its relationship to other testicular steroids (author's transl.). Fortschr Tierphysiol Tierernahr 10:1–136

Claus R, Hoffman B, Karg H (1971) Determination of 5α-androst-16-en-3-one, a boar taint steroid in pigs, with reference to relationships to testosterone. J Anim Sci 33:1293–1297

Denny PC, Ball WD, Redman RS (1997) Salivary glands: a paradigm for diversity of gland development. Crit Rev Oral Biol Med 8:51–75

Dorries KM, Adkinsregan E, Halpern BP (1995) Olfactory sensitivity to the pheromone, androstenone, is sexually dimorphic in the pig. Physiol Behav 57:255–259

Gower DB (1972) 16-Unsaturated C 19 steroids. A review of their chemistry, biochemistry and possible physiological role. J Steroid Biochem 3:45–103

Grinwich DL, Cliplef RL, Mckay RM (1988) Measurement of 16-androstenes (5-alpha-androst-16-en-3-one/5-alpha-androst-16-en-3-alpha-ol) in saliva of mature boars of 2 breeds following castration. Can J Anim Sci 68:969–972

Hogg CJ, Vickers ER, Rogers TL (2005) Determination of testosterone in saliva and blow of bottlenose dolphins (*Tursiops truncatus*) using liquid chromatography-mass spectrometry. J Chromatogr B 814:339–346

Kaminski RM, Marini H, Ortinski PI, Vicini S, Rogawski MA (2006) The pheromone androstenol (5 alpha-androst-16-en-3 alpha-ol) is a neurosteroid positive modulator of GABAA receptors. J Pharmacol Exp Ther 317:694–703

Katkov T, Booth WD, Gower DB (1972) The metabolism of 16-androstenes in boar salivary glands. Biochim Biophys Acta 270:546–556

Kirkwood RN, Forbes JM, Hughes PE (1981) Influence of boar contact on attainment of puberty in gilts after removal of the olfactory bulbs. J Reprod Fertil 61:193–196

Kirkwood RN, Hughes PE, Booth WD (1983) The influence of boar-related odors on puberty attainment in gilts. Anim Prod 36:131–136

Macchi E, Cucuzza AS, Badino P, Odore R, Re F, Bevilacqua L, Malfatti A (2010) Seasonality of reproduction in wild boar (*Sus scrofa*) assessed by fecal and plasmatic steroids. Theriogenology 73:1230–1237

Melrose DR, Reed HC, Patterson RLS (1971) Androgen steroids associated with boar odour as an aid to the detection of oestrus in pig artificial insemination. Br Vet J 127:497–502

Park CS, Yi YJ (2002) Comparison of semen characteristics, sperm freezability and testosterone concentration between Duroc and Yorkshire boars during seasons. Anim Reprod Sci 73:53–61

Patterson RLS (1968) 5α-androst-16-ene-3-one: compound responsible for taint in boar fat. Chem Ind 17:548–549

Perry GC, Patterson RLS, Macfie HJH, Stinson CG (1980) Pig courtship behavior—pheromonal property of androstene steroids in male sub-maxillary secretion. Anim Prod 31:191–199

Reed HC, Melrose DR, Patterson RLS (1974) Androgen steroids as an aid to the detection of oestrus in pig artificial insemination. Br Vet J 130:61–67

Sannikka E, Terho P, Suominen J, Santti R (1983) Testosterone concentrations in human seminal plasma and saliva and its correlation with non-protein-bound and total testosterone levels in serum. Int J Androl 6:319–330

Sinclair PA, Squires EJ (2005) Testicular sulfoconjugation of the 16-androstene steroids by hydroxysteroid sulfotransferase: its effect on the concentrations of 5-alpha-androstenone in plasma and fat of the mature domestic boar. J Anim Sci 83:358–365

Stefanczyk-Krzymowska S, Krzymowski T, Grzegorzewski W, Wasowska B, Skipor J (2000) Humoral pathway for local transfer of the priming pheromone androstenol from the nasal cavity to the brain and hypophysis in anaesthetized gilts. Exp Physiol 85:801–809

Trudeau VL, Grinwich DL, Sanford LM (1988) Seasonal variation in the blood concentration of 16-androstenes in adult landrace boars. Can J Anim Sci 68:565–568

Wang C, Plymate S, Nieschlag E, Paulsen CA (1981) Salivary testosterone in men: further evidence of a direct correlation with free serum testosterone. J Clin Endocrinol Metab 53:1021–1024

Weiler U, Claus R, Dehnhard M, Hofacker S (1996) Influence of the photoperiod and a light reverse program on metabolically active hormones and food intake in domestic pigs compared with a wild boar. Can J Anim Sci 76:531–539

Chapter 31
The Influence of Experimentally Reduced Vaginal Flora in Oestrus Females on the Mating Behaviour of Male Domestic Dogs (*Canis familiaris*)

Michał Dzięcioł, Wojciech Niżański, Roland Kozdrowski, Małgorzata Ochota, and Ewa Stańczyk

Abstract This study assessed the effect of reduced vaginal flora during oestrus on the mating behaviour of four male dogs. The experiment included a total of 32 bitches in oestrus that were allocated to three groups. Group A included untreated bitches ($n=15$), group B, bitches treated by an intravaginal infusion of an antibiotic ($n=15$) and group C, bitches treated by an intramuscular injection of antibiotic ($n=15$). Bitches in oestrus (i.e. with progesterone concentrations of 10–15 ng/mL) were presented to the male dogs. Our results demonstrate that antibiotic-treated bitches with reduced vaginal bacterial flora (females in groups B and C) were less attractive to males than untreated females. There was no significant difference between the two methods of antibiotic administration. These results indicate that vaginal flora produces semiochemical substances (pheromones) during oestrus that induce sexual arousal in male dogs.

31.1 Introduction

Olfactory signals play a significant role in the induction of animal behaviours, including social and sexual behaviours (Kustritz 2005; Ungerfeld et al. 2006). We investigated whether chemical signals produced by vaginal flora in oestrus bitches induced sexual arousal in male domestic dogs. We predicted that the administration of an antibiotic that kills semiochemical producing bacteria to oestrus bitches would lead to a reduction in the production of semiochemical substances and result in

M. Dzięcioł (✉) • W. Niżański • R. Kozdrowski • M. Ochota • E. Stańczyk
Department of Reproduction and Clinic of Farm Animals, Wrocław University of Environmental and Life Sciences, Plac Grunwaldzki 49, 50-366 Wrocław, Poland
e-mail: michaldzieciol@wp.pl

M.L. East and M. Dehnhard (eds.), *Chemical Signals in Vertebrates 12*,
DOI 10.1007/978-1-4614-5927-9_31,

decreased sexual interest in treated oestrus females by male dogs. We evaluated this idea by comparing the behaviour of male dogs to bitches treated with antibiotics and untreated bitches.

31.2 Material and Methods

31.2.1 Study Animals

The experiment was carried out during a period of 36 months in three German shepherd dog breeding kennels located in Poland and in Germany. Study animals included four stud male dogs between 5 and 7 years of age and 32 bitches between 2 and 5 years of age.

31.2.2 The Experimental Procedure

The 32 bitches were divided into three groups. Group A included 15 bitches that received no antibiotic treatment. Group B included 15 bitches that received intravaginal antibiotic treatment. Group C included 15 bitches which received intramuscular injection of an antibiotic. Some bitches were used in more than one group in the different cycles (i.e. 1 year a female was in group A, and in the following year, she was part of group B or C). The same bitches were also presented to various males in the different cycles. The bitches from group B and C were patients of the Clinic of Reproduction (Wrocław University of Environmental and Life Sciences). The antibiotic treatment was performed in these bitches, according the individual medical indications (reduced fertility in previous cycles, but without any pathological discharge from vagina). The bitches in all cases were treated with a broad spectrum antibiotic, Synulox® (Amoxicillin with clavulanic acid, Pfizer) at a dose 12.5 mg/kg body weight in 24 h intervals for 7 days, and the effectiveness of the therapy was verified by the control bacteriological examination performed 2 days before the expected time of mating.

31.2.2.1 Test of the Sexual Attractiveness of Bitches

The same test procedures described below were applied to bitches in all three groups. The sexual attractiveness of the bitches was tested using an experienced stud dogs that had a history of many previous matings per year (in Germany, stud German shepherd dogs are allowed up to a maximum of 90 matings per calendar year). Their reactions to females were analysed, based on the modified method proposed by Goodwin et al. (1979). Bitches in this study were exposed to a male when their progesterone concentration reached 10–15 ng/mL, which has been reported as the optimum time for mating (Johnston et al. 2001; Manothaiudom et al. 1995). The bitches were transported to the place of residence of the stud dog. Initially, a female was let free in a restricted area where she could urinate and the male could observe her behaviour. Then, the bitch was situated in the confined space which was known to the male as a place for mating (copulative point). The male was introduced to the female from behind. To avoid an adverse reaction by the female, in all cases, bitches were gently restrained by their owners during the procedures of testing. During this contact period, the reaction of the male to the female was recorded according to the following criteria: Reaction type I—the male was immediately interested in the bitch, after short contact the male mounted the female and within 3 min mated the bitch without any complications. Reaction type II—the male showed little interest in the bitch, he smelled her vulva, but he did not mount the bitch (refused mating). At least six females from groups A, four females from group B and four from group C were presented to each stud dog.

31.2.3 Bacterial Analysis of Vaginal Samples

Samples from the vagina were taken by a direct contact, avoiding rubbing; from each bitch, starting on the first day of oestrus, followed by a control check conducted 2 days before the planned date for mating. Each sample was cultured by conventional technique on blood agar and MacConkey agar media (Kustritz 2006; Szewczyk 2005). Isolated bacteria were identified using the following commercial tests: Api Staph, Api 20 C Aux, Api 20 Strept, Api 20 E, Api 20 Ne and Api Coryne (all tests were obtained from bioMérieux SA, Lyon, France).

31.2.4 Progesterone Assay and Timing of Mating

Eight millilitres of blood were taken by venipuncture from the cephalic vein in heparinised tubes. Plasma was separated 60 min after blood collection by centrifugation for 15 min at 2,000 g. Progesterone concentration was determined on the same day with the use of a commercial radioimmunoassay (RIA) kit (progesterone Coat-a-Count kit, Diagnostic Products Corporation, Los Angeles, CA, USA) validated for dog plasma (Srikandakumar et al. 1986). Blood samples ($n = 120$) were taken from each bitch at 48–72 h intervals starting from the sixth day of the oestrus cycle.

The study was approved by the Local Ethical Committee for the Affairs of Experiments on Animals of Wrocław University of Environmental and Life Sciences.

31.2.5 Statistical Analysis

Sexual attractiveness was analysed by Mann–Whitney *U* test (Statistica 9.1).

Table 31.1 The reaction of the male domestic dogs (Reaction type I and II) during contact with female dogs not treated with the use of antibiotic (group A) and treated with antibiotic, either administrated intravaginally (group B) or by intramuscular injection (group C)

Males	Experimental groups	No. of females in groups	RI/RII
I	A	6	6/0
	B	4	0/4
	C	4	1/3
II	A	6	6/0
	B	4	0/4
	C	4	1/3
III	A	6	6/0
	B	4	1/3
	C	4	0/4
IV	A	6	5/1
	B	4	1/3
	C	4	0/4

R I–reaction type I
R II–reaction type II

31.3 Results

In group A the males mated with almost all bitches. Only in one case (dog no. IV) mating was not performed successfully. In both group B and C, males mated with only two females and in most cases they refused mating the females treated with antibiotics (Table 31.1). Statistical analysis showed differences in attractiveness of females in group A and B ($p<0.000001$) and A and C ($p<0.000001$), but there was no difference between groups B and C ($p=0.31$). As all bacteriological samples from treated bitches were negative, these results indicate that antibiotic treatment significantly reduced vaginal microflora.

31.4 Discussion

The influence of bacterial flora on the composition of aromatic substances often called copulins, present in the vaginal discharge has been described in several species including humans (Albone et al. 1978; Dravnieks et al. 1970; Gower et al. 1986, 1994; Kohl et al. 2001; Konopski and Koberda 2003). In dogs, modification of bacterial flora in anal glands caused changes of the composition of chemical signals that influenced the behaviour in a dog pack (Pageat and Gaultier 2003). In humans a reduction of the saprophytic bacteria on human skin can cause a reduction of gender-specific odours (Bird and Gower 1982).

The presence of bacteria in the vagina of female domestic dogs during the proestrus and oestrus is known (Allen and Dagnall 1982; Baba et al. 1983; Janowski et al. 2006; Watts et al. 1996; Zduńczyk et al. 2006). The increase in the number of leukocytes observed during the transition period from oestrus to dioestrus is the reason for the significant reduction of bacteria and is correlated with the reduction of attractiveness of the females to the males (Baba et al. 1983). Hormonal action is also important for specific female behaviours during oestrus and the results of our experiment suggested the importance of the presence of vaginal microflora for the creation of specific semiochemical signals detected by the male dogs. These results are consistent with the observations made by previous studies (e.g. Merkx et al. 1988; Ungerfeld and Silva 2005).

References

Albone ES, Gosden PE, Ware GC, MacDonald DW, Hough NG (1978) Bacterial action and chemical signalling in the red fox. Proc Am Chem Soc 67:78–91

Allen W, Dagnall GJR (1982) Some observations on the aerobic bacterial flora of the genital tract of the dog and bitch. J Small Anim Pract 23:325–335

Baba E, Hata H, Fukata T, Arakava A (1983) Vaginal and uterine microflora of adult dogs. Am J Vet Res 44:606–609

Bird S, Gower DB (1982) Axilary androstenone, cholesterol and squalene in men: preliminary evidence for androstenone being a product of bacterial action. J Steroid Biochem Mol Biol 17:517–522

Dravnieks A, Krotoszynski B, Keith L, Bush I (1970) Odor threshold and gas chromatographic assays of vaginal odors: changes with nitrofurazone treatment. J Pharm Sci 59:495–499

Goodwin M, Gooding KM, Regnier F (1979) Sex pheromone in the dog. Science 203:559–561

Gower BD, Nixon A, Jackman PJH, Malet AI (1986) Transformation of steroids by axilary coryneform bacteria. Int Cosmet Sci 8:149–158

Gower DB, Holland KT, Mallet AI, Rennie PJ, Watkins WJ (1994) Comparison of 16-androstene steroid concentrations in sterile apocrine sweat and axillary secretions: interconversions of 16-androstenes by the axillary microflora-mechanism for axillary odour production in man. J Steroid Biochem Mol Biol 48:409–418

Janowski T, Zduńczyk S, Jurczak A (2006) Bacterial flora of the genital tract in healthy bitches during the oestrous cycle and in bitches with vaginal discharge. Reprod Dom Anim 41(suppl 1):17

Johnston SD, Kustritz MVR, Olson PNS (2001) Canine and Feline Theriogenology. Saunders, Philadelphia

Kohl J, Atzmueller M, Fink B, Grammer K (2001) Human pheromones: integrating neuroendocrinology and ethology. Neuro Endocrinol Lett 22:309–321

Konopski L, Koberda M (2003) Feromony człowieka. Środki komunikacji chemicznej między ludźmi. Scholar, Warszawa, pp 129–167

Kustritz MVR (2005) Reproductive behavior in small animals. Theriogenology 64:734–774

Kustritz MVR (2006) Collection of tissue and culture samples from the canine reproductive tract. Theriogenology 66:567–574

Manothaiudom K, Johnston SD, Hegstad RL (1995) Evaluation of the accuracy of the ICAGEN–Target Canine Ovulation Timing Diagnostic Test in detecting canine plasma progesterone concentrations. J Am Anim Hosp Assoc 31:57–64

Merkx J, Slob AK, van der Werff ten Bosch JJ (1988) Vaginal bacterial flora partially determines sexual attractivity of female rats. Physiol Behav 44:147–149

Pageat P, Gaultier E (2003) Current research in canine and feline pheromones. Vet Clin North Am Small Anim Pract 33:187–211

Srikandakumar A, Ingraham RH, Ellsworth M, Archbald LF, Liao A, Godke RA (1986) Comparison of a solid-phase, no-extraction radioimmunoassay for progesterone with an extraction assay for monitoring luteal function in the mare, bitch, and cow. Theriogenology 26:779–793

Szewczyk EM (2005) Diagnostyka bakteriologiczna. PWN, Warszawa

Ungerfeld R, Silva L (2005) The presence of normal vaginal flora is necessary for normal sexual attractiveness of estrous ewes. Appl Anim Behav Sci 93:245–250

Ungerfeld R, Ramos MA, Möller R (2006) Role of the vomeronasal organ on ram's courtship and mating behaviour, and on mate choice among oestrous ewes. Appl Anim Behav Sci 99:248–252

Watts JR, Wright PJ, Whithear KC (1996) Uterine, cervical and vaginal microflora of the normal bitch. J Small Anim Pract 37:54–60

Zduńczyk S, Janowski T, Borkowska I (2006) Flora bakteryjna pochwy i macicy suk w stanach fizjologicznych i zapalnych. Medycyna Wet 62:1116–1119

Chapter 32
Socio-Sexual Stimuli and Reproductive Function: Emerging Perspectives of the Male Effect in Sheep and Goats

Penny A.R. Hawken and Graeme B. Martin

Abstract Producers are facing increasing public scrutiny of the methods used in the production of animal products. Concerns over hormone residues in meat and milk may lead to restricted use of exogenous hormones in the reproductive management of farm animals in the future, so it is important to develop non-pharmacological methods of oestrus synchronisation. The 'male effect' was first reported in sheep in the 1940s and describes the capacity of the male to increase the secretion of reproductive hormones and induce ovulation in the female. It is widely accepted that olfactory signals (ostensibly 'pheromones') are primarily responsible for the profound shift in the activity of the reproductive centres of the female brain, though the chemical nature of this signal is yet to be conclusively identified for either sheep or goats. On the other hand, our understanding of the mechanism through which socio-sexual stimuli stimulate the hypothalamic-pituitary-gonadal axis of females has been greatly improved through a combination of histological, electrophysiological and endocrinological techniques. It is clear that olfactory stimuli from males are primarily transmitted through the main olfactory system, with audio and visual stimuli playing a synergistic but relatively minor role. In spite of over 60 years of research in this field, there are several areas of current and potential research that should improve our understanding of this remarkable phenomenon and its application to farm animal management.

P.A.R. Hawken (✉) • G.B. Martin
School of Animal Biology, The University of Western Australia,
35 Stirling Highway, Crawley, WA 6009, Australia

UWA Institute of Agriculture (Animal Production), The University of Western Australia,
35 Stirling Highway, Crawley, WA 6009, Australia
e-mail: penny.hawken@uwa.edu.au

M.L. East and M. Dehnhard (eds.), *Chemical Signals in Vertebrates 12*,
DOI 10.1007/978-1-4614-5927-9_32,

32.1 Introduction

Many strategies for the reproductive management of farm animals rely on exogenous hormones to synchronise oestrus prior to natural mating or artificial insemination. However, consumers are becoming increasingly concerned about the use of chemicals in animal production (Martin et al. 2004; Martin and Kadokawa 2006; Scaramuzzi and Martin 2008) as evidenced by the banning of synchronisation of oestrus with oestradiol-17β in the European Union in 2006 (Lane et al. 2008). Public scrutiny of animal production systems is likely to grow (Scaramuzzi and Martin 2008), so it is imperative that we develop non-pharmacological methods for controlling reproduction. One alternative lies in exploitation of the social interactions between males and females, a phenomenon termed the 'male effect'. In sheep and goats, exposure of anovulatory females to males increases the pulsatile secretion of gonadatrophin releasing hormone (GnRH) from the hypothalamus, and thus luteinising hormone (LH) from the anterior pituitary gland, a response that can lead to a preovulatory surge of LH, ovulation and a resumption of cyclic activity (reviews; Martin et al. 1986; Rosa and Bryant 2002; Ungerfeld 2007; Delgadillo et al. 2009). This sequence of events can be induced when spontaneous ovulation is absent because of season, sexual immaturity or lactation (Delgadillo et al. 2009) and is heavily affected by a number of internal (genotype, sexual experience) or external factors (e.g. photoperiod, nutrition). Socio-sexual stimuli from males can be used to induce puberty and reduce postpartum anoestrus in swine and cattle (Rekwot et al. 2001) but, for the purpose of this review, we will focus on evidence from sheep and goats for which the male effect is best documented.

The term socio-sexual stimuli encompasses olfactory, auditory, visual and tactile stimuli from males, all of which combine to yield the optimum ovulatory response in anovulatory ewes exposed to rams (Pearce and Oldham 1988; Gelez et al. 2004a). This phenomenon is predominantly driven by a pheromone produced by the skin of the male (Knight and Lynch 1980a, b), with auditory and visual stimuli playing only minor roles (Hawken et al. 2009b). In this review, we will describe our current understanding of how socio-sexual stimuli affect the hypothalamic-pituitary-gonadal (HPG) axis of female sheep and goats and we will present emerging perspectives that are likely to play a major role in the development of our understanding and future application of the male effect in farm animal species.

32.2 Where Are We Now?

Changes in GnRH secretion in females are difficult to measure directly, but we can bioassay the effects of exposure to males by measuring changes in the concentration and characteristics of LH secretion in the peripheral circulation (review: Delgadillo et al. 2009). The strength of this approach is demonstrated by the relationship between pulses of GnRH in the hypophyseal blood and LH in the peripheral circulation, well established by direct studies in sheep (Clarke and Cummins 1982) and supported by histological (Gelez and Fabre-Nys 2006) and electrophysiological analyses of the male effect (Hamada et al. 1996; Ichimaru et al. 1999).

32.2.1 Histological Evidence

The immediate early gene c-Fos is a valuable marker of neural activation because the transient nature of the protein product Fos allows identification of cells activated within 2 h of exposure to a stimulus. Gelez and Fabre-Nys (2006) used this technique to detect regions of the female sheep brain activated by exposure to an intact male, male fleece or female fleece. They were able to use fleece because, in contrast to rodents where male urine can be used to induce puberty (Vandenbergh 1973) or alter oestrus activity (Whitten 1956), the male pheromone in sheep and goats originates from the sebaceous glands of the skin and ante-orbital wax of the eyes (Knight and Lynch 1980b; Cohen-Tannoudji et al. 1994). Thus exposure to male fleece alone can elicit a robust increase in LH secretion in sexually experienced female sheep and goats (Over et al. 1990; Cohen-Tannoudji et al. 1994; Gelez et al. 2004a). Using this approach, Gelez and Fabre-Nys (2006) were able to differentiate between regions of the brain only stimulated by olfactory stimuli from the male and those regions responding to the full compliment of socio-sexual stimuli. Exposure to an intact male increased the number of Fos-immunoreactive (Fos-IR) cells in key structures of both the accessory and main olfactory systems including the cortical (CoA) and medial nuclei (MeA) of the amygdala, multiple cortical regions, the ventromedial nucleus (VMN) and medial pre-optic area (mPOA) of the hypothalamus and the dentate gyrus of the hippocampus. Exposure to male odour increased Fos immunoreactivity but to a lesser extent than males in the CoA, basal amygdala, mPOA and VMN of the hypothalamus and only elicited more Fos activation than exposure to the female odour in the CoA, in the mitral layer of the main olfactory bulb, CoA and dentate gyrus (Gelez and Fabre-Nys 2006). In contrast, the percentage of GnRH cells double-labelled for Fos was similar in the organum vasculosum of the lamina terminalis (OVLT) and mPOA whether females were exposed to male odour or the intact male. Gelez and Fabre-Nys (2006) detected very few GnRH cells double-labelled for Fos in other regions of the brain, including the mediobasal hypothalamus. This observation contrasts with the response of male sheep to female sheep, where activation of a population of GnRH neurons in the mediobasal hypothalamus was associated with the pulsatile increase in LH secretion (Boukhliq et al. 1999). Sexual dimorphism in the olfactory system has been reported in many species (review; Keller et al. 2009), so it seems likely that there are differences between male and female sheep in the neural pathways activated during exposure to socio-sexual stimuli. Alternatively, this divergence could be an artefact of differences in the time course of euthanasia relative to exposure to the socio-sexual stimulus (2 h; Boukhliq et al. 1999; 1.5 h; Gelez and Fabre-Nys 2006). Activation of the main olfactory system by both males and male odour concurs with previous studies showing that this pathway is critical in mediating the LH response of sexually experienced female sheep to male odour

Table 32.1 Density of Fos-immunoreactive (Fos-IR) cells in the arcuate nucleus (ARC), ventromedial nucleus of the hypothalamus (VMH) and pre-optic area (POA) of female sheep exposed to rams for 2 h (+2 h male) or 6 h (+6 h male)

	ARC Fos-IR cells/mm^2	VMH Fos-IR cells/mm^2	POA Fos-IR cells/mm^2
Control	2.6 ± 0.4^a	6.1 ± 0.9^a	7.3 ± 0.2^a
+2 h Male	9.3 ± 0.8^b	14.8 ± 0.9^b	16.8 ± 1.5^b
+6 h Male	12.4 ± 1.9^b	18.4 ± 1.8^c	21.9 ± 1.3^c

Different superscripts indicate differences between treatments within a specific region $P<0.05$ (Hawken et al., unpublished data)

(review; Gelez and Fabre-Nys 2004). The role of the accessory olfactory system is less clear because it cannot substitute for the main olfactory system (Cohen-Tannoudji et al. 1989) and is not activated by male odour or males in sexually naïve female sheep (Chanvallon and Fabre-Nys 2009).

Less is known about the activation of brain regions after the first 1.5 h of the acute LH response, but preliminary data from our laboratory indicates that the number of Fos-IR cells increased between 2 and 6 h after initial exposure to males in the ventromedial hypothalamus and pre-optic area but not in the arcuate nucleus (Table 32.1; Hawken et al., unpublished data). This observation suggests that there is an increased neural activation over time, possibly associated with positive feedback by oestradiol on GnRH secretion preceding the pre-ovulatory surge of LH.

32.2.2 Electrophysiological Evidence

Mori and colleagues adapted the multi-unit activity (MUA) technique originally developed to assess neural activation in the rhesus monkey for use in the Shiba goat (review; Okamura et al. 2010). Using this technique, they were able to show, in real time, that periodic increases in electrical activity (MUA 'volleys') were associated with pulses of LH secretion in the peripheral circulation (Mori et al. 1991; Mori and Tanaka 1995). They were then able to show that exposure of female goats to male hair (Hamada et al. 1996), or an extract reputedly containing the male pheromone (Iwata et al. 2003), elicited an MUA volley within just a few minutes of exposure; in turn, the volley was associated with a pulse in LH secretion. Recent studies have shown that continued exposure to male hair reduces in the inter-volley interval compared to control females (Ichimaru et al. 1999) and that the timing of exposure relative to the last MUA volley determines whether the male odour induces an MUA volley and a pulse of LH (Murata et al. 2011).

The electrodes were originally implanted in close apposition to the arcuate nucleus and median eminence, where they clearly reflected changes in electrical activity stimulated by olfactory signals from the male (review; Okamura et al. 2010). However, the identity, nature and function of the cells responsible for this electrical activity are more controversial. In more recent studies, they targeted at the population of kisspeptin cells in the caudal aspect of the arcuate nucleus (Ohkura et al. 2009; Murata et al. 2011) and detected a close relationship between MUA activity and pulses of LH in the peripheral circulation. The existence of a GnRH pulse generator is a controversial topic far beyond the scope of this review (see review by Clarke 2011), but some authors argue that the MUA volleys reflect activity of the kisspeptin cells in the arcuate nucleus and that these cells constitute the 'GnRH pulse generator' (Ohkura et al. 2009; Okamura et al. 2010; Wakabayashi et al. 2010).

32.3 Where to in the Future?

32.3.1 Kisspeptin and the Male Effect

The product of the Kiss-1 gene, kisspeptin and the kisspeptin receptor (GPR54) play a major role in the control of GnRH secretion (Gottsch et al. 2004; Smith et al. 2006; Oakley et al. 2009; Caraty et al. 2010; Tsutsui et al. 2010). Clarke (2011) summarised some of the most convincing evidence for sheep: (1) over 90% of GnRH cells express the GPR54 receptor; (2) infusion of kisspeptin can induce ovulation in anovulatory female sheep; (3) expression of kisspeptin in the arcuate nucleus is increased during the follicular phase of oestrus ewes; (4) kisspeptin cells in both the arcuate nucleus and pre-optic area express oestrogen receptor α (ERα; approximately 100% vs. 50%) and appear to be involved in mediating positive feedback by oestradiol on GnRH secretion.

The link between kisspeptin and oestrogen positive feedback, well established in rodent species, involves up-regulation of kisspeptin cells in the AVPV/PeV and down-regulation of kisspeptin cells in the arcuate nucleus (review; Smith 2009). This interaction is less clear-cut in the ovine brain where the kisspeptin cells of the ARC nucleus are up-regulated during the late follicular phase and there are dynamic differences in the up-/down-regulation of kisspeptin cells in response to acute or chronic oestrogen treatment (review; Clarke 2011). Almost all kisspeptin cells co-express dynorphin (DYN) and neurokinin B (NKB) and the secretion of these neuropeptides, combined with the presence of both progesterone and oestrogen receptors, has led to the hypothesis that these neurons play an integral role in controlling GnRH secretion (Okamura et al. 2010; Wakabayashi et al. 2010). However, despite the wealth of evidence linking kisspeptin to GnRH secretion, no studies have yet directly correlated activation of kisspeptin cells with the increase in LH secretion characteristic of the male effect. Their role has been inferred in the MUA-goat model through placement of electrodes in close proximity to the arcuate population of kisspeptin neurons (Ohkura et al. 2009). The authors detected regular volleys in MUA activity that were accompanied by a pulse of LH and found that exogenous administration of kisspeptin increased LH secretion but did not affect the timing of the next MUA volley (Ohkura et al. 2009). Ohkura et al. (2009) interpreted this as further evidence that the MUA volley was reflecting electrical activity from neurons other than the GnRH neurons themselves. The characteristics of the MUA volley in females implanted with electrodes in this location were affected by gonadal steroids and direct infusion of NKB and DyN, further supporting the hypothesis that the MUA signal reflects the activity of the KNDy neurons of the arcuate nucleus, implicating an integral role in pheromone-induced changes in GnRH/LH secretion (Wakabayashi et al. 2010). However, it should be noted that the changes in MUA activity and corresponding changes in LH secretion were observed in female goats exposed to the male odour after implantation of electrodes in a more general location within the arcuate nucleus and median eminence (Hamada et al. 1996; Ichimaru et al. 1999). It is clearly difficult to conclusively identify the cells responsible for the changes in electrical activity and we propose that detailed histological and pharmacological studies are required to definitively determine the role of kisspeptin in mediating the GnRH/LH response of females to male socio-sexual stimuli.

32.3.2 Cell Proliferation and Male Novelty

Male novelty is an important factor in determining the physiological response of female sheep to males. Female sheep maintained in continuous contact with males will respond to novel males, but not those that they have been maintained with for several months (Hawken et al. 2009c; Jorre de St Jorre et al. 2012). Activation of the reproductive axis by novel males appears to continue through to ovulation in both sheep (Cushwa et al. 1992) and goats (Veliz et al. 2006). The acute LH response of female sheep to males is associated with a rapid and robust increase in the rate of cell proliferation in the dentate gyrus of the hippocampus (Hawken et al. 2009c). It is logical to assume that these cells play a role in allowing female sheep to distinguish between novel and familiar males because neurogenesis in this region in mice is linked to the ability of females to distinguish between subordinate and dominant males (Mak et al. 2007). The subventricular zone and rostral migratory stream are also regions of pheromone-induced neurogenesis in mice (Mak et al. 2007) and thus cell proliferation in this region may be expected to be up-regulated in female sheep exposed to males. Constitutive cell proliferation occurs in the subventricular zone and main olfactory bulb of female sheep (Brus et al. 2010), but we were unable to quantify the impact of male stimuli on cell proliferation in the subventricular zone due to technical issues (Hawken et al. 2009c) so the male effect on cell proliferation in this region is not yet known.

The magnitude and time course of the male-induced increase in cell proliferation in the dentate gyrus of females contrasts dramatically between sheep and mice. For example, exposure of female sheep to males doubled cell proliferation in the dentate gyrus within hours of initial exposure to the male (Hawken et al. 2009c) whereas, in mice, the rate of neurogenesis only increased by 38% in association with synchronisation of pro-oestrus 7 days after initial exposure to the male odour (Mak et al. 2007). In sheep, changes in the rate of cell proliferation in the dentate gyrus have only been studied relative to the acute LH response (<6 h after male exposure), but the proven relationship between neurogenesis, oestradiol and LH in rodents

(Fowler et al. 2002; Mak et al. 2007) suggests that the rate of cell proliferation may increase further between the acute LH response and the LH surge.

In contrast to rodents, exposure of female sheep to males has little effect on the concentration of prolactin (Poindron et al. 1980) so this pituitary hormone is unlikely to be as involved in neurogenesis in sheep, as has been reported in rodent species (Larsen et al. 2006, 2008). Mak et al. (2007) argue an obligatory role for LH in mediating pheromone-induced neurogenesis in the dentate gyrus of female mice, but this remains a contentious issue due to uncertainty over the ability of LH to cross the blood–brain barrier and changes in other hormones, including oestradiol, that parallel changes in LH secretion. The mechanism through which male stimuli induce such a rapid increase in cell proliferation in the dentate gyrus of female sheep is currently unknown, but the well-established relationship between oestrogen and pheromone-induced neurogenesis in prairie voles makes oestradiol an attractive candidate (Fowler et al. 2002).

For decades, a period of separation of females and males was deemed necessary for males to induce ovulation in anovulatory female sheep and goats (Delgadillo et al. 2009). We now know that separation is not necessary if the males are unfamiliar or novel (Hawken et al. 2009c; Jorre de St Jorre et al. 2012), but we do not fully understand what makes a male 'familiar' or 'novel'. If the cells that proliferate during the first few hours of male exposure are involved in forming a memory of the male, as appears to be the case in mice (Mak et al. 2007), what happens to the 'memory' of these males once they have been separated from the females. In an attempt to answer this question, Jorre de St Jorre et al. (2012) compared LH secretion in female sheep maintained with males for several months that were either re-exposed to the same 'familiar' males after 15 min, 1 day, 17 days or 1 month of separation or were exposed to novel intact or castrated males. There was no change in LH secretion in females re-exposed to the same males after 15 min, 1 day or 17 days of separation. However, after 1 month of separation, re-exposure of females to the once-'familiar' males induced a similar but reduced increase in LH secretion compared to female sheep exposed to novel males. Furthermore, only 25% of females had an LH surge within 60 h of progestagen removal, compared to 100% of females exposed to novel males. Clearly, the familiar males were only able partially activate of the reproductive axis of female sheep, even after 1 month of separation (Jorre de St Jorre et al. 2012). Continued male presence is necessary for sustained elevation of pulsatile secretion of LH in female sheep (Pearce and Oldham 1984) and may also be important for the survival of cells that proliferate in response to the male stimulus. In rodents, environmental enrichment increases the rate of cell survival (van Praag et al. 2000) and odour deprivation is associated with reduced survival of newly proliferated cells (Corotto et al. 1994). Cohen-Tannoudji and Signoret (1987) found that female sheep exposed to males for 2–3 h responded to males with increased LH secretion as little as 24 h after the first exposure, indicating that this short period was insufficient for the female to become refractory to the male. No studies to date have established the time taken for females to become refractory to males, but it is clear from studies in our laboratory that this period is less than 3 months (Jorre de St Jorre et al. 2012). Clearly, the period of habituation to and separation from the male is correlated with the capacity of males to stimulate GnRH/LH secretion in the female sheep, but how this interacts with cell birth or cell death in the hippocampus or subventricular zone is currently unknown.

Female sheep can clearly differentiate between familiar and novel males and tailor their neuroendocrine response accordingly (Hawken et al. 2009c; Jorre de St Jorre et al. 2012), but which stimuli do they use to do this? Olfactory stimuli are perhaps the most logical candidate because female sheep use olfaction to discriminate between their own and alien offspring (review; Kendrick et al. 1997) and because, in mice, exposure to the male-specific, major urinary protein 'Darcin' stimulates the female to associate the unique combination of volatile compounds in the male urine with individual males (Roberts et al. 2010). No studies to date have tested whether female sheep respond differentially to the odour of familiar or novel males, so it is currently impossible to determine whether olfactory stimuli play a principle role in mediating the capacity of female sheep to differentiate between individual males. The male pheromone in both sheep and goats has proved difficult to identify (review; Delgadillo et al. 2009), but this complexity in itself may mean that it could encode the identity of individual males. However, visual stimuli are also important for social communication in sheep (reviews; Kendrick 1991, 2008; Kendrick et al. 1995) and we recently found that projected images of male sheep elicit a small but significant neuroendocrine response in female sheep (Hawken et al. 2009b). It seems unlikely that the sheep would 'forget' an individual male in as short a space of time as 1 month since they can remember faces for over 2 years (Kendrick et al. 2001). We postulate that the network of cells that proliferate in response to the male deteriorate over time if the male stimulus is removed, enabling the male to again stimulate LH secretion in the female. In wild populations of sheep, the males and females separate outside the rutting season and rejoin for mating (review; Ungerfeld et al. 2004). Thus it is logical that the olfactory memory of the male is lost to allow for migration and reunion of the sexes. Clearly further investigation is required to determine the role of cell proliferation and male identity in the male effect.

32.3.3 Sexual Experience

Sexual experience does not appear to affect the incidence, timing, magnitude or characteristics of the acute LH response of female sheep to males (Gelez et al. 2004a). However, it does directly affect the proportion of females ovulating in response to the male effect (Murtagh et al. 1984; Chanvallon et al. 2010a) and their expression of sexual behaviour (Gelez et al. 2004b; Hawken et al. 2008a). Therefore, it appears that while males can elicit a similar initial increase in LH secretion in sexually naïve and experienced females, maintenance of the response through to a pre-ovulatory surge of LH and ovulation is at least partially dependent on sexual experience. In spite of the similarity of the LH response of sexually naïve and sexually experienced female sheep to males, the pattern of neural activation in the brain appears to be different (Table 32.2).

Table 32.2 Regions of the brain quantified for Fos immunoreactivity in sexually experienced female sheep by Gelez and Fabre-Nys (2006) and in sexually naïve female sheep by Chanvallon and Fabre-Nys (2009)

	Sexually experienced females + male (Gelez and Fabre-Nys 2006)	Sexually naïve females + male (Chanvallon and Fabre-Nys 2009)
Main olfactory bulb		
Glomerular layer	×	×
External plexiform layer	NM	×
Mitral layer	✓	✓
Granular layer	✓	✓
Anterior olfactory nucleus	✓	×
Accessory olfactory bulb		
Glomerular layer	×	×
Granula mitral layer	✓	×
Diencephalic and hypothalamic structures		
OVLT	×	×
Pre-optic area	✓	×
Ventromedial nucleus	NM	✓
Paraventricular nucleus	×	×
Amygdala		
Cortical nucleus	✓	✓
Medial nucleus	✓	✓
Basal nucleus	✓	×
Central nucleus	×	×
Lateral nucleus	×	×
Hippocampus		
CA1	NM	×
CA3	NM	×
Dentate gyrus	✓	×
Cortical structures		
Orbitofrontal cortex	✓	×
Frontomedial cortex	✓	×
Insular cortex	NM	×
Cingulate cortex	×	×
Anterior piriform cortex	✓	×
Enterorhinal cortex	✓	×
Visual cortex		
V1	×	×
V2	×	×
Brainstem		
Olive nucleus	NM	×
Locus coeruleus	NM	×
Inferior colliculus	NM	✓

The presence o✓ or × indicates whether females did (✓) or did not (×) have more Fos immunoreactive cells in that region than control females isolated from male socio-sexual stimuli. NM indicates that Fos immunoreactivity in that region was not measured in that study

For example, the granulosa mitral layers of the accessory olfactory bulb are activated in sexually experienced but not naïve female sheep exposed to males (Chanvallon and Fabre-Nys 2009). The functional relevance of this activation is unknown because the accessory olfactory system is not necessary for the male odour to elicit an LH response in female sheep (Cohen-Tannoudji et al. 1989). Chanvallon and Fabre-Nys (2009) postulated that this neural activation may be associated with investigation and sustained interest in the sexual stimuli, a theory that is supported by the distinct differences in sexual behaviour between sexually naïve and sexually experienced female sheep (Gelez et al. 2004b; Hawken et al. 2008a). Sexual experience appears to be necessary for females to express proceptive behaviour towards the male and is associated with dynamic changes in norepinephrine that are not evident in sexually naïve females (Gelez et al. 2004b). Furthermore, in contrast to sexually experienced females, the proportion of GnRH cells that were immunoreactive for Fos was typically lower in sexually naïve females and unaffected by exposure to male odour or males (Chanvallon and Fabre-Nys 2009).

Exposure to male odour does not elicit an LH response in sexually naïve females, indicating that the ability of female sheep to respond to male odour is completely dependent on sexual experience (Gelez et al. 2004a). Importantly, Gelez et al. (2004a) showed that it is the sexual experience rather than age of the female that dictates their response to male odour. The divergence in the LH response of sexually naïve and sexually experienced females to male odour is driven by incomplete activation of the olfactory pathways responsible for transmitting olfactory stimuli from males to the HPG axis of sexually naïve females (Chanvallon and Fabre-Nys 2009). Specifically, cells in the mitral layer of the main olfactory bulb were activated in sexually naïve females exposed to male odour, but no other structures of the main or accessory olfactory systems showed any increase in Fos immunoreactivity compared to control sheep (Chanvallon and Fabre-Nys 2009). Thus, in sexually naïve females, the olfactory signal seems to be arrested at the main olfactory bulb (Chanvallon and Fabre-Nys 2009). The concept of a learned element to the neuroendocrine response of female sheep to male odour is supported by their capacity to associate the scent of lavender oil with males and then respond to this odour with an increase in LH secretion (Gelez et al. 2004a). The response of female mice to male pheromones is said to be innate, but a recent review article acknowledged that the responses can be conditional on early experiences in life (Wyatt 2010). For example, female mice reared without contact with males are attracted to male bedding but do not respond to the volatile compounds associated with the male odour until they have been exposed to male bedding (Moncho-Bogani et al. 2002). Thus, the involvement of learning in the neuroendocrine response of females to male odour may not be restricted to ungulate species.

Less is understood about what constitutes 'sexual experience' in terms of improving the ovulatory response of sexually naïve female sheep to males. Murtagh et al. (1984) reported an increase in the proportion of 10–11-month-old female sheep ovulating in response to males after a 6-week 'pre-exposure' to males. However, Chanvallon et al. (2010b) found either no beneficial effect or an inhibitory effect of pre-exposure to males at 5 or 9 months of age on the neuroendocrine and/or ovarian response to the males at 12 months of age. They proposed that the stress associated with forced exposure to the males, the indoor housing and/or handling had a negative effect on either the reproductive development of the females or their perception of the male stimulus (Chanvallon et al. 2010b). The experimental conditions were indeed markedly different to Murtagh et al. (1984) where 10–11-month sexually naïve females were exposed to rams at pasture and at a lower ewe:ram ratio (3 rams per 100 ewes vs. 7 per 100 ewes). In another study, Hawken et al. (2008a) found that the behavioural interactions between 14-month-old, sexually naïve female sheep and males were improved by 7 days of exposure to males at pasture. The impact of this pre-exposure period on their subsequent neuroendocrine response was more marginal, but the authors did find that only female sheep with prior experience of males responded to males with an increase in the mean concentrations of LH and that an LH surge was detected in numerically more of these ewes (Hawken et al. 2008a). From these limited studies, we tentatively propose that age, duration of the pre-exposure period to males and reproductive status (cyclic, anovulatory due to season and/or pre-puberty) interact to determine the beneficial effect of prior exposure to males on the physiological responses of sexually naïve female sheep to males. The question of whether they need to experience one or multiple oestrus cycles or indeed pregnancy and/or lactation to become 'sexually experienced' requires further investigation.

32.3.4 Cyclic Females

The existence of a 'breeding season' male effect in sheep and goats was initially disregarded due to the strong inhibitory effect of progesterone on the reproductive axis (review; Delgadillo et al. 2009). However, females under the influence of both endogenous progesterone (Hawken et al. 2007, 2009a) and progesterone implants (Pearce and Oldham 1983) or medroxyprogesterone acetate (Evans et al. 2004) responded to males with an increase in pulsatile LH secretion similar to that observed in anovulatory female sheep. Sheep at all stages of the oestrous cycle show this LH response to males, but goats appear to have a threshold concentration of progesterone above which the LH response to males is blocked (Hawken et al. 2009a). The change in the hormonal milieu in female sheep under the influence of endogenous or exogenous progesterone does not directly induce ovulation but does appear to affect the timing of reproductive events in the subsequent oestrous cycle. For example, exposure of female sheep to males towards the end of a synchronisation protocol was associated with an earlier LH surge, ovulation and oestrus than females separated from males until 48 h after progestagen removal (Evans et al. 2004). This effect does not appear to be limited to male exposure during the luteal phase because male presence shortens the follicular phase both in spontaneous (Parsons and Hunter 1967; Fletcher and Lindsay 1971) and progestagen-induced oestrous cycles (Ungerfeld and Rubianes 1999; Romano et al. 2000, 2001; Evans et al. 2004). The implications of these changes in the endocrine milieu, particularly in females under the influence of progesterone or artificial progestagens, are not well understood, but there is evidence in both sheep and goats that exposure to males alters the distribution of oestrus (Ngere and Dzakuma 1975; Chemineau 1983; Hawken et al. 2008b; Hawken and Beard 2009). The mechanism through which males induce oestrus synchrony in cyclic females is unclear, but Hawken et al. (2007) suggested that changes in LH secretion have a luteotrophic effect during the early-mid luteal phase and a luteolytic effect during the late luteal phase that is combined with shortening of the follicular phase to synchronise oestrus among a group of cyclic females. Male novelty also appears to interplay with this phenomenon—replacement of males with novel males at 17-day intervals during the transition into the breeding season advanced the timing of oestrus in female sheep compared to continuous exposure to the same males for 50 days (Hawken and Beard 2009). Plasma progesterone indicated that 100% of females in both groups were cyclic within the first oestrous cycle, so the novel males were clearly having an effect on the distribution of oestrus in cyclic females (Hawken and Beard 2009). No work to date has quantified the brain regions activated by males in cyclic females and it is unclear how the socio-sexual stimuli overcome or bypass the suppressive effect of progesterone.

A possible concern with this 'breeding season male effect' is the potential impact on fecundity. Exposure of female sheep to males during the last 3 days of progestagen synchronisation was associated with a reduction in litter size (Evans et al. 2004; Hawken et al. 2005) that appeared to be mediated through a reduction in ovulation rate (Hawken et al. 2005). Embryo survival may also be affected by male presence because exposure of pregnant female sheep to novel males 10 days after mating reduced the proportion of twin births compared to control females (Al-Gubory 1998). In both females treated with artificial progestagens and pregnant females, exposure to males was associated with increased LH secretion, perhaps explaining the unexpected effects on litter size. Hawken (2005) reported a similar trend towards reduced litter size in female sheep exposed for males for 24 h at 17-day intervals over the transition into the breeding season over two independent studies (2003: Control: 2.05 ± 0.05 vs. Ram exposed: 1.75 ± 0.05 lambs per ewe; 2004: Control: 2.16 ± 0.07 vs. Ram exposed: 1.99 ± 0.07 lambs per ewe). The potential for males to have a negative impact on fecundity in cyclic females contrasts with the male effect in anovulatory females where the first ovulation is associated with an increase rather than decrease in ovulation rate (review; Martin et al. 1986).

32.4 Conclusions

The effect of socio-sexual stimuli from males on the HPG axis of female sheep and goats is complex and multi-factorial both in terms of the nature of the male stimulus and the response of the female. The sheer complexity of the perception and processing of the stimuli is evident in the multiple regions of the female brain that are activated by exposure to the male. In this review, we have focussed on areas of the male effect that are either emerging in importance or have received minimal attention in previous reviews of this field. However, there are many other factors including breed, seasonality and sexual behaviour of the male covered in detail elsewhere (Delgadillo et al. 2009) that are critical to our understanding of the male effect. There are also many pressing questions that still need to be answered if we are to optimise this phenomenon as a non-pharmacological method of controlling reproduction in sheep and goats. For example, is the olfactory signal from males a species-wide 'pheromone' combined with individual olfactory signatures? What are the key aspects of the acute LH response that lead to successful ovulation; is it LH pulse frequency, basal or mean concentrations of LH or a combination of all three? Can we condition young, sexually naïve females to males to improve their subsequent responses to male stimuli or are they inevitably limited by their stage of reproductive development?

Optimisation of the male effect in anovulatory females during seasonal and post-partum anoestrus is critical if this phenomenon is to provide a real alternative to pharmacological methods of oestrus synchronisation in sheep and goats. The potential use of socio-sexual stimuli to manipulate the distribution of oestrus in cyclic females requires further investigation to determine whether and to what extent males can effect the distribution of oestrus in female ungulates.

References

Al-Gubory KH (1998) Effects of the presence of rams during pregnancy on lambing performance in ewes. Anim Reprod Sci 52:205–211

Boukhliq R, Goodman RL, Berriman SJ, Adrian B, Lehman MN (1999) A subset of gonadotropin-releasing hormone neurons in the ovine medial basal hypothalamus is activated during increased pulsatile luteinizing hormone secretion. Endocrinology 140:5929–5936

Brus M, Meurisse M, Franceschini I, Keller M, Levy F (2010) Evidence for cell proliferation in the sheep brain and its down-regulation by parturition and interactions with the young. Horm Behav 58:737–746

Caraty A, Franceschini GE, Hoffman DL (2010) Kisspeptin and the preovulatory gonadotropin-releasing hormone/luteinising hormone surge in the ewe: basic aspects and potential applications in the control of ovulation. J Neuroendocrinol 22:710–715

Chanvallon A, Fabre-Nys C (2009) In sexually naive anestrous ewes, male odour is unable to induce a complete activation of olfactory systems. Behav Brain Res 205:272–279

Chanvallon A, Blache D, Chadwick A, Esmaili T, Hawken PA, Martin GB, Vinoles C, Fabre-Nys C (2010a) Sexual experience and temperament affect the response of Merino ewes to the ram effect during the anoestrous season. Anim Reprod Sci 119:205–211

Chanvallon A, Scaramuzzi RJ, Fabre-Nys C (2010b) Early sexual experience and stressful conditions affect the response of young ewes to the male. Physiol Behav 99:457–465

Chemineau P (1983) Effect on oestrus and ovulation of exposing creole goats to the male at three times of the year. J Reprod Fertil 67:65–72

Clarke IJ (2011) Control of GnRH secretion: one step back. Front Neuroendocrinol 32:367–375

Clarke IJ, Cummins JT (1982) The temporal relationship between gonadotropin releasing hormone (GnRH) and luteinising hormone (LH) secretion in ovariectomized ewes. Endocrinology 111:1737–1739

Cohen-Tannoudji J, Signoret JP (1987) Effect of short exposure to the ram on later reactivity of anoestrous ewes to the male effect. Anim Reprod Sci 13:263–268

Cohen-Tannoudji J, Lavenet C, Locatelli A, Tillet Y, Signoret JP (1989) Non-involvement of the accessory olfactory system in the LH response of anoestrous ewes to male odour. J Reprod Fertil 86:135–144

Cohen-Tannoudji J, Einhorn J, Signoret JP (1994) Ram sexual pheromone: first approach of chemical identification. Physiol Behav 56:955–961

Corotto FS, Henegar JR, Maruniak JA (1994) Odor deprivation leads to reduced neurogenesis and reduced neuronal survival in the olfactory bulb of the adult mouse. Neuroscience 61:739–744

Cushwa WT, Bradford GE, Stabenfeldt GH, Berger YM, Dally MR (1992) Ram influence on ovarian and sexual activity in anestrous ewes: effects of isolation of ewes from rams before joining and date of ram introduction. J Anim Sci 70:1195–1200

Delgadillo JA, Gelez H, Ungerfeld R, Hawken PA, Martin GB (2009) The 'male effect' in sheep and goats—revisiting the dogmas. Behav Brain Res 200:304–314

Evans AC, Duffy P, Crosby TF, Hawken PA, Boland MP, Beard AP (2004) Effect of ram exposure at the end of progestagen treatment on estrus synchronisation and fertility during the breeding season in ewes. Anim Reprod Sci 84:349–358

Fletcher IC, Lindsay DR (1971) Effect of rams on the duration of oestrous behaviour in ewes. J Reprod Fertil 25:253–259

Fowler CD, Liu Y, Ouimet C, Wang Z (2002) The effects of social environment on adult neurogenesis in the female prairie vole. J Neurobiol 51:115–128

Gelez H, Fabre-Nys C (2004) The "male effect" in sheep and goats: a review of the respective roles of the two olfactory systems. Horm Behav 46:257–271

Gelez H, Fabre-Nys C (2006) Neural pathways involved in the endocrine response of anestrous ewes to the male or its odor. Neuroscience 140:791–800

Gelez H, Archer E, Chesneau D, Campan R, Fabre-Nys C (2004a) Importance of learning in the response of ewes to male odor. Chem Senses 29:555–563

Gelez H, Archer E, Chesneau D, Lindsay D, Fabre-Nys C (2004b) Role of experience in the neuroendocrine control of ewes' sexual behavior. Horm Behav 45:190–200

Gottsch ML, Cunningham MJ, Smith JT, Popa SM, Acohido BV, Crowley WF, Seminara S, Clifton DK, Steiner RA (2004) A role for kisspeptins in the regulation of gonadotropin secretion in the mouse. Endocrinology 145:4073–4077

Hamada T, Nakajima M, Takeuchi Y, Mori Y (1996) Pheromone-induced stimulation of hypothalamic gonadotropin-releasing hormone pulse generator in ovariectomized, estrogen-primed goats. Neuroendocrinology 64:313–319

Hawken P (2005) The role of socio-sexual cues in sheep reproduction. PhD thesis, School of Agriculture Food and Rural Development. University of Newcastle, Newcastle upon Tyne, UK

Hawken PA, Beard AP (2009) Ram novelty and the duration of ram exposure affects the distribution of mating in ewes exposed to rams during the transition into the breeding season. Anim Reprod Sci 111:249–260

Hawken PA, Beard AP, O'Meara CM, Duffy P, Quinn KM, Crosby TF, Boland MP, Evans AC (2005) The effects of ram exposure during progestagen oestrus synchronisation and time of ram introduction post progestagen withdrawal on fertility in ewes. Theriogenology 63:860–871

Hawken PAR, Beard AP, Esmaili T, Kadokawa H, Evans ACO, Blache D, Martin GB (2007) The introduction of rams induces an increase in pulsatile LH secretion in cyclic ewes during the breeding season. Theriogenology 68:56–66

Hawken PAR, Evans ACO, Beard AP (2008a) Prior exposure of maiden ewes to rams enhances their behavioural interactions with rams but is not a pre-requisite to their endocrine response to the ram effect. Anim Reprod Sci 108:13–21

Hawken PAR, Evans ACO, Beard AP (2008b) Short term, repeated exposure to rams during the transition into the breeding season improves the synchrony of mating in the breeding season. Anim Reprod Sci 106:333–344

Hawken PAR, Esmaili T, Jorre de St Jorre T, Jorre T, Martin GB (2009a) Do cyclic female goats respond to males with an increase in LH secretion during the breeding season? Anim Reprod Sci 112:384–389

Hawken PAR, Esmaili T, Scanlan V, Blache D, Martin GB (2009b) Can visual or audio-visual stimuli of a prospective mate stimulate an LH response in sheep? Animal 3:690–696

Hawken PAR, Jorre de St Jorre T, Jorre T, Rodger J, Esmaili T, Blache D, Martin GB (2009c) Rapid induction of cell proliferation in the adult female ungulate brain (*Ovis aries*) associated with activation of the reproductive axis by exposure to unfamiliar males. Biol Reprod 80:1146–1151

Ichimaru T, Takeuchi Y, Mori Y (1999) Stimulation of the GnRH pulse generator activity by continuous exposure to the male pheromones in the female goat. J Reprod Dev 45:243–248

Iwata E, Kikusui T, Takeuchi Y, Mori Y (2003) Substances derived from 4-ethyl octanoic acid account for primer pheromone activity for the "male effect" in goats. J Vet Med Sci 65:1019–1021

Jorre de St Jorre T, Jorre T, Hawken PA, Martin GB (2012) Role of male novelty and familiarity in male induced LH secretion in female sheep. Reprod Fertil Dev 24(4):523–530

Keller M, Baum MJ, Brock O, Brennan PA, Bakker J (2009) The main and the accessory olfactory systems interact in the control of mate recognition and sexual behavior. Behav Brain Res 200:268–276

Kendrick KM (1991) How the sheep's brain controls the visual recognition of animals and humans. J Anim Sci 69:5008–5016

Kendrick KM (2008) Sheep senses, social cognition and capacity for consciousness. In: Dwyer CM (ed) The welfare of sheep. Springer, Dordrecht, pp 135–157

Kendrick KM, Atkins K, Hinton MR, Broad KD, Fabre-Nys C, Keverne B (1995) Facial and vocal discrimination in sheep. Anim Behav 49:1665–1676

Kendrick KM, Da Costa AP, Broad KD, Ohkura S, Guevara R, Levy F, Keverne EB (1997) Neural control of maternal behaviour and olfactory recognition of offspring. Brain Res Bull 44:383–395

Kendrick KM, da Costa AP, Leigh AE, Hinton MR, Peirce JW (2001) Sheep don't forget a face. Nature 414:165–166

Knight TW, Lynch PR (1980a) Source of ram pheromones that stimulate ovulation in the ewe. Anim Reprod Sci 3:133–136

Knight TW, Lynch PR (1980b) The pheromones that stimulate ovulation in the ewe. Proc Aust Soc Anim Prod 13:74–76

Lane EA, Austin EJ, Crowe MA (2008) Oestrous synchronisation in cattle—current options following the EU regulations restricting use of oestrogenic compounds in food-producing animals: a review. Anim Reprod Sci 109:1–16

Larsen CM, Kokay IC, Grattan DR (2006) The presence of a male mouse initiates mitogenesis in the subventricular zone in virgin C57B6 mice. Front Neuroendocrinol 27:80–81

Larsen CM, Kokay IC, Grattan DR (2008) Male pheromones initiate prolactin-induced neurogenesis and advance maternal behavior in female mice. Horm Behav 53:509–517

Mak GK, Enwere EK, Gregg C, Pakarainen T, Poutanen M, Huhtaniemi I, Weiss S (2007) Male pheromone-stimulated neurogenesis in the adult female brain: possible role in mating behavior. Nat Neurosci 10:1003–1011

Martin GB, Kadokawa H (2006) "Clean, green and ethical" animal production. Case study: reproductive efficiency in small ruminants. J Reprod Dev 52:145–152

Martin GB, Oldham CM, Cognie Y, Pearce DT (1986) The physiological response of anovulatory ewes to the introduction of rams—a review. Live Prod Sci 15:219–247

Martin GB, Milton JT, Davidson RH, Banchero Hunzicker GE, Lindsay DR, Blache D (2004) Natural methods for increasing reproductive efficiency in small ruminants. Anim Reprod Sci 82–83:231–245

Moncho-Bogani J, Lanuza E, Hernandez A, Noverjarque A, Martinez-Garcia F (2002) Attractive properties of sexual pheromones in mice: innate or learned. Physiol Behav 77:167–176

Mori Y, Tanaka T (1995) Electrophysiological approach to the hypothalamic GnRH pulse generator. J Reprod Fertil Suppl 49:231–243

Mori Y, Nishihara M, Tanaka T, Shimizu T, Yamaguchi M, Takeuchi Y, Hoshino K (1991) Chronic recording of electrophysiological manifestation of the hypothalamic gonadotropin-releasing hormone pulse generator activity in the goat. Neuroendocrinology 53:392–395

Murata K, Wakabayashi Y, Sakamoto K, Tanaka T, Takeuchi Y, Mori Y, Okamura H (2011) Effects of brief exposure of male pheromone on multiple-unit activity at close proximity to kisspeptin neurons in the goat arcuate nucleus. J Reprod Dev 57:197–202

Murtagh JJ, Gray SJ, Lindsay DR, Oldham CM (1984) The influence of the ram effect in 10–11 month old Merino ewes on subsequent performance when introduced to rams again at 15 months. Proc Aust Soc Anim Prod 15:490–493

Ngere LO, Dzakuma JM (1975) The effect of sudden introduction of rams on oestrous pattern of tropical ewes. J Agric Sci 84:263–264

Oakley AE, Clifton DK, Steiner RA (2009) Kisspeptin signaling in the brain. Endocr Rev 30:713–743

Ohkura S, Takase K, Matsuyama S, Mogi K, Ichimaru T, Wakabayashi Y, Uenoyama Y, Mori Y, Steiner RA, Tsukamura H, Maeda KI, Okamura H (2009) Gonadotropin-releasing hormone pulse generator activity in the hypothalamus of the goat. J Neuroendocrinol 21:813–821

Okamura H, Murata K, Sakamoto K, Wakabayashi Y, Ohkura S, Takeuchi Y, Mori Y (2010) Male effect pheromone tickles the gonadotropin-releasing hormone pulse generator. J Neuroendocrinol 22:825–832

Over R, Cohen-Tannoudji J, Dehnhard M, Claus R, Signoret JP (1990) Effect of pheromones from male goats on LH-secretion in anoestrous ewes. Physiol Behav 48:665–668

Parsons SD, Hunter GL (1967) Effect of the ram on duration of oestrus in the ewe. J Reprod Fertil 1967:61–70

Pearce DT, Oldham CM (1983) 'Ram effect' in the breeding season. Proc Aust Soc Reprod Biol 15:49

Pearce DT, Oldham CM (1984) The ram effect, its mechanism and application to the management of sheep. In: Lindsay DR, Pearce DT (eds) Reproduction in sheep. Cambridge University Press, Cambridge, pp 27–34

Pearce DT, Oldham CM (1988) Importance of non-olfactory stimuli in mediating ram induced ovulation in the ewe. J Reprod Fertil 84:333–339

Poindron P, Cognie Y, Gayerie F, Orgeur P, Oldham CM, Ravault J-P (1980) Changes in gonadotropins and prolactin levels in isolated (seasonally or lactationally) anovular ewes associated with ovulation caused by the introduction of rams. Physiol Behav 25:227

Rekwot PI, Ogwu D, Oyedipe EO, Sekoni VO (2001) The role of pheromones and biostimulation in animal reproduction. Anim Reprod Sci 65:157–170

Roberts SA, Simpson DM, Armstrong SD, Davidson AJ, Robertson DH, McLean L, Beynon RJ, Hurst JL (2010) Darcin: a male pheromone that stimulates female memory and sexual attraction to an individual male's odour. BMC Biol 8:75–96

Romano JE, Christians CJ, Crabo BG (2000) Continuous presence of rams hastens the onset of estrus in ewes synchronized during the breeding season. Appl Anim Behav Sci 66:65–70

Romano JE, Fernandez Abella D, Villegas N (2001) A note on the effect of continuous ram presence on estrus onset, estrus duration and ovulation time in estrus synchronized ewes. Appl Anim Behav Sci 73:193–198

Rosa HJ, Bryant MJ (2002) The ram effect as a way of modifying the reproductive activity in the ewe: a review. Small Rumin Res 45:1–16

Scaramuzzi RJ, Martin GB (2008) The importance of interactions among nutrition, seasonality and socio-sexual factors in the development of hormone-free methods for controlling fertility. Reprod Domest Anim 43(suppl 2):129–136

Smith JT (2009) Sex steroid control of hypothalamic Kiss1 expression in sheep and rodents: comparative aspects. Peptides 30:94–102

Smith JT, Clifton DK, Steiner RA (2006) Regulation of the neuroendocrine reproductive axis by kisspeptin-GPR54 signaling. Reproduction 131:623–630

Tsutsui K, Bentley GE, Kriegsfeld LJ, Osugi T, Seong JY, Vaudry H (2010) Discovery and evolutionary history of gonadotropin-inhibitory hormone and kisspeptin: new key neuropeptides controlling reproduction. J Neuroendocrinol 22:716–727

Ungerfeld R (2007) Socio-sexual signalling and gonadal function: opportunities for reproductive management in domestic ruminants. Soc Reprod Fertil Suppl 64:207–221

Ungerfeld R, Rubianes E (1999) Estrus response to the ram effect in Corriedale ewes primed with medroxyprogesterone during the breeding season. Small Rumin Res 32:89–91

Ungerfeld R, Forsberg M, Rubianes E (2004) Overview of the response of anoestrous ewes to the ram effect. Reprod Fertil Dev 16:479–490

Van Praag H, Kempermann G, Gage FH (2000) Neural consequences of environmental enrichment. Nat Rev Neurosci 1:191–198

Vandenbergh JG (1973) Acceleration and inhibition of puberty in female mice by pheromones. J Reprod Fertil Suppl 19:411–419

Veliz FG, Poindron P, Malpaux B, Delgadillo JA (2006) Maintaining contact with bucks does not induce refractoriness to the male effect in seasonally anoestrous female goats. Anim Reprod Sci 92:300–309

Wakabayashi Y, Nakada T, Murata K, Ohkura S, Mogi K, Navarro VM, Clifton DK, Mori Y, Tsukamura H, Maeda K, Steiner RA, Okamura H (2010) Neurokinin B and dynorphin A in kisspeptin neurons of the arcuate nucleus participate in generation of periodic oscillation of neural activity driving pulsatile gonadotropin-releasing hormone secretion in the goat. J Neurosci 30:3124–3132

Whitten WK (1956) Modification of the oestrous cycle of the mouse by external stimuli associated with the male. J Endocrinol 13:399–404

Wyatt TD (2010) Pheromones and signature mixtures: defining species-wide signals and variable cues for identity in both invertebrates and vertebrates. J Comp Physiol 196:685–700

Part VIII
New Directions in Semiochemistry

Chapter 33
A Reverse-Engineering Approach to Identifying Which Compounds to Bioassay for Signalling Activity in the Scent Marks of African Wild Dogs (*Lycaon pictus*)

Peter Apps, Lesego Mmualefe, and J. Weldon McNutt

Abstract Scent marks can potentially be used to manage the movement patterns of free-ranging mammals. We are investigating the possibility that using synthetic scent marks to create artificial range boundaries will help to keep African wild dogs (*Lycaon pictus*) inside the safety of wildlife conservation areas. For this we need to produce synthetic scent marks, and so we need to identify which compounds among the hundreds in real scent marks are semiochemically active so that they can be formulated into the synthetic marks. Approaches that have previously been applied to locate and identify signalling compounds in mammal scent marks cannot be used on African wild dogs and so we are using "reverse-engineering" to locate and identify which compounds we should prioritise in bioassays for spatial semiochemical activity. For a compound, or a suite of compounds to be considered to be a good candidate for a spatial semiochemical role, it has to be species-specific, potentially airborne, persistent, and present in urine and not in faeces. We have used GC-MS to identify 102 components of African wild dog urine, faeces, and preputial tufts. Most of the major components are carboxylic acids and other compounds that are ubiquitous in mammal scents. Eighty-three of the identified components are poor candidates as single compound spatial semiochemicals because they also occur in other mammals that are sympatric with wild dogs. Four components from wild dog urine, pyridine, *N,N*-dimethylacetamide, propanamide, and 1-methylimidazole-5-carboxa ldehyde, meet the design criteria for spatial semiochemicals. These compounds are priority candidates for testing in artificial scent marks.

P. Apps (✉) • L. Mmualefe • J.W. McNutt
Paul G. Allen Family Foundation Laboratory for Wildlife Chemistry, Botswana Predator Conservation Trust, Private Bag 13, Maun, Botswana
e-mail: peterjapps@gmail.com; lmmualefe@gmail.com; predatorconservation@gmail.com

M.L. East and M. Dehnhard (eds.), *Chemical Signals in Vertebrates 12*, 417
DOI 10.1007/978-1-4614-5927-9_33, © Springer Science+Business Media New York 2013

33.1 Introduction

Semiochemicals could provide powerful tools to manage the behaviour of mammals (Campbell-Palmer and Rosell 2010; Linklater 2004; Schulte et al. 2007; Wyatt 2003, pp. 251–255). Translocated scent marks have been used to manipulate the movement patterns of free-ranging mammals (e.g. Jackson et al. 2012; Parsons and Blumstein 2010; Welsh and Müller-Schwarze 1989), and we are investigating the possibility that using synthetic scent marks to create artificial range boundaries will help to prevent African wild dogs (*Lycaon pictus*) from coming into fatal conflict with livestock farmers by deterring the dogs from leaving the safety of wildlife conservation areas. There are formidable challenges to the collection of sufficient natural marks for large-scale use, and the removal of large numbers of natural marks from one area for use to another might be expected to disrupt the signalling system in the source area. Because the supply of natural African wild dog scent marks is inadequate for large-scale manipulations, it is necessary to produce synthetic marks, and so we need to identify the compounds in real wild dog scent marks that need to be formulated into the synthetic ones. This requires the chemical characterisation of the active ingredients of natural marks, and because mammal scent marks contain hundreds of compounds, it is practically necessary to refine the formula of synthetic marks to include only those components that are semiochemically active.

The extreme complexity of mammal scents makes it impractical to identify all of their components and to reconstruct facsimile scents by mixing individual compounds, or to separately bioassay all the components and their possible combinations. Therefore, to create synthetic wild dog spatial scent marks, we need to locate which components among the hundreds emitted by wild dog scent marks inform wild dogs that an area is occupied by other wild dogs. Approaches that have been successful in other species and circumstances cannot be applied to free-ranging, endangered, large carnivores. Experimental manipulations such as castration, adrenalectomy, artificial crowding of single sex groups, and staged aggression (see for example Novotny et al. 1986, 1990) are self-evidently inappropriate. Both response-guided fractionation and serial bioassays require a rapid, straightforward, well-characterised bioassay (Albone 1984; Müller-Schwarze and Houlihan 1991; Rasmussen et al. 1997). Neurochemistry (Ferrero et al. 2011; Lin et al. 2005) requires a well-characterised animal model in which uniformly inbred individuals can be killed. Multivariate data analyses have not, as yet, located a signalling compound in any species. Because these approaches are either inapplicable or not likely to succeed, we propose to reverse-engineer the scent signals. To reverse-engineer is "to examine (a product) in order to determine its construction, composition, or operation, typically with a view to manufacturing a similar product" (Oxford English Dictionary 2011). The features that an effective signal of area occupation would need to exhibit generate a set of design criteria (Alberts 1992; Bossert and Wilson 1963) for a signal that an area is occupied by wild dogs, and the compounds or mixtures of compounds from wild dog scent marks whose properties and patterns of occurrence match the criteria are considered to be priority candidates for testing for semiochemical activity. The reverse-engineered design criteria for wild dog spatial semiochemicals are as follows. Scent marks have to be detectable against the complex chemical background that Wyatt (2003, p. 227) calls the chemical cacophony, and in particular, because scent marks operate when the animal that deposited them is absent, the signals that scent marks emit must be species-specific. African wild dogs share their habitats with dozens of other species, and if the wild dogs' spatial semiochemicals were also produced by any of these, the wild dogs' signals would be disrupted. Marks must persist at least long enough to be encountered by neighbours and potential intruders, which implies persistence for a few weeks. Marks are detected at a distance, and so the molecular weights and structures of the signalling compounds must allow a sufficiently high vapour pressure for them to be airborne, or the semiochemicals must be adsorbed onto airborne particles. Wild dogs respond to the urine of strangers by prolonged sniffing (Parker 2010, pers. obs.) and counter-marking (pers. obs.), but are less responsive to strangers' faeces (Parker 2010), which suggests that urine is semiochemically more active than faeces, and that the signalling compounds, or combination of compounds, are to be found in urine and not in faeces. In addition to these a priori expectations, deviations from simple chemistry may point to the existence of specific biological mechanisms that may in turn be linked to signalling.

Male wild dogs have a large glandular field at the tip of the prepuce covered with tufts of longer hair (Van Heerden 1981) that are frequently wetted with urine. The gland is much larger in dominant males than in subordinates. No other canids have such a preputial gland, making it a likely source for species-specific compounds.

It is as well to stress that, at short range, scent marks do not act as repellents and would be expected to attract animals to investigate them in detail (Gosling and Roberts 2001; Roberts and Gosling 2001).

African wild dogs are highly social predators that live in packs, usually with 4–9 adults, 2–6 yearlings, and up to a dozen pups. Their densities are naturally low even in healthy populations, and their numbers are declining due to loss of habitat and anthropogenic mortality (Creel et al. 2004; Woodroffe et al. 2004, 2007). Most of the protected areas in which wild dogs occur are not effectively fenced off from livestock areas, and dogs that move into livestock areas fall foul of lethal control by farmers. They are classified as endangered by the IUCN (McNutt et al. 2008). Packs have large home ranges (Woodroffe et al. 2004), and although ranges may overlap, packs avoid one another, contact between packs is rare, and when it does occur fatal conflict may ensue (Creel and Creel 2002; Parker 2010; Woodroffe 2011; pers. obs.). Home ranges are marked with urine and faeces (Parker 2010, pers. obs.). Although there is only limited evidence on how packs use scent marks (either their own or from other packs) to regulate their movements, the movements of free-ranging dogs can be experimentally manipulated on a scale of kilometres by deployment of urine and faeces from other packs (Jackson et al. 2012). If the presence of other packs' scent marks deters wild dogs from entering an area, then it is possible that replica wild dog scent marks could be used to limit the extent to which dogs move out of protected areas into livestock areas where they are at risk.

We focus here on the identification by gas chromatography—mass spectrometry of organic compounds from African wild dog urine, faeces, and preputial tufts, and whether the compounds are species-specific and present in urine and preputial tufts and not in faeces. Using gas chromatography for analysis intrinsically limits identifications to those compounds that have sufficient vapour pressure to be airborne.

33.2 Methods and Materials

Samples of voided urine and faeces were collected from African wild dogs in a study population that ranges over an area of approximately 2,600 km^2 in the eastern Moremi Game Reserve and adjacent Wildlife Management Areas of northern Botswana. Samples were collected from five packs in which all the dogs were recognisable from coat patterns and had known individual life histories (Table 33.1).

After dogs were seen to urinate or defecate and had moved away, faeces or patches of urine-soaked soil were collected and wrapped in aluminium foil or placed into specially cleaned glass jars with lids lined with aluminium foil. Samples were deep frozen on the day of collection until they were thawed for analysis. Tufts of urine-soaked hair from the preputial gland were collected, into glass vials with foil-lined caps when four dogs were immobilised to have radio-collars fitted. Tuft clippings were frozen within 2 h of collection.

Wild dog volatiles were compared with those from sympatric species reported in the literature, and in addition, urine and/or faeces were collected from elephants (*Loxodonta africana*), Cape buffalo (*Syncerus caffer*), impala (*Aepyceros melampus*), plains zebra (*Equus quagga*), lion (*Panthera leo*), leopard (*Panthera pardus*), cheetah (*Acinonyx jubatus*), and spotted hyaena (*Crocuta crocuta*) and anal gland secretion from lion, cheetah, spotted hyaena, and black backed jackal (*Canis mesomelas*).

Samples were analysed by gas chromatography on 30 m polar polyethyleneglycol capillary columns (Varian VF-WAXms, Restek Stabilwax, Restek RtxWax) in Varian (now Bruker) 450 GCs with FID, PFPD, and 320 MS detectors. Identities based on MS library searches were confirmed by co-injections of standards and by nitrogen- and sulphur-specific responses of the PFPD. Full details of sample preparation and analysis are given in Apps, Mmualefe, and McNutt (2012).

33.3 Results

One hundred and two compounds were identified from wild dog urine, faeces, and preputial tufts (Table 33.2) (see also Apps et al. 2012).

Dominant pair urine overmarks are consistently simpler in composition than is predicted from the composition of male and female urines separately (Table 33.3), and their composition differs according to whether the male or the female urinated first (Table 33.4).

Table 33.1 Number of urine and faeces samples of each type from each African wild dog pack that were analysed by GC-MS

Pack	Dominant male urine, not overmarked	Dominant female urine not overmarked	Dominant pair urine overmarks	Dominant female faeces overmarked by dominant male	Subordinate male faeces	Subordinate female faeces	Subordinate male urine	Subordinate female urine
Mathews	4	1	12	1	4	–	2	4
Hawkins	1	–	1	–	1	2	2	–
Kekaneka	–	–	1	–	–	–	–	–
Mankwe	–	3	1	–	2	–	2	1
Chitabe	–	–	1	–	1	–	–	–
TOTALS	5	4	16	1	8	2	6	5

Table 33.2 Compounds identified in the urine, faeces, and preputial tufts of free-ranging African wild dogs (*Lycaon pictus*), and their occurrence in allopatric and sympatric species

Compound	Identification criteria[a]	Material[b]	Potentially airborne	Species-specific to wild dogs	Not found in sympatric mammals	Present in urine or preputial tuft and not present in faeces	Molecular weight
Formamide	MS RI N	U F	Y				45.0
Formic acid	MS RI std	U F P	Y				46.0
Acetamide	MS RI std N	U F P	Y				59.3
Acetic acid	MS RI std	U F P	Y				60.1
Pyrrole	MS RI std	U F	Y				67.1
Propanamide	MS RI std N	U	Y		Y	Y	73.1
2-Methyl-1-propanol	MS RI std	F	Y				74.1
Propanoic acid	MS RI std	U F P	Y				74.1
1,2-Propanediol	MS RI std	U F	Y				76.1
1,3-Propanediol	MS RI std	U F	Y	Y	Y		76.1
Dimethyl sulfoxide	MS RI std S	U F	Y				78.1
Pyridine	MS RI std	P	Y		Y	Y	79.1
2-Pyrrolidinone	MS RI std N	U F P	Y				85.1
1-Penten-3-ol	MS RI std	F	Y				86.1
4-Hydroxybutanoic acid lactone	MS RI std	U F P	Y				86.1
2-Methyl-2-propenoic acid	MS std	U F P	Y		Y		86.1
N,N-dimethylacetamide	MS RI std	U	Y	Y	Y	Y	87.1
Butanamide	MS RI std N	F	Y	Y	Y		87.1
1-Pentanol	MS RI std	F	Y				88.1
2-Methylpropanoic acid	MS RI std	U F P	Y				88.1
Butanoic acid	MS RI std	U F P	Y				88.1
N,N-Dimethylurea	MS std N	U P	Y			Y	88.1

2-Methylpyrazine	MS RI std	F	Y				94.1
Dimethyl sulfone	MS RI std S	U F P	Y				94.1
Phenol	MS RI std	U F P	Y				94.1
Dimethyl disulphide	MS RI std S	F	Y				94.2
Furfural	MS RI std	U	Y			Y	96.1
2-Furanmethanol	MS RI std	U F	Y				98.1
1-Methyl-2-pyrrolidinone	MS RI std	U P	Y			Y	99.1
2-Piperidone	MS RI std N	U F P	Y				99.1
Hexanal	MS RI std	U F	Y				100.2
3-Methylbutanamide	MS RI std	U F	Y				101.1
Acetic anhydride	MS RI std	U F	Y				102.1
2-Methyl butanoic acid	MS RI std	U F	Y				102.1
3-Methyl butanoic acid	MS RI std	U F P	Y				102.1
Pentanoic acid	MS RI std	U F P	Y				102.1
1-Hexanol	MS RI std	F	Y				102.2
N,*N*-dimethylglycine	MS std	U F P	Y	Y	Y		103.1
Styrene	MS RI std	U F	Y				104.1
Benzaldehyde	MS RI std	U F	Y				106.1
2,6-Dimethylpyridine	MS RI std	U F	Y		Y		107.2
2,5-Dimethylpyrazine	MS std	F	Y		Y		107.2
2,6-Dimethylpyrazine	MS RI std	F	Y		Y		107.2
Benzyl alcohol	MS RI std	U F	Y				108.1
4-Methyl phenol	MS RI std	U F	Y				108.1
3-Methyl phenol	MS RI std	U F P	Y				108.1

(continued)

Table 33.2 (continued)

Compound	Identification criteria[a]	Material[b]	Potentially airborne	Species-specific to wild dogs	Not found in sympatric mammals	Present in urine or preputial tuft and not present in faeces	Molecular weight
1-Methylimidazole-5-carboxaldehyde	MS std N	U P	Y	Y	Y	Y	110.1
1-Methyl-2,4-imidazolidinedione	MS std N	U P F	Y				114.1
5-Methyl-2,4-imidazolidinedione	MS std N	U	Y			Y	114.1
2-Heptanone	MS RI std	F	Y				114.2
1-Heptanol	MS RI std	U F	Y				116.2
4-Methylpentanoic acid	MS RI std	U F P	Y				116.2
Hexanoic acid	MS RI std	U F P	Y				116.2
Indole	MS RI std N	U F P	Y				117.1
Acetophenone	MS RI std	U F	Y				120.1
3-(Methylthio)-propanoic acid	MS std S	F P	Y		Y		120.2
Benzoic acid	MS RI std	U F P	Y				122.1
2-Ethyl-6(5)-methylpyrazine	MS RI std	F	Y	Y	Y		122.2
Phenylethyl alcohol	MS RI std	U F	Y				122.2
Dimethyl trisulphide	MS RI std S	U F P	Y				126.3
Octanal	MS RI std	U F	Y				128.2
2-Octanone	MS RI std	F	Y				128.2
1-Octen-3-ol	MS RI std	F	Y				128.2
2-Octen-1-ol	MS RI std	U F	Y				128.2
Quinoline	MS RI std	U F P	Y				129.2
Quinazoline	MS std	U F P	Y	Y	Y		130.1
Pantolactone	MS RI std	U P	Y			Y	130.1

4-Methylhexanoic acid	MS std	P	Y			Y	130.2
Heptanoic acid	MS RI std	U P	Y			Y	130.2
Benzeneacetic acid	MS RI std	U F P	Y				136.1
Benzenepropanol	MS RI std	F	Y	Y	Y		136.2
2-Pentylfuran	MS RI std	U F	Y				138.2
2-Phenoxyethanol	MS RI std	U F	Y		Y		138.2
2-Nonanone	MS RI std	F P	Y				142.2
Nonanal	MS RI std	U F	Y				142.2
2-Ethyl-3-methylpyrazine	MS std	F	Y		Y		142.2
2-Ethylhexanoic acid	MS std	U	Y			Y	144.2
Octanoic acid	MS RI std	U P	Y			Y	144.2
Benzenepropanoic acid	MS RI std	U F P	Y				150.2
Decanal	MS RI std	U F	Y				156.3
Nonanoic acid	MS RI ret	U P	Y			Y	158.2
1,3-Propanediol diacetate	MS RI	U F	Y	Y	Y		160.2
Undecanal	MS RI std	U	Y			Y	170.3
n-Decanoic acid	MS RI std	U	Y			Y	172.3
Dodecanal	MS RI std	U	Y			Y	184.3
2-Phenylethylbutyrate	MS RI std	F	Y	Y	Y		192.3
n-Dodecanoic acid	MS std	U P	Y				200.3
Tetradecanol	MS RI std	F	Y				214.4
Pentadecanal	MS RI ret	U	Y			Y	226.4
Pentadecanol	MS RI std	F	Y				228.4
Tetradecanoic acid	MS RI std	U F P	Y				228.4
Hexadecanol	MS RI std	U F P	Y				242.4

(continued)

Table 33.2 (continued)

Compound	Identification criteria[a]	Material[b]	Potentially airborne	Species-specific to wild dogs	Not found in sympatric mammals	Present in urine or preputial tuft and not present in faeces	Molecular weight
Pentadecanoic acid	MS RI std	U F P	Y				242.4
Hexadecanoic acid	MS RI std	U F P	Y				256.4
Octadecanoic acid	MS RI std	U F P	Y				284.5
Nonadecanoic acid	MS std	P	Y			Y	298.5
Eicosanoic acid	MS std	P	Y			Y	312.5
Docasanoic acid	MS std	P	Y			Y	340.6
Tetracosanoic acid	MS std	P	Y			Y	368.6
Cholesterol	MS std	P	Y				386.7
Squalene	MS std	U F P	Y				410.7

Compounds are ranked by molecular weight, all else being equal compounds with higher molecular weights will have lower vapour pressures and be more persistent in scent marks

[a]*MS* mass spectral library match; *RI* retention index; *std* comparison with authentic material; *S* pulsed photometric detector in sulphur mode; *N* pulsed photometric detector in nitrogen mode; *ret* retention time increment from a homologue with a confirmed identity

[b]*U* urine; *F* faeces; *P* preputial tuft

Table 33.3 The percentage of urine samples from alpha-pair African wild dogs that contained each compound above trace levels, and the expected occurrence of each compound in overmarks calculated from their occurrence in non-overmarked urine. Analysis by GC-FID

	% Actual occurrence				
Compound	Female not overmarked ($N=6$)	Male not overmarked ($N=13$)	Male and female overmarked ($N=50$)	% Expected occurrence in overmarks	Expected/actual occurrence in overmarks
Acetic acid	100	100	100	100	1
Propanoic acid	50	54	22	76.9	3.5
2-Methylpropanoic acid	17	15	12	29.5	2.5
Butanoic acid	50	15	10	57.7	5.8
3-Methylbutanoic acid	67	61	40	87.2	2.2
Acetamide	83	54	42	92.3	2.2
Formamide	67	38	30	79.5	2.6
Dimethyl sulphone	67	69	60	89.7	1.5
Phenol	0	46	16	46.2	2.9
2-Piperidinone	33	23	10	48.7	4.9
Benzoic acid	100	77	82	100	1.2
Benzeneacetic acid	50	53.8	44	76.9	1.7
1-Methyl-2,4-imidazolidinedione	0	15	4	15.4	3.8

Table 33.4 The percentage of dominant-pair African wild dog urine overmarks in which each compound occurred above trace levels, according to which sex urinated first. Analysis by GC-FID

	% Occurrence		
Compound	Female first (*N*=28)	Male first (*N*=22)	Ratio
Acetic acid	100	100	1
Propanoic acid	36	11	0.3
2-Methylpropanoic acid	13	11	0.8
Butanoic acid	14	7	0.5
3-Methylbutanoic acid	36	43	1.2
Acetamide	50	36	0.7
Formamide	36	25	0.7
Dimethyl sulphone	59	61	1
Phenol	23	11	0.5
2-Piperidinone	9	11	1.2
Benzoic acid	91	75	0.8
Benzeneacetic acid	41	46	1.1
1-Methyl-2,4-imidazolidinedione	0	7	∞

33.4 Discussion

In all mammals for which semiochemicals have been chemically characterised, the active semiochemicals are a very small minority of the total chemical mixture—urine and faeces in particular contain metabolic waste products for which no semiochemical role has been demonstrated or is to be expected. Consequently, in addition to the analytical challenges posed by mammal scents of complexity, physico-chemical diversity, and low concentrations, the principal challenge in identifying mammal semiochemicals is to locate the signalling compounds among the hundreds of compounds in a mammal odour.

The concept of chemical signals fitting "design criteria" dates back at least as far as Bossert and Wilson's (1963) paper on insect pheromones, which was extended by Alberts (1992) to include the effects of environmental conditions on the properties of vertebrate chemical signals. Whether or not a compound (or set of compounds) from a mammalian odour is likely to be sending a signal can be assessed by whether it fits the design criteria for that type of signal. Reverse-engineering can proceed *a priori* on the basis of deductions about how a semiochemical needs to operate; for example, alarm pheromones require low molecular weight for rapid diffusion, localisation, and rapid fade out, while trail pheromones require low vapour pressures for persistence (Bossert and Wilson 1963). A compound that is sensitive to oxygen or water vapour is not a candidate semiochemical (or a candidate member of a semiochemical mixture), neither is a compound that is so widespread that its presence conveys no information.

Some criteria are uniform across species, for example, all spatial scent marks have to be species-specific and persistent, and any chemical signal that is detected at a distance has to include components that are volatile enough to be airborne.

Other criteria are species-specific—for example, which material (urine, faeces, glandular secretions, etc.) is used to produce scent marks shows which material contains the active semiochemicals. Compounds (or combinations of compounds) that occur in inactive materials can be eliminated from the list of potential semiochemicals derived from the analysis of the active material.

All compounds that can be analysed by GC are volatile enough to meet the criterion of wild dog spatial signalling compounds needing to be airborne. Of the 102 compounds identified from African wild dog urine, faeces, and preputial tufts, only 10 have not been reported from other mammals, and only 19 have not been reported from, or found in our analyses of, mammals that are sympatric with wild dogs. Sixteen compounds occur in urine and not in faeces. Only two compounds; *N,N*-dimethylacetamide and 1-methylimidazole-5-carboxaldehyde, are completely species-specific in the sense of not having been found in any other mammal and are present in urine or preputial tuft and not in faeces. Compounds that occur only in allopatric species cannot interfere with African wild dogs' scent marking, and if the species-specificity criterion is relaxed to include only species that are sympatric with African wild dogs, then two additional compounds, pyridine and propanamide, are added to the list of potential spatial semiochemicals. Only one of these compounds, 1-methylimidazole-5-carboxaldehyde, has the low vapour pressure that would be expected for a persistent signal component, but propanamide in the vapour phase could be produced persistently by dehydration of ammonium propanoate formed from ammonia and propanoic acid.

If the signal is composed of two or more compounds, then various combinations that meet the reverse-engineering criteria can be identified: for example, a combination of 1,3-propanediol with either benzoic acid or 1-methyl-2,4-imidazolidinedione is both species-specific and present only in urine and not in faeces. In addition to the two compounds that meet all the criteria, there are eight species-specific compounds, and 14 that are found in urine and not in faeces. Any of the 112 pair-wise combinations of one species-specific compound and one urine compound meets the criteria for species-specificity and occurrence in urine and not faeces. As the number of compounds in the signal increases the constraints on the occurrence of each compound relax, but the number of such mixtures very rapidly becomes too large for practical bioassays. Two or more widespread compounds in specific ratios of concentration could also generate a species-specific signal, although it must be noted that such specific ratios are not likely to be robust to weathering and environmental variations.

Species-specificity might be conferred by the absence of particular compounds, and it is notable that ketones, amines, and sulphur compounds that are common in mammals are absent or are at very low concentrations in African wild dogs. In a more specific example, among the three wild carnivore species in which we found methyl-2,4-imidazolidinedione isomers in urine, wild dogs are distinguished from leopards and spotted hyaenas by the absence of the 5-methyl-isomer which occurs in both leopards and spotted hyaenas.

If overmarks were simply mixtures of excretory urine from two animals, the composition of an overmark would be predictable from the composition of its component urines. In contrast to the predictions of straightforward chemistry the results in Tables 33.3 and 33.4 show that urine deposited on top of other urine is different in composition from urine that is not deposited on top of other urine, and also that urine which will be overmarked is different to urine that will not be overmarked. This implies that the composition of the urine in overmarks is controlled facultatively. Although overmarking in one form or another is nearly ubiquitous among mammals (Ferkin and Pierce 2007), there are no equivalent chemical results for other species, and their significance for scent marking by wild dogs is not obvious. It can be argued that the reason for compounds such as butanoic acid, 2-piperidinone, and 1-methyl-2,4-imidazolidinedione not according with simple predictions is that they are being facultatively released as signals. It is entirely possible that unpredictable compounds (and others) are involved in signalling only within the pack, and that they do not signal occupation of an area.

Reverse-engineering is as dependent on bioassays as any other approach to identifying semiochemicals; its aim is to decrease the number of compounds that need to be subjected to bioassay and to assign a priority order in which compounds will be bioassayed. To reduce the number of bioassays and to assign priorities to different compounds are necessary because African wild dogs are particularly challenging experimental subjects. How wild dogs respond when they naturally encounter neighbours' scent marks is not known in detail. Dogs that are outside their core range move further, and closer to their range centre, after encountering experimentally translocated marks from unknown dogs (Jackson et al. 2012), but this response has not been shown to be applicable when dogs are in their core range. Time spent investigating by sniffing is not a measure of territorial significance; dogs have been observed to spend more than 5 min investigating spotted hyaena faeces for example. We are currently investigating whether overmarking of scents is a viable primary screening test.

33.5 Conclusion

The reverse-engineering approach has identified four compounds that are priority candidates for bioassay as spatial semiochemicals in African wild dog urine scent marks. The biological activity of these candidates, and the viability of the reverse-engineering approach to locating and identifying mammal semiochemicals, can both be tested only by the responses of free-ranging wild dogs to artificial scent marks that contain these candidate compounds.

Acknowledgements We are extremely grateful to the Paul G. Allen Family Foundation for its funding of this research; the Botswana Department of Wildlife and National Parks, for permission to conduct research in Botswana under research permit number EWT 3/3/8 XXIV (71), and to the Restek Corporation for supplying columns and consumables. Wild dog samples were collected by research assistants at BPCT: L. Pfefo, M. Sarefo and B. Sango; leopard, cheetah, and hyaena samples were collected by A. Stein, F. Broekhuis, G. Cozzi, and R. Jackson.

References

Alberts AC (1992) Constraints on the design of chemical communication systems in terrestrial vertebrates. Am Nat 139:S62–S89

Albone ES (1984) Mammalian Semiochemistry. The Investigation of Chemical Signals Between Mammals. Wiley, Chichester, p 5

Apps PJ, Mmualefe L, McNutt JW (2012) Identification of volatiles from the secretions and excretions of African wild dogs (*Lycaon pictus*). J Chem Ecol 38: online DOI 10.1007/s10886-012-0206-7

Bossert WH, Wilson EO (1963) The analysis of olfactory communication among animals. J Theor Biol 5:443–469

Campbell-Palmer R, Rosell F (2010) Conservation of the Eurasian beaver *Castor fiber*, an olfactory perspective. Mammal Rev 40:293–312

Creel S, Creel NM (2002) The African Wild Dog: Behaviour, Ecology, Conservation. Princeton University Press, Princeton

Creel S, Mills MGL, McNutt JW (2004) African wild dogs. Demography and population dynamics of African wild dogs in three critical populations. In: Macdonald DW, Sillero-Zubiri C (eds) Biology and Conservation of Wild Canids. Oxford University Press, Oxford, pp 337–350

Ferkin MH, Pierce AA (2007) Perspectives on over-marking: is it good to be on top? J Ethol 25:107–116

Ferrero JDM, Lemon K, Fluegge D, Pashkovski SL, Korzan WL, Datta SR, Spehr M, Fendt M, Liberles SD (2011) Detection and avoidance of a carnivore odor by prey. Proc Natl Acad Sci U S A 108:11235–11240

Gosling LM, Roberts DSC (2001) Scent-marking by male mammals: cheat-proof signals to competitors and mates. Adv Stud Behav 30:169–217

Jackson C, McNutt JW, Apps PJ (2012) Managing the ranging behaviour of African wild dogs (*Lycaon pictus*) using translocated scent marks. Wildl Res 39:31–34

Lin DY, Zhang S-Z, Block E, Katz LC (2005) Encoding social signals in the mouse main olfactory bulb. Nature 434:470–477

Linklater WL (2004) Wanted for conservation research: behavioural ecologists with a broader perspective. Bioscience 54:352–360

McNutt JW, Mills MGL, McCreery K, Rasmussen G, Robbins R, Woodroffe R (2008) *Lycaon pictus*. In: IUCN 2011. IUCN Red List of Threatened Species. Version 2011.1. www.iucnredlist.org. Downloaded 6 Oct 2011

Müller-Schwarze D, Houlihan PW (1991) Pheromonal activity of single castoreum constituents in beaver, *Castor canadensis*. J Chem Ecol 17:715–734

Novotny M, Jemiolo B, Harvey S, Wiesler D, Marchlewska-Koj A (1986) Adrenal-mediated endogenous metabolites inhibit puberty in female mice. Science 231:722–725

Novotny M, Jemiolo B, Harvey S (1990) Chemistry of rodent pheromones: molecular insights into chemical signalling in mammals. In: Macdonald DW, Müller-Schwarze D, Natynczuk SE (eds) Chemical Signals in Vertebrates 5. Oxford University Press, Oxford, pp 1–22

Oxford English Dictionary (2011) http://www.oed.com/viewdictionaryentry/Entry/246393. Accessed 6 Jan 2012

Parker MN (2010) Territoriality and scent marking behavior of African wild dogs in northern Botswana. PhD thesis, University of Montana, Missoula

Parsons MH, Blumstein DT (2010) Familiarity breeds contempt: kangaroos persistently avoid areas with experimentally deployed dingo scents. PLoS One 5(5):e10403. doi:10.1371/journal.pone.0010403

Rasmussen LEL, Lee TD, Zhang A, Roelofs WL, Daves GD Jr (1997) Purification, identification, concentration and bioactivity of (Z)-7-dodecen-1-yl acetate: sex pheromone of the female Asian elephant, *Elephas maximus*. Chem Senses 22:417–437

Roberts SC, Gosling LM (2001) The economic consequences of advertising scent mark locations on territories. In: Marchlewska-Koj A, Lepri JJ, Müller-Schwarze D (eds) Chemical Signals in Vertebrates 9. Kluwer Academic, New York, pp 11–18

Schulte BA, Freeman EW, Goodwin TE, Hollister-Smith J, Rasmussen LEL (2007) Honest signalling through chemicals by elephants with applications for care and conservation. Appl Anim Behav Sci 102:344–363

Van Heerden J (1981) The role of integumental glands in the social and mating behaviour of the hunting dog *Lycaon pictus* (Temminck, 1820). Onderstepoort J Vet Res 48:19–21

Welsh RG, Müller-Schwarze D (1989) Experimental habitat scenting inhibits colonization by beaver, *Castor canadensis*. J Chem Ecol 15:887–893

Woodroffe R (2011) Ranging behaviour of African wild dog packs in a human-dominated landscape. J Zool 283:88–97

Woodroffe R, McNutt JW, Mills MGL (2004) African wild dog *Lycaon pictus*. In: Sillero-Zubiri C, Hoffmann M, Macdonald DW (eds) Canids: Foxes, Wolves, Jackals and Dogs—2004 Status Survey and Conservation Action Plan. IUCN, Gland, Switzerland, pp 174–183

Woodroffe R, Davies-Mostert H, Ginsberg J, Graf J, Leigh K, McCreery K, Mills G, Pole A, Rasmussen G, Robbins R, Somers M, Szykman M (2007) Rates and causes of mortality in endangered African wild dogs (*Lycaon pictus*): lessons for management and monitoring. Oryx 41:1–9

Wyatt TD (2003) Pheromones and Animal Behaviour. Communication by Smell and Taste. Cambridge University Press, Cambridge, UK

Chapter 34
The Application of Proteomics to the Discovery and Quantification of Proteins in Scent Signals

Robert J. Beynon, Stuart D. Armstrong, Sarah A. Roberts, Amanda Davidson, Deborah M. Simpson, Michael J. Turton, Jennifer Unsworth, and Jane L. Hurst

Abstract The increasing awareness of the occurrence and role of proteins in scent secretions has created a requirement for analytical approaches that can define the protein complexity of a scent secretion and identify and quantify those proteins. In this brief review, we summarise current approaches to the characterisation of proteins in scent secretion. By applying analytical methodologies of progressively increasing complexity, it is possible for behavioural labs to complete analytical surveys of scent complexity and protein component prior to the move to more specialised methods based on protein and peptide mass spectrometry. Whilst protein mass spectrometry is most easily delivered when the cognate gene or mRNA sequence is known, it is possible to recover most or all of the protein sequence by mass spectrometry, improving the chances of a cross-species identification. Finally, once the identity of the protein(s) is known, new analytical approaches permit high-quality quantification of the proteins, an essential pre-requisite to assessment of the regulation of the protein in scent secretions.

34.1 Introduction

We are increasingly aware of the molecular complexity of chemical communication in vertebrates and the subtlety of information that different components convey. In particular, it is now clear that proteins play a multitude of roles in the synthesis,

R.J. Beynon (✉) • S.D. Armstrong • D.M. Simpson • M.J. Turton • J. Unsworth
Protein Function Group, Institute of Integrative Biology,
University of Liverpool, Liverpool L69 7ZB, UK
e-mail: r.beynon@liv.ac.uk

S.A. Roberts • A. Davidson • J.L. Hurst
Mammalian Behaviour and Evolution Group, Institute of Integrative Biology,
University of Liverpool, Leahurst Campus, Neston CH64 7TE, UK

M.L. East and M. Dehnhard (eds.), *Chemical Signals in Vertebrates 12*,
DOI 10.1007/978-1-4614-5927-9_34, © Springer Science+Business Media New York 2013

propagation and perception of scent signals. Fortunately, the emergence of the need for analysis of protein components of scent marks has been coincident with the rapid development of the methodologies and instrumentation in the field of proteomics. In this brief review, we discuss how such methodologies can be applied to the study of vertebrate semiochemistry.

The term 'proteomics' was first coined in 1995 (Wilkins et al. 1996) and can be defined as the study of the entire complement of proteins in a particular biological sample, at a specific physiological or pathophysiological state. A proteome analysis is achieved predominantly by the combination of high-quality separation technologies coupled to mass spectrometers of remarkable sensitivity and resolution. It is not the purpose of this chapter to provide a detailed review of proteomics methodologies—the reader is referred to many recent publications. Rather, we will show how the application of proteomics to mammalian semiochemistry research can provide new insights. The stated aim of proteomics—the description of a 'complete' proteome—has rarely been met, primarily for two reasons. First, the 'protein world' is exquisitely multidimensional in its complexity, and we, as yet, lack the technology to tease apart all the dimensions of this complexity. Secondly, whereas every cell has a single copy of a gene, the number of protein molecules encoded by that gene can range from zero (no expression of that gene) to hundreds of millions of copies. This large dynamic range can compromise proteome analyses, because methods that attempt to reach deeply into the low abundance proteome can be compromised by dominant, high abundance proteins. Notwithstanding these deficiencies, there are two ways in which proteomics might be applied in vertebrate semiochemistry. *Tissue-level proteome analyses* might allow profiling of the protein response to specific semiochemical signals, an approach that is fundamentally no different to many other ligand:receptor signalling analyses and which will not be discussed further here. More specifically, proteome characterisation can be applied to the analysis of *scent signals themselves*, and it is this topic that we address here.

34.2 The Complexity of Scent Mark Proteomes Dictates the Analytical Strategy

It is helpful to have an approach that permits a quick assessment of the overall protein complexity of the secretion, most easily assessed by visual inspection of a gel electrophoretogram of denatured proteins (sodium dodecyl sulphate polyacrylamide gel electrophoresis, abbreviated SDS-PAGE), in which proteins are separated in a clear, transparent gel according to their molecular weight and subsequently visualised by a stain (Fig. 34.1). This is a straightforward methodology, readily deployed in practically every laboratory with minimal instrumentation outlay, and all solutions and even pre-cast gels can be purchased, minimising the laboratory overhead (Garfin 2009).

As can be seen in Fig. 34.1a–c, the SDS-PAGE analysis can give a quick visual representation of the overall complexity of the scent sample (number of bands),

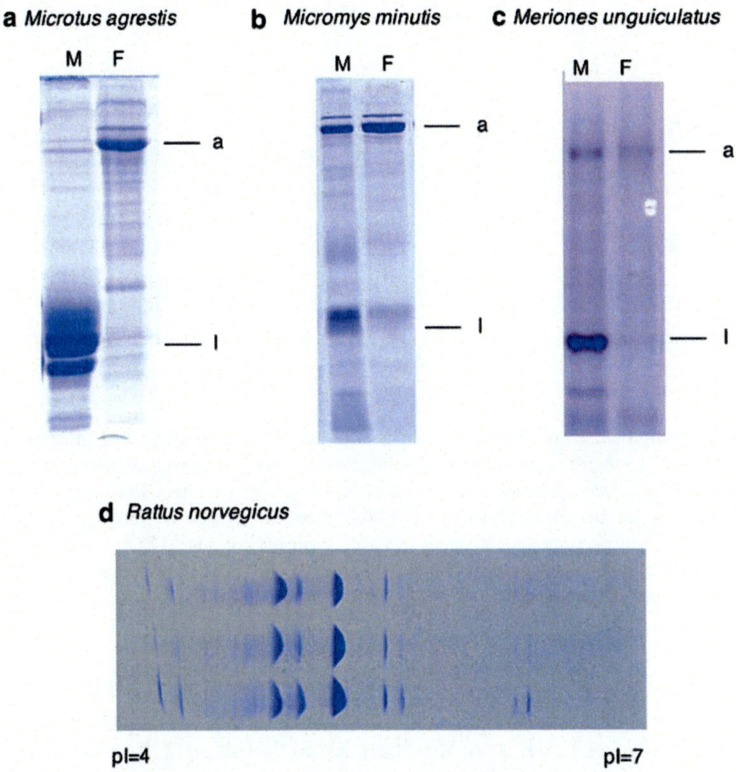

Fig. 34.1 Rapid evaluation of protein complexity in scent marks. A rough estimate of the number of proteins, and their abundances, can be obtained from sodium dodecyl sulphate polyacrylamide gel electrophoresis (SDS-PAGE). This methodology separates proteins approximately according to their size, such that smaller proteins move towards the bottom of the gel faster than large proteins. A set of calibrant proteins can provide a rough estimate of the size of each protein in the scent sample. This set of lanes, cropped from images of individual gels run at different times, display the urinary proteins from a range of species. (**a**) Adult male and female urine from *Microtus agrestis* (field vole), (**b**) adult male and female urine from *Micromys minutis* (harvest mouse) and (**c**) adult male and female urine from *Meriones unguiculatus* (gerbil). The band on some lanes in the region of marker 'a' is of the size expected for serum albumin, and the band in the region of marker 'l' is the size expected for a semiochemical lipocalin, such as the major urinary proteins (MUPs) in the house mouse. In addition to separation by molecular weight it is possible to separate proteins in gels by virtue of their charge (isoelectric focusing, IEF) or by a native gel electrophoresis in which mobility is the outcome of a combination of overall charge and molecular weight of the protein. In this example (**d**), *Rattus norvegicus* (Norway rat) urinary proteins are resolved according to their isoelectric point (pI)

their relative abundance (by intensity of staining and thickness of bands) and the overall protein complement (integrated stain intensity over entire lane). Moreover, several samples can be run in parallel on the same gel, allowing a quick comparison of different individuals, or of animals of different sex or social status. The simplest analysis is an SDS-PAGE gel analysis, but there are other gel methods that separate

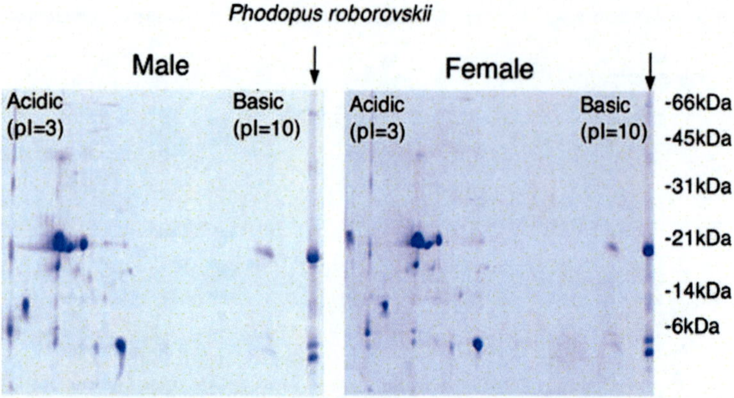

Fig. 34.2 Two-dimensional gel electrophoresis of proteins in a scent mark. Two-dimensional gels resolve proteins by their charge (*horizontal dimension*) and molecular weight (*vertical dimension*). In this example, urinary proteins from an adult male and an adult female Roborovski hamster (*Phodopus roborovskii*) are resolved by 2D-PAGE. Note the overall similarity of the pattern of proteins, confirmed by more detailed analyses. The *arrow* indicates a 1D SDS-PAGE analysis run at the same time. It should be evident that the 1D analysis concealed considerable complexity that is revealed by the 2D separation

proteins by, for example, net charge—isoelectric focusing (Towbin et al. 2001; Friedman et al. 2009) or native gel electrophoresis (Wittig and Schagger 2008; Drab et al. 2011) (Fig. 34.1d). It is worth remembering that some of these gel approaches only resolve proteins that are positively charged (cationic) or negatively charged (anionic) and thus, some proteins of the 'wrong' charge might not be seen as they will not enter the gel. A further exploration of the complexity of a protein samples is a two-dimensional separation (two-dimensional polyacrylamide gel electrophoresis (Marcus et al. 2009), or 2D PAGE), in which proteins are first resolved by their charge, and then, in a second dimension, by their molecular weight in an SDS-PAGE separation. A well-run 2D PAGE separation can resolve many thousands of discrete proteins, and because of the extreme subtlety of the charge-based separation, it is often possible to resolve charge-varying polymorphisms (Garfin 2003). Two-dimensional gels can be technically challenging and require rather more protein than one-dimensional gels, but are excellent at visualising complexity at the protein level (Fig. 34.2).

For proteins that are water-soluble (the majority in scent secretions), it may also be possible to explore complexity by column chromatography. A mixture of proteins can be bound to an absorptive matrix, and individual proteins are then released by progressively weakening the interaction between the proteins and the chromatographic matrix. Separation is usually on the basis of charge (anion or cation exchange chromatography), but can also, using a different separation principle, be according to molecular weight (size exclusion, or gel filtration chromatography). The proteins that are eluted from the column are usually visualised by their ability to absorb light at 280 nm, but it is possible to collect fractions from the elution profile and analyse fractions independently (Robertson et al. 1996, 2007, 2008; Armstrong et al. 2005; Cheetham et al. 2009). Unlike 1-D or 2-D gel separations, column chromatography will usually retain the proteins in their native state, and they may be used for subsequent behavioural or other functional analyses.

Some caveats are appropriate. First, two proteins that migrate to the same part of the gel or elution profile will be superimposed, and it would not be possible to assess the relative contribution of the different proteins contributing to that band or peak. Secondly, polymorphic variations within the same protein are often undetectable by this method. Finally, the most abundant proteins might not be the semiochemically active proteins. For example, serum albumin is often a component of secretions and whilst this protein may have a role in propagation of scent signals, this has only been overtly explored in the elephant (Lazar et al. 2004). This type of analysis can therefore be used to discover the overall protein complement and its complexity, but care must be taken against over-interpretation in the absence of more detailed behavioural experiments. Nonetheless, these methods are relatively routine approaches to the demonstrable presence of proteins in a scent sample, which is an important step that prefaces more detailed characterisation.

34.3 Identification of Proteins in Scent Marks

The logical next step in characterising the proteins in scent marks is to gain their identity. The capability of proteomics methodologies permits analysis of single protein bands or spots from 1-D or 2-D gels, or of proteins in individual chromatography fractions. However, it is rare that any protein sample is analysed at the level of protein—'top down' analysis. Much more commonly, the proteins are first converted to smaller peptide fragments, usually by a digestive enzyme, and most commonly with trypsin ('bottom-up' proteomics). Trypsin is the preferred enzyme because it cuts a protein into short peptides (average 15 amino acids long) in a highly predictable way. If the protein is pure, then, implicitly, all trypsin-generated peptides must have been derived from that protein. In this situation, it can be enough to measure the masses of the peptides by mass spectrometry, and the peptide mass 'fingerprint' thus obtained can be sufficient information to match the fingerprint to a database of fingerprints from all known proteins (Perkins et al. 1999). The database search engines give a score that reflects the probability that the protein is a true match—it is, however, not enough to state that the highest scoring match is 'correct'—this protein also has to meet the threshold score for a high probability match (Fig. 34.3). The two implicit requirements of this method of identification are first, that the protein is pure, so that all peptides are 'connected' by virtue of having been derived from the same protein (it will be appreciated that connectivity does not include the order of the peptides in the protein) and secondly, that the protein, or a closely similar protein, already exists in the database of known protein sequences. The second assumption is often challenged when working with exotic species that do not qualify as 'model organisms'. Moreover, proteins associated with scent signalling might be expected to

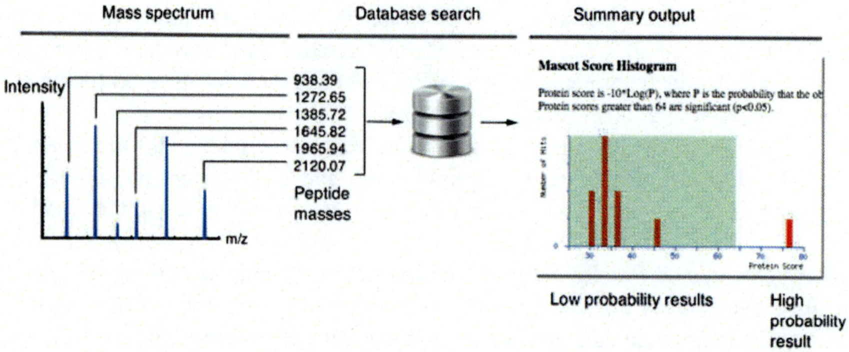

Fig. 34.3 Identification of a protein by peptide mass fingerprinting. A mass spectrum of a digest of a protein yields a series of peptide masses that are a 'fingerprint' of the protein. The masses of these proteins can be used to search the database of all known proteins (possibly restricted to specific taxonomic groups). The search engines (e.g. Mascot; www.matrixscience.com) return a score that reflects the probability that the protein is similar or identical to the top-matched targets in the database. The identification is never certain, and care must be taken in interpretation of the probabilistic scoring systems. All hits in the shaded, green area are below the threshold for a $p<0.05$ identification. The *y*-axis defines the number of hits at each score value (tick are single integers)

undergo rapid evolution (Karn and Laukaitis 2009; Ramm et al. 2009), and thus, diminish the probability of a cross-species match in the database search.

In the absence of a high probability peptide mass fingerprint result, the next logical step is to gain further information about each peptide in a digest of a protein. This is achieved by diagnostic fragmentation of the peptide in the gas phase, inside the mass spectrometer, using the approach of tandem mass spectrometry. Tandem mass spectrometry (MS) couples two stages of mass spectrometry (MS/MS). In the first stage, the masses of the peptides are measured, and in the second stage, a single peptide ion is isolated and subjected to collisions with inert gases at sufficiently high energy that the peptide ion is broken into smaller fragments. In the most commonly used forms of tandem MS, the peptide is fragmented at the peptide bond between two amino acids, generating a series of overlapping product ions that differ from the next fragment ion in the series by the mass of an amino acid (Fig. 34.4). The fragmentation of a peptide in the instrument leads to diagnostic ion series from which it is possible to recover full or partial sequence information for that peptide (see later). However, when working with a species for which the genome, and hence proteome, is available, it is rare for this ion series to be interpreted, whether by inspection or automatically. Rather, the pattern of fragment ions (Fig. 34.5) is used as a pattern that is searched against all patterns of peptides that can be generated by the proteome of the organism (Cottrell 2011). So, for species such as human and mouse, tandem-MS-based searches are relatively simple. If the protein is highly conserved, the fragment ion search algorithms can bring up cross-species matches that can be of high confidence. This tends to be optimal with proteins that are evolving slowly, such as core metabolic enzymes such as those of glycolysis (Ramm et al. 2009).

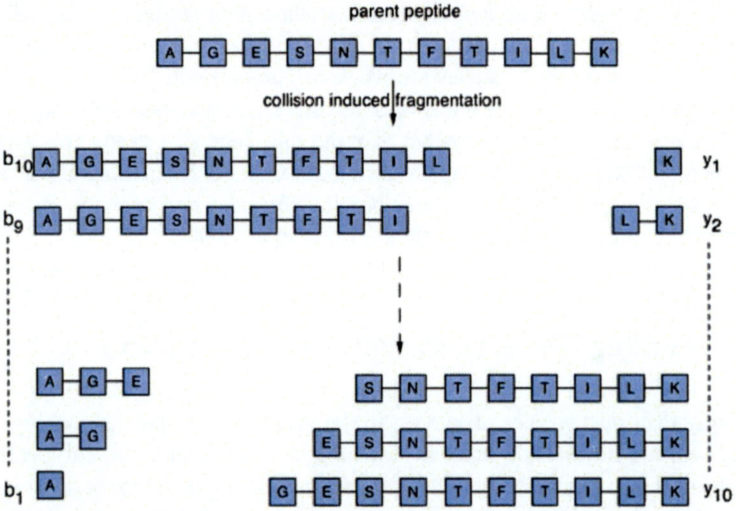

Fig. 34.4 Principles of tandem mass spectrometry peptide fragmentation. In tandem mass spectrometry, a peptide is subjected to collision-induced decomposition to generate a series of fragments (in this example, the *y* and *b* series of ions) that differ in mass by one amino acid. By measurement of the mass difference between the product ions, it is possible to reconstruct the amino acid sequence of a peptide

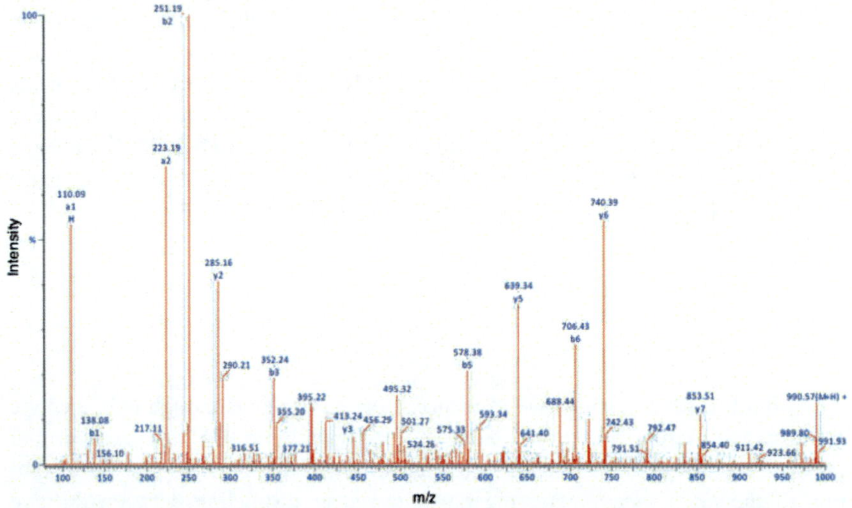

Fig. 34.5 Example tandem mass spectrum of a peptide. This is the product ion spectrum from collision-induced breakdown of a single peptide. The series of diagnostic ions (labelled b_n and y_n) can be used to gain the sequence of the peptide or can be used as a pattern to search a database of peptide fragmentation patterns. Other ions, labelled a_n, are the products of different fragmentation events

Proteins with a high rate of evolution, such as those that are driving speciation or which are under strong selective pressures (Karn and Laukaitis 2009), will evolve so rapidly that cross-species matches are weak or non-existent. Proteins involved in chemical communication fall under this second class, and reliable cross-species identifications are the exception rather than the rule. Indeed, a strong cross-species match is more likely for conserved proteins that might not be an integral part of the semiochemical process, such as serum albumin, which can be a common component of almost any biological secretion (but see Lazar et al. 2004).

34.4 Working Without a Cognate Genome Sequence

It is likely that most species of interest to the vertebrate chemical signalling community will not have a full genome sequence, or even limited protein level data. This makes the challenge of identification of a protein in a scent mark considerably greater. A mass fingerprint might not reveal any high probability hits, even though the mass spectrum of the peptides indicates multiple peptides. Under these circumstances, it is necessary to gain further information about the protein, by detailed analysis of the peptides. Simple mass spectrometry can only reveal the mass of the peptides, but a more sophisticated approach, defined generically as 'tandem mass spectrometry' (tandem MS or MS/MS), can greatly increase the information about a single peptide. There is nothing to be lost by a broad species search of MS or MS/MS data, but it is likely that the ion fragmentation patterns might need to be interpreted manually (Ma and Johnson 2012)—a slow and challenging process that requires considerable experience, although software tools are improving rapidly (Waridel et al. 2007; He and Ma 2010; Allmer 2011). This will yield short segments of protein sequence (typically 10–15 amino acids, Fig. 34.6) that can be used to search protein databases for similar proteins using tools such as BLAST (basic local sequence alignment tool; http://blast.ncbi.nlm.nih.gov). To illustrate, the sequence interpreted in Fig. 34.6 yields this match in a BLAST search:

```
Query    2     LTPEKHE    8

               LTPE HE

Subject 132    LTPEQHE    138
```

Within this seven amino acid sequence, it would seem as though only a single amino was different. However, lysine (K) and glutamine (Q) have very similar masses (128.09496 and 128.05858, respectively) and it is possible that the resolution of the mass spectrometer did not allow clear distinction between the two possibilities.

Short sequence matches can give some indication of the nature of the protein; it is preferable to obtain several peptide sequences and to run them independently and explore similarities in the search results. Although there are theoretically 20^{10} different 10-amino acid sequences (approximately ten million million), it does not follow that

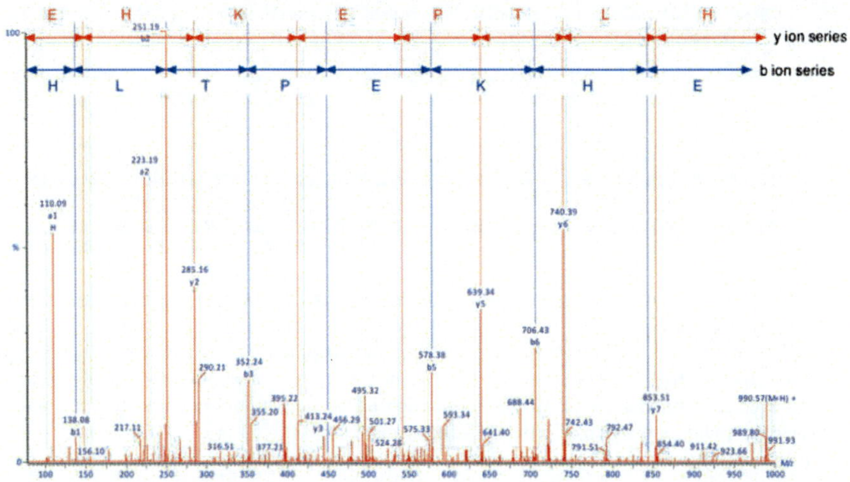

Fig. 34.6 Manual interpretation of an MS/MS product ion spectrum of a peptide to obtain the amino acid sequence. In a tandem mass spectrometer, a peptide can be subjected to sufficient energy to allow it to fragment, predominantly at the peptide bond. Thus, the mass difference between each successive decomposition product can be used to recover the amino acid sequence. In this example (from a bank vole urinary lipocalin), there are two sequences of ions (*b* and *y*), reading from either end of the peptide. Because the *b* and *y* ions are derived from either end of the peptide, the sequences should be complementary

there will be a high probability match to any known protein sequence. Under these circumstances, a more focused study of a specific protein would be required. If the protein is digested by a single proteolytic enzyme, the sequences of individual peptides can be obtained, but it would not be possible to reconstruct their order. To resolve this, the protein can be fragmented with enzymes of different specificity to generate overlapping fragments (Fig. 34.7)—an approach we used in recovering virtually all of the protein sequence of roborovskin (Turton et al. 2010). With longer sequence runs, the probability of a meaningful match in BLAST searches greatly increases.

34.5 Accurate and Absolute Quantification of Proteins in Scent Marks

It is a reasonable starting assumption that proteins that are directly involved in the chemical communication process have their expression regulated, whether during maturation, through seasons or by sex. To test this demands quantitative assessment of the absolute amount of each protein in a scent mark. In simple scent secretions that contain a single or few resolvable proteins, it might be possible to determine, to modest accuracy, the amount of the protein by quantitative densitometry of stained bands on a gel (Fig. 34.8). If an antibody to the protein is available, the proteins in the gel

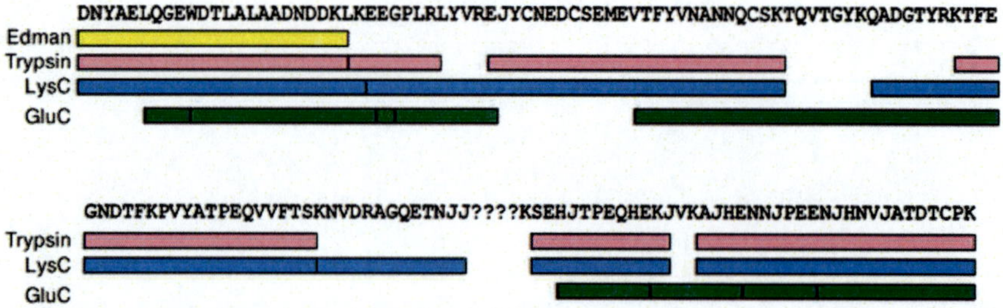

Fig. 34.7 Construction of more complete protein sequences by overlapping peptides. In this example (the urinary lipocalin roborovskin from *Phodopus roborovskii*, see text), three different proteolytic enzymes were used in separate digests, and peptides from each were sequenced by tandem mass spectrometry. The overlap in the sequences from the different sets of proteolytic fragments, even if incomplete, permits the construction of longer runs of amino acids that greatly enhance the ability to perform effective BLAST searches

can be transferred to a stable membrane and located and quantified by the technique of western blotting, in which the antibody is employed as a specific probe to target and stain the protein of interest. Gel-based quantification can give some initial indication of protein output, but there are times when more precise measurements are needed. This can be particularly so when a polymorphic group of proteins co-migrate on one- or two-dimensional separation, because if two proteins are overlaid in 1D or 2D separations, their relative contributions to the total band or spot intensity cannot be resolved. For such quantification, more complex approaches are required. Moreover, the common strategy of digestion of the mixture of proteins to peptides may conceal small polymorphic differences, and there is a strong case to be made for development of analytical strategies at the protein level (Meyer et al. 2011).

It may also be feasible to analyse the profile of proteins after measurement of their intact masses by mass spectrometry. In this methodology, the protein mixture is introduced directly into the mass spectrometer, and the profile of mass can be recovered to a precision of about 0.01 %—a MUP, of approximate mass 20,000 Da can be mass measured to ±2 Da. Profiles from genetically identical individuals are remarkably consistent and can clearly show sex differences (for example, see Armstrong et al. 2005). This approach may only be suitable for relatively concentrated proteins in scent secretions (such as MUPs) but can give a valuable rapid and quantitative survey of the protein profile. Moreover, small mass shifts can hint at the presence of polymorphic variants (Robertson et al. 1996, 2007).

The highest quality approach to MS-based quantification is to use stable isotope labelled standards as a reference against which the protein of interest can be assessed (Becker and Bern 2011). Stable isotope standards are chemically identical to the analyte of interest, but some of the atoms are substituted with stable isotope ('heavy') equivalents: carbon-13 for carbon-12, nitrogen-15 for nitrogen-14, deuterium for hydrogen. Modern mass spectrometers have high enough performance that they can readily discriminate between natural compound and synthetic 'heavy' standards. The analyte and the standard are mixed and analysed simultaneously in

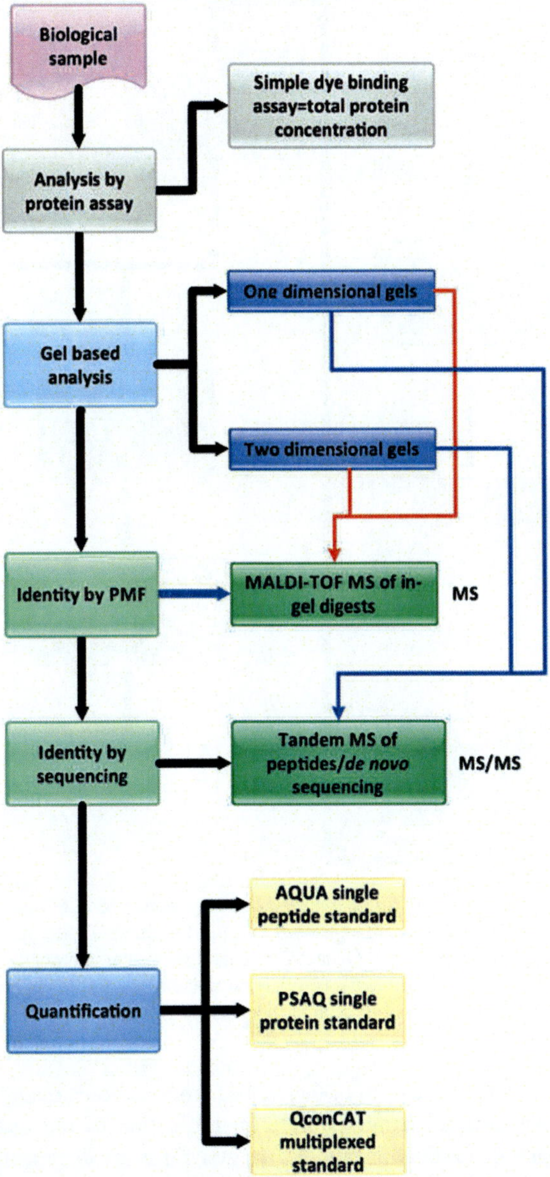

Fig. 34.8 Strategies for protein characterisation and quantification in scent secretions. A logical strategy for the characterisation of the protein content of scent secretions begins with relatively simple analyses and gel-based display methods and can develop through complete identification and characterisation to precise and accurate quantification of specific proteins in scents. Most effort is likely to be invested in gaining the identity of the protein(s), especially if the genome sequence of the animal is not known. Eventually, proteomics workflows can deliver high-quality quantification using a range of approaches, defined by the complexity of the protein mixture and the availability of different types of mass spectrometer. PSAQ: protein standard for absolute quantification, AQUA: absolute quantification peptide standard

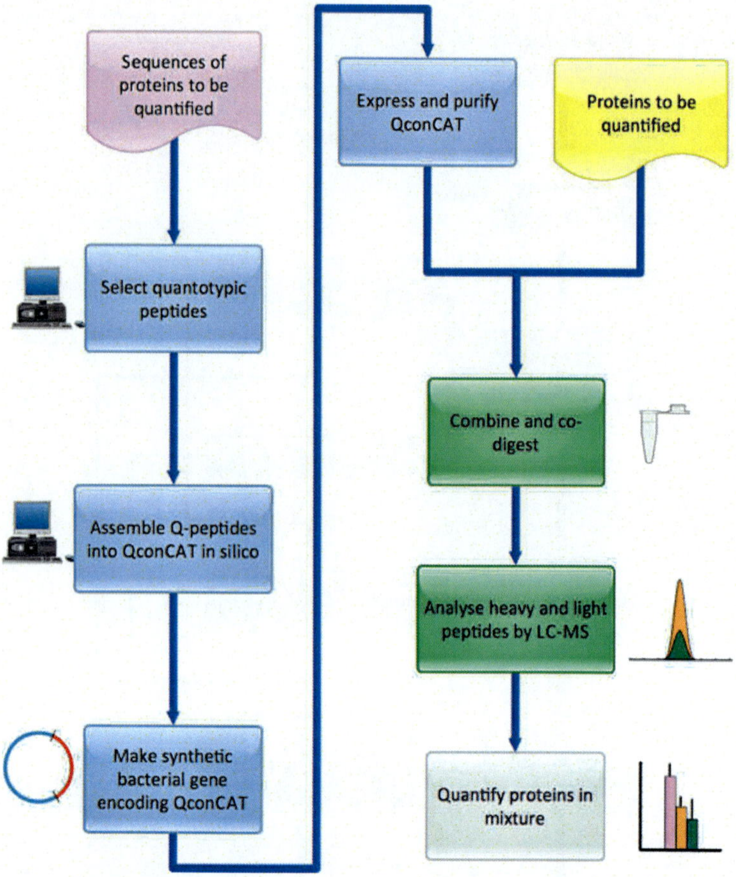

Fig. 34.9 The QconCAT strategy for multiplexed quantification. QconCATs are proteins encoded by artificial 'designer' genes that are expressed and purified in stable isotope labelled form in bacteria. When digested with trypsin, QconCATs generate a stoichiometrically equal mixture of peptides, each of which can be used as a quantification standard for a different protein, although it is more common to rely on at least two peptides per analyte protein

the mass spectrometer—the ratio of the signals from a known quantity of standard and an unknown analyte permits simple quantification of the unknown natural sample. For many quantification studies, the standards are chemically synthesised peptides that are heavy, termed AQUA, or 'absolute quantification' peptides (Gerber et al. 2003; Kirkpatrick et al. 2005), used as a reference against which a digested protein is compared (Fig. 34.9). Under some circumstances, the standard can be the entire protein (PSAQ, Dupuis et al. 2008; Brun et al. 2009; Adrait et al. 2012), but this requires that the entire protein be cloned and expressed in bacteria or a cell-free labelling system—a not insignificant technical challenge. For small proteins (of the order of 20–30 kDa), it is probably more cost effective to purchase an optimised synthetic gene that can be expressed in bacteria and synthesised in a stable isotope labelled form. If multiple proteins are to be quantified, it may be preferable to build a QconCAT.

QconCATs are artificial proteins, encoded by 'designer genes' that are made chemically, and which are designed to ensure the production of large amounts of the encoded protein when expressed in bacteria (Beynon et al. 2005; Pratt et al. 2006; Rivers et al. 2007). The protein is designed such that when it is digested by trypsin, it creates a mixture of standard peptides (stoichiometrically 1:1) that can be used to achieve quantification of several proteins simultaneously (Fig. 34.9). This approach is of particular value for quantification of polymorphic variants on different proteins, provided at least one peptide has a polymorphic amino acid change that elicits a detectable mass change. For example, we have designed and used a MUP QconCAT to generate standards for quantification of MUP variants (*manuscript in preparation*). This protein standard allowed us to quantify different MUP forms in a single experiment, even separating two MUPs that differ in a single amino acid out of 162. The use of a QconCAT has to be justified in that it is only cost-effective if more than about 5–10 proteins require quantification (Brownridge et al. 2011). However, this approach might be of value in determining the complex changes in a scent mark protein complement, even in the absence of polymorphic variants.

34.6 Conclusions

Our increasing awareness of the role of proteins in the generation, mediation and perception of scent signals has introduced opportunities for new analytical strategies for protein discovery, characterisation and quantification. At the same time, the developments in instrumentation and methodologies in proteomics are greatly enhancing the scope and scale of such analyses. It is not appropriate to enter into a deep discussion of proteomics workflows in this chapter. However, a few general comments are appropriate. Most scent secretions are likely to be uncharacterised and there are optimal sequences of experiments of increasing complexity that can be used to build a picture of the protein complement. The workflow for quantification (Fig. 34.8) does in fact embody the methodologies of increasing complexity that can be brought to bear as the understanding of the scent proteins develops. We are primed for a fuller understanding of the role of proteins in the semiochemical process. The real challenge will then be the reconstruction of complex, protein containing scents in the endeavour to 'close the loop' and prove that the analytical results have correctly identified semiochemically active components.

The Chemical Signals in Vertebrates (CSIV) community is a small one and comprises a broad range of specialisms from field ecology and behaviour to analytical chemistry and biochemistry. There is scope for interaction whereby analytical labs commit limited resources to preliminary analyses of scent mark composition, paving the way to more detailed and complex analyses. The ideal medium to bring together the communities is through the CSIV web presence (www.csiv.org) and it seems timely to develop the site to encompass such collaborations, in the spirit of the community.

Acknowledgements This work has been supported by grants from the Biotechnology and Biological Sciences Research Council (BBSRC, BB/J002631/1, BB/26/S13738) and Natural Environment Research Council (NE/G018650/1). Jenny Unsworth is grateful to BBSRC for provision of a studentship.

References

Armstrong SD, Robertson DH, Cheetham SA, Hurst JL, Beynon RJ (2005) Structural and functional differences in isoforms of mouse major urinary proteins: a male-specific protein that preferentially binds a male pheromone. Biochem J 391:343–350

Allmer J (2011) Algorithms for the de novo sequencing of peptides from tandem mass spectra. Expert Rev Proteomics 8:645–657

Adrait A, Lebert D, Trauchessec M, Dupuis A, Louwagie M, Masselon C, Jaquinod M, Chevalier B, Vandenesch F, Garin J, Bruley C, Brun V (2012) Development of a protein standard absolute quantification (psaq) assay for the quantification of *Staphylococcus aureus* enterotoxin A in serum. J Proteomics 75(10):3041–9

Becker CH, Bern M (2011) Recent developments in quantitative proteomics. Mutat Res 722:171–182

Brun V, Masselon C, Garin J, Dupuis A (2009) Isotope dilution strategies for absolute quantitative proteomics. J Proteomics 72:740–749

Brownridge P, Holman SW, Gaskell SJ, Grant CM, Harman VM, Hubbard SJ, Lanthaler K, Lawless C, O'Cualain R, Sims P, Watkins R, Beynon RJ (2011) Global absolute quantification of a proteome: challenges in the deployment of a QconCAT strategy. Proteomics 11:2957–2970

Beynon RJ, Doherty MK, Pratt JM, Gaskell SJ (2005) Multiplexed absolute quantification in proteomics using artificial QCAT proteins of concatenated signature peptides. Nat Methods 2:587–589

Cottrell JS (2011) Protein identification using MS/MS data. J Proteomics 74:1842–1851

Cheetham SA, Smith AL, Armstrong SD, Beynon RJ, Hurst JL (2009) Limited variation in the major urinary proteins of laboratory mice. Physiol Behav 96:253–261

Drab T, Kracmerova J, Ticha I, Hanzlikova E, Ticha M, Ryslava H, Doubnerova V, Manaskova-Postlerova P, Liberda J (2011) Native red electrophoresis—a new method suitable for separation of native proteins. Electrophoresis 32:3597–3599

Dupuis A, Hennekinne JA, Garin J, Brun V (2008) Protein standard absolute quantification (psaq) for improved investigation of staphylococcal food poisoning outbreaks. Proteomics 8:4633–4636

Friedman DB, Hoving S, Westermeier R (2009) Isoelectric focusing and two-dimensional gel electrophoresis. Methods Enzymol 463:515–540

Garfin DE (2009) One-dimensional gel electrophoresis. Methods Enzymol 463:497–513

Garfin DE (2003) Two-dimensional gel electrophoresis: an overview. Trends Anal Chem 22:263–272

Gerber SA, Rush J, Stemman O, Kirschner MW, Gygi SP (2003) Absolute quantification of proteins and phosphoproteins from cell lysates by tandem MS. Proc Natl Acad Sci USA 100:6940–6945

He L, Ma B (2010) Adepts: advanced peptide de novo sequencing with a pair of tandem mass spectra. J Bioinform Comput Biol 8:981–994

Karn RC, Laukaitis CM (2009) The mechanism of expansion and the volatility it created in three pheromone gene clusters in the mouse (*Mus musculus*) genome. Genome Biol Evol 1:494–503

Kirkpatrick DS, Gerber SA, Gygi SP (2005) The absolute quantification strategy: a general procedure for the quantification of proteins and post-translational modifications. Methods 35:265–273

Lazar J, Rasmussen LE, Greenwood DR, Bang IS, Prestwich GD (2004) Elephant albumin: a multipurpose pheromone shuttle. Chem Biol 11:1093–1100

Ma B, Johnson R (2012) *De novo* sequencing and homology searching. Mol Cell Proteomics 11:O111.014902

Marcus K, Joppich C, May C, Pfeiffer K, Sitek B, Meyer H, Stuehler K (2009) High-resolution 2DE. Methods Mol Biol 519:221–240

Meyer B, Papasotiriou DG, Karas M (2011) 100% protein sequence coverage: a modern form of surrealism in proteomics. Amino Acids 41:291–310

Perkins DN, Pappin DJ, Creasy DM, Cottrell JS (1999) Probability-based protein identification by searching sequence databases using mass spectrometry data. Electrophoresis 20:3551–3567

Pratt JM, Simpson DM, Doherty MK, Rivers J, Gaskell SJ, Beynon RJ (2006) Multiplexed absolute quantification for proteomics using concatenated signature peptides encoded by QconCAT genes. Nat Protoc 1:1029–1043

Robertson DH, Cox KA, Gaskell SJ, Evershed RP, Beynon RJ (1996) Molecular heterogeneity in the major urinary proteins of the house mouse *Mus musculus*. Biochem J 316:265–272

Robertson DH, Hurst JL, Searle JB, Gunduz I, Beynon RJ (2007) Characterization and comparison of major urinary proteins from the house mouse, *Mus musculus domesticus*, and the aboriginal mouse, *Mus macedonicus*. J Chem Ecol 33:613–630

Robertson DH, Wong SC, Beynon RJ, Hurst JL, Gaskell SJ (2008) Observation of heterogeneous gene products by FT-ICR MS. J Am Soc Mass Spectrom 19:103–110

Ramm SA, McDonald L, Hurst JL, Beynon RJ, Stockley P (2009) Comparative proteomics reveals evidence for evolutionary diversification of rodent seminal fluid and its functional significance in sperm competition. Mol Biol Evol 26:189–198

Rivers J, Simpson DM, Robertson DH, Gaskell SJ, Beynon RJ (2007) Absolute multiplexed quantitative analysis of protein expression during muscle development using qconcat. Mol Cell Proteomics 6:1416–1427

Towbin H, Ozbey O, Zingel O (2001) An immunoblotting method for high-resolution isoelectric focusing of protein isoforms on immobilized pH gradients. Electrophoresis 22:1887–1893

Turton MJ, Robertson DH, Smith JR, Hurst JL, Beynon RJ (2010) Roborovskin, a lipocalin in the urine of the Roborovski hamster, *Phodopus roborovskii*. Chem Senses 35:675–684

Wilkins MR, Sanchez JC, Gooley AA, Appel RD, Humphery-Smith I, Hochstrasser DF, Williams KL (1996) Progress with proteome projects: why all proteins expressed by a genome should be identified and how to do it. Biotechnol Genet Eng Rev 13:19–50

Wittig I, Schagger H (2008) Features and applications of blue-native and clear-native electrophoresis. Proteomics 8:3974–3990

Waridel P, Frank A, Thomas H, Surendranath V, Sunyaev S, Pevzner P, Shevchenko A (2007) Sequence similarity-driven proteomics in organisms with unknown genomes by LC-MS/MS and automated *de novo* sequencing. Proteomics 7:2318–2329

Index

A
Accessory olfactory system, mice, 28
Acid methyl ester, 215, 225, 226
Activity range
  home range, 289
  lactating and pregnant females, 289
  predator treatment, 284, 285
  size, predator odour, 288
  vegetation, 288–289
Addo Elephant National Park (AENP), 138
AENP. *See* Addo Elephant National Park (AENP)
African clawed frog *(Xenopus laevis)*
  larval development, 47
  middle chamber epithelium, 38
African elephants
  adolescence, male Asian elephants, 137
  AENP, 138
  chemosensory behaviour, 137, 139
  communication, 142
  developmental patterns and chemical stimuli, 138, 140
  dispersal, 136
  enantiomers ratio, 142
  implementation, double-blind procedure, 139
  and insects, 139
  investigation, urine and faeces, 142
  maintenance, photographic database, 138
  male elephants, 136–137
  matriarchal societies, 136
  observations, waterholes permit, 139
  oestrous pheromone, 139
  pheromone, 141
  reproduction, 136
  SCPF responses rates, 141
  social environment, 136
  urine and potential signals, 139–140
  vegetation, 138
African wild dogs *(Lycaon pictus)*
  alpha-pair, 427
  dominant-pair, 428
  free-ranging, urine, faeces and preputial tufts, 420, 422–426
  livestock areas, 419
  male, 419
  scent marks, 418–419
  urine and faeces, 420, 421
Age
  behavioural assay, 220
  effects, relative abundance, 216, 218
  mammalian urine, 208
  mating season and sub-adult males, 219–220
AGS. *See* Anogenital gland secretions (AGS)
Ailuropoda melanoleuca. See Giant panda *(Ailuropoda melanoleuca)*
Air-borne compounds transmitted into egg, 372–373
Alarm cue
  animal collection and maintenance, 297
  antipredator behaviours, 296
  aquatic organism, 296
  description, 307
  experimental protocol, 299–300
  fathead minnows (*see* Fathead minnows)
  grey tree frogs, 305–306
  HPLC analysis, 298–299
  hypoxanthine-3-N-oxide, 306–307
  identity, Ostariophysan, 306
  predation, 295
  statistical analysis, 300–301

M.L. East and M. Dehnhard (eds.), *Chemical Signals in Vertebrates 12*,
DOI 10.1007/978-1-4614-5927-9, © Springer Science+Business Media New York 2013

Alarm cue (*cont.*)
  stimuli preparation, 297–298
  wood frogs *(Lithobates sylvaticus)*, 307
Albatrosses, 329
Alloanointing
  ectoparasite-repelling substances, 330
  ruff-sniff display, 330
American wood stork, *Mycteria Americana*, 354
Amethystine python *(Morelia amethistina)*, 270, 273
Amniotic fluid (AF)
  biological substrates, 79
  initial suckling, 79–80
  maternal label source, sheep, 62–63
Amygdaloid area (AA), 4
Anal gland scent, spotted hyena
  degree of mixing, over-pasting, 151
  dry-paste animals, 158
  grass stalk plus, 151
  marks, natal and immigrants, 152, 158
  natal female clan members, 158
  remote mechanism, 150
  and urine, 159
  vegetation, Hyaenidae 'paste, 150
Anal gland secretion, 420
16-Androstenes measurement, saliva of GLB and GMB
  breeding season, 388
  chemical substances, intraspecies communication, 382
  chromatograms, 385, 386
  collection system, 384
  concentrations, 383
  description, 382
  and estosterone concentrations, 385, 387
  GC-MS analysis, 384–385
  human, 383
  literature, datas, 388–389
  musk-and urine-like smelling steroids, 386, 388
  odorous substance, 388
  olfactory system, pigs, 383
  pheromone synthesis, 383
  production, oral cavity, 386, 388
  quantification, 385
  testis, male pig produces, 386
  testosterone, 385
  variation, gonadal function, 383
Androstenol
  gas chromatography, 383
  ratio, 3α-androstenol, 382, 388
  salivary concentrations, 383
Androstenone. *See* 16-Androstenes measurement, saliva of GLB and GMB
Anogenital gland secretions (AGS), 208, 218, 219
Anointment of fur with the scent, 153
Antarctic prions, *Pachyptila desolata*
  blue petrels, 328
  burrow-nesting petrels, 332
Antibiotic treatment, domestic dogs, 394
Antipredator behavior
  finescale dace *(Phoxinus neogaeus)*, 306
  fishes and amphibians, 296
  grey tree frog, 306
  wood frogs *(Lithobates sylvaticus)*, 307
Anti-predatory behaviour, 4, 251, 252, 269, 273, 274
Antorbital gland, 198
Arboreal and terrestrial predator, 270
ARC. *See* Arcuate nucleus (ARC)
Arcuate nucleus (ARC), 400
Artificial insemination, 382
Asian elephants *(Elephas maximus)*
  epidermal structures, mammals, 230–231
  osmetrichia, 231–232
  'scent flagging' behaviour, 228
  sexual communication, 228
  sheep wool, 232
  tail hairs (*see* Tail hairs, Asian elephants *(Elephas maximus)*)
Associative learning, aggression, 129–130
Auklet odour
  accumulation, headspace, 367
  behaviours, E18 embryos, 368, 369
  detection and discrimination, 368–373
  E18 egg, laboratory hood, 366–367
  embryo scents, 367
  incubator treatment, 367
  responses, 367
  statistical analysis, 368, 369
  treatment, 371, 372

B

Bacteria
  herbs and
    colonies, 354, 355
    description, 354
    herbless nests, 354
    starling nestlings, 354
  vagina, female domestic dog (*see* Vaginal bacterial flora, oestrus females)
Bank vole *(Myodes glareolus)*, 122

Bats
- chemical signatures, 332
- 'osmetrichia', 231

Behavioural modification. *See* Bird development
Behavioural responses, predator odours, 251–252
Bengal tiger *(Panthera tigris tigris)*
- FAME, 225
- fatty acids, 225–226
- hair ultrastructure, 225
- methods and materials, 224–225
- pheromones, 223–224

Bioassays
- African elephants
  - development, chemosensory behaviour, 137, 142
  - double-blind procedure, 138
  - non-oestrous urine, 139
  - waterholes, 138
- Klipspringer antelope *(Oreotragus oreotragus)*, 198–199

Bird/avian olfaction
- chemical signals, 326
- passerines (*see* Passerine/songbird)
- *Sturnus vulgaris*, 326

Bird development
- avian behavioural ecology, 364
- domestic chickens, see
- early exposure chemical stimuli, prior birth, 364
- *Gallus gallus domestics* (*see* Domestic chickens *(Gallus gallus domestics)*)
- volatile odorants, shell, 364

Black backed jackal *(Canis mesomelas)*, 420
Black-tailed deer *(Odocoileus hemionus columbianus)*, 197–198, 203
Blue petrels, *Halobaena caerulea*
- chemical composition, wax, 329
- preen waxes, 328

Blue tits, *Cyanistes caeruleus*
- anti-parasitic purposes, 327
- antipredatory behaviours, 327

Body odour, human, 163–164
Body odour-perfume coevolution
- cognitive biases, 192–193
- communities, 192
- description, 191–192
- patterns, 193
- scent ingredients, 192

Body odour, tiger. *See* Bengal tiger *(Panthera tigris tigris)*
Brain activity/multi-unit activity (MUA) technique
- characteristics, 401–402
- exposure, 400
- kisspeptin cells, 400–401
- LH secretion, 400, 401
- rhesus monkey, 400

Brown antechinus *(Antechinus stuartii)*, 273
Brushtail possums *(Trichosurus vulpecula)*, 90
Budgerigars, *Melopsittacus undulates*, 331
Bugle *(Ajuga reptans)*, 356
Burrow-nesting/burrow odour
- chemical divergence, 328
- procellariiform species, 329

Buzzards, *Buteo buteo*, 354

C

Cane toad *(Bufo marinus)*. *See* Recessus olfactorius
Cape buffalo *(Syncerus caffer)*, 420
Cape sparrow, *Passer melanurus*, 354
c-Fos gene/Fos-immunoreactive (Fos-IR) cells
- brain activation, 400
- density, ARC, VMH and POA, 400
- exposure, male, 399

Cheetah *(Acinonyx jubatus)*, 420
Chemical signals
- African elephants
  - communication, 142
  - identity, 142
  - oestrous, 137
  - SCPF responses, 141
  - urine and faeces, 137
  - waterholes, 138
- Asian elephants, 137
- sexual displays, 163

Chemical Signals in Vertebrates (CSIV), 445
Chemical signatures
- "biological olfactometer", 333
- chemical analytical methods, 325
- seabirds, 334
- uropygial waxes, 332

Chemical stimuli cause neuroanatomical change
- auklet odour treatment, 371
- chemosensory responses, 365, 373
- E13-16 egg development, 371–372
- egg's aqueours environment, 371
- embryonic exposure, 364–365
- orders, nest materials, 366

Chemosignals
  mate choice, 333–334
  origin, 332–333
Chromatography, 436–437
Chrysocyon brachyurus. *See* Maned wolves (*Chrysocyon brachyurus*)
Circadian activity, 278–279, 291
Common brushtail possum (*Trichosurus vulpecula*), 273
Common voles (*Microtus arvalis*)
  activity range (*see* Activity range)
  adaptations, predation risk, 278
  capture location data, 280, 282–283
  crepuscularity and diurnality indexes, 282, 286–287, 291
  description, 291–292
  DOS, 281, 285–286, 290
  foraging ranges, 278
  home range, 283
  identification, 279
  mean relocation distance, 281, 285–286, 290–291
  odour sources, 279–280
  prairie deer mice (*Peromyscus maniculatus bairdii*), 278
  predator evasion, 278–279
  radio-tracking (*see* Radio-tracking)
  statistical analysis, 282
Crested auklets (*Aethia cristatella*), 327, 366, 371
CSIV. *See* Chemical Signals in Vertebrates (CSIV)
Cultural evolution. *See* Body odour-perfume coevolution

D

Damselflies
  chemical cues, 312–313
  comparative test, 317
  visual cues, tadpole models, 313, 314
Damselfly larvae of the species *Microstigma rotumdatum*, 316
Darcin, 127–128
Dark-eyed junco (*Junco Hyemalis*), 343
Detection strategies, 315, 316
Development
  cat, 54
  domestic dog, 55
  rabbit, 52–53
Dietary odour cues, human sexual selection
  adaptationist approach, 162
  chemical cues, 167
  chemosignalling, 161
  communication and formation, attributions, 161
  effects, 163–164
  garlic effects, axiallary odour, 164–166
  health benefits, 166
  hormonal contraception, females, 165
  impacts, 164
  mean ratings, pleasantness, attractiveness and intensity, 165, 166
  meat/non-meat consumption, male odour, 164
  nutrition signals, 162–163
  starving individuals, 167
  STATISTICA 7.0, 165, 166
Dimethyldisulfide (DMDS)
  nipple grasping, 83
  sexual attraction, male and female, 82–83
Distance to odour source (DOS), 281, 285–286, 290
Domestic cat (*Felis silvestris catus*), 54–55
Domestic chickens (*Gallus gallus domestics*)
  Auklet odour, 366
  avian model development, 364
  behavioural modification, 371, 373–374
  chemosensory responses, 365, 373
  chemosensory stimuli, amniotic fluid (*see* Chemical stimuli cause neuroanatomical change)
  fertile eggs, E9, 365–366
  odour detection and discrimination, 368–373
  orders, nest materials, 366
  testing, odour (*see* Auklet odour)
  undergraduate development, feritle eggs, 365–366
Domestic dog (*Canis lupus familiaris*)
  arboreal predator, 270
  control and exposure trials, 271, 272
  European colonisation of Australia, 270
  tree-kangaroos, 274
DOS. *See* Distance to odour source (DOS)
Dry-pasting, spotted hyenas
  female rates, 154–155, 157
  functions, 152
  males, 156
  and pasting, 158
  and pawing, 156
  rates, 153, 154, 157
  scent deposition, 150
Dual-inheritance theory, 188–189
Dwarf hamster (*Phodopus roborovskii*), 119, 121
DYN. *See* Dynorphin (DYN)
Dynorphin (DYN), 401, 402

E

Early chemosensory exposure and behavioural modification, aviation system, 374–375
Eastern quoll *(Dasyurus viverrinus)*
development, neural connections, 93
nerve layer and neuroepithelium, 92
young deposition, 88
Egg, development. *See* Domestic chickens *(Gallus gallus domestics)*
Elder *(Sambucus niger)*, 356
Electrophoresis, SDS-PAGE, 434–435
Elephants. *See* Asian elephants *(Elephas maximus)*
Elephants *(Loxodonta africana)*, 420
Elephas maximus. *See* Asian elephants
Embryonic amphibians
flatworms, 260
hatching (*see* Hatching)
larval, 259–260
non-predator recognition (*see* Non-predator recognition)
plasticity, 266
post-hatching behaviour, 262
predator recognition (*see* Predator recognition)
Embryonic detection of/ response to odours/ embryo behaviour, See
Embryonic learning, 263
Enantiomers, 137, 142
Enantioselective gas chromatography. *See* Gas chromatography–mass spectrometry (GC-MS)
Estradiol
diester waxes, 330
female mice
evaluation, puberty onset, 19, 20
hormonal responses, 21, 23
neuroendocrine mechanisms, 18–19
hormonal control, 330
European rabbit *(Oryctolagus cuniculus)*, 52–53
European starlings, *Sturnus vulgaris*
avian olfaction, 353–354
description, 354
fresh plants, nests, 354
green nest material (*see* Green nest material)
nest box colony, 354
parasites and herbivores, 354
European storm-petrel, *Hydrobates pelagicus*
chicks, 332
Y-maze experiments, 333

F

Faeces, African wild dogs. *See* African wild dogs *(Lycaon pictus)*
FAME. *See* Fatty acid methyl ester (FAME)
Fathead minnows *(Pimephales promelas)*
dashing and freezing, 301, 303, 304
experiment, 300
filtered and nonfiltered tree frog AS, 301, 305
HPLC, 298
hypoxanthine, 298
Ostariophysans, 306
ponds, collection, 297
time spent moving, 301, 302
Fatty acid methyl ester (FAME), 225
Fatty acids
chloroform, 229
FAME, lipid fraction, 230, 231
pentadecanoic acid, 232
sheep wool, 232
Female social integration, spotted hyena. *See* Spotted hyena *(Crocuta crocuta)*
Field voles *(Microtus agrestis)*, 278, 290
Fire-bellied toad *(Bufo orientalis)*. *See* Recessus olfactorius
Flehmen, 139, 141, 142
Follicle stimulating hormone (FSH)
male urinary odours, female mice, 21, 23
puberty onset, female mice, 18, 19
Fragrance
dual-inheritance theory, 188
social effects, perfumes, 188
status markers, 189–190
universal use, 187–188
FSH. *See* Follicle stimulating hormone (FSH)

G

Garlic consumption, axiallary odour, 164–166
Gas chromatographuic-high resolution electron impact mass spectrometry (GC-HRMS), 64, 65
Gas chromatographuic-low resolution electron impact mass spectrometry (GC-LRMS), 64, 65
Gas chromatography–mass spectrometry (GC-MS)
analysis, 64
enantioselective, 64, 65
FAME, 230, 231
fatty acids, elephant hair, 232
identification, wool VOCs, 64
instrument, 64
ketones, 201

Gas chromatography–mass spectrometry (GC-MS) (*cont.*)
  LRMS and HRMS analysis, 64, 65
  maned wolf scent, 173
  multivariate scatterplot, 69, 71
  polydimethylsiloxane (PDMS), 64
  VOCs wool, Döhne Merino lambs, 64, 66–69
  volatile compounds, urine, 210
GC-LRMS. *See* Gas chromatographuic-low resolution electron impact mass spectrometry (GC-LRMS)
GC-MS. *See* Gas chromatography–mass spectrometry (GC-MS)
German landrace boars (GLB) (domestic pig, *Sus scrofa domesticus*). *See* 16-Androstenes measurement, saliva of GLB and GMB
German shepherd dogs, 392
Giant panda *(Ailuropoda melanoleuca)*
  AGS and urine, 208
  animals, 209
  behavioural assay, 220
  chemical communication, 209
  developmental effects, 219
  GC-MS analysis, 210
  identification, urinary constituents, 208
  mammalian urine, 208
  mating season (*see* Mating season)
  relative abundance (*see* Relative abundance)
  sample collection and solvent extraction, 209–210
  seasonal changes, sex differences, 219
  sexual dimorphism, 219
  statistical analyses, 210–211
  TIC, 211–215
GnRH neurons, 18–19
GnRH secretion. *See* Gonadatrophin releasing hormone (GnRH) secretion
Goats *(Capra aegagrus hircus)*. *See* Male effect, sheep and goats
Gonadatrophin releasing hormone (GnRH) secretion
  cells double-labelled Fos activation, 399
  females, 398
  immunoreactive, Fos, 406
  LH secretion, 401–402
  neurons, 401
  pulsatile secretion, 398
  pulse generator, 401
Gonadotropins, 18, 19
Göttingen minipig boars (GMB). *See* 16-Androstenes measurement, saliva of GLB and GMB
Goutweed (*Aegopodium podagraria*), 356
Green nest material
  choice of nest plants, 358–360
  herbs and
    bacteria, 354, 355
    female attraction, 355–356
    fledging mass, 356, 357
    immunity, 357
    nest temperature, 357–358
    red blood cells, 356
  OB, 360
  volatile compounds, nest herbs, 356
Grey tree frog *(Hyla vesicolor)*, 296, 305–306
Group odour, spotted hyena
  badge, 152
  clan, 152
  cues, 152

H

Habenula and lateral preoptic area, 92
Hatching
  alarm cues, 260–261
  mean time, size and developmental stage, 260, 261
  predators and non-predatory, 260
Health
  benefits, garlic consumption, 166
  and diet, 161
  endogenous chemosignals, 333
  guide, 237
  herbs and red blood cells, 356
  human, 163
  obscure poor skin, 190
Hemiterpenoids
  S-adenosylmethionine, 179
  hydrogen sulphide (H_2S), 179
  isomeric alcohols, 178
  3-methyl-2-butene-1-thiol, 175, 177
  3-methyl-1-methylthiobut-3-ene, 175, 177
  sulphide, 178
Herbs and fledgling mass, 356, 357
Herbs and immunity
  basophil leucocytes, 357
  drug hypothesis, 357
  immuno-stimulants, 357
  malnutrition, 357
Herbs and nest temperature, 357–358
Herbs and red blood cells, 356
Heterozygosity, 126–127

High performance liquid chromatography (HPLC)
fathead minnow, 298, 299, 303–305
grey tree frog, 298, 299
red-legged salamander, 102, 104
Varian PrepStar 218 system, 298
wood frogs *(Lithobates sylvaticus)*, 307
Histology, *B. marinus* and *B. orientalis*
cellular ultrastructure, 45
nuclei location, 45
sensory epithelia, vomeronasal and main olfactory organ, 40–41
transverse sections, nasal cavity, 39–40
uniform epithelial structure, 41
Hogweed *(Heracleum spondylum)*, 356
Home range
computation, 280
dataset, observations, 282
herbivore and neutral groupings, 283
non-predator treatments, 283
size, microtines, 288
Homing behaviour
avian chemical communication, 331
olfactory navigation, 326
petrel seabirds, 325
Hormones
domestic cat, 55
domestic dog, 55–56
rabbit, 52
House finches, *Carpodacus mexicanus*, 162, 327
House mouse *(Mus musculus domesticus)*
AOB and MOB projections, 5
individual polymorphism, 122
MUP genes, 119
suckling odours (*see* Suckling odours, rats and mice)
HPG axis. *See* Hypothalamic-pituitary-gonadal (HPG) axis
HPLC. *See* High performance liquid chromatography (HPLC)
Human
behaviour, dual-inheritance theory, 188–189
body odour, 163–164
ethnic group markers, 189
milk, breastfeeding, 193
semiochemistry, 186–187
sexual selection (*see* Dietary odour cues, human sexual selection)
status markers, 189–190
Human semiochemistry, 186–187
Humboldt penguins *(Spheniscus humboldti)*, 343
Hyaenidae, 150
Hypothalamic-pituitary-gonadal (HPG) axis, 398, 406, 408
activation, female mice, 18, 19
female mice, 18
kisspeptin neurons regulation, 30, 32
precocious activation, 29
puberty acceleration, 29
Hypothalamus
chemosignal information, 28
electrical activity, 29
integration, 29
kisspeptin neurons regulation, 30, 31
neural signal, 24
ventromedial, 28

I

IACUC. *See* Institutional Animal Care and Use Committee (IACUC)
Impala *(Aepyceros melampus)*, 420
Indian house sparrow, *Passer domesticus*, 354
Individual recognition
chemical communication and, 331–332
mate choice, 326
MUPs, 124–125
Individual variation in pheromone isoform ratios
among males, PRF isoforms
chromatograms, 105, 106
individual distribution, PRF-B, PRF-C1 and-C2, 106, 107
quantitative analyses, 106
ternary plot, 106, 108
animal collection and protein extraction, 101
behavioural and biochemical analysis, 101
bioassays, 100
biochemical and data analysis, 102
chemical signal, 100
definition, red-legged salamander, 100
mental gland (*see* Mental gland (proteinaceous secretions of))
minor components, 104–105
neurophysiological modes, 101
pheromone composition
diversity, 103
proportional composition, PRF and PMFs, 104
representative chromatogram, mental gland profile, 102–103
PRF and PMF, 100
proportional abundance, major components, 107–109

Individual variation in pheromone isoform ratios (*cont.*)
  structural diversity, 100–101
  vomeronasal organ (VNO), 100, 111, 112
Injections
  AOB
    amygdala area, 6
    anterogradely and anterior Pir labelled fibres, 6
    dorsolateral olfactory tract, 6
    frontal sections, 6, 8
    glomerular and mitral layer, 6
    parasagittal and frontal sections, 6, 7
  MOB
    cortical amygdala, 9
    lateral olfactory tract, 8–9
    medial amygdala, 9–10
    posterior medial amygdala, 10
    ventral, 7, 9
Institutional Animal Care and Use Committee (IACUC), 176
Inter-digital gland, spotted hyena
  front feet, pawing sites, 150
  secretions, individuals, 151
Intraspecific pheromone variation, Red-legged salamander
  communication scenarios, 112
  distributions, individual l isoform proportions, 107, 110
  efficacy, interactive and function, 111–112
  males and females, 112
  molecular evolution, vomeronasal receptors, 112
  PRF and PMF, 112
  sensory stimulation, female, 112
  ternary plot depiction, PRF isoform ratios, 108, 110
Isoamylamine, 25–27
Isobutylamine, 25–27

J

Japanese quails, *Coturnix japonica*
  brain activation, male, 330
  female signal, 331
Juvenile red-sided garter snakes, methyl ketone production
  chemical analysis, 237–238
  concentration, skin lipids, 242
  description, 237
  fractionation method, 238–239
  hibernacula, 241–242
  hormone treatment and analysis, 237
  identification, snakes, 240, 241
  oestradiol levels, 238
  radioimmunoassay, 242
  sample sizes, 242
  saturated and monounsaturated compounds, 238–239
  sex pheromone (*see* Sex pheromone)
  statistical analysis, 238
  SVL, mass and skin lipids, 239, 240

K

Kairomones, 252
Kangaroo *(Macropus rufus)*, 88
Kin recognition
  description, 343
  inbreeding, 343
  long-tailed tit nestlings, 344
  mate choice, 344, 345
  MUPs, 125–126
  olfactory cues, 346
  peacock, 343
  time frame, sensitive phases, 345
  visual and acoustic cues, 345
  zebra finches, 344
Kisspeptin
  male urinary odours regulation, mice, 30–31
  population, neurons expression, 29
Kisspeptin/kisspeptin receptor (GPR54), 401
Klipspringer antelope *(Oreotragus oreotragus)*
  analytical techniques, 198
  bioassays, 198–199
  black-tailed deer, 197–198, 203
  fruity smell, 201
  ketones and esters, identification, 201–203
  steps, general workup scheme, 199–200
  tarsal and antorbital glands, 198
  workup and purification procedure, 198, 199
Kowari *(Dasyurides byrnei)*
  olfactory bulb, neonate, 92
  pouch and teat, 93
  VNO, 93

L

Larval grey tree frogs *(Hyla vesicolor)*, 296
Learned recognition, predator odours
  alarm and predator cues, 253
  anti-predator response, 253
  diet, minnows, 253
  intensity, 253
  minnows reap enormous fitness, 249
  single conditioning event, 254

Lectin histochemistry
animal material, 39
binding characteristics, DBA, 42
binding patterns, 46
distribution, labelled secretory vesicles, 46
HRP-and the rhodamine-labelled lectin, 44–45
HRP-molecule, 46
N-a-D-galactosaminyl-residues, 46
sensory epithelium, main olfactory organ, 42–43
vomeronasal organ, 43–44
Leopard *(Panthera pardus)*, 420, 429
LH. *See* Luteinising hormone (LH)
Lion *(Panthera leo)*, 420
Lipocalins, 435, 441, 442
Lordosis reflex, 382
Loxodonta africana. See African elephants
Lumholtz' tree-kangaroo *(Dendrolagus lumholtzi)*
"alertness", 273
amethystine python, 270
behavioural studies, 270–271
bipedal hopping locomotion, 270
control and exposure trials, 271, 273
dog odour sources, 270–271
ethogram, 271
marsupials, 273
odours, 273–274
olfactory cues, 274
predator odours, 270
predator–prey relationships, 269
Luteinising hormone (LH)
acute response, 402–403
anterior pituitary gland, 398
characteristic, male effect, 401
divergence, 406
female sheep, 403, 404
GnRH secretion, 402
MUA activity, 401
neurogenesis, 403
peripheral circulation, 398, 400, 401
sexually experienced female sheep and goats, 399
sexually naïve and experienced females, 404
surge, 403, 406

M

Main and accessory olfactory bulbs, mice
amygdala, 14
convergence, 13
convergent projections, 4
injections (*see* Injections)
interaction, 4
layers labeling, rats and mouse, 11–12
long-distance chemosensory organs, rodents, 4
male pheromones, 4–5
materials and methods, 5–6
olfactory recipients, AOB projects, 11
Pir, 11
in 1970s, 4
sexual differences, 14
tract-tracing experiments, 4
vomeronasal and olfactory systems, 10–11
Main olfactory epithelium (MOE)
olfactory sensory systems, mammals
bowman glands, 93
chemosensory organs, nasal cavity, 90
description, 91
neonatal tammars and axons, 92
surface, 92
and VNO, 4, 91
by zinc sulphate, 28
Major histocompatibility complex (MHC)/ MHC-dependent mating preferences
chemical assessment, 334
chemical signals and, 333
individual recognition and mate choice, 333
sexual traits, 334
Major urinary proteins (MUPs)
house mouse, 435
male mouse
chemosignal(s), 27
purification and lyophilisation, 25
synthesis, 25
volatile ligands, 25, 27
QconCAT, 445
scent signals, genetic identity
heterozygosity, 126–127
individual recognition, 124–125
Kinship signals, 125–126
Male effect, sheep and goats
breeding season, 408
kisspeptin, 401–402
reproductive function (*see* Socio-sexual stimuli and reproductive function)
Male-male competition
communication channels, 328
nesting sites, 328
spotted hyena, 150
Male novelty, cell proliferation, 402–404
Male pheromone, mice, 18

Male urinary odours
  hormonal responses, 21–23
  magnitude effects, 21
  stimulation, 20
Mammalian semiochemicals, 180, 428, 430
Mammals, chemical signals and sucking
  cat, 54–55
  dog, 55–56
  newborn and learned chemical cues, 52
  rabbit, European, 52–53
  similarities and differences, 56–57
  survival and lactation, young species, 52
Maned wolves *(Chrysocyon brachyurus)*
  hemiterpenoids (*see* Hemiterpenoids)
  IACUC, 176
  increase, abundances, 180
  information, 172
  pyrazines, 179–180
  reproduction, 172–173
  sample collection and preparation, 174
  scent, 173
  thiol and sulphide, 173, 176
  urine analysis, 174–176
Marsupial and monotreme mammals, birth
  altricial young, 88
  chemical cues, survival, 89–91
  description, 88
  detecting pheromonal cues produced by mother
    description, 91
    differences, chemical signals, 94
    exception, 91
    main olfactory system development, 91–92
    milk source, 94
    olfactory bulb, 92–93
    tammars and axons, 92
    VNO and nasopalatine duct, 93–94
  lay eggs, 88–89
  management, 89
  milk source, 88
  red kangaroo and eastern quoll, 88
Marsupials
  behavioural patterns, 270
  predator-experiences, 273
  rainforest habitats, 274
Mass spectrometry
  fingerprint, 437
  peptides, 440
  remarkable sensitivity and resolution, 432
  tandem, 437, 439, 441, 442
Mate attraction, 355
Mate choice/mate attraction
  avian chemical communication, 325
  implications, 333–334
  olfactory ornament, 328
Mating behaviour, 382, 383
Mating season
  animals, 209
  chemical communication, 209
  gas chromatographs, urine, 216, 217
  relative abundances
    seasonal differences, 216, 217
    sexual differences, 211, 216
  scent marking, 208
  sexual dimorphism, 219
  TIC, 211–215
Maturity, sexual, 121
Meadow voles *(Microtus pennsylvanicus)*, 163, 164, 290
Mental gland (proteinaceous secretions of)
  analyses
    intraspecific diversity, 110–112
    PRF range, minor components, 109–110
    proportional abundance, major components, 107–109
  animal collection and protein extraction, 101
  secretion, 100
2-Methylbut-2-enal, 53
Methyl ketones. *See* Juvenile red-sided garter snakes, methyl ketone production
Milfoil, *Achilea millefolia*, 356
Milk
  biological substrates, 79
  birth survival, 89
  breast, 164
  cat, 54
  cuing value, 82
  dog, 55–56
  ducts, 90
  finding process, 89
  human odour, 193
  lamb, 62
  mother's, 52
  newly born rat pups, 81
  patches, 94
  pheromonal and learned chemical cues, 56
  rabbit, 52, 53
  and saliva share volatiles, 82
  source, 88, 90, 93, 94
Monkeys, 332, 400
Monotreme. *See* Marsupial and monotreme mammals, birth
Mother-infant bonding. *See* Suckling odours, rats and mice
Mother-offspring bond formation, sheep, 61

Mother-offspring bonding. *See* Marsupial and monotreme mammals, birth
Mozambique rain frog *(Breviceps mossambicus)*, 46
Multicomponent chemical signal
  communication, 111
  courtship duration reduction, 101
  functionality, 113
MUPs. *See* Major urinary proteins (MUPs)

N

Nasal cavity
  anuran amphibians, 38
  transverse section, recessus olfactorius, 39–40
Nasopalatine duct, 93
Neonate
  chemical cues, 89–90
  developmental stage, 88
  guiding cue, 90
  human, 89
  marsupial families, 88
  milk source, mammalian, 88, 94
  olfactory cues, 90
  rats and mice
    AF, 80
    birth right, 83
    care and protection, mother, 77
    fresh milk, 81
    motivating and guiding, initial behaviour, 78
    own fluid, 80
    suckling performance, 79
    suckling response, 82
  tammar wallabies, 93
  teats, mother, 91
  VNO, 92
Nest herbs/choice of
  immuno-stimulants, 357
  red fowl mite, 354
  starling nest box colony, 354
  volatile compounds, 356
Nest odour, 359
Neuroendocrine mechanisms, 18–19
Neurokinin B (NKB), 401, 402
Neuropeptides, 401
Neurosteroid, 383
Nipple preference/order
  ensure regular sucking, 56–57
  establishment, rabbits and kittens, 54
  lack of order, 56
  postnatal months, 53, 54
Nipple-related odour cues, 81–82
Nipple-search pheromone, 53
NKB. *See* Neurokinin B (NKB)
Noenatal recognition, sheep
  analytical methods, 64
  bioassays, 65, 74–75
  bond formation, mother-offsprings, 62
  description, 64–65
  ewe implantation, 63
  GC-LRMS and GC-HRMS analysis, 65–67
  hypothesis, twins lambs, 65
  individual odour profile, 63
  observations, 63
  optimal 2D presentation, PCA biplot, 71–73
  post-partum lamb deaths, 62
  prevention, odour diffusion, 74
  sample collection and preparation, 63
  SEP-enriched wool volatile
    day-old twin lambs, 70
    unrelated lambs, 70
  SEP technique, 65
  statistical analysis, 64–65
Non-predator recognition, 265
Northern pike *(Esox lucius)*, 248
Nursing, 52, 54, 56

O

OB. *See* Olfactory bulb (OB)
Odorous steroids, 382
Odour, avian communication
  anosmic/microsmatic birds, 326
  chemosignals and implications (*see* Chemosignals)
  description, 326
  intraspecific interactions (*see* Sexual behaviours)
  land birds, 326
Odour communication, 188
Odour learning
  by accessory olfactory system, 53
  nipple-search sequence, 53
  suckling, 53
Odours in nest materials, 366
Oestradiol
  levels, treatment, 238
  peanut oil, 237
  radioimmunoassay, 242
Oestrogen receptors
  acute/chronic treatment, 401
  kisspeptin, 401
  pheromone-induced neurogenesis, 403
  progesterone, 401

Oestrus females. *See* Vaginal bacterial flora, oestrus females
Olfaction
biological and cultural chemosignalling, 190–191
body odour-perfume coevolution, 191–193
dual-inheritance theory, 188–189
ethnic group markers, 189
fragrances, 187–188
human semiochemistry, 186–187
recessus (*see* Recessus olfactorius)
social effects, perfumes, 188
status markers, 189–190
visual cues (*see* Dietary odour cues, human sexual selection)
Olfactory bulb (OB)
description, 360
non-breeding season, 360
olfactory receptor genes, 343
Passeriformes, 342
Procellariiformes, 341
songbirds, 341
song nuclei, songbirds, 360
Olfactory cues that direct young to pouch and teat
brushtail possums, 90
chemical structure, 90
guiding cue, 90
macropodids, 90
marsupial young, 89
monotreme females, 90
neonatal quokka, 89
odour cues, 89
polytocous species, 90
sensory cues, 89
tammar wallaby, 90–91
Olfactory epithelium, 2
Olfactory identity cues, spotted hyena
den-bound juvenile, 149
juvenile females, 153
scent marks, 157
Olfactory receptor genes, 343
Olfactory social learning
benefits, 347
description, 346
males and females, 346
olfactory kin recognition, 347
parent offspring communication, 346
Oreotragus oreotragus (Artiodactyla). *See* Klipspringer antelope *(Oreotragus oreotragus)*
Organum vasculosum of the lamina terminalis, 91
Orientation
and foraging, 347
nest construction and, 343
Orkney Islands vole *(Microtus arvalis orcadensis)*, 278
Overmarking
African wild dog urine, 427, 428
composition, 429–430
dominant pair urine, 420, 421
Over-pasting, spotted hyena
behaviour, 150, 152
degree of mixing, anal gland scent, 151
rates, 152
scent marks, clan members, 152

P

Pandas. *See* Giant panda *(Ailuropoda melanoleuca)*
Panthera tigris tigris. *See* Bengal tiger *(Panthera tigris tigris)*
Parental care, 311
Passerine/songbird
blue petrels, 343
mechanisms, kin recognition (*see* Kin recognition)
predator avoidance, 342
preen oil secretion, 343
Passerines *Passeriformes*
olfaction (*see* Passerine/songbird)
sense of smell, 341–342
Pasting, spotted hyena
behaviour, 153
description, 150
dry-pasting (*see* Dry-pasting, spotted hyenas)
female rates, 154–155, 157
functions, 152
males, 156
mean rates, 154, 158
over-pasting (*see* Over-pasting, spotted hyena)
scent, 157
social interactions, 153
Pawing, spotted hyena
behaviour, 156–157
calculation, 154
event, 153
functions, 152
inter-digital gland secretions, 151
male-male competition, 150
males, 158
mean rates, 154

olfactory display, 158
scent marking, 150
Peacock *(Pavo cristatus)*, 343
Peacock gudgeons *(Tateurndina ocellicauda)*, 249
Pheromaxein, 382–383
Pheromones
aggression, 129–130
copulins, 394
Darcin, male mouse sex, 127–128
description, 127
GnRH/LH secretion, 402
individual responses, 128–129
induced neurogenesis
mice, 402, 403
prairie voles, 403
mammals (*see* Mammals, chemical signals and sucking)
meaning, 127
mice (*see* Main and accessory olfactory bulbs, mice)
production, 398
sexual interest, oestrus females by male dogs, 391–392
sheep and goats originates, 399
Phylogenetic effects
MUPs, 127
olfactorius and epithelium, 47
Pimephales promelas, 248
Piriform cortex, 92
Pitohui/Ifrita genera
feathers, 327
sour odour, 327
Plains zebra *(Equus quagga)*, 420
Platypus *(Ornithorhynchus anatinus)*, 88
Plethodontid modulating factor (PMF). *See* Individual variation in pheromone isoform ratios
Plethodontid receptivity factor (PRF). *See* Individual variation in pheromone isoform ratios
POA. *See* Pre-optic area (POA)
Poison frog *(Ranitomeya variabilis)*
anuran tadpoles, 317
biological communication, 309–310
chemical recognition, phytotelmata, 315
comparative test, 317
damselfly larvae, chemical cues, 312–314
different cues, tadpole depositions, 315, 316
life-stages, anuran amphibians, 310
parental care, 311–312
predators, 310–311
primary defence mechanisms, 310
statistical analyses, 314
visual and chemical detection, 315, 316
visual cues, tadpole models, 313, 314
Polygyny, mammals, 136
Predation. *See* Poison frog *(Ranitomeya variabilis)*
Predation risk
adaptations, 278
diel activity, 291
foraging ranges, voles, 278
levels, animal and non-animal odours, 279–280
mustelid, 288, 289
Predator avoidance, 342
Predator introductions
minnows reap enormous fitness, 249
trials with dashing response, 249, 251
Predator odour
feeding behaviour, 270
Microtus arvalis (*see* Common voles *(Microtus arvalis)*)
trapping methods, 273
Predator odours recognition by minnows
alarm and predator cues, 253
antipredator behaviour, 248–249
anti-predator response, 249
chemical ecology, predator/prey interactions, 254
dashing behaviour, 249, 251
diet-based cues, 255
Esox lucius, 248
laboratory conditions, 253
model fish system, 248
model species, 247
molecules, 254–255
olfactory system, 249, 250
pike odour concentration documentation, 251
predation pressure exertion, 248
predator introductions (*see* Predator introductions)
predator–prey interactions, 247
predatory pike/non-predatory suckers, 254, 255
prey animals, 248
proportion, fish dashing, 252
respondent, 252–253
risk-averse strategy, 254
single conditioning event, 254
Predator recognition
aquatic animals, 262
newt odour, 262–263
tadpoles, 264
Pre-optic area (POA), 400

Preputial gland, 419, 420
Procellariiform
chemical composition, wax acids, 329
olfactory bulb size, 341
phylogenetic effects, 328
seabirds, 326
uropygial secretion, 328
Progesterone, 53
Progesterone assay, 393
Progesterone receptors
GnRH secretion, 401
inhibitory effect, 407
LH response, 407
medroxyprogesterone acetate, 407
plasma, 408
Prolactin/lactation
cat, 54
rabbit, 53
Protein, scent secretions
accurate and absolute quantification, 441–445
characterisation, 434
chemical communication, 433–434
cognate genome sequence, 440–441
complexity, 434–437
identification, 437–440
'proteomics', 434
strategies, characterisation and quantification, 443, 445
Proteomics
'bottom-up', 437
description, 434
workflows, 443
Puberty acceleration/puberty-accelerating chemosignals, mice
identification
female sexual maturation, 24
HMW and LMW, male urine fraction, 24
MUPs, 25
in 1990s, 25
solid phase extraction, 27–28
volatile amines, 28
volatile ligands, MUPs, 25–28
integration
chemosensory perception, 28
hypothalamic, 29
kisspeptin neurons regulation, 30–31
male urinary odours (*see* Male urinary odours)
production mechanisms
extraction, 24
hormonal regulation, 23
social influences, 24
sources, 23
Puberty onset, female mice by male urinary odours
ecological validity and potential applications, 31
evaluation, 19–20
neuroendocrine mechanisms, 18–19
puberty acceleration (*see* Puberty acceleration/puberty-accelerating chemosignals, mice)
Purple martin, *Progne subis*, 354
Pyrazines, 175, 176, 179–180

Q
QconCAT strategy, 444, 445
Quokka *(Setonix brachyurus)*, 89

R
Radio-tracking
activity range, 288
vs. capture location data, 284, 289–290
datasets, 284
48 h sessions, 280–281
monitoring and treatment phases, 282
session, 18 females, 283–284
Rat *(Rattus norvegicus)*
AOB and MOB projections, mice, 11–12
individual polymorphism, 124
males invest, urinary MUPs, 121
suckling odours (*see* Suckling odours, rats and mice)
Recessus olfactorius
B. marinus and *B. orientalis*
histology (*see* Histology, *B. marinus* and *B. orientalis*)
lectin histochemistry (*see* Lectin histochemistry)
definition, 38
description, 38
Red-backed salamander *(Plethodon cinereus)*, 163
Red-billed woodhoopoe, *Phoeniculus purpureus*
foul-scented uropygial secretion, 327
Pitohui feathers, 327
Red-legged salamander *(Plethodon shermani)*. *See* Individual variation in pheromone isoform ratios
Red-sided garter snakes *(Thamnophis sirtalis parietalis)*. *See* Juvenile red-sided garter snakes, methyl ketone production

Relative abundance
age effects, urinary compounds, 216, 218
seasonal differences, 216, 217
sexual differences, males and females pandas, 211, 216
sexual dimorphism, 219
Reproduction
activity range, 288
bank voles *(Myodes glareolus)*, 290
capture records, 280
predation risk, 278
spatial and circadian activity, 279
Reproduction and social behaviour, Plethodontid salamanders, 112
Reproductive hormones. *See* Socio-sexual stimuli and reproductive function
Reproductive management of farm animals, 398
Reverse-engineering approach
African wild dogs
alpha-pair, 427
dominant-pair, 428
free-ranging, 420, 422–426
urine and faeces samples, 420, 421
bioassays, 430
chemical signals, 428
compounds, 428, 429
gas chromatography and mass spectrometry, 420
male wild dogs, 419
mammals, 429
mammal scents markes, 417–418
MS library, 420
neurochemistry, 418
overmarking, 430
packs, 419
semiochemicals, 417
signalling compounds, 419
spatial scent marks, 428–429
species-specific compound, 429
urine and faeces, collection, 420
wild dog volatiles, 420
Risk assessment
predator odour, 247, 248
predator recognition
behavioural response, 262–263
newt odour, 263
temporal pattern, 263
Rodent urinary proteins
communication, 118
complexity, volatile molecules, 118
description, 118
individual polymorphism
bank vole, 122
characteristics, species and sex, 122
gene cluster, MUP, 124
isoforms express, 122
MUP patterns, 122–124
investment
creatinine control, 120–121
lipocalins, 120
MUPs expression, 121
social condition, male and female house mice, 121–122
wild-stock house mice, 120
ligand binding, 118–119
mice and rats, 118
pheromones (*see* Pheromones)
scent signals, genetic identity (*see* Major urinary proteins (MUPs))
vertebrate scents, 118
Root voles *(Microtus oeconomus)*, 290

S

Saliva, rats and mice
amniotic fluid (AF), 79
changes, 81
collection, 80
cueing value, 82
lactating conspecifics, 80–81
maternal, 81
maternal and pup, 79
PND 1, 81–82
post-parturient females, 80
predisposed mechanisms, 83
Scanning electron micrographs. *See* Tail hairs, Asian elephants *(Elephas maximus)*
Scanning electron microscopy, *B. orientalis* and *B. marinus*
coated samples, 39
dehydration, 39
investigation, epithelial surface structures, 39, 43
material, 38–39
surface differentiations, 45
Scent flagging, 228
Scent mark. *See* Reverse-engineering approach
Scent marking. *See* Spotted hyena *(Crocuta crocuta)*
Seabirds
chemical communication, 334
chemical signatures, 334
procellariiform, 326
sexual behaviours, 330
Season, domestic boars, EU, 383
Semiochemical, mammals, 180, 428, 430

Sex
AGS and urine, 208
alcohols, phenols and ketones, 218
Asian elephant *(Elephas maximus)*, 208
differences, males and females pandas, 211, 216
dimorphism, 219
mammalian urine, 208
pheromones, 219
relative abundances, 210–211
Sex difference in body odour
bird species, 343
visual and acoustic signals, 342
visual plumage traits, 342
Sex pheromone
aggression, 128, 129
androgens, 236
darcin, 128, 130
male mouse, 127–128
oestrogen, 236
radioimmunoassay, 242
sexual attractiveness, 236
Sexual attraction
adult male *vs.* female, 127–128
aggression, 129
individual male, 128
instinctive, 127, 130
targets and reinforces, 130
Sexual attractiveness of bitches, 392–393
Sexual behaviours
chemical communication and Avian, 330–331
chemical communication and individual recognition, 331–332
Sexual dimorphism
abundances, urinary chemical compounds, 219
AOB and MOB, 14
avian chemical, 330, 331
budgerigars, 331
coded, 219
communication proteins, 130
olfactory system, 399
protein investment, 120
sensitivity, female pigs, 383
Sexual dimorphism in the olfactory system, 399
Sexual displays, 162–163
Sexual experience
mice
and antipredatory behaviour, 4
pheromones detection, VNO, 4
reproductive function (*see* Socio-sexual stimuli and reproductive function)
Sexual selection, 111, 112
Sheep *(Ovis aries)*
noenatal recognition (*see* Noenatal recognition, Sheep)
reproductive function (*see* Male effect, sheep and goats)
Short-beaked echidnas *(Tachyglossus aculeatus)*, 88
Small volatile ligands, mice, 25, 27
Smell
ancestral spirits, 187–188
'anxious' odours, 186
body odour-perfume coevolution, 191–193
cumin, 164
musk-and urine-like smelling steroids, 386, 388
red clay, 187
starving, 167
Snakes. *See* Juvenile red-sided garter snakes, methyl ketone production
Social chemosignals, avian embryos, 364, 366
Social status, spotted hyena
age, 157
communal den, 158
correlation, female, 158
immigrant males and females, 151
increases, 152
median, 149, 156, 158
reproductive active males, 156
standardised, 153, 154
statistical analyses, 154
Socio-sexual behaviour, 4
Socio-sexual stimuli and reproductive function
cell proliferation and male novelty, 402–404
consumers, 398
cyclic females, 407–408
electrophysiological evidence, 400–401
GnRH, 398–399
histological evidence, 399–400
HPG, 398
kisspeptin and male effect, 401–402
LH response, 409
males, 398
olfactory signal, 409
seasonal and postpartum anoestrus, 409
sexual experience, 404–407
sheer complexity, 408
Sodium dodecyl sulphate polyacrylamide gel electrophoresis (SDS-PAGE)
analysis, 434–435
separation, 436
Solid phase dynamic extraction (SPDE), 173, 174, 176

South American monk parakeet, *Myiopsitta monachus*, 354
Southern brown bandicoot *(Isoodon obesulus)*, 273
Spatial placement of scent marks, 150
SPDE. *See* Solid phase dynamic extraction (SPDE)
Spiny mouse *(Acomys cahirinus)*, 163
Spotless starlings, *Sturnus unicolor*, 355
Spotted hyena *(Crocuta crocuta)*
  behaviours, 152
  communal den, 158
  communal den area, 152–153
  communication, clan members, 150
  dry-pasting, 153
  dyadic interactions, 153
  female rates, pasting and dry-pasting, 154–155, 157
  Hyaenidae 'paste' scent, 150
  identification, 153
  marking rates, reproductively active males, 155
  mean behaviour rates, 154, 157
  median status and dry-paste, 158–159
  over pasting (*see* Over-pasting, spotted hyena)
  pasting behaviour, 153
  pawing (*see* Pawing, spotted hyena)
  scent mark, 157–158
  scent marks, natal and immigrant members, 152
  social status, 151–152
  trade-off, young males, 159
  visual contact, 151
Starlings, *Sturnus vulgaris*, 326–327
Suckling odours, rats and mice
  biological substrates, 79
  chemosensory systems, 78
  lactating females, 79
  neonates survival, 77–78
  nipple-related odour cues, 81–82
  nipple search and suckling, newborns, 78
  postnatal odour cues, 80–81
  predisposed mechanisms, 82–83
  prenatal odour cues, 79–80
  1970s and 1980s, 78–79
Sulphides
  hydrogen sulphide (H_2S), 179
  isomeric, 176
  isopentenyl diphosphate, 178
  mass spectral ions, abundances, 178
  S-methylsulphonium ion, 179

T

Tactile stimuli, 52
Tail hairs, Asian elephants *(Elephas maximus)*
  cross section, 229, 230
  extraction, lipid, 229
  FAME, 229
  fatty acid profile, 230, 231
  GC-MS analysis, fatty acids, 232
  preparation, samples, 229
  surface image, 229, 230
Tammar wallaby *(Macropus eugenii)*
  climb, 90
  knobs olfactory, transmission electron micrograph, 92
  nose, 91–92
  olfactory epithelium, 93
  VNO, 93
Tentatively identified compounds (TIC), 211–215
Territorial marking pheromones. *See* Klipspringer antelope *(Oreotragus oreotragus)*
Territory, 430
Testosterone
  and 16-androstenes, 386–388
  Asian and African elephants, 137
  chemosignal(s), 24
  concentrations, 388–389
  distinct seasonal course, serum, 388
  domestic boars, EU, 383
  formation, young pigs, 383
  injection, 23
  male pig, 386
  olfactory bulb, 360
  salivary, 385
  sexual maturity, 121
  wild boar, 383
Thamnophis sirtalis parietalis. *See* Juvenile red-sided garter snakes, methyl ketone production
TIC. *See* Tentatively identified compounds (TIC)
Tiger. *See* Bengal tiger *(Panthera tigris tigris)*

U

Urinary lipocalins
  excrete, 118
  protein investment, 120
  rodent tertiary structure, 119
  vole and hamster sequence, 118
Urinary metabolites, 174–176
Urinary volatile, giant pandas. *See* Giant panda *(Ailuropoda melanoleuca)*

Uropygial secretions
blue petrels, 332
procellariiforms, 328
red-billed woodhoopoe, 327
sexual behaviours, 330

V
Vaginal bacterial flora, oestrus females
analysis, samples, 393
animal behaviours, 391
copulins, 394
experimental procedure, bitches, 392
hormonal action, 395
male domestic dogs, 391
Mann–Whitney U test, 393
progesterone assay and mating, 393
reaction, mating male domestic dogs, 394
reduction bacteria and attractiveness, 395
saprophytic bacteria, humans, 394
semiochemical substances production, 391–392
sexual attractiveness, bitches, 392–393
study animal, German shepherd dog, 392
Vandenbergh effect, 18
Ventromedial nucleus of the hypothalamus (VMH), 400
VMH. *See* Ventromedial nucleus of the hypothalamus (VMH)
Volatile compounds in nest herbs
choice experiment, 356
description, 356
"drug hypothesis", 356
starling nests, 356
Volatile ligand(s), 25–28
Volatile male odours, 4
Volatiles amines, 28
Vomeronasal organ (VNO)
and main olfactory epithelium, 90, 91
and main olfactory organ, 38
mice
and MNO, 2
sexual pheromones detection, 4
nasopalatine duct, 93–94
Plethodon shermani, 100, 111, 112
recessus olfactorius
cilia, 40
lower parts, nasal cavity, 39
main olfactory, 40
posteriorly located section, 40
tammar wallaby olfactory knobs, 92
Vomeronasal receptors, MUPs, 119

W
Weaning, African elephants, 136, 138
Wilson's storm petrels, *Oceanites oceanicus*, 332
Wintergreen essential oil (98% methyl salicilate), 367
Wood chat shrike *Lanius senator*, 354
Wood frogs *(Lithobates sylvaticus)*, 307

Z
Zebra finch *(Taeniopygia guttata)*
olfaction, passerines (*see* Passerine/songbird)
olfactory social communication (*see* Olfactory social learning)
smell, passerines, 341–342